Occupational Health Practice
Second Edition

Occupational Health Practice

Second Edition

Edited by
R.S.F. Schilling, CBE,MD,DSc,FRCP,FFCM,FFOM,DPH,DIH
Emeritus Professor of Occupational Health
University of London

BUTTERWORTHS
London – Boston
Sydney – Wellington – Durban – Toronto

First published 1973
Reprinted 1975
Reprinted 1977
Italian translation 1978
Second edition 1981

© **Butterworths & Co (Publishers) Ltd. 1981**

British Library Cataloguing in Publication Data

Occupational health practice. – 2nd ed.
  1. Industrial hygiene
  I. *Schilling,* Richard Selwyn Francis
  613.6'2   RC967   80–41044

  ISBN 0–407–33701–6

Typeset by Butterworths Litho Preparation Department
Printed and bound by Mackays of Chatham

Foreword to the First Edition

Occupational health is a field which is growing rapidly all over the world. Its objectives have gradually broadened from dealing mainly with occupational hazards causing accidents and occupational disease to include all kinds of factors at work or related to working conditions that may cause or contribute to disease or deviation from health.

The activities of occupational health services have changed and their scope is now much wider than before, not only preventing occupational hazards but also promoting the general health of the worker and the adjustment of work to man and of man to work. Occupational health services are now expanding from industrial undertakings to all places of employment, such as offices, agriculture, forestry and transport.

The basic problems of occupational health and of occupational health services are the same or very similar in different countries. The practical solutions, however, may be different as many local factors have to be considered, such as the administration, the development of medical care and public health in the country concerned.

In many countries there is a great need for practical information on occupational health practice. Professor Richard Schilling and his staff at the TUC Centenary Institute of Occupational Health in the London School of Hygiene and Tropical Medicine, have international experience in this field from working in different countries as WHO consultants and as members of international research groups; they also have experience derived from the training and education of postgraduates from many countries. This book will meet the great need for information in this rapidly expanding field all over the world, and will be of greatest value in promoting the development of occupational health.

Sven Forssman

Preface to the Second Edition

Since the first edition was published in 1973, there have been many developments in occupational health. Some of these I have attempted to cover in this expanded edition by including seven new chapters, by having 13 others rewritten and the remaining five chapters revised and updated.

The greatest challenge on the international scene is to raise the standards of health and safety in the work places of developing countries. Its importance is emphasized in the introduction by Dr H. Mahler, Director-General of the World Health Organization. The challenge is outlined in more detail in a new chapter on occupational health in developing countries and in another on migrant workers whose health gives rise to world-wide problems.

Two new chapters describe the response to toxic agents of particular organs (liver, kidney, skin, blood, lung and nervous system). This approach to industrial toxicology is unusual, but it is of value to health workers in hospital and general practice and it enables students to grasp general principles. There are three new chapters on the screening of organ systems and toxicity testing. These are increasingly important methods of health protection.

Chapters which appeared in the first edition, but which have been completely rewritten, include those on work and health, screening well people, uses and methods of epidemiology, environmental measurement and control, personal protection, occupational safety and accident prevention, mental health and the prevention of occupational disease and ill health.

Throughout the book greater emphasis is given to the promotion of health and well-being. At the workplace this is done by encouraging workers to participate in forming and executing health and safety policies, which enables them to influence their own work situation. Not only does this achieve higher standards of health and safety but it also improves job satisfaction.

This edition is written primarily for occupational physicians and nurses, either in training or practising in the field. However, the more extensive coverage of environmental measurement and control, and the chapters on ergonomics and accident prevention, extend its usefulness to occupational hygienists, safety officers and others.

Once again I express my indebtedness for the help received with the first edition. I am indebted to those who have helped to recast and produce the new edition, especially the contributors, the publishers and the members of my family: my wife Heather for improving and typing manuscripts, my daughter Erica Hunningher for her editorial expertise and my son Christopher Schilling for valuable suggestions and improvements to manuscripts in early draft.

I should like to thank the Information and Advisory Service of the TUC Centenary Institute of Occupational Health for their constant help in providing reference material. I am grateful to those who supplied new illustrations: some are acknowledged in the text; others include Mr C. J. Webb and his staff of the Visual Aids Department, London School of Hygiene and Tropical Medicine, and Mr Joost Hunningher.

I acknowledge with thanks the help from many other prople who read the manuscripts or who made valuable suggestions for adding new material. They include Mr David Hodge Brown, Dr J.C. Graham, Dr P.G. Harris, Dr James P. Hughes, Dr C.P. Juniper, Dr A. S. McLean, Surgeon Lieut. Commander G.H.G. McMillan, Dr Siza Mekky, Dr A. Newton Taylor, Dr John M. Peters, Dr C.A.C. Pickering, Professor Geoffrey Rose, my daughter Margie Schilling, Dr M.L. Snaith and Dr Michael Williams.

Richard Schilling

Preface to the First Edition

The need for a book describing what the physician, hygienist and nurse actually do to protect and improve the health of people at work has become increasingly obvious to the staff of this Institute. Although many books have been written on occupational health, there are none in English which deal comprehensively with its practice. We teach the principles of occupational health practice to postgraduate students in occupational medicine, nursing and hygiene, and the lack of a standard work of reference has made the task of both teaching and learning more difficult.

Our academic staff and visiting lecturers have attempted to fill this gap, which is repeatedly brought to our notice by students. While our primary aim is to meet a need in formal course programmes it is hoped that the book may also be useful to the many whose interests encompass occupational health but who cannot attend a course, and that it will be of some value to medical and non-medical specialists in related fields.

Our students come from all over the world, many from countries undergoing rapid industrialization. We have therefore tried as far as possible to offer a comprehensive, up-to-date, account of occupational health practice, with some emphasis on the special needs of work people in developing countries. Eastern European countries attach great importance to occupational health and provide comprehensive occupational health services and training programmes. We refer to their methods of practice and training as well as to those of the western world because we believe both East and West have much to learn from each other, and the developing countries from both. Terms such as occupational health, medicine and hygiene often have different meanings, particularly in the eastern and western hemispheres. Occupational health in the context of this book comprises two main disciplines: occupational medicine, which is concerned primarily with man and the influence of work on his health; and occupational hygiene, which is concerned primarily with the measurement, assessment and control of man's working environment. These two disciplines are complementary and physicians, hygienists, nurses and safety officers all have a part to play in recognizing, assessing and controlling hazards to health. The terms industrial health, medicine and hygiene have a restricted meaning, are obsolescent, and are not used by us.

The three opening chapters are introductory; the first gives an account of national developments, contrasts the different forms of services provided by private enterprise and the State; and discusses factors which influence a nation or an industrial organization to pay attention to the health of people at work. The second is about a man's work and his health. Everyone responsible for patients needs to realize how work may give rise to disease, and how a patient's ill health may affect his ability to work efficiently and safely. It is as important for the general practitioner or hospital consultant as it is for the occupational physician to be aware of the relationship between work and health. The third chapter outlines the functions of an occupational health service. The chapters which follow describe in more detail the main functions, such as the provision of treatment services, routine and special medical examinations, including 'well-person' screening, psychosocial factors in the working environment and the mental health of people at work. There are chapters on occupational safety and the prevention of accidents and occupational disease which are often the most important tasks facing an occupational health service. Methods used in the study of groups of workers are outlined in sections on epidemiology, field surveys and the collection and handling of sickness absence data; these chapters are of special importance, as it is essential that those practising occupational health think in terms of 'groups' and not just of the individual worker. Epidemiological expertise enables this extra dimension to be added to the investigation and control of accidents and illness at work.

One chapter is devoted to ergonomics while five on occupational hygiene deal with the physical and thermal environments, airborne contaminants, industrial ventilation and protective equipment and clothing. There are concluding chapters on ethics and education in occupational health. Undergraduates in medicine and other sciences frequently lack adequate teaching on this subject and we hope that this book may be useful to them and their teachers.

Although it is not possible to cover fully the practice of occupational health in 450 pages, we hope to convey the broad outlines of the subject to a wide variety of people.

I owe many thanks to many people for help in producing this book, especially to the contributors and to those who assisted them in preparing their manuscripts, and to the publishers for their patience and understanding. For the illustrations I am particularly grateful to Mr. C.J. Webb, Miss Anne Caisley and Miss Juliet Stanwell Smith of the Visual Aids Department at the London School of Hygiene and Tropical Medicine, and also to the Wellcome Institute of the History of Medicine, to the Editors of many journals, and to Professor Kundiev and Sanoyski of the USSR.

Manuscripts were read by members of the Institute staff and others who made valuable suggestions; the latter include Professor Gordon Atherley, Professor R.C. Browne, Dr J. Gallagher, Dr J.C. Graham, Dr Wister Meigs, Mr Wright Miller, Mr Andrew Papworth, Miss Brenda Slaney, Professor F. Valić, my wife and my daughter Mrs Erica Hunningher – I am indebted to them all; I am also grateful to Dr Gerald Keatinge and Dr Dilys Thomas for reading proofs, and to my secretary, Miss Catherine Burling for her help and enthusiasm throughout the long period of preparation.

Richard Schilling

List of Contributors

Alan Bailey, MB BS, MRCP
Director of Research,
BUPA Medical Centre,
Battlebridge House,
300 Gray's Inn Road,
London WC1X 8DU

M.A. El-Batawi, MD, DSc, MPh
Chief Medical Officer,
Occupational Health Section,
World Health Organization,
1211 Geneva 27,
Switzerland

Professor E. Boyland, DSc, PhD
Emeritus Professor of Biochemistry,
TUC Centenary Institute of Occupational Health,
London School of Hygiene and Tropical Medicine,
Keppel Street,
London WC1E 7HT

Alexis Brook, MA, MB BChir, FRCPsych, DPM
Consultant Psychiatrist,
Tavistock Clinic,
120 Belsize Lane,
London NW3 5BA

E.N. Corlett, PhD
Professor of Production Engineering,
Head, Department of Production Engineering and Production
Management,
University of Nottingham,
University Park,
Nottingham NG7 2RD

G.W. Crockford, BSc, MIBiol,
Senior Lecturer,
TUC Centenary Institute of Occupational Health,
London School of Hygiene and Tropical Medicine,
Keppel Street,
London WC1E 7HT

D. Else, PhD,
Senior Lecturer,
Department of Safety and Hygiene,
University of Aston in Birmingham,
Gosta Green,
Birmingham B4 7ET

A. Ward Gardner, MD, FFOM, CPH, DIH
Esso Medical Centre,
Fawley,
Southampton SO4 1TX

J.M. Harrington, MD, MSc, MRCP, FFOM,
Professor of Occupational Health,
University of Birmingham,
PO Box 363,
Birmingham B15 2TT

J.C. McDonald, MD, MS(Harvard), FRCP(Lond),
FRCP(Can.), FFCM, FFOM, DPH, DIH,
Director and Professor of Occupational Health,
TUC Centenary Institute of Occupational Health,
London School of Hygiene and Tropical Medicine,
Keppel Street,
London WC1E 7HT

H. Mahler, MD
Director-General of the World Health Organization,
1211 Geneva 27,
Switzerland

M.K. Molyneux, MSc, PhD, DipOH, FIOH
Senior Occupational Hygienist,
Shell UK Ltd.,
Carrington,
Urmston,
Manchester M31 4AJ

R. Murray, OBE, BSc, MB ChB, FRCP, FFOM, DPH, DIH, DTech
Robert Murray Associates,
Quality House,
Quality Court,
Chancery Lane,
London WC2A 1HP

M. L. Newhouse, MD, FRCP, FFCM, FFOM
Senior Research Fellow,
London School of Hygiene and Tropical Medicine,
Keppel Street,
London WC1E 7HT

S. J. Pocock, MA, MSc, PhD
Senior Lecturer in Medical Statistics,
Department of Clinical Epidemiology and Social Medicine,
Royal Free Hospital,
21 Pond Street,
London NW3 2PN

P. A. B. Raffle, KStJ, MD, FRCP, FFOM, DPH, DIH
Chief Medical Officer for London Transport,
280 Old Marylebone Road,
London NW1 5RJ

R. S. F. Schilling, CBE, MD, DSc, FRCP, FFCM, FFOM, DPH, DIH
Emeritus Professor of Occupational Health,
University of London,
46 Northchurch Road,
London N1 4EJ

P. J. Taylor, MD, BSc, FRCP, FFCM, FFOM, DIH
Chief Medical Officer, the Post Office,
23 Howland Street,
London W1P 6HQ
and
Vice Dean to the Faculty of Occupational Medicine,
Royal College of Physicians,
London NW1 4LE

H. A. Waldron, MD, PhD, MRCP, FFOM, MIBiol, DHMSA
Senior Lecturer,
TUC Centenary Institute of Occupational Health,
London School of Hygiene and Tropical Medicine,
Keppel Street,
London WC1E 7HT

Joan Walford, MIS
Statistician,
TUC Centenary Institute of Occupational Health,
London School of Hygiene and Tropical Medicine,
Keppel Street,
London WC1E 7HT

T. J. Wilmot, MS, FRCS, DLO
Consultant Ear, Nose and Throat Surgeon,
West Tyrone and Fermanagh Hospital Group,
Omagh,
County Tyrone BT79 OAN

H. Beric Wright, MB BS, FRCS, MFOM
Deputy Chairman,
BUPA Medical Centre,
Battlebridge House,
300 Gray's Inn Road,
London WC1X 8DU

Contents

xvii

Introduction

'Health for all by the year 2000' has become a main social goal of governments and WHO in recent years. As part of attaining this goal, efforts will be made to make available to all workers, including those employed in remote areas, preventive health care based on convenient and appropriate technology and workers' participation.

Recent international surveys in occupational health have demonstrated the occurrence of a wide variety of health problems affecting the working populations, particularly in the developing world. In addition to prevailing tropical diseases and malnutrition, workers in developing countries are increasingly exposed to new health hazards associated with industrialization and mechanization of agriculture. Large numbers of people working in agriculture, small industries, mining and construction are not provided with the essential protective measures against occupational health risks. This presents a serious challenge to national health services and reflects on socioeconomic development. It also calls for vigorous concerted efforts at the international level, to emphasize the need for technical co-operation among the developing countries and for the development of appropriate occupational health technology relevant to the real needs in various countries. To fulfil these needs it is of fundamental importance to increase the number of adequately trained health personnel in the field of occupational health with a broad understanding of epidemiology and preventive medicine.

Occupational health is no longer a narrow field of public health that merely concerns itself with the identification and control of specific occupational diseases. It aims at the health protection and promotion of health of workers, and the identification and control of health hazards in workplaces, including those not only of a physical, chemical and biological nature, but also psychosocial factors that may have harmful effects on the health and productivity of workers. It also aims at the recognition and control of work-related diseases where the working conditions, together with the general environment of living, play a role in causation, and which are susceptible to control by preventive measures taken at an early stage through appropriate occupational health practice.

Proper occupational health utilizes work as a factor in health promotion. Once occupational hazards have been placed under control, work processes and tools adapted to human capacities and limits can play an effective role in promoting mental and physical health. This phenomenon has not as yet been used to the advantage of health for people.

This new edition takes account of these modern trends, and particularly highlights the needs in developing countries.

H. Mahler

1

Developments in Occupational Health

R.S.F Schilling

Indifference to the health and safety of work people has been a feature of both ancient and modern societies. It is only since the Second World War, from the 1940s onwards, that there has been a rapid growth in occupational health on a worldwide scale. A brief historical review helps to reveal those factors that retard and those that accelerate developments (Sigerist, 1943). Important factors are the humanity of a society, its wealth and the economic need to conserve a healthy and efficient work force, the status of its workers, and a knowledge of occupational risks. These have all influenced countries, industries and individual workplaces to control hazards and promote well-being.

AGE OF ANTIQUITY, MIDDLE AGES AND RENAISSANCE

Mining is one of the oldest industries and has always been a hazardous occupation. Conditions in the gold, silver and lead mines of ancient Greece and Egypt reveal an almost complete disregard for miners' health and safety (Rosen, 1943). Since the miner of antiquity was a slave, prisoner or criminal, there was no reason to improve his working conditions because one of the objectives of this employment was punishment, and there were ample reserves of manpower to replace those who were killed or maimed.

Agricola and Paracelsus

The first observations on miners and their diseases were made by Agricola (1494–1555) and by Paracelsus (1493–1541). During the Middle Ages the status of the miner had changed. From being a feudal enterprise, manned by serf labour, mining in Central Europe had become a skilled occupation, which led to the emancipation of the miner. The growth of trade had created a demand for currency and

3

capital which was filled by increasing the supply of gold and silver from these mines. Mines got deeper and conditions worsened. When, in 1527, Agricola was appointed official town physician to Joachimstal, a flourishing metal mining centre in Bohemia, he described in *De Re Metallica* (Agricola, 1556) the diseases that prevailed in the mining community. At that time mortality from pulmonary diseases was not recorded, nor were the causes known, but they would have included deaths from silicosis and tuberculosis, and from lung cancer due to the mining of a radioactive ore in siliceous rock. Mortality must have been high, judging by the evidence of Agricola's statement that 'in the mines of the Carpathian mountains, women are found who have married seven husbands, all of whom this terrible consumption has carried off to a premature death'. Apart from improvements in ventilation, miners remained without any significant means of protection. However, they organized themselves into societies which provided sickness benefit and funeral expenses, giving them some security and preventing the extremes of social misery (Rosen, 1943). Such improvements as there were followed the changed social status of the miner, and the recognition by outstanding physicians like Agricola and Paracelsus of the extent and severity of occupational disease. Paracelsus based his monograph on occupational diseases of mine and smelter workers (Paracelsus, 1567) on his experience as town physician in Villach, Austria, and later as a metallurgist in the metal mines in that area. He relates, 'We must have gold and silver, also other metals, iron, tin, copper, lead and mercury. If we wish to have these, we must risk both life and body in a struggle with many enemies that oppose us.' Paracelsus realized that the increasing risk of occupational disease was a necessary and concomitant result of industrial development.

Bernardino Ramazzini

During the sixteenth and seventeenth centuries mining, metal work and other trades flourished in Italy following the Renaissance, which had encouraged the transition from feudalism to capitalism (Bernal, 1969). In 1700 Bernardino Ramazzini, 1633–1714 (*Figure 1.1*), physician and professor of medicine in Modena and Padua first published *De Morbis Artificum Diatriba*. It was the first systematic study of trade diseases and in it he put together the observations of his predecessors and his own, based on visits to workshops in Modena. Rightly acclaimed the father of occupational medicine, he showed an unusual sympathy for the less fortunate members of society. He was the first to recommend that physicians should enquire about a patient's occupation (*see* Chapter 3).

Ramazzini's interest in occupational medicine appears to have been inspired by the opportunities it offered to make new observations as well as by his sympathy for the common people. As a physician of that time he was probably unique. 'I hesitate and wonder whether I shall bring bile to the noses of doctors – they are so particular about being elegant and immaculate – if I invite them to leave the apothecary's shop which is usually redolent of cinnamon and where they linger as in

Figure 1.1 Bernardino Ramazzini (1633–1714)

their own domain, and to come to the latrines and observe the diseases of those who clean out the privies.' Neither his medical colleagues nor other people of standing had any strong humanitarian sense to inspire them to heed his words, nor at that time was there any economic necessity to protect the life and health of workmen. This is in direct contrast to the improved conditions of miners which had taken place in Central Europe a century earlier as a result of the combined effects of an awareness of their hazards, and a change in their social status.

THE INDUSTRIAL REVOLUTION AND NINETEENTH CENTURY IN BRITAIN

Towards the end of the sixteenth century the manufacture of cotton textiles came to England with the religious refugees from Antwerp. Spinning and weaving thrived as a cottage industry until the latter half of the eighteenth century, when mechanization transferred the making of textiles from people's homes to the new factories. Later the factory system spread to other industries in Europe and North America. This change in method of manufacture so unsettled the traditional routine of family and community life that it became known as the Industrial Revolution. Several forces led to these fundamental changes in the methods of manufacture. Science and technology had enabled the use of steam to be developed for motive power. There were large increases in the population of England and Wales. The breakdown of the strong central government of the Tudors and Stuarts, which had attempted to keep society geographically and socially static, allowed people to move from the country to towns, to man the new factories. From her commercial and banking enterprises oveseas, Britain had accumulated the financial resources to build new factories, and towns to house the people.

Thus the eighteenth century brought great technological inventions and laid the foundations, in Europe and North America, of modern society with its factory system. It exposed workers of all grades to the pressures of increasing production and associated physical and psychosocial hazards of work.

Forces which are not dissimilar from those preceding the Industrial Revolution have enabled rapid industrialization to take place in developing countries, where work people have been exposed to the same pressures and hazards. Such forces are the development of hydroelectric and other forms of power, increases in population, the end of colonialism and the financial and technical assistance made available to them by developed countries.

Effects of industrialization on community health

The more serious effects of health which followed the Industrial Revolution were not directly occupational in origin. Family life was disrupted when men moved into new industrial areas leaving their families behind. This was a situation which encouraged alcoholism and prostitution. Epidemics followed as a result of overcrowding in insanitary conditions. The change from peasant to town life led to malnutrition, made worse by the poverty and unemployment caused by fluctuations in the economy.

Workpeople moved from the countryside of rural England to the squalor and ugliness of the new industrial towns. The Hammonds (1917) described them as 'bare and desolate places without colour, air or laughter where man, woman and child ate, worked and slept'. There were a few sympathetic employers, such as Robert Owen, 1771–1858, and Michael Sadler, 1780–1833, who provided good working and housing conditions for their employees. But poor housing, overcrowding and lack of sanitation caused by the concentration of an expanding population around the new factories led to the development of the public health services which were designed to control disease and improve the health of these communities.

The health problems arising from industrial progress in developing countries today are, in many aspects, similar to those during industrialization in the nineteenth century. In addition, tropical countries have to face major threats of disease which are not directly related to work (*see* Chapter 2).

Effects of industrialization on workers' health

Inside the factories and mines of the nineteenth century the workers were exposed to hazards of occupational disease and injury and the adverse effects of excessively long hours of work.

As manufacturing techniques improved, machines became speedier and more dangerous. Little attention was paid to safety devices and workers were often simple people untrained to handle the new machinery. Toxic hazards increased by reason of prolonged exposure to a wider range of new chemicals introduced without considering their possible effect on workers. In the milieu of his cottage industry, the handloom weaver or the spinner had worked by the rule of his strength and convenience. He could take a break to cultivate his plot of land. In the factory these rules no longer applied, and he became exposed to the pressure of continuous work at a speed imposed by the needs of production – a pressure which dominates our society today and against which man so often rebels.

Humanists and public opinion

Man's indifference to his less fortunate fellow men was perhaps assuaged during the eighteenth century by the liberal ideas of men like Rousseau, Voltaire, Kant and Thomas Jefferson. Society was also influenced by the action of humanists like John Howard, 1726–1790, who led the reform of British prisons; William Tuke, 1732–1822, who set an example for the humane treatment of the mentally sick

(Glendenning, 1960); and William Wilberforce, 1759–1833, who started the campaign for the abolition of the slave trade. Later, the seventh Earl of Shaftesbury, 1801–1885 (*Figure 1.2*), evangelist and aristocrat, spent most of his life trying to relieve the conditions of the desititute and deprived in Victorian England. As a Member of Parliament he helped to promote legislation which reduced the hours and improved the conditions of work of women and young persons

Figure 1.2 Anthony Ashley Cooper, seventh Earl of Shaftesbury (1801–1885)

employed in mines, factories and other workplaces. His reforms were bitterly opposed by employers, but his powerful influence as an acknowledged man of integrity and a leading member of the aristocracy did much to relieve the oppressive conditions created by the Industrial Revolution in Great Britain.

In the nineteenth century, manufacturers generally believed it was economically important to keep their new machines running continuously with cheap labour. Even though the philosophy of the government at that time was to let the people be free and society would take care of itself, it was obliged to interfere because of public reaction to the adverse working conditions of women and children who were subjected to long and arduous hours and were unable to look after themselves. These conditions were compared with those of the Negro slaves whose lives had been made more tolerable by the abolition of slavery in 1833.

Insanitary conditions in factories and the dormitories attached to them introduced a risk of infectious disease. Influential people in the vicinity of the factories feared that the mills might be a source of contagion to themselves. Thomas Percival, 1740–1804, a Manchester physician called in by the people of Ratcliffe in Lancashire to investigate an epidemic of typhus, went beyond his remit and produced a report on hours of work and conditions of young persons. This report influenced Sir Robert Peel (Snr), a mill-owner, to introduce into the British House of Commons the first Factory Bill which became the famous Health and Morals of Apprentices Act of 1802. It limited hours of work to 12 a day, provided for their religious and secular education, and demanded the ventilation and lime-washing of workrooms. The Act was meant to be enforced by visitors appointed by Justices of the Peace, but it was ineffective. Nevertheless, the principle of Government interference was established and this early legislation culminated in the Ten Hour Act of 1847 which restricted the hours of work of women and young persons in factories to 58 in the week. This marked the beginning of the Welfare State – the principle of looking after those who were unable to look after themselves, such as the young, the old, the indigent and the sick.

The factors which encouraged more state control and made a generation of hard bitten employers give way to men who were socially more responsible, were the influence of enlightened employers and humanitarians, of medical men, and later of trade unions.

Enlightened employers

A small group of perceptive employers, like Sir Robert Peel, Robert Owen, and Michael Sadler, influenced Parliament to introduce new legislation to control hours of work of women and young persons.

Robert Owen became a mill manager in Manchester at the age of 20. Later he moved to Scotland to manage the New Lanark Mills where be became famous for his good management and humane treatment of work people. He refused to employ young persons of under 10 years of age. He shortened hours of work, provided for adult and child education, improved the environment, and still was financially successful. He persuaded Sir Robert Peel, the architect of the 1802 Act, to introduce legislation to protect young persons in all types of textile mills, to prohibit employment for those under 10 years of age, and to limit their hours to 10 a day. The Bill of 1819 was passed in the House of Commons but was emasculated in the House of Lords. Robert Owen's influence was limited by his professed atheism and socialism, which were wholly unacceptable to his manufacturing colleagues.

Medical influence

During the eighteenth and nineteenth centuries, facts about the ill effects of work on health emerged from the observations of a few physicians who followed the example of Ramazzini and took an active interest in the diseases of occupations. In 1775 Percivall Pott, 1713–88, had drawn attention to soot as a cause of scrotal cancer in chimney sweeps. The reports of Dr Thomas Percival on conditions in the mills of Ratcliffe influenced Sir Robert Peel to get the Act of 1802 passed through Parliament. Later, Charles Turner Thackrah, 1795–1833, (*Figure 1.3*), a Leeds physician, published the first British work on occupational diseases. Thackrah died of pulmonary tuberculosis at the

Figure 1.3 Charles Turner Thackrah (1795–1833)

age of 38, but not before he had made his mark, which earned him recognition as one of the great pioneers in occupational medicine. In 1832, Michael Sadler introduced in the House of Commons a new Factory Bill, later to become the Act of 1833 which created the Factory Inspectorate. During his speech he said: 'I hold in my hand a treatise by a medical gentleman of great intelligence, Mr Thackrah of Leeds.' He then quoted from the text of this book, entitled *The Effects of the Principal Arts, Trades and Professions and of Civic States and Habits of Living on Health and Longevity* (Thackrah,1832).

Measurements of occupational mortality were first introduced in England and Wales in the middle of the nineteenth century by Dr William Farr of the General Register Office. He used census population figures, and recorded deaths in certain occupations to calculate mortality rates. This drew attention to the gross risks of injury and disease in factory workers and miners at that time.

Edward Headlam Greenhow, 1814–88, one of the outstanding epidemiologists of the nineteenth century, used the unpublished records of the General Register Office to examine occupational mortality in more detail. He compared crude death rates from pulmonary disease in the lead mining towns of Alston and Reith in the North of England, with those of nearby Haltwhistle which had no lead mines (*Table 1.1*). He did not consider the possibility of differences in age distributions affecting the rates, but his conclusion that the near fourfold mortality excess in Alston and Reith was associated with heavy exposure to dust in the lead mines was almost certainly correct. Greenhow reached a general conclusion that much of the very high mortality from pulmonary disease in the different districts of England and Wales was due to the inhalation of dust and fumes arising at work. Under the influence of his reports, Factory Inspectors were given powers in the Factory Acts of 1864 and 1867 to enforce occupiers to control dust by fans or other mechanical means (Ministry of Munitions, 1918).

Table 1.1. AVERAGE ANNUAL DEATH RATES PER 1000 FROM PULMONARY DISEASE FROM 1848 TO 1854 IN MEN AND WOMEN AGED OVER 20 YEARS, IN THE MINING TOWNS OF ALSTON AND REITH AND IN THE NON-MINING TOWN OF HALTWHISTLE

	Men	*Women*
Alston	14.4	7.8
Reith	13.0	7.2
Haltwhistle	3.7	5.8

William Farr's successors at the General Register Office have made occupational mortality data more valuable by using purer occupational groups, standardizing for age, comparing the mortality of workers with that of their wives, in order to distinguish between occupational and socio-economic risks, and most recently, by standardizing for social class.

Thus, medical intelligence stemming from national and local mortality data (*see* Chapter 13), and from the testimony of individual physicians, had an early influence on the development of health and safety measures.

Medical pioneers have on occasions inspired the hostility of both their colleagues and manufacturers. John Thomas Arlidge (1822–99), an outstanding physician in the pottery district of North Staffordshire, devoted himself to the study of potters' diseases. A medical colleague wrote of Arlidge: 'He made an unfortunate beginning of his career by compiling statistics of the people working in the potteries which gravely reflected on the humanity of the manufacturers. He was instrumental in the appointment of factory surgeons for earthenware and china manufacturers, upon whom this entailed much expense. His medical friends were against him, and up to his death this feeling never died out' (Posner, 1973).

Early influence of trade unions

The French Revolution, at the end of the eighteenth century, had a profound effect on Britain (Fay, 1945). The government, confronted by war abroad and the threat of Jacobite revolt at home, adopted a policy of repression. The Combination Acts of 1799 and 1800, which made trade unions illegal, were not repealed until 1824. Their restraining effect is evident from the large number of unions which then came to life and enabled organized labour to exert its influence to obtain improvements in working conditions. At that time the unions were concerned with reducing hours of work and raising wage levels. Their interest in occupational health and safety came much later.

Development of industrial medical services

Government service

The Factory Act of 1833 introduced two fundamental innovations: the appointment of Factory Inspectors and the necessity of certification by a medical man that a child seemed by its strength and appearance to be a least nine years old, the age below which employment was prohibited in textile mills. Later the Act of 1844 gave Inspectors powers to appoint certifying surgeons in each district to introduce more uniformity into certification and prevent parents taking their children from one doctor to another until they got a certificate. With the advent of birth registration in 1837, age certification by the surgeons became redundant. The Factory Act of 1855 gave them new duties: to certify that young persons were not incapacitated for work by disease or bodily infirmity, and to investigate industrial accidents. Thus a rudimentary industrial medical service, the first of its kind, was introduced by law in Great Britain. Towards the end of the nineteenth

Figure 1.4. Sir Thomas Morison Legge (1863–1932)

century, workers in certain dangerous trades were required by regulations to be examined periodically by the certifying surgeons. These regulations applied to those making lead paints, lucifer matches and explosives with dinitro-benzene; and to those vulcanizing rubber with carbon disulphide and enamelling iron plates (Schilling, 1963).

To obtain knowledge of important industrial diseases like lead, phosphorus or arsenic poisoning, and anthrax, the principle of notification was introduced in 1895. The investigation of notified cases of occupational disease was added to the duties of these surgeons, who also had powers to suspend sufferers from work. The high prevalence of lead poisoning in the potteries and in white lead works, and the incidence of phossy jaw among the match makers, received a great deal of publicity. These events, and the need to deal with notifications and reports from certifying surgeons, led to the appointment in 1898 of Thomas Morison Legge, 1863–1932 (*Figure 1.4*) as the first Medical Inspector of Factories.

By his own researches, and through his axioms for preventing occupational disease, which he evolved during his 30 years of unique experience as a factory inspector, Legge made an outstanding contribution to occupational medicine (*see* Chapter 25). He ended his distinguished career in the civil service by resigning when the government refused to ratify an international convention prohibiting the use of white lead for the inside painting of buildings. For a few years before his death he became Medical Adviser to the Trades Union Congress and wrote his classic work *Industrial Maladies* (Legge, 1934).

Employers' services

Even before the Industrial Revolution there were isolated examples of occupational health services. In the eighteenth century the Crawley Iron Works in Sussex retained the services of a doctor, clergyman and schoolmaster for the benefit of employers and employees. In the nineteenth century a factory near Stirling in Scotland employed 'a medical gentleman to inspect work people and prevent disease'. The report of Michael Sadler's Select Committee (1832) on the employment of children is one of the main sources of knowledge of factory conditions at that time. It recalls the unusually enlightened actions of John Wood, a Bradford mill-owner, who employed a doctor and sent his work children to Buxton, or other health resorts, when they were 'overdone'. He had baths on the premises and his works had high standards of ventilation and cleanliness. The motives for setting up the small number of medical servicies in industry were at that time almost entirely humanitarian.

The first real impetus to the voluntary appointment of doctors by employers came after the passing of the first Workmen's Compensation Act in 1897. The larger firms appointed physicians as a means of protecting themselves against claims for compensation, rather than as a measure to protect employees. Unfortunately the industrial medical officer was often regarded by workmen as the employer's man – a suspicion which, however unfounded it may have been, has died hard (Meiklejohn, 1956).

THE TWENTIETH CENTURY TO THE OUTBREAK OF THE SECOND WORLD WAR

Great Britain

In Britain, the state gradually built up a statutory medical service for factory workers, provided by about 1800 part-time certifying factory surgeons (later called appointed factory doctors). Most of them were general practitioners, who were supervised in their work by the Medical Inspectors of Factories. They had three main tasks: to examine young persons under the age of 18 for fitness for work when starting employment and at annual intervals thereafter; to undertake periodic medical examinations of persons employed in certain dangerous trades; and to investigate and report on patients suffering from any of the notifiable industrial diseases, or injured by exposure to noxious substances. The limitations of this type of statutory service were obvious to the more enlightened employers who made provisions for the medical care of their employees. Later the government recognized the shortcomings of this Appointed Factory Doctor Service. It was abolished by Act of Parliament in 1972 and replaced by the Employment Medical Advisory Service.

Influence of war

The First World War (1914–18) introduced important changes in outlook towards the health of people at work in Britain. The desperate shortage of munitions in 1915 was not made good by long hours of work. This led to the appointment of the Health of Munition Workers' Committee which sponsored scientific investigations into the effects of work on health and efficiency. It studied new toxic hazards from handling explosives such as trinitrotoluene (TNT) and solvents used in making aircraft. There followed a rapid growth in first aid, and in industrial medical and nursing services. National survival was the motive for this new interest in occupational health.

The economic slump which followed the war slowed down developments. Nevertheless, the more enlightened and wealthy industries provided their own health services, because they realized that legislation laying down minimum standards of health and safety, and the statutory medical examinations of the certifying factory surgeons were inadequate. These services broadened in their scope and, generally speaking, their aim was to improve and maintain the employees' health, and not merely to protect the employers from compensation claims. As there was no systematic training of doctors and nurses in occupational medicine and the practice of occupational hygiene was almost non-existent, the achievements of these services were limited.

Developments in other countries

Developments in Great Britain illustrate those factors which have stimulated changes in attitudes towards the health of people at work. Changes in other countries followed similar patterns for much the same reasons. Developments in the United States of America are contrasted with those in the Union of Soviet Socialist Republics.

United States of America

The vastness of the USA and the wide range in the origin and culture of its settlers, produced a federation of States in which there was considerable freedom for each State to pursue its own policies for dealing with the problems of rapid industrialization. The State of Massachusetts passed the first Child Labour Law in 1835, and by 1867 had appointed a special police officer to enforce the law prohibiting the employment in factories of children under 10 years of age. Massachusetts was the first State to establish a Bureau of Labor Statistics. Other States followed suit, and these Bureaux eventually became State Departments of Labor with responsibilities for enacting and enforcing a growing range of codes to protect workers from long hours, hazardous processes and adverse environmental conditions (McKiever, 1965). The Federal Government dealt only with the control of working conditions for persons employed by or on behalf of the United States Government. By its Constitution, the main responsibility had to be left to individual States, which varied considerably in the standards of health and safety they demanded for people at work. The Federal Government created a Bureau of Labor in 1884, a Bureau of Mines in 1910, and the Office of Industrial Hygiene, as part of the United States Public Health Service, in 1914. These did much to encourage the promotion of occupational health, by undertaking research, by their education programmes and by giving advice to individual States on specific problems. The Federal Government also had an important influence on the development of occupational health through the funds it made available to the various States for setting up occupational hygiene programmes. As a result, in the three years before the Second World War 30 units were established to provide medical and hygiene services for the control of occupational disease. This federal activity created a body of occupational hygienists and enabled the United States to lead the field in environmental measurement and control in the workplace. Railroad, steel and mining companies were among the first industries to set up industrial medical services. Their in-plant health services provided by employers followed much the same pattern as those in the United Kingdom, with many of the large firms employing full-time medical officers.

Among the great pioneers in occupational health is Alice Hamilton, 1869–1970 (*Figure 1.5*). She spent 40 years of her life searching for occupational hazards which had been overlooked by industry and plant physicians (Hamilton, 1943). In 1910 she began her crusade with a survey of poisoning in the lead industries. She had to face opposition both from employers and from members of her own profession, one of

Figure 1.5 Alice Hamilton (1869–1970)

whom described her report on lead poisoning as false, malicious and slanderous. Nevertheless, her investigations led to improvements in working conditions and higher standards of medical surveillance. Working for Federal and State governments, and finally in the University of Harvard, she continued her investigations so that workmen might be protected against serious risks, such as silicosis in the Arizona copper mines, carbon disulphide poisoning in the viscose rayon industry, and mercurialism in the quicksilver mines of California. In 1919 Harvard paid Dr Hamilton the great compliment of appointing her Assistant Professor of Industrial Medicine. She was the first woman to be a member of the academic staff and one of the first ever to hold a university post in occupational health. She travelled widely and was able to compare the provisions made for the health of work people in many countries. During her visits to Europe in the 1920s she was surprised by the elaborate provisions for the study and treatment of occupational diseases in the USSR, which she rated as better than in any country she had visited (Grant, 1967).

Union of Soviet Socialist Republics

The first important phase in the development of occupational health in Eastern Europe began in the USSR after the October Revolution of 1918, and the second took place in other countries, such as Bulgaria, Czechoslovakia, Poland, Romania and Yugoslavia after the Second

Figure 1.6 F.F. Erisman (1842–1915)

World War (*see* page 22). Before the revolution Russia, like other European countries, at that time had no organized occupational health services, and generally there was little or no interest in this subject among members of the medical profession (Ministry of Health USSR, 1967). An exception was F.F. Erisman 1842–1915 (*Figure 1.6*), one of the founders of the science of hygiene in Russia. Erisman pressed hard for improvements in environmental conditions in factories, but his views were not acceptable. Before 1917 the Bolshevik party had formulated a health policy with two cardinal principles. Health services were to be free and to concentrate particularly on prevention. Alexander Semashko, who became the first Commissar of Health in the Russian Soviet Federative Socialist Republic, was one of the architects of this policy. One of his first actions as Commissar was to separate the medical schools from the universities, with the result that the content of teaching programmes was decided at a politcal level and not by physicians. The first Medical Institute outside the universities for training undergraduates was set up in Moscow shortly after the Revolution. In 1922 this Institute, later named after Semashko, established the first Chair of Hygiene of Labour. A year later a Research Institute* of Occupational Health and Safety was set up in Moscow. Health services in workplaces were organized as an integral part of all medical care in the USSR.

FROM THE SECOND WORLD WAR ONWARDS

During this period, developments in the provision of health care at work have been accelerated by war and economic expansion, and by changing patterns of work which can have adverse effects on mental health and well-being. These latter developments have emphasized the importance of the health of the organization itself, as well as that of its workers. The belief that occupational health services are economically worthwhile and the demands by workers for better conditions have further stimulated their growth.

Standards of service have been raised by the increasing number of professionally trained health workers, by major advances in the techniques for health and environmental monitoring and by improved methods of collecting and distributing knowledge about work hazards and their control through national and international agencies.

War and economic expansion

While the First World War had a positive, but somewhat ephemeral influence on developments in occupational health in Europe and North America, the Second World War and the economic expansion

---

*The first Institute of Occupational Health, the Clinica del Lavoro, was founded in 1904 in Milan. Later, in the 1930s, other European countries – Czechoslovakia, France, Germany and Romania – established academic departments for research and teaching.

which immediately followed, provided a stronger impetus for developments in countries all over the world. During the war industry had to employ the disabled, as well as the fit, due to the manpower shortages caused by the demands of the armed services. In Western countries, the emphasis on the functions of occupational health services changed from detecting unfitness to assessing ability for work.

The armed forces made special contributions by developing techniques for selecting personnel. They adapted military equipment to suit the soldier, sailor and airman in order to increase his fighting efficiency, and so gave a boost to ergonomics (*see* Chapter 20). Similarly, the need to get highly trained personnel, such as air or tank crews, back to active service as soon as possible led to substantial improvements in methods of rehabilitating the injured and sick (*see* Chapters 3 and 8).

A sustained period of high employment has further motivated expansion in occupational health. In many countries, however, with the slowing down of economic expansion due to the need for energy conservation, unemployment has risen. It is likely to rise still further because of restricted fuel supplies and the introduction of microprocessing systems, which will eliminate many repetitive jobs. Both the numbers employed, and the hours worked on production processes, may be reduced, but there will be an increase in service industries. More occupational health services are being provided in shops, offices, hospitals and universities.

In developing countries health services for the gainfully employed have been given high priority to improve their work efficiency and to deal with the major health problems associated with industrial progress. Persons who have been accustomed to the slow tempo of rural life and are ignorant of new processes, are especially vulnerable to the hazards of modern industry (*see* Chapter 2).

The health of the organization

Management, trade unions and health workers have become more aware of the importance of the relationship between the individual and the organization, and the manner in which this may affect health and well-being. Organizational factors such as role ambiguity, overload and underload, which adversely affect the mental health of the individual worker, are discussed in Chapter 23. It is, however, difficult to alleviate or control sources of organizational stress.

The traditional approach is to hope that problems will either go away or resolve themselves. A more constructive approach is to recognize that an organization has a personality of its own, with its own behaviour and state of health. This may be studied by looking at

In the USSR the trade unions have extensive responsibilities for health and safety through their technical factory inspectors who are broadly equivalent in their functions and powers to inspectors in Labour Ministries in Western countries (WHO, 1963).

Workpeople

In many countries it is now obligatory for workers to be represented at workplace level on safety committees and for them to appoint their own safety representatives. In Britain, through regulations which came into force in October 1978, safety representatives appointed by unions have the legal right to investigate potential hazards, dangerous occurrences, accidents and complaints, and to carry out periodic plant inspections (*see* Chapter 25).

In whatever way trade unions and workers participate in forming and executing health and safety policies, the common objective is that workers may be able to influence their own work situation towards higher standards of safety, health and improved job satisfaction.

ROLE OF GOVERNMENTS

The state's role in providing occupational health and safety varies enormously in different countries. It ranges from a complete state service to minimal provision where the government does little more than set standards through statutory laws. In the nineteenth century and early years of the twentieth, minimal provisions prevailed and were almost completely inadequate because of difficulties in enforcing the law.

Extensive state involvement has become necessary because of the increasing costs of providing health care for workpeople and the demands for higher standards. Within the last decade the United Kingdom's occupational health programme has been drastically reorganized as a result of the Health and Safety at Work, etc., Act 1974. Health and safety legislation is stricter and more comprehensive, covering almost all workers including the self-employed. There is an Employment Medical Advisory Service (EMAS) to advise the government, management, unions and staff of occupational health services. Similarly the Federal Government of the USA, through its Occupational Safety and Health Act 1970, has set up a new administration to enforce the law and a new Institute, NIOSH, to recommend hygiene standards, sponsor and undertake research and to finance training programmes for health workers.

There is no single government system applicable to all countries, though the aims of occupational health may be similar. Methods of achieving these aims will vary according to the form of government and the type of health service provided outside the workplace. Two distinct systems exist. In one the state provides the service, as in Eastern Europe. In the other, the state plays an advisory and supervisory role, encouraging or making it a statutory obligation for employers to provide their own services. This second method has been adopted by most countries in the European Economic Community (EEC).

Comprehensive services in Eastern Europe

In Eastern European countries where governments are responsible for providing comprehensive health care, services at the workplace are planned as an integral part of all medical care. This helps to avoid unnecessary overlap, to use resources where they are most needed, and to encourage the exchange of information about the health of the individual and the community. There may be difficulties in maintaining standards, of encouraging flexibility and adapting the system to meet local needs. Within Eastern Europe there are at least two different systems for providing occupational health services.

In the USSR, Czechoslovakia, Romania and Bulgaria, health services are organized into separate branches of therapeutic and prophylactic medicine. Therapeutic services are provided by the hospitals, polyclinics and medical departments of large plants, and preventive services by the sanitary and epidemiological stations (sanepids), which are located in towns, rural areas and in large workplaces. The physicians in the hospitals, polyclinics and large plants are responsible for all forms of medical treatment, including the diagnosis and treatment of occupational diseases; whereas the physicians in the sanepids assess and control the working and general environment and are in charge of the prevention of communicable and non-communicable disease. This system encourages integration of preventive medical services and, presumably, avoids duplication of treatment. The separation of treatment and preventive services has a possible disadvantage in that treatment may show where prevention is needed.

In Yugoslavia the organization differs from the above in two respects. The aim is to decentralize responsibility, giving cities more freedom to develop occupational health services according to local needs. Industrial plants are free to set up their own type of health service provided they stick to certain basic standards. At all levels there is closer integration of therapeutic and preventive services.

Services in EEC countries

Countries belonging to the EEC differ fundamentally from the socialist countries of Eastern Europe in that no government provides a comprehensive health care system, which embraces occupational health, although Italy plans such provision through its Servizio Sanitario Nazionale (EEC, 1977). Most of the countries in the EEC have accepted a policy based on the International Labour Conference's blueprint for health services in places of employment (Recommendation 112, ILO, 1959). This recommends that occupational health services should eventually be provided in all industrial, non-industrial and agricultural undertakings and public services. It lists three priorities: undertakings where health risks appear greatest, undertakings where workers are exposed to special hazards, and undertakings which employ more than a prescribed number of workers. The EEC (1962) recommended that services must be based on statutory requirements and not on voluntary efforts. The aim is state controlled, but not state run, occupational health services. It embodies a compulsory 'do it yourself' system.

CONCLUSION

Since the end of the Second World War there have been rapid and extensive developments. Group occupational health services have been set up to meet the needs of small plants. Most of the heavy industries and public organizations have established services. Many shops, offices, universities and hospitals have followed suit. Services which had previously provided only primary care, for example, those for hospital staff and students, now include an occupational health element. The scope of occupational health services has widened to meet health care needs which are not directly related to the effects of work on health. This is for the convenience of workpeople for whom community health services are absent or inadequate.

Changes for the better have not been universal. In developing countries, through a combination of failure to recognize needs and inadequate resources to meet them, services are often either insufficient or non-existent. Even in highly industrialized countries examples of good and bad practice may be seen side by side. Standards are dictated as much by attitudes of management and workers as by statutes. An important factor is the extent to which employers and workers want to have safer and healthier workplaces.

Where there is a health service much depends on the training and experience of its staff. In the past, when so little was known about the ill effects of work on health, a few pioneers encouraged or cajoled governments and employers to make improvements by revealing facts about loss of life and limb, and of sickness caused by disregarding working conditions. Standards of health and safety now depend on a more positive approach and more widely available expertise in occupational health.

REFERENCES

Agricola, G. (1556) *De Re Metallica*, trans. H.C. Hoover and L.H. Hoover (1912). *Mining Magazine, London*

Bernal, J.D. (1969) *Science in History* II *The Scientific and Industrial Revolutions.* Harmondsworth: Penguin

EEC: European Economic Commission (1962) 'Recommendations from the Commission regarding occupational health services in places of employment.' *Official Journal of the Common Market.*

EEC (1977). Standing Committee of Doctors of the EEC, GP77/113, Copenhagen

Fay, C.R. (1945) *Life and Labour in the Nineteenth Century.* Cambridge University Press

Glendenning, Logan (1960) *Source Book of Medical History.* New York: Dover

Grant, Madeline P. (1967) *Alice Hamilton, Pioneer Doctor in Industrial Medicine.* New York and London: Abelard-Schuman

Hamilton, Alice (1943) *Exploring the Dangerous Trades.* Boston: Little, Brown

Hammond, J.L. and Hammond, Barbara (1917) *The Town Labourer* London: Longmans Green

Hughes, James P., ed (1974) 'Cost effectiveness of occupational health programmes.' *Journal of Occupational Medicine*, **16**, 153–186

ILO: International Labour Office (1959). International Labour Conference, Recommendation 112. Geneva: ILO

Jaques, Peter (1977) 'Trade unions and the working environment'. *Journal of the Royal Society of Arts* **125**, 672–683

Lane, R.E. (1949) 'The care of the lead worker.' *British Journal of Industrial Medicine*, **6**, 125–143

Legge, T.M. (1934) *Industrial Maladies*, ed. S.A. Henry, London: Oxford University Press

McKiever, J. (1965) *Trends in Employee Health Services.* US Department of Health, Education and Welfare

Meiklejohn, A. (1956) 'Sixty years of industrial medicine in Great Britain.' *British Journal of Industrial Medicine, 13*, 155–162

Munitions, Ministry of (1918) Health of Munition Workers Committee. *Final Report*, Cmnd 9065, London: HM Stationery Office

Paracelsus (1567) *On the Miners' Sickness and Other Miners' Diseases*, in *Four Treatises of Paracelsus*, ed. H.E. Sigerist (1941). Baltimore: Johns Hopkins Press

Posner, E. (1973) 'John Thomas Arlidge (1822-99) and the Potteries,' *British Journal of Industrial Medicine* **30**, 266–270

Ramazzini, B. (1713) *De Morbis Artificum*, 2nd edition trans. W.C. Wright (1940). Chicago: University Press

Rosen, George (1943) *The History of Miners' Diseases: a medical and social interpretation.* New York: Schuman

Sadler Committee: Select Committee of House of Commons (1832) on Bill relating to labour of children in mills and factories

Schilling, R.S.F. (1963) 'Developments in occupational health during the last thirty years.' *Journal of the Royal Society of Arts*, **111**, 933–984

Sigerist, Henry E. (1943) Introduction to *The History of Miners' Diseases. See* Rosen, George (1943)

Thackrah, C.T. (1832) *The Effects of Arts, Trades and Professions and of Civic States and Habits of Living on Health and Longevity.* 2nd ed. London: Longmans. Reprinted in *The Life, Work and Times of C.T. Thackrah* by A. Meiklejohn. Edinburgh and London: E. and S. Livingstone, 1957.

USSR: Ministry of Health (1967) *The Systems of Public Health Services in the USSR*

WHO: World Health Organization (1963) *Occupational Health in Four Countries, Yugoslavia, the USSR, Finland and Sweden.* Geneva: WHO

WHO (1979) *Report on Study of Periodic Health Examinations of Workers exposed to Industrial Hazards*, ed. R. Murray. Copenhagen: WHO

WHO (1980) *Health Aspects of Well-being in Working Places; Euro Reports and Studies 31.* Copenhagen: WHO

2

Special Problems of Occupational Health in the Developing Countries

M.A. El-Batawi

With rapid industrialization and mechanization in developing countries, occupational health problems are becoming more prominent. Governments are paying more attention to them through the development of workers' health services and occupational health institutes and by training of personnel and the pursuit of field research. In some countries, new legislation is being enacted and new systems for occupational health care have been initiated.

This chapter reviews occupational health problems in developing countries, assesses the conditions of the services available and describes the trends in national and international developments.

GENERAL HEALTH PROBLEMS OF WORKERS

The predominant occupation in developing countries is agriculture. This is followed by small-scale industries, construction and extraction of mineral resources. Although industry is a new development in industrializing societies, work was an important activity of human beings in ancient civilizations and, as in Egypt, the interaction between work performance and health was graphically depicted in the script of Egyptian monuments several thousand years ago (*Figure 2.1*; Noweir, 1979). Some of the problems mentioned below may not be new, but recent changes have made many of them more apparent.

With the prevalence of endemic diseases and malnutrition and in the absence of medical screening, workers in developing countries are at risk of aggravating general health problems by inadequately controlled occupational hazards. This is an important feature in developing countries and requires a special approach.

Few reports are available on the type and magnitude of 'non-occupational diseases' among workers in these countries. The main

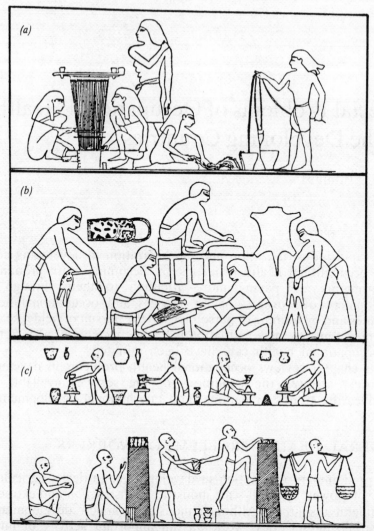

Figure 2.1 Industry in ancient Egypt. (a) Dyeing and hand weaving of textiles. (b) Leather manufacturing from hides including tanning. (c) Pottery and firing Source: Noweir (1979)

sources of such information have therefore been limited to the results of field studies. These reveal, for example, high rates of parasitic diseases, including malaria in Africa (Hall, 1970), respiratory and gastrointestinal diseases resulting from dual exposure to inadequate living conditions and various work hazards (El-Batawi, 1972).

Although statistical information is lacking, it is possible to speculate that liver damage caused by parasitic diseases may be aggravated by chemical exposures, particularly hepatoxic industrial solvents (*see* Chapter 5).

Malnutrition in working populations in many developing countries is a problem reflected in health and work output. As most occupations in these countries require a good deal of physical effort, there is a special need to deal with nutrition.

For these reasons, occupational health practitioners should always pay attention to the total health of workers in relation both to the work demands and environment and to overall living conditions, and public health officers should take account of the occupation of people in analysing the causes of morbidity.

Agriculture and small industries

Agricultural workers have a multitude of health problems, some of which are the result of work hazards. These problems are often forgotten because of the widespread misconception that occupational health is concerned mainly with industry and industrialized countries, and because of the lack of adequate data about agricultural workers.

'Small industries' often have uncontrolled health hazards, and the workers in them do not often have any health supervision. Workers in agriculture and in small industries form together the major part of the productive workforce in developing countries. They are inadequately served and their health and safety problems are often complex. They are less well paid than other workers and usually unorganized in labour unions. Many of them are not protected by occupational health and safety legislation. In most countries national health services, provided by rural and urban centres, pay little attention to occupational disease.

Workers in agriculture

Zoonotic diseases are acquired by workers handling animals or animal products. Such diseases as brucellosis, anthrax, leptospirosis and tetanus have, in some countries, been included in the compensatable occupational diseases (*Table 2.1*).

The use of pesticides is increasing, particularly as developing countries are concerned about higher agricultural yield. The main problems arise from poisoning by the more toxic organo-phosphorous and carbamate compounds. Under-reporting occurs because symptoms at the onset of poisoning are mild and shortlived, particularly in early cases, and are frequently ignored by employees. Serious cases of intoxication have been reported, however, from hospitals located in rural and suburban areas. Thousands poisoned by carbamates were reported in 1978 by the Ministry of Health of Egypt. Sri Lanka too has

Table 2.1. INFECTIOUS DISEASES PRINCIPALLY CONTRACTED THROUGH AGRICULTURAL OCCUPATIONS

Viral

Viral encephalitis, tick-borne (Russian spring–summer, louping ill, diaphasic meningo-encephalitis, European tick)
Viral haemorrhagic fever, tick-borne (Crimean, Omsk, Kyasanur forest disease)

Rickettsial

Q fever

Bacterial

Anthrax
Brucellosis
Erysipeloid
Glanders
Melioidosis
Leptospirosis
Tetanus
Tuberculosis, bovine
Tularaemia

Parasitic

Ancylostomiasis
Schistosomiasis

Source: WHO, (1962)

identified a serious problem. Reports of intoxication have also been made in countries in Africa, Asia and Latin America. Spraying without adequate protection appears to be the major cause of poisoning.

In agricultural work, exposure to pesticides is often associated with other factors, such as heat in the tropics, and in some instances malnutrition and parasitic infestation, which may increase susceptibility to poisoning. Protein deficiency, for example, was found to enhance the toxicity of most pesticides, particularly organophosphates and carbaryl (Shakman, 1974). Numerous factors, including quantity of body fat, levels of methionine, carbohydrates, riboflavin and nicotinic acid in the diet may offer partial explanation for increased susceptibility. Experimental studies on combined effects, as carried out by a WHO Collaborating Centre in Bulgaria, suggest the possible gravity of combined exposures. This is a problem that needs further studies and an epidemiological trial is currently in progress in the Sudan.

With the increasing use of agricultural machinery in developing countries, occupational injuries are becoming more frequent. The trend can be demonstrated by experience in some industrialized countries where the occupational accident rate for agriculture now ranks second to mining. The rapid spread of mechanization in agriculture in developing countries has reduced human effort and increased yield, but has brought with it hazards that had been confined to manufacturing industry.

Exposure of agricultural workers to vegetable and other organic dusts is widespread. Several occupational diseases have been described and some are included in the statutory lists of notifiable diseases in a number of countries, for example, byssinosis, farmer's lung and occupational asthma. Many dusts and their health effects have not been systematically investigated. Exposure to dusts of grains, rice, cocoa, coconut fibres, tea, kapok, tobacco and wood is common in countries where these products are grown. There is evidence that such exposures may cause obstructive respiratory disease and asthma.

Byssinosis is known to be caused by exposure to dusts of cotton, flax, sisal and soft hemp. In several countries where fibre processing is still a 'home industry', significant exposure occurs in homes, barns and farms among rural populations. Respiratory disability has been observed among young villagers exposed since childhood to flax dusts at home (El-Batawi and Hussein, 1964). Investigations in almost all countries producing and processing textile materials made of cotton or flax have revealed significantly high rates of byssinosis. In some instances byssinosis affected most of the workers exposed, and in others, depending on the dust concentration and duration of exposure, a prevalence of from 20 to 40 per cent has been observed, and permanent pulmonary disability has been found in some of the affected workers (Khogali, 1969).

Respiratory diseases have been observed during the author's field visits among workers in African countries. Respiratory allergy in handling cocoa beans following storage before export is one example. There have been various reports on different types of obstructive respiratory disease and asthma from exposure to vegetable and other organic dusts. There is, however, limited epidemiological information on prevalence and the relationship between airborne concentrations and health effects.

Workers in small industries

The size of the industry plays an important role in dealing with health and labour problems. It influences, for example, production, wages, work stability and turnover, and general conditions of work. Small industries are described as having the following characteristics (WHO, 1976):

1. A closer personal or face to face relationship between the owner/manager and workers;

2. Little, if any, specialization in the performance of management functions;
3. Smaller economic compass expressed in terms of number employed, capital, products, power, etc.

Examples are shops for the preparation, mixing and packing of crops such as tea, cocoa and spices; cotton ginneries, handicraft shops, stone crushing plants, printing works, electroplating, small foundries (*Figure 2.2*) and blacksmiths.

Figure 2.2 Preparing a mould in a foundry; no dust control

The small industries have been estimated to constitute almost 70 per cent of the work force in manufacturing and related trades (with a range of from 45 to 95 per cent) in developing countries. In spite of their large number and importance to the national economy, the standard of working conditions has tended to depend on economic and technical competence as well as on the attitude of employers or owners towards their workers. Many small industries are operated in the owner's own house or backyard.

Although the health problems in small industries are not unique to developing countries, it is likely that in these countries such problems are more serious. This may be attributed to a number of factors, including limited financial resources, a lower level of education and skill among the workers and owners, and insufficient health and safety care. Hours of work are often unregulated, wages generally low, welfare facilities almost absent and work premises may, in some instances, be dusty and overcrowded.

In some countries the legal provisions for protecting workers against occupational diseases and injuries do not apply to workplaces employing less than a certain number. Where there are legal provisions, enforcement of the law may be difficult because of the large number of work places to be supervised and their economic and technical inability to comply. They usually contain old machinery that is housed in primitive buildings where it may be difficult to introduce changes. At the same time, health supervision may be absent or limited to part-time visits by medical or nursing practitioners.

Small industries usually hire workers indiscriminately from many vulnerable groups, including the very young, the old and the partially handicapped. The lack of social amenities required to facilitate the employment of women imposes a strain on working mothers in connection with the care of their children (*Figure 2.3*).

Figure 2.3 Working mother with her children in a small glass factory, South-East Asia

Temporary work by contract may be encountered, for example in seasonal operations such as ginneries and in manufacturing certain products for large establishments. In such cases, workers may not be registered and their diseases and injuries may go unreported.

The World Health Organization made a review in 1976 of various health problems of workers in small industries based on field studies carried out in a number of countries. Workers were found to have a greater risk of suffering from toxic effects or fully developed occupational disease than those in large industrial concerns (WHO, 1976).

In 1975 a pilot study of health problems of small industries in the Republic of Korea (Cho and Lee, 1975) reported that among 3600 workers, employed in 60 small work places, the numbers exposed to potential health hazards were:

'to noise, 870 (24.2 per cent); siliceous dusts, 427 (11.9 per cent); lead, 268 (7.4 per cent); organic solvents, 563 (15.6 per cent); to various chemical substances – chlorine compounds, carbon monoxide, sulfur dioxide – 297 (8.3 per cent)'.

Out of the 268 workers in lead smelter and accumulator factories, there were 112, or 41.8 per cent, with signs of lead poisoning. Furthermore, there were 56 suspected cases of intoxication by solvents such as benzene, toluene and xylene from the 563 workers exposed in rubber and machinery workshops; a prevalence of 9.9 per cent. The overall prevalence of occupational diseases among 2630 workers at risk was 480 or 18.2 per cent.

Large industries and mines

The last two decades have witnessed a rapid expansion of manufacturing industries and mining. The change in some countries has been so sudden that there has not always been corresponding development in health care, particularly for the gainfully employed.

The rapid change from manual to mechanized agriculture and from traditional agricultural employment to industry simulates to some extent the Industrial Revolution in Europe in the eighteenth and nineteenth centuries, when occupational diseases and accidents started to appear more frequently. Although there are differences between the two eras, industrialization in developing countries at the present time is, like the Industrial Revolution, associated with changes in living and working patterns of people. These cause a variety of health problems. The migration of workers from their original habitat to industrial areas adds further health and social problems.

The first of these changes is manifested in the form of psychosocial stress in the adaptation of workers to various types of mechanized processes and to occupational and work-related disability. These are briefly dealt with below. The problems associated with labour migration are dealt with in Chapter 4.

OCCUPATIONAL AND WORK-RELATED DISEASE

Schedules of compensatable occupational diseases exist in most developing countries. Some schedules date from colonial times and may not represent the existing and prevailing conditions. Many do not contain certain diseases that are known to be occupationally caused, for example, noise-induced hearing loss and chronic respiratory diseases due to inhalation of organic dusts.

In developing countries the official reporting of occupational diseases is limited and the information available mainly comes from a number of field studies by different institutions. These studies provide evidence that the data on officially reported occupational diseases usually represent a fraction of their real occurrence. The main reasons for under-reporting are as follows:

1. Medical units in large industries and mines, often owned by multinational corporations, may be technically unable to establish a diagnosis or may be discouraged from doing so because of the possibility of legal action and costly preventive measures.
2. There are considerable difficulties in evaluating the role that occupational exposures have played in causing or aggravating disease. Such disease may be insidious in onset with a long latent period between exposure and demonstrable effects (this difficulty also occurs in developed countries).
3. There are the administrative and technical constraints of poorly developed national programmes for occupational health.

In spite of this under-reporting there is evidence that the major groups of occupational diseases occur frequently. They include the pneumoconioses and obstructive respiratory diseases caused by dusts; intoxications by various pesticides; poisoning by metals, particularly lead, and by solvents; occupational dermatitis, acute and chronic effects of respiratory irritant gases and vapours; and noise-induced hearing loss (WHO, 1979).

In addition there are work-related diseases in which work plays only a partial role in causation. Several examples of work-related diseases in developing countries have been reported. Chronic gastroenteritis and gastroduodenal ulcer have been described among workers in various occupations associated with inadequate dietary habits, irritants in the work and work stress, including alternating shifts (El-Batawi, 1972). Arthritis and locomotor disorders have been described as work-related diseases in persons with unduly heavy physical work, carrying heavy loads, inappropriate posture and differences in microclimatic conditions. There are also work-related syndromes such as silico-tuberculosis, low back pain and chronic bronchitis. For example, in the Republic of Korea the prevalence of tuberculosis may be as high as 26 per cent in pneumoconiotics, as against 3 per cent in the total working population (Cho and Lee, 1978).

OCCUPATIONAL ACCIDENTS

These present by far the most serious causes of disability due to work. They are more obvious and easier to report. In terms of loss of work days and payment of compensation they inflict heavy losses every year in most countries. In 1974 a survey in a number of countries in Latin America (WHO, 1974) showed that economic losses resulting from occupational accidents in terms of absenteeism and workmen's compensation were substantial. In Colombia, 900 000 cases of occupational accidents were reported in 1971 from among 4.5 million insured workers in one year. In Chile, in the same year, the occupational safety insurance schemes reported 74 800 disabling injuries from a population of 350 000 workers. In 1972 the corresponding number was 112 700 cases of injuries. In Bolivia in 1972 the population of 24 000 workers in large mines had 5430 injuries in one year. These figures give annual rates of accidents, causing disabling injuries ranging from 21 to 34 per 100 workers at risk. In other countries reports have shown similarly high rates. These are a few examples of accident rates in developing countries which are much higher than those in industrialized countries. For example, in the United Kingdom, the insured working population has an annual industrial injury rate of about 2.6 per 100 employed (Royal Commission on Civil Liability, 1978).

The three main factors that play a role in the high incidence of occupational accidents are inadequately controlled environmental factors, limited safety education and lack of protective equipment, and higher susceptibility attributable to difficulties in adapting to mechanized work and to low standards of general health.

PSYCHOSOCIAL FACTORS AT WORK

In developing countries psychosocial factors are important elements in health and productivity. The main problem is associated with the transfer of workers from traditional work methods to modern mechanization and assembly lines. On the other hand, there are other factors that help to improve the psychosocial environment.

An exact parallel cannot be drawn between the working conditions during the Industrial Revolution of the eighteenth century and those in developing countries at present. During the Industrial Revolution mechanization of processes was a gradual development associated with new inventions and discoveries which took a long time to implement, thereby providing a better opportunity for adaptation. The rapid introduction of complex work methods and tools in developing countries has been associated with psychosocial stress, and the main problem has been to adjust to rapid change. Workers have moved from their quiet and more intimate rural life to noisy impersonal factories; from a traditional dependence on natural methods and manual work, where the results of the labours can be seen, to standardized production, precise timing and dependence on energy. They are likely to move away from their friendly work associates to work with busy strangers, under different management and in a different hierarchy. The benefits expected, mainly from material rewards, may play an important role in motivation to meet such challenges. In many instances, however, this material reward may not be enough to compensate for the alienation and dissatisfaction, manifested in increased absenteeism and labour turnover, and by the occurrence of overt psychosomatic disturbances.

According to unpublished information, available to WHO, in one country in South-East Asia, the annual turnover for 1967 in a new steel mill was very high in the first two years. Most of the newly recruited workers were familiar with iron processing as blacksmiths in private shops, but the mill was forced to slow down production because of shortage of labour. This was explained by the difficulty in adapting to elaborate industrial operations, where pots of molten iron were handled, where workers were responsible for only limited parts of the production process and where communication was difficult and personal identification limited, despite the rewarding pay.

In another country, in the Western Pacific in 1970, where new coal mines had been set up, many of the miners used to stay in their jobs for an average of six months because of the remote location of mines, the high incidence of pneumoconiosis and casualties from underground collapses of roofs and walls. This resulted in a high labour turnover, despite the material attractions of pay, free housing, schooling and medical care for workers and their dependants.

Little study has been made of psychosomatic disorders in workers in developing countries, and no information is available on the occurrence of such behavioural problems as alcoholism or drug abuse. One study in Singapore (Chew, 1978), reported epidemics of 'hysteria' in young female workers from certain ethnic and cultural backgrounds. These episodes led to temporary suspension of work in the factories affected.

On the other hand, the positive effects on health associated with work should not be overlooked. There are employers in developing countries who maintain long-inherited traditions in dealing humanely with their workers. Religious traditions are upheld and employees take a break for worship in the premises of factories.

OCCUPATIONAL HEALTH SERVICES IN DEVELOPING COUNTRIES

Personal health and medical services

In large industrial and mining establishments and in some organized agricultural plantations there are medical units capable of dealing with out-patient and hospital cases of diseases and injuries. Many have preventive health services for workers and their families, for example, the Estate Health Schemes organized by the Planters Association in Sri Lanka, and the medical services in the plantations of Liberia and Malaysia. Health assistants provide health care in remote areas in these countries. Some of the medical units of big companies are quite elaborate in terms of personnel and equipment. Health insurance programmes and out-patient clinics and hospitals are commonly found in countries of Latin America and the Middle East.

In areas where farming is carried out at subsistence level and in small industries, rural and suburban health services belonging to government should be concerned with work hazards and should take responsibility for instituting preventive measures. Generally this provision appears to be lacking. In addition to primary care these services should be prepared to deal with the prevention of specific occupational health problems, such as toxicity of pesticides, respiratory diseases due to various dusts, and occupational accidents. This requires the training of health personnel, including health assistants, in simplified techniques for the evaluation and control of work hazards (WHO, 1979). Workers should be involved and encouraged to participate in initiating the services and stating their needs.

Legislation and administration

Legislation may cover occupational health and safety, workmen's compensation and, in many instances, social security. The minimum government role in this field is to set standards, inspect work places for hazardous exposures and require employers to comply with standards. This is the usual practice, particularly where work places and their health services are privately owned.

Several countries have already revised and updated their legislation to suit presentday requirements. In many developing countries, however, occupational health and safety legislation fails to regulate the provision of occupational health care to the underserved working population such as agriculture and small industries and makes inadequate provision for setting standards for health and safety services within workplaces. In addition, legislation in many countries fails to provide adequate guidelines for establishing occupation hygiene and safety standards for new workplaces.

Labour inspection

The number of trained inspectors is usually too small. Often they are not provided with the technical equipment required to detect and evaluate various hazards and they may have inadequate means of transport. On the whole, inspectors are underpaid. They are handicapped by bureaucracy and excessive delays in judicial proceedings. The effectiveness of inspection is often limited because the legal penalties for violations may be small or non-existent where the work establishments are owned by government. Inspection is incomplete, with very little attention paid to preventive medical and hygiene practice.

Co-ordination of national services

Workers' health may be the concern of several agencies in government and private sectors. For example, labour, and sometimes health, authorities are responsible for enforcing legislation on occupational safety, while social security covers therapeutic care in out-patient clinics and hospitals. Sometimes ministries, including those concerned with industry and transport, deal with various aspects of workers' health. In many countries the health authorities have an occupational health institute which provides training, research and advisory services. There are also groups of doctors, working in centres which provide treatment and rehabilitation for small and large industries, in return for payment by employers. Such overlap of services is generally associated with low effectiveness, which can be solved only by coordination or by integration.

Where there is limited involvement of national health services in occupational health, workers may be deprived of important aspects of health care. This can happen in the case of tuberculosis detection and control, immunization against communicable disease, health education and improvement of nutrition. Where labour ministries are handling the laws and inspection, with the weakness of enforcement referred to earlier, there can be little hope of dealing effectively with the prevention and control of health problems.

It is encouraging to see that in many countries health authorities have started to take action in occupational health. Some have developed central units, departments or institutes which are gradually developing their role in this field. It is essential, however, to co-ordinate the activities of government departments and to establish contact with industry and workers' organizations. Co-ordinating committees at the national level are useful, but coordination should be ensured at the periphery where health and labour officials should work together, providing health services and supervising workplaces.

Occupational health manpower

Health workers are in short supply everywhere in the developing world. According to the information available to WHO, the estimated proportion of medical and other health personnel employed by industry and other trades and/or in workers' social security and health insurance schemes, ranges from 2 to 35 per cent of the total health manpower in different countries. There is an outstanding shortage of persons trained in occupational hygiene, which leads to limited action in assessing and controlling hazards in the work environment. Training in occupational health nursing is also needed, as many nurses are employed at workplaces. Auxiliary personnel may be available in plantations and industries, and they too are in need of training.

Undergraduate teaching of occupational health to medical students is provided in many medical schools. The number of hours is limited. Postgraduate training institutions also exist in Egypt (both for physicians and hygienists), India, Singapore and the Sudan.

A large number of postgraduate fellows from developing countries, supported by WHO and other bodies, receive training in the United Kingdom, the United States and Eastern European countries. There are several short international courses on epidemiology and occupational toxicology, organized annually for physicians from developing countries. The main problem for those who get these fellowships is the absence in their countries of an established career in occupational health that would employ them effectively on their return.

It is not easy to assess occupational health manpower needs in developing countries. The magnitude of the problem can be measured by considering the nature and extent of occupational risks and other health problems, and the number of health personnel employed in the major types of industry. These numbers need to be related to the total of workers employed in different occupational sectors. It may be possible to set up hypothetical targets for occupational health manpower needs. Their achievement depends on the overall number of health personnel that may become available in the country and the allocations to other health fields. Another determining factor is the number that could be trained annually in occupational health by the available national institutions and other educational programmes. In some countries legislation stipulates a minimum number of workers in a workplace for whom the provision of part-time or full-time health personnel (physicians, nurses and safety supervisors) is obligatory. It is obvious that with the limitations in training facilities, these requirements can best be met with untrained personnel. In-service training, short-term courses, simplified guidelines and educational material may help to improve the situation. The fact that a substantial proportion of the nation's health personnel may be employed by industry to provide medical care offers an opportunity for government health departments to ensure that these medical services provide preventive health measures for workers and, where possible, for their families.

In some countries physicians employed by industry have been asked to comply with certain standards in their daily practice that would realize some of the aims of a national health programme in occupational health. At the same time, national health programmes should develop preventive health services for workers where services do not exist. One way of doing this is by occupational health services filling gaps in community services, and community services providing occupational health care where this does not exist.

Information and field studies

More information is needed on the nature and magnitude of workers' health problems in developing countries. Information is essential for programme planning and evaluation. Existing systems of reporting diseases are underdeveloped and fragmented and cannot help, either in planning or establishing priorities. Surveys and investigations of occupational health, and training of appropriate teams to carry them out, are important primary steps for initiating developments in occupational health. Information from field studies would also provide the justification for investing in workers' health programmes and would identify urgent problems.

SUMMARY OF NEEDS

There is a need to develop a new *understanding* of 'occupational health' in developing countries. A broader definition is required, which embraces general and occupational health problems, in order to make the best use of limited resources and to cater for the large numbers of underserved working populations.

Field surveys

In order to identify health problems and their magnitude, surveys need not be elaborate. They should aim at observing selected samples of workplaces, stratified by type and size, to evaluate environmental exposure and to examine workers for specified diseases in a standardized manner. Continuation of such studies would enable control measures to be assessed, would identify specific needs of countries and provide a flow of information for planning and evaluation.

Training of personnel

Training can be carried out at the national level or through systems of inter-country co-operation. Short courses have frequently been organized within the countries. In the design of training programmes, it is necessary to emphasize aspects that are of particular importance to developing countries, including related matters of general health such as nutrition, tropical diseases, health education and epidemiology in addition to the specific aspects of occupational health. Many countries have now reached the stage of being able to give training in various specific fields, such as occupational hygiene, ergonomics and toxicology. A few countries are developing their own institutions for research, training, advisory services and standard setting. It is necessary to train factory inspectors and health assistants.

Participation

Participation of *workers* in health programmes is highly desirable and should be initiated, stimulated and guided by health and safety education in work premises through labour unions and other media (*see* Chapter 25).

Equally important is the education of *employers* in both small and large workplaces.

Legislation

The introduction and *updating of legislation* in occupational health and safety is necessary in many developing countries. Account should be taken of past experience and of specific national priorities. Legislation should be comprehensive. The responsibilities and functions of government bodies, management and labour should be clearly defined and emphasis should be laid on coordination.

Legislation is of limited usefulness in the absence of an adequate and *well-equipped administrative body,* and it is essential to give special attention to enforcement.

Methods of evaluation and environmental exposure

Only few developing countries have the necessary facilities for evaluating workers' health and assessing standards of hygiene and safety in workplaces. The methods used for these purposes are described in later chapters. There are, however, several occupational health problems that are peculiar to developing countries. There is a need for both the application and *adaptation of technology in occupational health* so that these problems may be overcome. Some of the areas in which this is most necessary are:

1. Developing criteria, and methods of, periodic health examination and early detection of occupational and work-related diseases.

2. Developing simplified and more reliable techniques at low cost for detecting and evaluating occupational hazards.

3. Developing safety methods and personal protective devices.

4. Application of epidemiology in workers' health programmes, e.g. devising survey methods for the detection and control of psycho-social disturbances.

Occupational hygiene standards are mainly derived from the United States and European countries. They may not be appropriate for environmental and human conditions in developing countries, and do not cover certain occupational exposures which occur in rural communities, for example, wood and other vegetable dusts. Epidemiological studies would enable standards to be set for the commonly encountered occupational exposures in developing countries. They could also reveal some of the unknown effects of combined exposures, including endemic disease and high atmospheric temperatures and malnutrition, which may influence safety margins set for toxic exposures.

The role of health services

Health services in rural and suburban areas should concern themselves with the prevention of specific health problems related to occupations. Other national health resources available to industry should be used to the greatest possible extent for the development of preventive health care for workers and, wherever possible, their families. At the same time social security and health insurance schemes for industrial workers should develop preventive health services in the working environment and they should promote the suitable placement of workers.

Medical care units, provided by employers in work establishments, should also follow preventive programmes, in cooperation with national health services.

National occupational health institutes serve a wide variety of important functions, including the development of preventive measures to control hazards at workplaces, standard setting, research and training. Many developing countries have established institutes which have achieved notable progress in this field. There is a need for further development of such institutes and centres of research and teaching. Continuous contact and co-ordination at the international level are essential for the harmonization of methods, development of standards and co-ordination of research.

THE FUTURE

Countries undergoing development often suffer from poverty and disease and are learning that these can only be tackled by a co-ordinated attack on all the causative elements that are, to a great extent, interdependent. Health programmes have to be closely co-ordinated with all other measures intended to combat poverty, including industrial development, social enlightenment and education. Societies should not be carried away by the luring attraction of wealth through industrialization and the mounting figures of production of material elements. In the process they may forget the health and welfare of human beings. More precisely, they should not allow industry, production and work to be a source of illness, handicap and disability among their people.

Priority status for occupational health

The factors that are usually considered in setting priorities for different health services are the size of the population at risk and the degree of morbidity; the extent to which a particular programme contributes to overall socioeconomic development; its effectiveness, feasibility and cost; and its relative 'urgency'.

All these factors are relevant to a workers' health programme. The population at risk is substantial and growing, and its level of morbidity is close to that of the general population, quite apart from its vulnerability to additional health hazards at work. It is an area of health that directly reflects on human performance and on productivity, thereby contributing to socioeconomic development. The effectiveness and feasibility of the programme are enhanced by the fact that working populations are in most cases relatively easy to reach and many of their health problems are preventable. If effective action is not taken to set up workers' health programmes at the outset of industrialization, the long-term consequences to health and productivity may be serious and much more costly to society.

Technical co-operation among developing countries

Through international efforts of various kinds developing countries are increasingly aware of the value of technical co-operation whereby the resources available in one country may be shared and used in solving the problems in another. National inventories on occupational health, which were initiated in 1976 by WHO, have helped to identify resources, including personnel and their skills, laboratories and institutes. They have highlighted major problems and administrative constraints in dealing with them. Only a few of those inventories have been published, for example, that undertaken in Malaysia (Mahathevan, 1976). Nevertheless, Regional and Interregional Centres for research and development have been assigned to carry out surveys and training. Exchange of visits are taking place among countries and projects of long-term duration have been started. The World Health Organization, the International Labour Office and the United Nations Environment Programme are coordinating their efforts to raise standards of occupational health in developing countries by meeting outstanding needs.

Ultimate responsibility of management

There is no doubt that legislation is necessary, provided there is a reasonably effective machinery for implementing it. Nevertheless experience has shown that it cannot be the only, or even the major, means by which governments try to achieve objectives. There are essential components in human nature and behaviour. Conscience, understanding and co-operation help in the assessment of needs and the creation and development of effective health services where organizations are capable of establishing them on their own. Such a sense of responsibility for the health and safety of workers can be stimulated by education and training, and by demonstrating the economic value of a healthy work-force. It is undoubtedly as important as, or even more important than, legislation and enforcement, and more lasting and effective.

REFERENCES

Chew. P.K. 'How to handle hysterical factory workers.' *Occupational Health and Safety*, **47** (2), 50-54

Cho, K.S. and Lee, T.J. (1975) 'Une surveillance pilot en médicine du travail; problème des petites industries en Corée du Sud, ACMS.' *Cahiers de Médicine Inter-Professionelle*, **57** (1), 11–l2

Cho, K.S. and Lee, S.H. (1978) 'Occupational health hazards of mine workers.' *Bulletin of the World Health Organization*, **56** (2) 205–2l8

Egypt: Ministry of Health (1978) 'Poisoning by insecticide in Egypt', *Proceedings of the 19th International Congress on Occupational Health.* Yugoslavia (in press)

El-Batawi, M.A. (1972) 'Health problems of industrial workers in Egypt.' *Industrial Medicine,* **41** (2), 18–23

El-Batawi, M.A. and Hussein M.(1964 'Endemic byssinossis in an Egyptian village.' *British Journal of Industrial Medicine,* **21,** 121–138

Hall, S.A. (1970) 'The work of university based health advisory services for industry in East Africa, in a decade 1960-1969.' *Proceedings of the 16th International Congress on Occupational Health,* p.352–354, Tokyo

Khogali, M. (1969) 'Population study in cotton ginnery workers in the Sudan.' *British Journal of Industrial Medicine,* **26,** 308–313

Mahathevan, R. (1976) 'Occupational health in Malaysia.' *Medical Journal of Malaysia,* **30** (4), 273–278

Noweir, M.H. (1979) 'Highlights of broad-spectrum industrial hygiene research in a developing country.' *Journal of the American Industrial Hygiene Association* **40,** 839–859

Royal Commission on Civil Liability, United Kingdom (1978) *Statistics and Costings.* Cmnd 7054-11. London: HM Stationery Office

Shakman, R.A. (1974) 'Nutritional influences on the toxicity of environmental pollutants.' *Archives of Environmental Health,* **28,** 105–113

WHO: World Health Organization (1962) 'Occupational health problems in agriculture.' *Fourth of the Joint International Labour Organization/WHO Committee on Occupational Health. Technical Report Series,* **246,** 27

WHO (1974) 'Study of occupational health in the Andean countries.' Document, WHO: OCH/74.3

WHO (1976) Meeting on organization of health care for small industries, Geneva, 22–27 July 1975. Document, WHO: OCH/76.2

WHO (1979) *Occupational Health Programme, Progress Report* by the Director-General, Thirty-second World Health Assembly A32/WP/1

3

Work and Health

J.M. Harrington and R.S.F. Schilling

Those who are responsible for the care of patients need to be aware of the influence that work may have on health and health on work. An occupation may not only be the cause of disease or of physical and mental stress, it may also exacerbate non-occupational disease. Nevertheless, not all effects of work are adverse. It offers opportunities for creative and stimulating activity and frequently acts as the base for establishing social contacts and companionship. Doctors need to bear in mind these positive aspects of work when assessing a patient's fitness for employment.

Health may also have an influence on safety and capacity for work. Physical or mental limitations, either inherited or acquired through injury or illness may affect people's capacity to work efficiently and safely.

The staff of an occupational health service has a major responsibility for dealing with these aspects of medical care; but many workers are not covered by in-plant health services. Even where there is such a service, a sick worker will often seek advice first from his general practitioner or, in an emergency, from a hospital. Therefore the general practitioner and hospital doctor need to know how to ask questions about the patient's work and health, and when to consider the occupational factors that may have influenced the illness they are investigating.

THE OCCUPATIONAL HISTORY

Bernardino Ramazzini was the first physician to stress the importance of taking an occupational history. After quoting the advice of Hippocrates on the interrogation of the patient and relatives, Ramazzini adds another question – 'What occupation does he follow?' It is, he says, 'concerned with exciting causes and should be particularly kept in mind when the patient belongs to the common people. In medical practice attention is hardly ever paid to this matter though for effective treatment, evidence of this sort has the utmost weight' (Ramazzini, 1713). With his unusual experience of hazardous trades gained from visiting workshops in Modena, he realized that incorrect diagnosis may occur if the occupation is ignored and that inadequate or inappropriate treatment, based on such omissions, could prolong the patient's recovery or exacerbate his illness. A modern example illustrates this point. A farm worker handling mouldy hay which contains the spores of *Micropolyspora faeni* may develop an acute influenza-like illness and seek medical treatment. Failure to diagnose farmer's lung would mean bed rest for a few days until the symptoms subside and then the patient would return to work. This would allow further exposure to the fungus, which may render the worker permanently disabled by restrictive pulmonary disease. The correct diagnosis at the time of the first attack could prevent further exposure and the serious disability which may follow. On the other hand, the physician must consider the evidence carefully before ascribing the cause of a disease to the patient's occupation. A wrong assumption of this sort can harm the patient and may persuade him to change his job and take legal action which is almost certain to be unsuccessful.

Ramazzini's advice given over 250 years ago is even more relevant today with increasing industrialization and a wider use of toxic chemicals and physical agents. In addition, there are many more men and women, not just 'the common people' who may be exposed to such hazards. Furthermore, there are those in key jobs, such as higher executives, airline pilots and food handlers, whose illness, if they stay at work, may have serious or even disastrous consequences on the health and safety of fellow workers or the general public.

Donald Hunter, 1898 – 1978 (*Figure 3.1*), like Ramazzini, was a distinguished general physician who had a special interest in occupational disease, derived from his contact with patients who followed dirty and hazardous trades in the East End of London. In his classic work, *Diseases of Occupations* (1978) he gives the following useful advice:

'No doctor can be expected to be familiar with the details of all occupations and every working environment, but at least he should take the opportunity to study those industries which fall within the area of his practice ... there are few surer and quicker means of gaining a patient's confidence than the display of an intelligent knowledge of his job. Many workers are intelligent, cooperative and good witnesses. Although some may be deaf, disconsolate, forgetful, obtuse, garrulous, or monosyllabic, the worker is still the best witness of what happened.'

Where an occupational disease is suspected, Hunter's advice is to 'ask whether any similar illness has occurred in a fellow workman'. The answer to this question can provide valuable evidence of an occupational cause of disease and may be life-saving. All persons with responsibilities for the medical care of patients should use this simple epidemiological approach (*see* Chapter 13) to find out if others in the

Fig. 3.1 Donald Hunter, 1898–1978

same group are affected. Occupational physicians have the advantage here as it is relatively easy for them to talk to the patient's workmates, visit their workplace, assess their environment and review health records of the group concerned.

For example, tumours of the nasal mucosa are so unusual that their occurrence in clusters may indicate an occupational aetiology. In 1932 the occupational physician at a nickel refinery observed that in 11 years 10 cases of carcinoma of the nasal sinuses had occurred among workers at the plant. All but one had died. The risk of acquiring this type of cancer in this plant was later assessed by Doll (1958) as 150 times that of the general population. Changes in the plant process, made at the time of the original observation, eliminated the risk in this refinery. Failure to recognize the epidemiological warning signs in the 1930s could have led to the deaths of many more men.

Occupational history-taking

The type of occupational history to be taken depends on the detail required by the physician. In medical practice time is often precious, and unnecessary questioning has to be avoided. As a routine, the history recorded in hospital or general practice should include at least a description of the patient's present occupation. This is important in deciding if a patient's disease or disability may seriously impair work capacity or endanger his or her own health and safety, or that of other workers and the public (*see* page 62). When an occupational cause of the patient's illness is suspected a detailed history is necessary.

Present occupation

Questions about the present work situation include duration of job, hours of work, levels of responsibility, unusual job demands as well as types of exposures including psychosocial factors and other possible health risks (*Table 3.1*).

Table 3.1. *AIDE MEMOIRE* FOR QUESTIONING A PATIENT ABOUT HIS OR HER OCCUPATION

Duration, hours of work, shifts, level of responsibility, unusual job demands

Types of exposure

Chemical agents	– dust, fumes, gases and liquids
Physical agents	– high and low temperatures, noise, radiations, inadequate lighting
Biological agents	– e.g. infected materials
Ergonomic factors	– machine design, seating, etc.
Psychosocial factors	– methods of payment, joint consultation, lines of communication, work satisfaction

Any other health risks

Source: Based on World Health Organization (1973)

When the patient is suffering from an acute illness or symptoms of recent origin, and exposures at work are suspected of being the cause, the physician must methodically pursue all possible causes among the materials used in manufacturing processes or other types of work. Particular attention should be paid to changes in work practices which antedate the onset of the symptoms. Materials handled at work may be exotic and unusual. For example, one group of laboratory workers suffering from asthma and hay fever were eventually found to have become sensitized to a protein excreted in rat and mouse urine (Newman Taylor, Longbottom and Pepys, 1977).

Several problems arise in obtaining a history. The job title may not mean anything to the doctor. For example, in the brewing industry many job titles go back several centuries and a knowledge of the industry or an explanation from the worker is needed to decide what is involved in jobs such as masher, wort runner, bottoms presser, racker, titter, stripper, smeller and trouncer! It is, therefore, important to find out what the worker actually does.Even so, the risk may still be hidden because the hazard is not directly related to the occupation. For example, a cooper who repairs and makes barrels, could be exposed to lead dust contained in the barrels before repair.

Previous occupations

The present occupation is often not the one to cause the occupational disease. Previous occupations must be taken into account, especially in the case of a patient in an 'end occupation'. In the United Kingdom certain jobs such as car park attendants and passenger lift operators are primarily reserved for registered disabled people. If a patient is employed in such an 'end occupation', further enquiry into his or her previous work may elucidate the cause of disability.

Table 3.2. PNEUMOCONIOSIS DEATHS IN 'END OCCUPATIONS' IN ENGLAND AND WALES (1970/72) AT AGE 15–74

Occupation	Number of deaths
Warehousemen and storekeepers	28
Stationary engine drivers	23
Clerks	11
Guards and related workers	9
Caretakers	6
Salesmen	4

Source: Registrar General (1978)

The occupations of men dying of pneumoconiosis in England and Wales between 1970 and 1972 (Registrar General, 1978) reveal a substantial number occurring in those employed in 'end occupations' (*Table 3.2*). Such jobs as warehouseman, clerk and stationary engine driver are sedentary and require no special skills. Few deaths from pneumoconiosis occurred among those in more skilled occupations, such as proprietors and managers.

Detailed enquiry is particularly important in malignant disease since occupational cancers may have a latent period of up to 50 years. Where cancer or lung disease may have been caused by past exposures which took place many years ago it is essential to take the occupational history from the time the patient left school, and to record the dates and details of subsequent jobs. An investigation into the deaths among housewives from asbestos related diseases such as asbestosis and mesothelioma, revealed that during the Second World War they had worked in a gas-mask factory putting asbestos filters into masks (Knox *et al.*, 1968).

Careful histories taken by the hospital or general practice physician not only help individual patients, they can also play a part in preventing the recurrence of the disease when the findings are reported to the patient's occupational physician or to the government department responsible for health and safety.

<u>OCCUPATIONAL HISTORY</u>

A) FOR NEW EMPLOYEES

(i)   What job are you going to do for the Company? ....
....
....

(ii) When is the proposed starting date? ....

(iii) Please list your jobs, since leaving school, in the table below. The medical department is particularly interested to know of any hazardous materials you may have worked with in the past - such as dust, gases and toxic chemicals.

Date starting job	Date leaving job	Description of Job (including hazardous materials handled)	Employing company

B) FOR CURRENT EMPLOYEES

(i)   Give a short description of your current job...
....

(ii) How long have you been doing it? ....

(iii) How long have you worked for this Company? ....

Fig. 3.2 An occupational history questionnaire for use in an occupational health service

Work systematically from birth forwards for residence, and from leaving school for occupation, ensuring that no periods are omitted. Record ACTUAL YEARS job and residence started and stopped. Under 'Residence' record ACTUAL TOWN lived in (put 'OUTSKIRTS' if this applies). For villages and rural areas record county. For foreign residence, record only the country; seamen as "at sea".

Under "Exposures" give FULL details of any periods of work in coal and other mines, foundries, potteries, cotton/hemp, flax, asbestos and any other dusty jobs. Record also any exposures to irritating gas or chemical fumes.

DATES From To	INDUSTRY (inc. name of company)	JOB (Actual occupation)	EXPOSURES	RESIDENCE	Please leave blank

Fig. 3.3 A form used in taking a full occupational and residential history

Occupational health practice

In occupational health practice a history of past and present jobs is required as part of an employee's health record. An example is given in *Figure 3.2*. This should be kept up to date by recording subsequent job changes in the same organization. Previous employment may affect suitability for new work, past heavy exposure to asbestos, for example, being a reason for not employing a worker in a job where further exposure will occur. Where persons are to be employed in the manufacture or use of manmade mineral fibres (MMMF) previous exposures to asbestos should be recorded to avoid incriminating MMMF as the cause of the disease. Enquiries into a possible risk of occupational pulmonary disease among electric power plant workers, exposed to pulverized fuel ash, revealed that the few workers who had pneumoconiosis were ex-coal-miners (Bonnell, Schilling and Massey, 1980).

Special surveys

Where the investigator is searching for causes of disease (*see* Chapter 15), a history of all relevant previous exposures, occupational and environmental, has to be obtained. *Figure 3.3* is an example of the type of occupational and residential history to be taken in such a survey. It serves also as a guide for questioning a patient whose previous occupation may be responsible for his or her disease. In a survey where the cause of the disease is known, and the object is to examine other factors such as prevalence and severity, exposure and dose response effects, the past history may be simplified as, for example (*Figure 3.4*) in the medical surveillance of textile workers (British Occupational Hygiene Society, 1972). In routine health surveillance programmes, details of job changes and levels of exposure should be kept as an ongoing record for each worker.

DIFFERENTIAL DIAGNOSIS OF OCCUPATIONAL DISEASE

Occupational exposures can affect any of the organ systems of the body and may give rise to signs and symptoms simulating non-occupational disease. Wrist drop, abdominal colic and anaemia, when found in one patient, although they are not pathognomonic of plumbism, probably signify inorganic lead poisoning, but on their own could indicate other diagnoses. Most hospital physicians or general practitioners need to

Record on dotted lines number of years in which subject has worked in any of these industries.

	Yes	No
Have you ever worked in a dusty job?	☐	☐
At a coal mine	☐	☐
In any other mine	☐	☐
In a quarry	☐	☐
In a foundry	☐	☐
In a pottery	☐	☐
In a cotton, flax or hemp mill	☐	☐
With asbestos	☐	☐
In any other dusty job	☐	☐

If 'Yes', specify ....

....

Have you ever been exposed regularly to irritating gas or chemical fumes? ☐ ☐

If 'Yes', give details of nature and duration?

....

....

Space for additional questions on special risks or exposures.

Fig. 3.4 A simplified history of past occupations taken in the medical surveillance of textile workers

know the clinical manifestations of the more important toxic materials and their differential diagnoses; a detailed knowledge of industrial toxicology is beyond their scope. Many physicians are aware of the toxic compounds relevant to their particular field or specialization; although the standard medical texts are particularly deficient in listing toxic compounds, other than drugs, that can affect body systems. Psychiatrists should know of the commonly used chemicals which cause behavioural disorders (*see* page 139). A university lecturer who had consulted a psychiatrist on account of his anxiety and depression attended a course on health and safety in the laboratory. The lecture on metal poisons alerted him to the possibility that exposure to mercury had caused his illness. Examination of his laboratory revealed pools of mercury in the cracks between the floor boards (*Figure 3.5*), the airborne concentrations of mercury vapour being well above the threshold limit value.

The knowledge required by occupational physicians lies between that of a toxicologist and the general physician. The postgraduate training of occupational physicians and nurses requires them to know

Fig. 3.5 Mercury under the floor boards of a laboratory

in some detail the more common toxic compounds, as well as physical and infective agents, that can contribute to occupational disease. Awareness of the less common occupational risks will be required in specific industries. For example, a physician in the dye-stuffs industry needs an extensive knowlege of the symptomatology and treatment of a wide range of organic chemical poisonings, whereas the medical officer to a coal mine would be more profitably concerned with a detailed knowledge of dust related lung diseases.

It is useful for all physicians and nurses to keep in mind the occupational causes of the more common symptom complexes. The effects of exposures at work on the various organ systems are described in Chapter 5 (liver, renal system and skin) and Chapter 6 (central and peripheral nervous systems, blood, and respiratory tract).

The patient's pastimes and other exposures

The patient's pastimes may be an important factor in elucidating the cause of his or her illness. Bird fancier's lung, an extrinsic allergic alveolitis provoked by dust from pigeon, parrot and other bird droppings, will be missed if the patient's hobbies are not considered, and continued exposure will lead to severe and irreversible disability. An attack of asthma may follow the use of isocyanate foam by the do-it-yourself boat builder. People who paint their own houses and burn off old lead paint without taking adequate precautions have been known to suffer from acute lead poisoning. Many people are gardeners and are exposed to fertilizers, fungicides, insecticides and plants which may cause dermatoses or more serious poisonings.

Lead poisoning was a frequent hazard of the pottery and ceramics industry in Great Britain until the use of any but leadless or low solubility glazes was prohibited by law. Lead glazes continue to be used by the amateur potter because of ignorance of the risks involved. Such unwitting exposure to the insidious effects of cumulative lead poisoning is a risk not only to the potter but also to the recipients of the finished pottery. A woman visited her doctor showing signs and symptoms of lead poisoning which was confirmed by a high blood level, 5.9 $\mu mol/\ell$. The cause remained obscure until she told him that she was making her own lemonade in a lead-glazed jug bought in the 1920s (Williams, 1972).

Diseases such as haemolytic anaemia and agranulocytosis may be occupational in origin. On the other hand they are just as likely to be iatrogenic since numerous modern drugs may affect the blood and blood forming organs.

One area of potentially unrecognized occupational exposure is the risks a worker may take by doing an unauthorized 'second' job. These 'moonlighting' activities are not readily divulged to the enquiring doctor because they frequently involve tax evasion. However, engaging in uncontrolled practices in this way is often more hazardous than the admitted occupation, because of the clandestine nature of the employment.

An exceptionally rare element in the differential diagnosis of occupationally related disease was noted by one of us some years ago. It concerned an 'outbreak' of ascending peripheral neuropathy, alopecia and anxiety neurosis in an optical glass factory. Although thallium was eventually implicated, it came as some surprise to realize that those affected were being poisoned, not through occupational exposure, but through the activities of a psychopathic workmate (Harrington, 1972).

Exacerbation of non-occupational disease by work

There are diseases such as ischaemic heart disease, varicose veins, arthritis and peptic ulcer which have multifactorial aetiologies and may be work-related, but their control seldom depends on a single main causative agent. There are others such as chronic bronchitis, mental illness and the dermatoses which may be aggravated by work, although originally caused by non-occupational factors.

It may be difficult to determine the extent to which work is the cause or merely an aggravating agent. Nevertheless, patients with certain diseases should be advised against working in particular environments or be kept under regular surveillance if such work is essential.

Work-related diseases

Ischaemic heart disease

Much research has been done to investigate risk factors in ischaemic heart disease. There is some epidemiological evidence to suggest that the stress of high pressure occupations increases the worker's chance of developing a myocardial infarction (Russek, 1967). In a review of studies of psychological and social risk factors in coronary heart disease undertaken between 1970 and 1975, Jenkins (1976) found evidence of work overload and chronic conflict situations being related to risk of developing coronary heart disease.

The physician may have to decide what to do with the salesperson who has ischaemic heart disease. Attempts must be made to redeploy people so afflicted in an effort to diminish the stress. Much worry and overt illness might be saved if regular assessments of risk factors enabled the physician to concentrate on the high risk groups, such as cigarette smokers, hypertensives and those people with low density hyperlipidaemia or sedentary life styles (*see* Chapter 10). Physical exertion, either at work or at play, seems to exert a protective role in the pathogenesis of ischaemic heart disease.

Varicose veins

Varicose veins have been recognized for some time as being associated with hereditary and environmental factors. Women who stand at their work have a higher prevalence of varicosities than those who sit (Mekky, Schilling and Walford, 1969). The presence of symptomatic varicose veins in a shop assistant or waitress should be good grounds for suggesting that the condition be treated and more sedentary work be considered.

Arthritis and locomotion disorders

These are common diseases in working populations and have been extensively described as related to, or caused by work exposures such as joint strain from carrying heavy loads, excessive use of particular joints, work posture and changes in climatic conditions (humidity, heat and cold exposures). One of the commonest complaints – back pain – can be both occupationally induced or occupationally exacerbated. The incidences of these disorders and their severity can be reduced by changing work methods. Failure to make effective changes in work patterns may necessitate a change of job for those most seriously affected.

Peptic ulcer

Peptic ulcer has frequently been associated with occupational factors such as repeated anxiety, irregular shifts and meals. It has been found to be associated with the degree of responsibility that is carried by the job (Doll and Avery Jones, 1951). A study of health professions in the USSR showed that peptic ulcer occurred in 15 per cent of physicians, 8.6 per cent of nurses and 3.9 per cent of sanitary assistants. Higher work responsibility corresponded with a higher peptic ulcer rate (WHO, 1976). This could be explained by occupational stress, or equally well be the hypothesis that those with a conscientious, ambitious type of personality are particularly prone to develop the disease. Rigid rules should not be made for the employment of patients with peptic ulcer. However, shift work and work which entails irregular meal times are to be avoided.

Work aggravated diseases

Chronic bronchitis

There is an increased morbidity and mortality from chronic bronchitis in certain groups which have in common hard physical work and air pollution, both at work and in the general environment, as well as adverse socioeconomic factors. The available evidence suggests that occupational factors exacerbate rather than cause this disease (Gilson, 1970). Nevertheless, there can be little doubt that a chronic bronchitic should not be exposed to airborne dusts or irritant gases and fumes, nor be employed on heavy work, if respiratory symptoms are not to be worsened.

It is possible that chronic obstructive pulmonary disease caused by work may be misdiagnosed as non-occupational bronchitis. Much of the byssinosis in cotton workers in England was in the past regarded as Lancashire chronic bronchitis not related to work but to air pollution of the general environment.

Mental illness

There is no evidence that severe mental illness, schizophrenia or manic depressive psychosis are causally related to particular types of work. But there is evidence that social factors in the working group may precipitate neurotic illness in vulnerable persons and that their employment in certain occupations, particularly those entailing danger and responsibility, can lead to mental breakdown or psychosomatic illness (*see* Chapter 23).

Skin diseases

Certain non-occupational skin diseases such as psoriasis and eczema may be aggravated by work which is dirty or involves continual exposures to dust, liquids or heat. This may be a source of diagnostic confusion considering the high frequency of occupationally *induced* dermatoses.

Shift work

The siting of shift work in the section on work-aggravated diseases should not imply that the shift work is *necessarily* hazardous to health. Shift work is now an integral part of modern industrial society. About one-fifth of the working populations of most developed countries are engaged in such work schedules, and, although little increase is expected in the manufacturing industries, more people may be required to work shifts in service and computing operations in the forseeable future.

A review of the literature (Harrington, 1978) revealed few carefully planned and scientifically conducted studies. Entrenched views for and against shift work are widespread but are frequently based on inadequate scientific evidence and on studies of shift workers which are, at best, valueless, and at worst, frankly biased. The facts are that no evidence exists to suggest that shift workers experience excess mortality as a result of such work practices. Moreover, sickness absence is lower in shift workers than day workers (*see* Chapter 13).

Unfortunately, little work has been done on ex-shift workers who do seem to be more sickly than their colleagues who continue in such work. Whether this increased morbidity is caused by, or merely exacerbated by, shift work, remains to be determined. The problem is of sufficient magnitude, however, to cause 20 per cent of all persons starting shift work to give up such work and the commonest reason for quitting is a medical one.

There is no doubt that rotating shift work, particularly the night shift, causes marked disruption of circadian rhythm as well as sleep disturbances. Night work can cause considerable problems for the worker who has sociocultural activities which involve participation in groups outside the home. Subjective well-being, work performance and output per shift can all be detrimentally affected by night shift work, particularly if the work involved is boring and repetitive. Short-cycle working with frequent breaks in the work can vitiate any attempt at circadian rhythm adaptation but does seem to lessen the strain on workers of unpopular antisocial shifts.

So far as specific health effects are concerned, there is a balance of good epidemiological evidence linking gastrointestinal disorders, particularly gastric and duodenal ulcers, with shift work. Shift work seems to exacerbate peptic ulcers but there is insufficient evidence to state that it can *initiate* them.

The almost universal dislike of night work seems to stem from its 'unnaturalness' and the general disruptive effect it exerts on the worker's life. It is particularly disliked for its effect on social, familial and sexual activities. These effects are made worse by the poor working conditions frequently encountered at night, with minimal catering and medical facilities. Such conditions are potentially dangerous for the shift worker who has diabetes, heart disease or epilepsy. Management and organized labour rarely appear to work out optimal shift-working schedules appropriate to the workforce at a given plant, though this lack of communication appears to be becoming less common.

EFFECT OF HEALTH ON WORK

The effects of health on work may be considered at three levels. First, there are many people, particularly the young and middle-aged, who have no obvious health impairment. Nevertheless, they vary in their biological capacity for work. Their efficiency, and indeed their well-being, depend to a large extent on successful matching of human capabilities and job demands. This is often left to the individual, who may try several jobs before finding the one which seems to suit his or her abilities. Increasingly, applicants for jobs are helped in this selection by medical screening and vocational guidance, which are now regarded as essential in the selection and training of management and skilled personnel.

Secondly, there are those whose work capacity and efficiency have been impaired by illness or injury. Provided this impairment is not progressive, nor likely to be made worse by work and is not affecting the health and safety of others, it is better that such people should work. This is particularly so if work capacity is not substantially impaired. In time, many adapt to their limitations, or their disability gets less, so that their effectiveness is not noticeably diminished. Indeed, such people often compensate for their disability by being conscientious workers and good timekeepers.

Thirdly, and most important, there are those whose health may adversely influence the health and safety of fellow workers and the community. Examples of those who *have* to be fit for their job are airline pilots, vehicle drivers, railway workers and food handlers. Hospital and other health care workers, as well as senior executives who are responsible for managing people, should also maintain a high degree of fitness for their job (*Table 3.3*).

Table 3.3. OCCUPATIONS IN WHICH ILLNESS OR DISABILITY IS A RISK TO HEALTH AND SAFETY OF OTHERS

Airline pilots	
Drivers	– Public service vehicles, heavy goods vehicles, other motor vehicles (including chauffeurs)
	Cranes
	Fork-lift trucks
Railway workers	– Drivers, guards, signalmen
Food handlers	– Persons in contact with food in its manufacture, storage and sale; catering staff
Hospital and health care staff	
Executives	

All airlines and most transport undertakings have an occupational health service which applies rigorous pre-employment, routine periodic and post-sickness examinations (*see* Chapter 9) to air pilots and drivers of public vehicles. For example, the policy of London Transport is to preclude a man from being a bus driver if he has had a myocardial infarction, or has hypertension, diabetes mellitus, or epilepsy. Ischaemic heart disease provides the greatest risk. Over a period of 20 years, 32 London bus drivers became acutely ill at the wheel from ischaemic heart disease. Eight of these men collapsed and were unable to stop, causing serious accidents in six instances. Bearing in mind the number at risk, this is a low figure which might have been higher had it not been for strict medical surveillance (Raffle, 1974). Drivers in undertakings other than passenger transport need surveillance to reduce the risk of accidents to themselves and others (*see* Chapter 9).

Food handlers are a potential danger to the public. The main hazard is the transmission of communicable disease, especially the gastroenteritides. One of the functions of an occupational health service in the food industry is to screen staff for alimentary tract and skin infections, thereby protecting the food. All responsible food manufacturers regularly inspect plant to ensure that the environment, where product handling occurs, is kept scrupulously clean. Strict medical standards and close surveillance are necessary for all employees who have direct or indirect contact with food products (*see* Chapter 9). Such precautions apply also to those who prepare and serve food within a catering service. They are especially important in hospital catering.

Occupational health services are being increasingly provided in hospitals. As hospital staff are often recruited from immigrants and low socioeconomic groups there is a need for careful pre-employment screening for pulmonary tuberculosis and other infectious diseases, and for psychiatric illness. Hospitals also have a curious attraction, as a place of employment, for the physically and mentally sick. The occupational health service should be aware of these phenomena and plan their health surveillance programmes accordingly. Hospital staff are immunized against tetanus, diphtheria, rubella, poliomyelitis, tuberculosis and smallpox. They undergo continuing surveillance for infectious disease to protect them and their patients, particularly children, and those with lowered resistance to intercurrent infection.

Screening of executives is now common practice (*see* Chapter 10). One reason for its introduction has been the need for fitness among those who have to make decisions and supervise others. A psychiatric illness, alcoholism or cerebrovascular disease may cloud and distort judgement. The doctor may be in a dilemma where a clinical decision could have unpleasant consequences for the patient and possibly jeopardize the doctor's status with the firm. Nevertheless, failure to cope with such problems may be disastrous to the efficiency of the organization. Lord Moran describes graphically his difficulties as a doctor in caring for Winston Churchill when he was ailing but holding on to office as Prime Minister (Moran, 1966).

General practitioners and hospital doctors have to make decisions about the fitness of their patients on their own, especially where there is no occupational health service to turn to for advice. They need, especially, to bear in mind the occupation of patients whose illness or disability can be a risk to the health and safety of others (*see Table 3.3*).

MANAGEMENT AFTER RECOVERY FROM SICKNESS

If the worker has not fully recovered it is the physician's and nurse's job to help such disabled people to adjust to their incapacities and to guide them back to normality by recommending modification of their work. For some, shorter working hours may be enough. For others it may be necessary to reduce their work load permanently or to recommend retraining and resettlement in a new job, or even premature retirement.

Some principles of rehabilitation

An essential prerequisite of a doctor, who is rehabilitating a patient, is a thorough understanding of the patient's condition. This supports the argument that it is the job of specialists, but how effectively it can be done depends on their commitment and work load. If, for example, orthopaedic surgeons were to spend a lot of time on the minutiae of rehabilitation, they would have less time for consultative work and surgery. This is less true of rheumatologists whose patients often suffer from chronic diseases, the effects of which are continually changing and which, therefore, need to be kept constantly under review.

The opposing view is that rehabilitation can be applied in general to patients whose diagnosis and treatment are not undergoing constant revision and whose disorder is essentially stable.

Both these principles belong to rehabilitation medicine, which developed as a speciality because of the need to restore the physical and social functions of large numbers of young men injured in the war. The resources which met those needs are being adapted to cope with chronic disease, which is a major problem in many countries. These resources vary from simple common sense to the full-scale rehabilitation unit.

The rehabilitation of patients with coronary thrombosis has improved enormously now that they are encouraged to return to progressively greater physical activity within a matter of a week or two following an uncomplicated myocardial infarction. This rarely requires any facilities or therapists at all, although gymnasia are often useful. Atheromatous vascular disease is progressive, but for the majority of patients a further infarction may be a long time coming and it is never too late to lose excess weight, stop smoking and become much fitter.

The rehabilitation of the traumatic paraplegic requires a concentration of resources, usually in a specialized centre. Only those patients who can be shown to benefit should be treated in this way. Initial assessment of a patient's disability, aspirations and potential for improvement is fundamental. Continual reassessment, to decide when these resources are no longer required, is also necessary. Such specialized centres are often needed to rehabilitate people with comparatively minor impairment who, if managed differently at an earlier stage, might not have become so disabled. In the Camden Medical Rehabilitation Centre in North London the average duration of disability, following uncomplicated fractures of the tibia and fibula, is 8.7 months.

Stages in rehabilitation

The stages in rehabilitation consist of diagnosis, likely prognosis and follow-up. Prognosis tends to evolve with time and may become self-evident to all, including the patient. Where an accurate diagnosis has been made, it is sometimes thought that the observed variations in the duration of disability are almost entirely out of the physician's hands and are in those of the patient. The conventional view of capsulitis of the shoulder tends to be that, in spite of whatever therapy is used, the condition will inevitably and spontaneously resolve somewhere between nine and eighteen months from the onset. There is little data concerning the natural history of such common rheumatic problems, and controlled trials of the various treatments are even more lacking. Clinical experience indicates that the progress of what is apparently the same lesion can differ remarkably from one patient to another both in its response to treatment and in whether it is an isolated lesion or part of a syndrome of neck, shoulder and arm pain. With a better understanding of the condition, a more accurate prognosis could be given. Recent developments in fields as widely dispersed as immunogenetics and nuclear physics, indicate that the pathology of the 'frozen shoulder' is heterogenous. Treatment and prognosis, therefore, may need to be varied, depending on the underlying pathology.

There is an association between a narrow neural canal and the likelihood of developing back pain and sciatica. Perhaps people with narrow spinal canals should not work at the coal face or in other jobs which carry a high risk of spinal injuries. This may provide a fresh set of dilemmas, but also opportunities to reduce spinal injuries.

The reasons for follow-up vary from the need for continuous surveillance, as in monitoring penicillamine treatment for rheumatoid arthritis, to the 'special interest' follow-up of patients with asymptomatic or untreatable disorders. The patient's needs and competence are reassessed at intervals to find out if there has been a change which requires alteration in the provision of care. Much of this reassessment need not be done by doctors but can be the concern of therapists or social workers. A change in the patient's ability to cope may indicate a need for reviewing the diagnosis and prognosis.

Employment is dealt with more fully in Chapter 8. Co-operation is essential between the clinician, whether in family or hospital practice, and the place of work. Too often the clinician is expected to shield the patient from an unsympathetic employer. The fact that the hospital doctor is less often confronted with employment problems does not indicate that they are rare, but merely that they are not sought. When

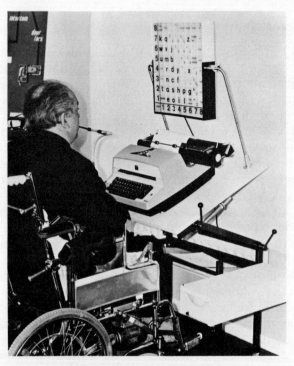

Fig. 3.6 Typewriter used by man with no hand movements due to spinal injury. It is operated by gentle sucking and blowing down a tube. (Reproduced by permission of Possum Controls Ltd. Middlegreen Trading Estate, Slough, Berks.)

such employment difficulties become apparent, the hospital doctor may discover that the patient has lost his job and even may have been put on the permanent disability list. Early consultation between the hospital doctor, the patient's general practitioner and the occupational physician or nurse often prevents such unfortunate occurrences and helps to get those who are permanently disabled resettled and, where necessary, retrained to do suitable work.

Recent developments in electronic aids enable people with very severe physical disability to exercise efficient and effortless control over electromechanical devices. For example, there are typewriter control systems which make it possible for a disabled person to type at speeds in excess of 40 words per minute. These systems can be used for employment or education, or for communication by persons who are unable to speak or who are paralysed. They can be operated by residual movements, such as the flicker of a finger or gentle mouth suction (*Figure 3.6*).

Specialist advice

The majority of rehabilitation problems do not need the attention of a specialist in rehabilitation medicine. The responsible clinician should consider them from the outset. The specialist should be consulted, as any other specialist is consulted, when his opinion, or the facilities he can offer, are needed. The case, for example, of a traumatic paraplegia needs no discussion. This patient will be passed on to the occupational therapists and physiotherapists as soon as possible and may need to go to a specialist rehabilitation unit. The stroke patient, on the other hand, is probably not referred soon enough and will undoubtedly need the help of therapists as soon as possible following the stroke. Most fit young people with injuries rehabilitate themselves within a few months, provided the motivation is there and the diagnosis and therapy have been adequate. The responsible clinician should consult a rehabilitation medicine specialist when progress has been too slow. Equally the physician in rehabilitation medicine must not take on patients if he has nothing more to offer than the referring clinician. This avoids wasting scarce resources.

VULNERABLE GROUPS

Disabled persons are a vulnerable group, but three other types of vulnerability need to be considered: age, sex and the unborn child.

Age

The extremes of the working age groups in most developed countries are the 15–20 year olds and the 60–65 year olds. These age strata are not universally applicable. Child labour still continues in many countries and the employment of persons past the formal retiring age is increasingly prevalent.

It used to be compulsory in the United Kingdom for all young persons entering employment from school to be examined by the

Appointed Factory Doctor (AFD). The Employment Medical Advisory Service Act (1972) abolished the AFD system and with it went the routine examination of school leavers. Although only a minority of young people have significant disease, much of the responsibility for its detection and management now falls on the occupational physician or nurse undertaking the pre-employment health screening. Even in the absence of specific disease, adolescence is a time of traumatic change. It involves a major adjustment to the social, physical and psychosexual mores of the adult population. These adjustments may be made more difficult by the dramatic change from school to factory. Sympathy and understanding may be needed to ease the late adolescent through this period and the occupational health practitioner should be prepared, and able, to counsel the young person on a wide variety of matters from adjustment to the new job and place of work, to advice on drugs, sex and complaints such as acne.

Physical development may be slow in some school leavers, and the work may need to be modified to cope with the still growing employee. The physician or nurse must be aware of the limitation on hours of work for persons under a particular age (18 years in the United Kingdom) and the concomitant ban on night work. There are anomalies in the law, however, as these restrictions tend to apply to factories but not necessarily to other places of employment.

In older age, the disparity between physiological and chronological age may be even more marked than in young people. Some 70-year-old people are still physically and mentally capable of a full 40 hour working week, while for others the ageing process is taking its toll before they reach their sixtieth birthday. The matter is complicated by the confusion between the ageing process itself, which proceeds at a variable rate in different organ systems, and ageing accompanied by disease.

Normal old age brings with it a variety of physiological changes, such as a fall in lean body mass, intracellular water content, cardiac output, urine concentrating ability, glomerular filtration rate, vital capacity and locomotor function. Repair of damaged tissues takes longer and mental and bodily function may recover less well from injury. Decreased cardiovascular capacity in older subjects reduces their ability to withstand the continued effects of exposure to physical and chemical agents such as a high physical work load and carbon monoxide exposure (WHO, 1975).

Certain diseases, such as pernicious anaemia, rheumatoid arthritis and acute leukaemia, are now common in the elderly. Some illnesses present in atypical forms. Neurological and cerebrovascular diseases such as parkinsonism and stroke are most common in this age group. The occupational health practitioner has to bear in mind the increased vulnerability of older people to toxic hazards and their inability to cope with normal work loads because of chronic disease and hormone imbalance, and the wide range of medicaments which many of them take.

Sex

Campaigns for equal rights for women have emphasized the unique problems that may be faced by women at work. Some of these problems are related to the need to protect the unborn child (*see* page 70). Others relate to the special employment problems of women (Bingham, 1977). More and more women are seeking regular employment outside the home and their needs have to be considered. In many ways, men might be more appropriately called a vulnerable group compared with women, in view of diseases which selectively afflict men. Nevertheless, women are traditionally viewed as weak and frail compared with men and, it is argued, that they therefore lack the stamina, physical strength and manual dexterity required to perform many jobs previously restricted to men.

There is little doubt that women are less physically strong than men but there are always exceptions to this rule and the mean difference in muscular strength of comparable groups of men and women is not more than 30 per cent. It makes more sense to fit the job to the capabilities of the most appropriate applicant of either sex. Another argument supporting the exclusion of women from certain jobs is that they are more 'susceptible' to certain toxic substances, for example lead and benzene. The scientific evidence for such a statement is meagre. Excluding childbearing, men and women are more alike than different in terms of biochemical and pathophysiological processes.

It is in their life styles that a major difference exists. Working women frequently remain the primary caretaker of the home and family. Their working day is often four to six hours longer than that of their husbands or work partners. Another relevant aspect of life style is that women tend to drink less alcohol and smoke fewer cigarettes than men, and therefore may be less at risk from a number of diseases which may affect work performance.

For most women, the ability to perform mental and muscular work is not essentially altered by the menstrual cycle. In general, only a small percentage of women suffer from dysmenorrhoea sufficiently to cause lost time. The type of work may effect menstruation. Irregularity of the menstrual cycle appears to be associated with the stress of repeated time zone changes among air stewardesses (Preston, 1976). Thus women are not necessarily a vulnerable group, if childbearing is excluded. They are however, frequently discriminated against on unscientific evidence. The need to fit the job to the person and not the person to the job is probably the most important consideration, overriding as it does any consideration of sex.

The unborn child

The mutagenic, teratogenic and childhood cancer risks associated with occupational hazards are outlined in Chapter 6. Males and females are both at risk to substances which can alter reproductive function. However, in addition to the mutagenic hazards, women need protection by virtue of their role as the carrier of the unborn child. Transplacental passage of toxic substances can cause teratogenesis and childhood neoplasia; these risks are difficult to quantify.

A simple solution is to exclude women of childbearing age and who are likely to become pregnant from exposure to substances potentially harmful to the fetus. This is difficult if not impossible to achieve in practice. It is not easy to differentiate women who wish to become pregnant from those who do not. Even if it were feasible at one point in time, the whole situation might have altered in six months as partners decide on the need or otherwise for a family; the clinical manifestations of pregnancy occur too late to protect the fetus at its most vulnerable period – the first trimester. Thus, in order to protect the fetus, it would be necessary to exclude women between the ages of 15 – 45 from an increasingly large range of jobs. A possible alternative is to alter the working environment so as to produce minimum toxic risks and maximum protection for the *whole* workforce. This still begs the question of what is a safe *fetal* level for various toxic substances.

Until more precise data are available on the best way of protecting the fetus, each hazard and process must be considered individually and everything possible should be done to ensure maximum protection against occupational disease (*see* Chapter 25).

CONCLUSION

The interactions between work and health can be complex. While the occupational health practitioner has a central role to play in ensuring that the worker and the job are appropriately matched, much also depends on hospital doctors, general practitioners and nurses being fully aware of these interactions.

REFERENCES

Bingham, E., ed. (1977) *Proceedings of Conference on Women and the Work Place.* Society for Occupational and Environmental Health, Washington DC

Bonnell, J.A., Schilling, C.J. and Massey, P.M.O. (1980) *Annals of Occupational Hygiene,* **23,** 159–164

British Occupational Hygiene Society (1972) 'Hygiene standards in cotton dust.' *Annals of Occupational Hygiene* **15,** l65–192

Doll, R. (1958) 'Cancer of the lung and nose in nickel workers.' *British Journal of Industrial Medicine,* **15,** 217–223

Doll, R. and Avery-Jones F. (195l) *Occupational Factors in the Etiology of Gastric and Duodenal Ulcers.* MRC Special Report Series 276 London: HM Stationery Office

Gilson, J.C. (1970) 'Occupational bronchitis.' *Proceedings of the Royal Society of Medicine,* **63,** 857–864

Harrington, J.M. (1972) 'How to miss a murderer in one uneasy lesson.' *World Medicine,* **7,** 25–27

Harrington, J.M. (1978) *Shift Work and Health. A critical review of the literature.* Health and Safety Executive, London: HM Stationery Ofice

Hunter, D. (1978) *The Diseases of Occupations.* 5th edition. London: English Universities Press

Jenkins, C.D. (1976) 'Psychological and social risk factors in CHD.' *New England Journal of Medicine,* **294,** 987–1033

Knox, J.F., Holmes, S., Doll, R. and Hill, I.D. (1968) 'Mortality from lung cancer and other causes among workers in an asbestos textile factory.' *British Journal of Industrial Medicine,* **25,** 293–303

Mekky, S., Schilling, R.S.F. and Walford, J. (1969) 'Varicose veins in women cotton workers: an epidemiological study in England and Egypt.' *Brtish Medical Journal,* **2,** 59l–595

Moran, Lord (1966) *Winston Churchill: the struggle for survival.* London: Constable

Newman Taylor, A., Longbottom, J.L. and Pepys, J. (1977) 'Respiratory allergy to urine proteins of rats and mice.' *Lancet,* ii 847–849

Preston, F.S. (1976) Health of female air cabin crew. Symposium on Health of Women at Work (April 1976) organized by the Society of Occupational Medicine, 11 St Andrew's Place, London, NW1

Raffle, P.A.B. (1974) 'Fitness to drive.' *Medical Society of London,* **90,** 197–205

Ramazzini, B. (1713) *De Morbis Artificum.* Geneva. Trans. W.C. Wright (1940). Chicago University Press

Registrar General (1978) *0ccupational Mortality.* Decennial Supplement England and Wales 1971. London: HM Stationery Office

Russek, H.L. (1967) 'Emotional stress and coronary disease.' *Diseases of the Chest,* **52,** 1–6

Williams, M.K. (1972) Letter to the *Lancet,* ii, 480

WHO: World Health Organization (1973) *Environmental and Health Monitoring in Occupational Health.* Technical Report Series 535 ,WHO, Geneva

WHO (1975) *Early Detection of Health Impairment in Occupational Exposure to Health Hazards.* Technical Report Series 571, WHO, Geneva

WHO (1976) *Occupational Health Programme.* Report by the Director General. 29th World Assembly

4

Migrant Workers

J.C. McDonald

INTRODUCTION

Article 10, International Labour Organization Convention 143 (1975)

'Each Member for which the Convention is in force undertakes to declare and pursue a national policy to promote and to guarantee, by methods appropriate to national conditions and practice, equality of opportunity and treatment in respect of employment and occupation, of social security, of trade union and cultural rights and of individual and collective freedoms for persons who as migrant workers or as members of their families are lawfully within its territory.'

Beginning in 1949, member nations of the International Labour Organization agreed on a series of Conventions and Recommendations to secure and protect the human, civil and political rights of migrants. Although there can be few countries where the letter, let alone the spirit, of these obligations has been fulfilled, the existence of this international moral pressure is a notable step forward. Migration to find work and a better way of life, even survival, is as old as mankind. Unfortunately, industrialization and other economic developments are determined by many factors other than the availability of labour. The economic and political desirability of encouraging the transfer of capital and technology rather than the transfer of workers is beyond the scope of this book. Both processes are going on and are likely to continue. As poverty is a great evil, there will always be men and women who are ready, indeed happy, to seek other lands to avoid it. Nevertheless, labour is not a commodity to be bought and sold or moved without regard for emotional and cultural ties or basic requirements for health and safety.

Migration is not confined to those who move from one country to another for the purpose of employment. There are other and more complex social, political and psychological reasons for migration; moreover, the problems associated with internal migration may be just as serious, although of less international concern. But the major problems of migration affect those men and women, predominantly poor and unskilled, who seek work in another country, usually more industrialized and economically advanced than their own. By focusing in this chapter on their needs and the role to be played by occupational health services in recognizing and alleviating them, many of the less serious needs of other migrants will be covered. There are, for example, the less numerous skilled migrants, but both they and those who employ them are already well placed to deal with their problems. The special needs of refugees and clandestine migrants will not be considered here, because, although these groups are large and face serious hazards, the services and procedures required by the legal and voluntary migrant are neither sufficient nor wholly appropriate for them. Also excluded from consideration are such categories as frontier workers, seasonal workers, seamen, students, and persons with professional training or other skills admitted to perform specific tasks usually defined by contract.

DEMOGRAPHIC FEATURES

The dimensions and patterns of international migration are extremely difficult to assess because of problems over definition of terms, lack of appropriate statistics and the massive proportion which is illegal, though often tolerated and sometimes even encouraged. In order to describe the phenomenon of migration adequately the following basic items of information would be required, at least annually, for all countries:

1. Number of persons entering by age, sex, family status, education, occupational skills, country of birth and of previous residence.
2. Number of persons leaving, similarly classified, together with additional information for expatriates on destination and duration of stay.
3. Additional data estimated from surveys on the illegal or clandestine components.

Except occasionally and in a few countries, these data are clearly not obtainable; statistics on immigration and emigration could be collected systematically, but at present only a handful of countries do this. Some information can be derived for a larger number of countries which conduct periodic censuses, but at best these describe the situation at points in time, at intervals of 10 years or so. Even were all the relevant facts accurately recorded in national censuses, it would be virtually impossible to derive from such data alone a satisfactory picture of the character, rate and direction of migratory ebb and flow. In the absence of adequate demographic facts, one has to rely on incomplete, sometimes unreliable, information and on estimates and even frank guesses.

During the Second World War, vast sections of industry were destroyed in Europe and Asia, and many millions of persons fled, were displaced or were transported. In subsequent decades of recovery and industrial expansion many returned home, but others did not return and never will. Meanwhile other more localized hostilities and persecutions continue with similar effects on a lesser scale. Other determining factors of importance have been the changing geographical patterns of industrialization and exploitation of energy and raw materials, in part associated with the dissolution of former colonial systems and the appearance of other forms of economic domination. Fear, want and hope in varying proportions have always forced men to emigrate. *Plus ça change, plus c'est la même chose.* Only in absolute terms are the dimensions of the problem greater today, and perhaps only in recent years has there been much concern to reduce its ill effects.

The term 'migrant worker' implies that those who migrate for the purpose of employment can readily be distinguished from persons who do so for other reasons. One thinks at once of men, usually without families and typically from Mediterranean countries, now so evident in the industrial cities of northern Europe. There are parallels in other continents, such as Mexicans in the United States, Bolivians in Argentina and Lesothans in South Africa. Clearcut though these appear, they remain part of the spectrum of population movement, variously motivated and largely unmeasured.

Most migration is between countries of the same continent but, for many years, the intercontinental drift has been away from Asia, Africa, Latin America and Europe and towards North America and Oceania (Australia and New Zealand). Europe now contributes much less than it did, and the rest of the world correspondingly more. Individual countries seldom publish more than the number and source of their expatriates and little or nothing on rates of arrival and departure. The general geographical picture, based on a report by Lasserre-Bigorry (1975) and on fragmentary data from various other sources, is presented in *Figures 4.1* and *4.2*. For the most part, the pattern of emigration is the reverse of that for immigration but there are a few exceptions (e.g. Canada, Great Britain, Uruguay and some West African countries), where there is substantial traffic in both directions. There are also those countries, shown white on both maps, which report little movement in or out. For some of these countries information is lacking, in others, notably China, India and USSR, the populations are so enormous that any movement across their frontiers is proportionately small and internal migration is of more relevance. Conversely, it is the smaller

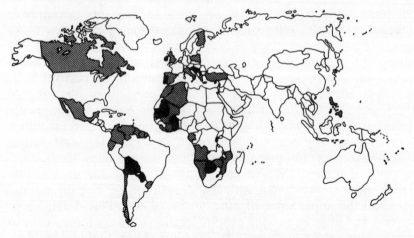

Figure 4.1 World pattern of emigration. Proportion of population living abroad: black — more than about 10 per cent; striped — between about one per cent and 10 per cent; white — less than about one per cent or no information

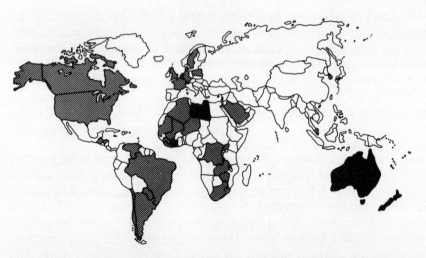

Figure 4.2 World pattern of immigration. Proportion of population foreign born : black — more than about 10 per cent; striped — between about one per cent and 10 per cent; white — less than about one per cent or no information

countries which tend to have the most dramatically high proportion of foreign nationals (e.g. Hong Kong, Singapore, Ivory Coast, Switzerland, Israel), or of their citizens living abroad (e.g. Ruanda, Ireland, Paraguay, Mauretania, Togo, Upper Volta).

Migration, for whatever reason, may entail social and health problems of concern to occupational medicine but these are liable to be most acute when it is determined by the demand for unskilled labour. Some differences therefore exist between families and individual persons who emigrate in hope of finding a better or more prosperous way of life, and men and women brought in for jobs which the local population is reluctant to do. The latter situation is typical of the wealthier countries of northern Europe today where (excluding Britain) some 6 million foreign workers now make up almost 10 per cent of the total work force. About half these workers are from southern Europe and the rest from North Africa or elsewhere. It is estimated that in addition there are almost 3 million women, some of whom are also employed, and over 2.5 million children; a third of the adults are unmarried. This postwar European phenomenon has no close equivalent elsewhere.

THE POTENTIAL RISKS

The migrant worker's problems have much in common with other work hazards which many men and women often accept without complaint. At first sight they make their choice voluntarily and for their own benefit, or at least for that of their families. It can be argued that, as with most other occupational risks, regulation and control will force the 'industry' out of the market and, both directly and indirectly, deprive workers of their only opportunities and discourage the enterprising. This is not the place to explore these arguments which contain some short-term truths, but rather to recognize that migrant workers do tend to share certain basic characteristics.

The migrant is keen to work, even if forced to break the law to do so. He, or sometimes she, is typically a young person in the prime of life, physically fit, enterprising and highly motivated. Although often from a predominantly rural and economically backward environment, characterized by high levels of infant and child mortality, the migrant is triply selected. First, he is a survivor, which is no mean achievement in some of the countries which provide most migrant labour. Second, he is one of a minority of his fellows who have chosen to emigrate, to leave the security of family and friends, and to persevere sufficiently to succeed in this intention. Finally, he has usually met the entry requirements of the host country and has the skills or other requirements of the foreign employer. Why then should such generally healthy and enterprising young men and women be at special risk? What in particular should an occupational health service be looking for in pre-employment and other examinations of these workers?

Most of the reasons for their vulnerability are obvious enough and are essentially either personal or environmental. The migrant may look healthy, but his nutritional status is often of marginal adequacy with little reserve. Although a survivor of many infections, some, such as tuberculosis, remain latent and ready to be activated by undue physical or mental stress. He may well have acquired immunity to many infections but these reflect the ecology of his own land and not that of a more industrialized country with possibly quite different climate and population density. In an analogous way, other patterns of behaviour and defence mechanisms accquired from past experience constitute a threat to the migrant. He has learned to deal with one set of problems at home or at work in ways which do not necessarily apply in the new country; he meets unfamiliar hazards which he may not recognize or know how to avoid. At home, when sick or in trouble he knows what to do; in the new land he neither knows nor can he easily find help. To all this must be added the personality and psychology of the migrant. It is a big step, requiring emotional strength, to try one's luck in an unknown country about which one may even have been seriously misinformed. Many of those who take it are driven either by bad, even desperate, economic circumstances, or by more than average ambition and adventurousness. For either of these reasons, anxiety, emotional instability and unrealistic optimism are common in migrants. There is also loneliness due to the fact that they are often not allowed to bring their families with them. Thus they are specially vulnerable when, as may happen, they meet disappointment, frustration, discrimination and hostility.

The new environment in which the migrant finds himself is potentially hostile in several ways. Jobs taken by migrants are generally the least attractive and poorest paid. Many work in mines and quarries, agriculture, forestry or construction, and a variety of industries where local labour is lacking, most of which carry specific hazards of occupational disease and accidents against which particular preventive measures may not be properly taken. Ignorance, unfamiliarity, language difficulties and sometimes climatic and nutritional factors or latent infections put the migrant at risk, especially during the first few months of employment. The migrant's domestic arrangements may be equally unsatisfactory since accommodation provided by the employer or obtained by the worker himself is often substandard. Adequate opportunities for sleep, recreation and cultural activities are frequently lacking. Because of low income and the necessity to save and send money home, there may be insufficient left to maintain a reasonable diet and living standard. Finally, and perhaps most seriously, the migrant may find the local population and native fellow workers unfriendly and without sympathy for his difficulties. The social isolation, loneliness and stress which result from this complex of factors may lead to depression, mental breakdown and psychosomatic disorders, and indirectly to alcoholism and even suicide.

EPIDEMIOLOGY OF MIGRATION

The implementation of effective measures to protect and improve the health of migrant workers requires information on the prevalence and causes of diseases which affect them, not easily obtained without well-designed surveys. Since migration is a world-wide phenomenon, health patterns are bound to vary and reliable data are scanty. In general, however, it cannot immediately be assumed that migration *per se,* or the new working environment, are necessarily the cause of illness or injuries, even those occurring at high frequency among migrants. The strong selective forces already described and the effects of adverse social circumstances in early life could well be the major determinants. Even so, it would be equally unsound to conclude that there was any less need for intervention; the priority should be to ensure that whatever is done is epidemiologically reasonable and the results evaluated. First, data are needed on the frequency of disease and injury among migrant groups and their families, defined by place of work and type of industry and, second, clues to the cause of unduly high rates. Action would then depend on whether those affected were already at high risk before leaving home, and whether migrants subsequently fared worse than comparable men and women who stayed behind. Again, there is a dearth of appropriate statistics, not only for migrants but also for indigenous populations with which they can be compared. Some of this information will have to be collected by government agencies. Occupational health services, however, in firms employing substantial numbers of migrants may well provide more precise data on work accidents, communicable disease and psycho-social disorders (*see* below). It has been argued that it would be discriminatory to separate the data for foreign and national workers and, perhaps with more justification, that migrant workers are understandably reluctant to disclose information which may be used against them. Occupational health services, if alive to this problem, should be able to avoid it.

Work accidents

It seems very probable that migrant workers suffer considerably more accidents at work than their fellows, although this is not well documented. It is even less certain that the excess remains after proper allowance is made for age, duration of employment and type of work. These corrections are necessary because it is generally true that certain jobs are more dangerous than others, that young workers have more, though perhaps less serious, accidents than old workers and that the first few months in any job constitute the period of greatest risk. All these three factors tend to operate against the migrant.

Data presented by various authors (ILO–WHO, 1977; Djordjevic and Lambert, 1977; Djordjevic, 1979; Opferman, 1979) suggest that crude accident rates among migrants in several European countries are two to three times greater than for the native-born workers, This applies particularly to minor accidents, less to the more serious and disabling accidents and not, apparently, to fatalities. There is no convincing evidence of any important difference between the various racial groups nor that these differences between accident rates of migrant and native workers are consistent in various types of industry.

The surveys reported by Opferman (1979), of the German Federal Ministry of Labour and Social Affairs, were more searching than most and correspondingly informative. Three points emerge: first, that the more closely the national and migrant workers are matched, job for job, and in other ways, the more similar are their accident rates and, second, that such differences as there are reflect mainly the experience of newcomers to the job. For a variety of reasons, migrants are more mobile than local workers and thus more frequently find themselves in new and unfamiliar work situations; after several years the differences in accident frequency disappear.

Communicable diseases

Three groups of infections consitute a potential threat to migrants and to the local populations with which they come in contact: mycobacterial disease (particularly tuberculosis), tropical parasitic infections and sexually transmitted diseases. Migrants, and indeed any travellers, are at special risk of contracting infections against which they lack immunity or of having a latent infection activated by adverse social circumstances; they are also capable of carrying infection to others. Examples can be cited of all these occurring, so they are of more than academic interest, but in practice the number of disease outbreaks attributable to migrant workers has been relatively small and only tuberculosis and venereal disease constitute a serious threat to the migrants themselves.

In the wealthy industrial countries to which most migrant workers go, tuberculosis is now uncommon in the indigenous population and a high proportion of all clinically active cases are in immigrants (Khogali, 1979). In Britain, for example, 32 per cent of all notifications in 1971 were of persons from abroad (British Thoracic and Tuberculosis Association, 1973). The extent to which the infections are imported and then activated, or acquired in the host country, is unknown. Among adults, the former possibility might well predominate, whereas among young children it may be the latter. Stress, overwork, poor nutrition and bad housing are likely to be the factors responsible. In countries, such as South Africa, Zimbabwe and Zambia, which employ migrant labour in mining, silicotuberculosis is a special hazard.

The occurrence of venereal disease among migrant workers inevitably reflects an unnatural way of life imposed on young persons, predominantly male, separated for long periods from their families and friends. Only by removal of this basic cause, is the problem likely to be reduced.

Psychological disorders

Mental illness, behavioural disorders and psychosomatic complaints are all most difficult to quantify. In part, this is because they are definable only in cultural terms which have defied satisfactory standardization. This is particularly true for population groups removed from their normal social environment. Thus, although it is repeatedly stated that psychiatric disorders are much more common in recent migrants than in the local population (see, for example, ILO–WHO, 1977), this view is based more on general opinion than on valid statistics. In some situations, indeed, there is evidence that the reverse is the case (Burke, 1976a and 1976b; Murphy, 1973).

For many years, both voluntary and enforced migration have been of special interest to social psychiatrists and have been the subject of many books, papers and conferences (Zwingman, 1978). The basic issue is whether mental illness in migrants is primarily determined by the various associated stresses or whether migration is in part a manifestation of mental instability. The nature of this still unanswered question is well described by Kuo (1976) who attempted to evaluate the relationship of four main factors – social isolation, culture shock, goal-striving stress, and cultural change – to measured symptoms of psychological stress among the Chinese population of Washington DC. The first two factors appeared the more important but still of low predictive value. The mental health of migrant workers deserves research of similar quality which some occupational health services are well placed to undertake.

It seems reasonable to believe that migrant workers may often have more than their share of restless, foolhardy and other unstable personality types. It is equally clear that if isolation, loneliness, worry and cultural confusion are capable of causing or aggravating mental illness, migrants must be considered at special risk. Either way, protective measures are needed.

SOME REMEDIES

The main problems experienced by migrant workers are primarily social, psychological and political. They have, therefore, to be attacked in many ways, in most of which there is a role, sometimes a major responsibility, for the occupational health service. The various measures required can be discussed under a number of headings:

International agreement

Convention 143 (International Labour Conference, 1975) adopted by the International Labour Organization, called on member countries to legislate and take other appropriate steps:

1. To ascertain and suppress clandestine movement and illegal employment of migrants, with severe penal sanctions against organizations responsible.
2. To ensure equality of treatment for migrants and nationals in respect of security of employment, remuneration, social security and other benefits.
3. To inform migrants fully of their rights, including equality of opportunity, working conditions and treatment.
4. To facilitate the reunification of the families (spouse, children and parents) of the migrant worker and encourage them to preserve their ethnic identity, language and cultural ties.

The implementation of this Convention was discussed more fully in the ILO Migrant Worker Recommendations (International Labour Conference, 1975). This urges the need for both the country of origin and the country of employment to take account of long-term social and economic consequences for all concerned. Equality of opportunity and treatment, which are the right of migrants and their families, apply to all aspects of employment, including fair advancement, membership of trade unions and co-operatives, and access to all amenities enjoyed by nationals, such as holidays with pay. As a matter of social policy, countries should aim at providing the migrant family with accommodation of reasonable local standard or, where this is not possible, to grant paid home leave for the migrant or to cover the cost of visits from his family. The special needs of migrants for informal advice and additional social services are to be recognized, as are security of employment and residence and the rights of legal and illegal migrants regarding pay and accumulated benefits, on their return home.

Social change

International conventions, even when fully implemented by national laws, have little effect unless accompanied by a change of heart among the people of the host country. Racial, religious, colour and cultural prejudices, largely based on ignorance and tradition, are widespread. These are re-enforced by economic insecurity, particularly among the poorer and less skilled sections of the population which are most vulnerable to foreign competition over work and pay. Above all, there is lack of understanding: most people simply do not appreciate the feeling and needs for friendship of the strangers in their midst. It follows that determined efforts must be made by leaders and makers of public opinion to set an example. This responsibility falls most heavily on politicians, journalists (in various news media), trade unionists, employers, professional people and the police. Among these categories are the staff of health services within industry, who are often well placed to demonstrate in a practical fashion both an understanding and a willingness to give that little extra help to migrant employees and to collaborate fully with other agencies which have similar objectives. It is easy to feel frightened and unwanted; it is more difficult to prescribe the recipe for a genuine welcome.

Community services

The scope of the community services required to deal with migration is much wider than is usually appreciated. In most countries, migrants and their dependents may need help and advice, not only on arrival but also during their stay and before returning home. In addition, services are wanted in the home country to prepare prospective migrants and to assist in their settlement on return. In some countries, for example Yugoslavia and the Philippines, the departure and return of citizens are well organized and co-ordinated, so far as possible, with the countries of employment. In neither of these countries can labour be recruited except though a national agency which ensures that the workers are properly qualified, physically suitable and adequately informed about the place and jobs which await them. Vocational and language courses are also available.

A reasonable balance has to be kept in countries of employment between:

1. The development of separate services and facilities for migrants to a level which may keep them apart from the local population and even increase discrimination and resentment.

2. The need for employees in the general social services to possess the linguistic and other abilities necessary to deal adequately with the special difficulties experienced by migrants.

These services must be able to offer effective assistance, information and advice on all aspects of life in the new country. Help such as is available to all at Citizens' Advice Bureaus in Britain, should cover:

1. Housing and related questions.
2. Employment problems and opportunities.
3. Family problems, specially related to education and health.
4. Legal advice.

Loneliness and isolation, which can become dangerous if excessive, are the common lot of strangers in any society. Social occasions and recreational activities with compatriots on the one hand but also with people from the new country are both important. Language is an obvious bar to communication, so every opportunity should be afforded for the migrant to learn at least sufficient for simple conversation and some appreciation of cultural activities in his or her new homeland.

Health services

The migrant and his family must of course have the same rights of access to medical care, including the occupational health services, as normal residents. However, this may be no great assurance or privilege. The health care for working people, inside or outside industry, is far from satisfactory in many countries which accept migrant labour, and foreign workers seldom make full use of what is available. The reasons for the latter may be simply ignorance, reticence and inability to communicate adequately with doctors and nurses, who understand neither the language nor cultural attitudes to illness of the migrant. The migrant worker may also fear that if found sick he may lose his job. It is important that immigrants should be fully covered for medical care from time of arrival, either by national or privately arranged insurance. The need is particularly great during the first few weeks. The major employers and host governments share a responsibility to provide this cover and also appropriately staffed occupational health services.

Although limited in their extent and availability, occupational health services are particularly well suited to the task of protecting the health of foreign workers. They have a responsibility for all employees of the enterprise or industry according to individual need, and in this sense are not discriminatory. Their role is primarily preventive, so they can seek out workers in real or potential trouble before serious damage is done. It is normal that they should conduct comprehensive health screening (including X-ray or other special tests) at or before employment, and periodically thereafter. It is their job to keep track of the attendance, accident and sickness records, to arrange whatever educational courses are required in the interests of health and safety, to advise management on questions of suitability for employment in certain jobs and to provide information and counselling for all employees. Good occupational services would greatly reduce the health risks of migrant work; a case therefore exists for requiring that any employer of migrant labour should provide them. Most countries which receive migrant workers are relatively wealthy and benefit from their presence financially and in other ways. Provision of adequate health surveillance and preventive health services would not appear to be an unreasonable price (Montoya-Aguilar and El-Batawi, 1979).

Health information

Some of the difficulties and adverse effects experienced by migrant workers are the direct result of ignorance on questions of health and safety. This is not surprising, as many are poorly educated and from countries where life and its hazards are quite different. Accurate and well-balanced information is needed in five main areas:

Medical and social services The newcomer should know exactly what agencies exist to help him when sick or in need of advice and how to use them; he should understand his rights and both what he can and cannot reasonably expect.

New environment Life in a large industrial city may expose the worker and his family to risks of accident, infection and temptation. They should be made aware of these risks and how to avoid them.

Food and nutrition Unaccustomed foods, the need to save money and ignorance of basic dietary requirements, commonly undermine the migrant family's health and strength.

Safety at work Every job has its risks to which the newcomer, especially one ignorant of the language, safety procedures and the use of protective equipment, is most vulnerable. Special safety training is required for migrant workers.

Social adaptation   Social isolation, loneliness and severe nostalgia may be reduced if the migrant has insight into reasons for any unfriendliness from the local population. He needs information and guidance on use of leisure, on how to enjoy life in his new home and on longer-term employment prospects.

The educational task is thus a large one and must be approached in a variety of ways. Ideally, the process should begin long before the migrant leaves home; the responsibility then lies with both countries and perhaps also with other workers after their return from abroad. Intensive instruction may be indicated during the first few weeks after arrival in the new country and at the new place of employment. Educational opportunities of a more relaxed and enjoyable kind are required indefinitely thereafter. Clearly it is for the government, the employers, and the trade unions of the host country to see that these facilities exist.

Research

The range of epidemiological studies which could be designed to explain the causes of migrants' problems is almost infinite and, for practical purposes, it may be best to concentrate on the evaluation of factors which are amenable to change. In particular, there are such questions as type of work and promotion prospects; attitude of local workers and unions; and access to medical and social services. Migration itself can be considered in terms of duration and distance. Work abroad for varying periods, with or without home leave or family visits, warrants study, as do some questions of linguistic and cultural 'distance'. The identification of personal characteristics of the migrant – physical, psychological and social – which appear associated with unacceptable risks might be useful. However, the use of such information in counselling or selection, would raise ethical questions.

An urgent requirement is to identify *effective* programmes for the protection of the migrant. Priority should be given to forms of intervention which are well-defined, straightforward, ethically acceptable and widely applicable. These might include:

1. Steps to reduce separation from home and family, for example by limiting duration of time abroad, frequent home leave, family visits and free telephone calls home.
2. Educational courses for the migrant before and after leaving the country of origin.
3. Vocational training courses for the specific work which is to be done.

4. Participation of labour unions in defined aspects of migrant welfare.
5. Programmes in which occupational health services should participate for the early detection of behaviour suggestive of undue stress.

CONCLUSION

Towards the end of the 1977 International Symposium on Safety and Health of Migrant Workers one of the discussants (Schou, 1979) made a statement at the conclusion of his paper which well summarizes the situation:

'. . . we have mostly heard about the poor, sometimes illiterate, unskilled migrant workers going to richer industrialized countries to take up work which nobody of the indigenous population wishes to carry out any longer. The migrant worker is, to put it bluntly, the voluntary slave of our time and personally I find it high time that international bodies like the ILO and WHO try to lay down some international rules and regulations which can safeguard the humanitarian rights of these people because the problem of migrant workers is not a temporary one.'

He listed twelve major points which should be considered if migrant workers and their families are to have the safety and happiness to which they are entitled:

1. Information, counselling and language training before leaving.
2. Medical and psychological screening of all family members contemplating migration.
3. Families kept united and separation to an absolute minimum.
4. Preparation of local population.
5. Information, advice and adequate accommodation on arrival, with social workers available.
6. Same right to social benefits as local population.
7. Safety instructions in language of migrant workers.
8. Encouragement to keep traditions and practise their religion.
9. Opportunities for vocational training.
10. Children to maintain language and traditions of parents.
11. Special centres to give advice on financial matters.
12. Aid centres in home country to help returning migrants to readapt.

This paper deserves to be read in its entirety as it describes succinctly the actions required, not just from ILO and WHO, but from the many persons and agencies responsible in most industrialized countries.

REFERENCES

British Thoracic and Tuberculosis Association (1973) 'A tuberculosis survey in England and Wales 1971: The influence of immigration and country of birth upon notifications.' *Tubercle,* **54,** 249–260

Burke, A.W. (1976a) 'Attempted suicide among the Irish-born population in Birmingham.' *British Journal of Psychiatry,* **128,** 534–537

Burke, A.W. (1976b) 'Socio-cultural determinants of attempted suicide among West Indians in Birmingham; ethnic origin and immigrant status.' *British Journal of Psychiatry,* **129,** 261–266

Djordjevic, D. (1979) 'Les accidents du travail et la morbidité des travailleurs migrants.' In *Safety and Health of Migrant Workers - International Symposium,* Occupational safety and Health Series, No. 41, pp. 19–30, Geneva: International Labour Organization (ILO)

Djordjevic, D. and Lambert, G. (1977) 'Migrant workers in the construction industry.' In *Migrant Workers – Occupational Safety and Health.* Occupational Safety and Health Series, No. 34, pp. 55–62, Geneva: ILO

International Labour Conference (1975) *Convention 143:* Convention concerning migration in abusive conditions and the promotion of equality of opportunity and treatment of migrant workers. Geneva: ILO

International Labour Conference (1975) *Recommendation 151:* Recommendation concerning migrant workers. Geneva: ILO

ILO–WHO: International Labour Organization and World Health Organization (1977) Joint Committee on Occupational Health. In *Migrant Workers - Occupational Safety and Health,* Occupational Safety and Health Series, No. 34, pp. 1–50, Geneva: ILO

Khogali, M. (1979) 'The role of occupational medical services in the United Kingdom in promoting the health of migrant workers.' In *Safety and Health of Migrant Workers - International Symposium,* Occupational Safety and Health Series, No. 41, pp. 171–177, Geneva: ILO

Kuo, W. (1976) 'Theories of migration and mental health: an empirical testing on Chinese-Americans.' *Society for Science and Medicine,* **10,** 297–306

Lasserre-Bigorry, J.H. (1975) 'Panorama des principales migrations internationales contemporaires aux fins d'emploi.' *Informations sur les conditions générales de travail,* No. 34, Geneva: ILO

Montoya-Aguilar, C. and El-Batawi, M.A. (1979) 'An introduction to the planning and organization of preventive and curative services for migrant workers.' In: *Safety and Health of Migrant Workers - International Symposium,* Occupational Safety and Health Series, No. 41, pp. 157–163, Geneva: ILO

Murphy, H.B.M. (1973) 'The low rates of mental hospitalization shown by immigrants to Canada.' In *Uprooting and After,* ed. C. Zwingman and M. Pfister-Ammende, pp. 221–231, Berlin, Heidelberg, New York: Springer Verlag

Opferman, R. (1979) 'Epidemiology and statistics of occupational accidents and morbidity in migrant workers.' In *Safety and Health of Migrant Workers – International Symposium,* Occupational Safety and Health Series, No. 41, pp. 31–42, Geneva: ILO

Schou, C. (1979) 'Practical prevention measures for safety and health of migrant workers: Contribution to Discussion.' In *Safety and Health of Migrant Workers – International Symposium,* Occupational Safety and Health Series, No. 41, pp. 279–283, Geneva: ILO

Zwingman, C. (1978) *Uprooting and Related Phenomena.* A descriptive bibliography MNH/78.23, Geneva: WHO

5

The Effects of Work Exposures on Organ Systems I — Liver, Kidney and Bladder, and Skin

J.M. Harrington and H.A. Waldron

In this chapter and the next the effects of work exposure on the organ systems of the body are described. It is not possible in a book on occupational health practice to include an exhaustive account of industrial toxicology with detailed references to studies of toxicology and occupational diseases, nor is it the place for a detailed resumé of clinical medicine. For reference to such texts, the reader should consult the books listed at the end of Chapter 6. However, good occupational health practice does require a working knowledge of clinical toxicology, and the purpose of these two chapters is to review the ways in which organ systems of the body respond to toxic influences that may be encountered in the workplace.

There are two ways of viewing the effects of toxic substances on the human body: more commonly, by considering each chemical or physical agent in turn and describing its effects on the body. Less frequently, an organ's responses to differing noxious stimuli are described. The former method is easier to write whereas the latter is of more practical value as it is the way that most patients will present to the examining physician or nurse. Patients do not usually complain that they are suffering from mercury intoxication; they are more likely to recount symptoms associated with the effect of mercury on certain target organs such as the central nervous system.

These two chapters consider industrial toxicology from the perspective of the patient with a demonstrable organ dysfunction. The 'target' organs reviewed are arbitrarily divided into two groups: liver, kidney, bladder and skin (Chapter 5) and blood, lungs and nervous system (Chapter 6). Each organ is dealt with in a similar manner; a brief review of structure and function is followed by an account of the range of dysfunctional response. Examples of physical and chemical substances which can produce such dysfunction are given. Only brief reference is made to screening the patient for exposure and absorption of such materials, a subject dealt with in Chapter 11. Chapter 6 concludes with a short account of other target organs, the cardiovascular and endocrine systems which are not, at present, well described for the effects of occupationally related toxic influences.

THE LIVER

Structure and function

The liver is the largest of the visceral organs and weighs about 1500 g in the adult. It consists of a mass of parenchymal cells with little supporting collagen. The liver would, therefore, be easily injured by direct trauma but for its protection under the right lower ribs.

The parenchymal liver cells are arranged in lobules and are well supplied with blood, from both the hepatic arterial and portal venous systems. The hepatic artery, portal vein and bile duct travel together as the portal triads. They branch together and blood flows from the portal vein, which drains the gastrointestinal tract, and hepatic artery via the sinusoids and between the parenchymal cells, to the central hepatic vein in the middle of the lobule. The hepatic veins drain into the inferior cava. The lobule is easy to visualize in the pig but is not clearly defined in human microscopical sections. In functional terms, a group of sinusoids running between a terminal portal tract and a few terminal hepatic venules comprise the acinus. The liver cells (hepatocytes) near the hepatic venules differ from those near the portal tracts as they receive blood lower in oxygen content as a result of passage through the acinus. These centrilobular hepatocytes are therefore more prone to toxic or anoxic conditions than the periportal cells. In addition to such parenchymal cells the liver also contains important reticuloendothelial cells, the Kupffer cells, which play a major role in storage and immune processes.

The hepatocytes lie one cell thick along the sinusoids and are the main functional unit of the liver 'factory'. Electronmicroscopy reveals that these cells contain a number of important organelles which seem to perform differing functions. The mitochrondria are concerned with energy supply through the citric acid cycle, and the synthesis of adenosine triphosphate. In addition, mitochondrial activity is important in the synthesis of urea, haemoglobin and fatty acids. The endoplasmic reticulum may be rough or smooth depending on the presence or absence, respectively, of ribosomes. The rough variety is concerned with protein synthesis and glucose phosphatase activity whereas the smooth endoplasmic reticulum is involved in bilirubin conjugation, steroid biosynthesis and many chemical detoxification mechanisms.

The liver, therefore, plays a central role in metabolic processes and is susceptible to the effects of absorbed toxic substances, especially if they are fat soluble. The normal function of the liver may be summarized as follows:

1. The synthesis of all the plasma proteins except immunoglobulins as well as the synthesis and metabolism of non-essential amino acids. The breakdown of amino acids to alphaketo acids and ammonia can be followed by the conversion of ammonia to urea.
2. The metabolism of glucose and galactose to glycogen for storage in the liver and the subsequent breakdown of glycogen to glucose when required.
3. The utilization of free fatty acids for energy production or conversion to triglyceride and other lipids. Neutral fat can be converted to glycerol and free fatty acid.
4. The production of bilirubin from old red cells and the production and assembly of the other constituents of the bile such as bile salts.
5. The breakdown of steroid hormones, such as cortisol, oestrogens and testosterone. In addition, the liver alters many drugs and foreign chemicals to make them more acidic or more polar, thereby increasing their water solubility and facilitating renal excretion.
6. The liver acts as a storage organ for iron, vitamin B_{12} and folic acid.

Disordered function

There are two main aspects of disordered hepatic function. First, the hepatocytes may be damaged; and second, the transport mechanisms to or from the hepatocytes may be blocked. Both dysfunctions can lead to jaundice which is an important sign of hepatic damage.

Hepatocellular damage

The liver has a remarkable facility for regeneration. This ability is partly linked to the slow but constant regeneration which goes on in the normal state. Insult from some external source can accelerate regeneration. For example, hemi-hepatectomy is followed by a remarkable acceleration in mitotic rates. After a few weeks the organ has returned to its preoperative size. Such an ordered return to normal size with the concomitant development of the correct architecture of the sinusoids is not necessarily seen when the damaging influence is chemical. This is particularly so if the toxic exposure is continued during the regenerative process.

For example, ethyl alcohol is a well-known liver toxin and continued absorption of excessive quantities will preclude the regeneration of organized liver lobules. The patterns of liver cell damage and repair are illustrated in *Figure 5.1*. Following ethyl alcohol absorption the liver regenerates in a disorded fashion. There is excessive proliferation of the stroma and the hepatocytes that survive the initial onslaught multiply to produce nodules. New acini are not formed and the failure

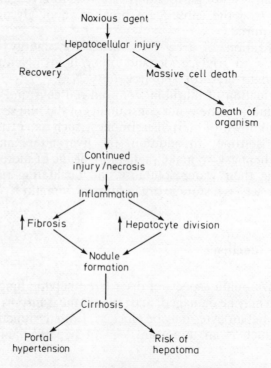

LIVER CELL DAMAGE AND ITS SEQUELAE

Figure 5.1 Liver cell damage and its sequelae

of effective vascular and biliary systems to develop, means that these foci of hepatocytes cannot function normally. The inflammatory stage which precedes fibrosis and nodular formation is also ineffective in re-establishing normal liver function, and the end result of the whole process is permanent, irreparable damage, which if severe or widespread enough, can lead to overt clinical evidence of diminished liver function, cirrhosis, malignancy and death.

Not all livers respond equally to the same degree of chemical insult. Pre-existing liver disease and the host response account for a large measure of the individual variation. These hepatic reactions have been best described for alcoholic liver disease, but there is every reason to believe that similar clinicopathological mechanisms occur with other hepatoxic substances.

Cholestasis

Cholestasis can occur as a side effect of disrupted hepatic architecture which may follow hepatocellular damage. Cholestasis can occur following an occupationally induced exposure to a noxious agent. It may be extrahepatic (prehepatic and posthepatic) or intrahepatic (*Figure 5.2.*)

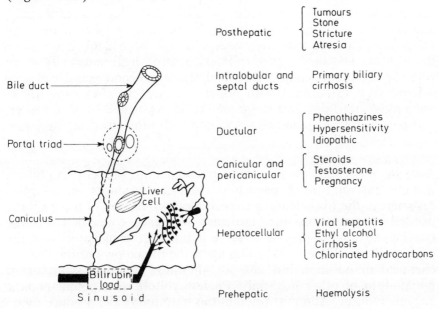

Figure 5.2 Cholestasis – its sites and some of the causes of jaundice

   In normal circumstances bilirubin is taken up by the liver through the sinusoids, conjugated in the smooth endoplasmic reticulum and excreted into the bile caniculi, whence it proceeds via the biliary tract to the gut. This mechanism can be swamped prehepatically by an excessive production of unconjugated bilirubin from haemolytic processes, or, rarely, by one of the inborn errors of metabolism. Hepatic causes include hepatocellular damage, conjugation process defects, defective canicular secretion and obstruction to the intrahepatic biliary system. Posthepatic cholestasis occurs when the biliary system is obstructed between the liver and the duodenum. Obstruction is either due to something in the lumen, in the wall or outside the wall of the bile tracts such as stone, strictures, carcinoma or inflammatory disease.

Other dysfunctions

Although hepatocellular damage and interference with bilirubin metabolism are the commonest types of hepatic dysfunction associated with hepatotoxic drugs, two other minor mechanisms are worth noting. These are:

1. Hypersensitivity
2. Enzyme induction.

Hypersensitivity   reactions occur primarily following drug absorption and are ill-understood. Genetic factors, immunological responses and variable reactions to toxic metabolites have all been invoked to explain the effects. The most serious effects of drug hypersensitivity occur when an hepatic reaction occurs. Halothane is a good example of such a hepatoxin. Cholestasis with hepatitis can occur after exposure to phenothiazine drugs and is frequently accompanied by skin rashes, and past history of jaundice associated with previous drug ingestion.

Enzyme induction   is a strange phenomenon whereby continued hepatic exposure to a particular chemical produces an adaptive response in the liver, leading to increased production of the enzymes needed to metabolize the particular substance concerned. As the enzymes so induced may be involved in other metabolic processes, the effect will not only be a speeding up of the breakdown of the drug or chemical in question but also an alteration in other synthesis or metabolism of other materials. Phenobarbitone is a potent hepatic enzyme inducer. Administration of this barbiturate can produce folate deficiency and hypocalcaemia. The altered metabolism of drugs, such as oral anticoagulants or anticonvulsants may cause serious side effects and if the metabolites of a given chemical are as toxic (or more toxic) than the original material, for example carbon tetrachloride, then enzyme induction will increase the effective hepatotoxicity of a given dose of these substances.

Pre-existent liver disease and hepatotoxic chemicals

The pre-existence of liver disease would be expected to alter a patient's reponse to a hepatic toxin. In experimental animals acute liver disease has been shown to alter hepatic handling of drugs. Although cirrhotic patients do not seem to fare any worse than normal people in their response to the drugs, it would be prudent to exclude persons with liver disease from exposure to hepatotoxins in the workplace, in case they are unable to cope with doses not shown to be hazardous to workers with normal livers.

Occupational hepatotoxins

The potential list of such substances is enormous and growing. The more important ones are listed in *Table 5.1,* according to the type of damage they produce. Some are capable of producing more than one hepatic response: where this is possible, the agent is listed under its more important hepatic effects. Most produce hepatocellular damage.

Table 5.1. SOME OCCUPATIONAL HEPATOTOXINS

Centrilobular necrosis ± fatty change

Carbon tetrachloride	Antimony
Chlorinated naphthylenes	Arsenic
Chlorodiphenyls (PCB's)	Boranes
Dimethyl hydrazine	
Dimethyl nitrosamine	Selenium
Ethyl alcohol	Thallium
Halothane	
Nitrobenzene	
Tetrachloroethylene	Phosphorus (yellow)
1.1.1. trichloroethane	
Trichloroethylene	
Trinitrotoluene	

Hepatitic effect

Halothane
Viral hepatitis
Leptospirosis

Cholestatic-cholangiolitic effect

Organic arsenicals
Toluene diamine
4–4' diaminodiphenyl methane

Hepatic venous thrombosis

Urethane

Neoplastic agents

Aflatoxin
Androgens
Arsenic
Nitrosamines
Oral contraceptives
Thorium dioxide
Vinyl chloride

Organic compounds

Aliphatic halogenated hydrocarbons   are capable of producing severe acute centrilobular necrosis. Carbon tetrachloride (CCl_4) is the classic example of a hepatotoxin in this group. The first damage occurs in the endoplasmic reticulum with vacuolation and ribosome detachment. Whilst the mitochondria remains intact, lipid droplets appear intracellularly and gross fatty changes occur in the lobular mid-zone. The toxicity of this chlorinated hydrocarbon is enhanced by prior treatment with enzyme inducing drugs but diminished by previous low protein diet. Death induced by carbon tetrachloride is usually caused by renal failure rather than hepatic failure.

Although the hepatotoxic effects of carbon tetrachloride are particularly well recognized, all the aliphatic halogenated hydrocarbons are capable of some degree of liver damage, though the fat solubility also ensures high uptake by the central nervous system and bone marrow where the clinical effects of toxicity may be more obvious or the long-term sequelae more serious. In this group 1,1,1-trichloroethane is probably the least toxic of the commoner solvents. Halothane can produce a hepatic reaction in which hypersensitivity to the substance may be involved.

Alcohols   are fat-soluble and have a predilection for the liver; the effects of ethyl alcohol are the best known and understood. Exposure is primarily a problem of social drinking and seldom one of occupational inhalation. Certain occupational groups are at high risk – licensed traders, actors, brewers and doctors.

Aromatic halogenated hydrocarbons   can also be potent hepatotoxins; with the exception of the chlorinated benzenes, the more highly chlorinated the compound, the greater the toxicity. Chlorinated naphthylene and the chloro diphenyls are particularly potent causes of liver damage.

Nitro-compounds,   particularly the aromatic ones, can cause hepatocellular damage. Trinitrotoluene has caused numerous toxic fatalities, as opposed to explosive fatalities. The cause of death is usually toxic hepatitis or aplastic anaemia. Nitrobenzene is also described as causing jaundice, both from hepatocellular damage and from excessive haemolysis.

Hydrazines and nitroso-compounds Dimethylhydrazine and dimethylnitrosamine can cause severe centrilobular necrosis. Both are suspected of being human hepatic carcinogens.

Metals

The metals are not, in the main, considered as hepatocellular poisons. They tend to exert their toxic effects elsewhere, but some metals do have a major hepatic effect.

Antimony and arsenic are very much alike toxicologically, and whereas both elements can produce haemolytic anaemia through the effect of the metal hydrides, they are also both capable of direct hepatocellular damage. Nevertheless, the risk of liver damage from these metals and their compounds is low.

Boranes, particularly the pentaboranes, can produce a wide range of toxic effects. The most serious sequelae are on the kidney, lungs and central nervous system, but prolonged exposure can cause liver damage.

Selenium and thallium are also reported to cause liver damage. Thallium salts are very hepatotoxic.

Yellow phosphorus can cause acute yellow atrophy of the liver. Exposure is usually high, short lived, and the effect occurs within a few weeks. Massive hepatic necrosis is the usual cause of death. Lower, more prolonged, exposure to yellow phosphorus, though liable to produce 'phossy jaw', seems rarely to lead to chronic liver disease.

Compounds producing liver effects other than hepatocellular damage

These are relatively few in number and include the rare causes of cholangiolitic-cholestatic jaundice such as organic arsenicals, toluene-diamine and the celebrated cause of 'Epping jaundice' – 4,4′ diaminodiphenyl methane, which was ingested in bread. It is used as an epoxy resin hardener.

DDT and Dieldrin can both cause hepatic enzyme induction and aflatoxin toxicity is enhanced by protein deficiency. An outbreak of porphyria has been described following exposure to hexachloro-benzene.

Table 5.2. SOME DRUGS CAUSING LIVER DAMAGE

Altered bilirubin metabolism

   Methyl dopa
   Methyl testosterone
   Norethandrolone
   Novobiocin
   Phenacetin
   Phenylhydrazine
   Rifampicin
   Salicylates
   Sulphonamides

Hypersensitivity reactions

   Isoniazid
   Methyl dopa
   Monoamino oxidase inhibitors
   Phenyl butazone
   Phenothiazines
   Most first line anti-tubercular drugs
   Phenindione
   Tolbutamide

Enzyme induction

   A long list of sedatives, anti-convulsants and steroids. Phenobarbitone is particularly potent.

Direct hepatocellular damage

   Cyclophosphamide
   Ferrous sulphate
   6 mercaptopurine
   Methotrexate
   Paracetamol
   Tetracycline-s (i/v)
   Vitamin A

Drugs In general medicine, the hepatic response to drugs forms the most important part of any article on hepatotoxic substances. None of the previously cited compounds – with the possible exception of halothane and chloroform – are mentioned. Drugs can affect liver function either by altering bilirubin metabolism, causing hypersensitivity reactions, induction of liver enzymes, or by direct cellular damage. The latter is the least important pharmacological effect, but the most important in occupational health practice.

The physician or nurse in industry should be aware of the possible interaction between drug therapy and occupational exposure to hepatotoxins. The commoner drugs that can cause liver damage are listed in *Table 5.2*.

All the drugs listed have to be synthesized or extracted for therapeutic use. All are, therefore, potential occupational exposures for workers in the pharmaceutical industry.

Infectious agents　The infections of prime importance in occupational health practice are serum hepatitis and leptospirosis. Serum hepatitis is a hazard of medical laboratory workers. Other medical and paramedical personnel are also at risk, especially in renal dialysis and blood-transfusion units. The disease has an incubation period of 7–23 weeks, and can be rapidly fatal. Recovery is more usual but chronic active hepatitis may also occur. Leptospirosis is a spirochaetal infection spread by rodents. Jaundice occurs in about half the patients with overt illness. Recovery is the rule but if renal failure and haemolysis are severe, the disease may be fatal.

Physical agents　The liver is automatically well protected from physical agents. Ionizing radiation can, however, cause fibrotic changes and, conceivably, could induce neoplastic change.

Hepatic tumours　Primary neoplasms of the liver are arbitrarily separated from hepatocellular toxic effects of occupationally induced hepatomas. About one fifth of cirrhotic patients develop hepatomas and in the United Kingdom approximately 70 per cent of all hepatomas develop in cirrhotic livers. The incidence of cirrhosis varies greatly from country to country, so these figures should only serve to illustrate the fact that previous liver damage appears to be an important precursor of many primary liver tumours. In Central Africa, for example, where hepatomas are more common, aflatoxin may be a more important aetiological factor than ethylalcohol. Experimental animal hepatomas can be produced by a variety of compounds obtained from moulds, particularly *Aspergillus flavus*. Another aetiological factor that has recently come to light is an increased incidence of hepatomata in patients who had recently undergone contrast angiography with thorium dioxide. These patients also have a higher incidence of angiosarcoma of the liver. Occupational health practitioners are more familiar with the recent publicity surrounding the relationship between previous occupational exposure to vinyl chloride monomer and angiosarcoma of the liver, although arsenic has also been implicated in the aetiology of this extremely rare tumour. Other suspected causes of hepatomata include persons given androgen therapy. Benign hepatic adenomata appear to be associated with previous use of oral contraceptives.

Screening for occupational liver disease

The laboratory tests that may be of value in investigating the patient suspected of occupationally induced liver disease fall into two groups: those related to the measurement of exposure to, and absorption of, the suspected hepatotoxin, and those which monitor general hepatic function.

There is an enormous number of individual compounds which could be measured in blood or urine to estimate occupational exposure to the suspected hepatotoxic agent. For monitoring devices for these compounds, the reader is referred to the standard toxicological texts. Liver function tests are described in Chapter 11. Other methods of investigating the cause of suspected liver damage include the bromosulphthalein test. It is very sensitive but cannot be carried out in the presence of jaundice. Radiological investigations include cholecystography and intravenous cholangiography. Needle biopses of the liver can provide specific histological confirmation of a generalized hepatic disease. Three methods of scanning the liver are currently available in some countries: whole body computerized traverse axial tomography; isotope scanning, with such isotopes as Technetium-99m or Gallium-67 and gray-scale ultrasonic scanning.

THE KIDNEY AND URINARY BLADDER

Structure and function

The kidney plays a central role in the detoxification mechanisms of the body. When a toxic substance is absorbed, the liver will frequently alter its chemical composition. Many such substances are relatively water insoluble and the liver increases the polarity and/or acidity of the absorbed compound. Both these processes tend to increase its water solubility, thereby rendering it more suitable for excretion by the kidney.

Some toxic materials will reach the kidney via the blood in an unaltered state. Limited metabolic transformation will occur pre-renally if the substance concerned is already highly polar, water soluble or acidic.

The kidney's main role in excretion is the production of approximately 1500 ml of urine per day. This function is vital to the maintenance of fluid and electrolyte balance in the body, and urine also serves as the most important vehicle for the excretion of unwanted waste products. These unwanted materials can be the byproducts of normal metabolic processes, or the results of the organism's attempt to detoxify absorbed toxic compounds.

To achieve these ends the kidney must be capable of altering the concentration of various solutes in the urine depending on the solute and water concentrations in body tissues. It has to adjust the volume excreted to the amount ingested and also has to reabsorb some material back into the blood from the glomerular ultrafiltrate. In addition, the kidney has the capability of secreting substances into this ultrafiltrate. This adds a further fine adjustment to salt and water balance.

The kidney has an outer cortical layer and an inner medulla. The medulla, with its papillary projections into the calyces of the renal pelvis, is concerned primarily with urine concentration and waste product excretion. The cortex has both excretory and non-excretory functions; they include erythropoiesis, calcium metabolism and blood pressure regulation.

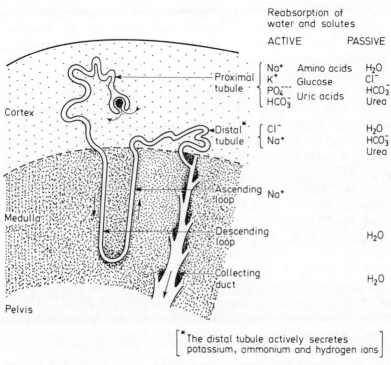

Figure 5.3 The nephron – its anatomy and the sites of active and passive renal handling of water and solutes

The function unit of the kidney is the nephron (*Figure 5.3*). There are one million present in the neonatal kidney but, by old age, 40 per cent have been lost. Most of the nephron lies in the cortex. It starts with the glomerulus, a blind-ending dilatation, shaped like a goblet. Its inner wall is in intimate contact with a tuft of capillaries from the renal artery. Diffusion of fluid from the capillaries through the glomerular wall and into the nephron is facilitated by the limited cellular barrier that exists between the lumen of the nephron and the blood in the capillary. The fluid leaving the glomerulus for its journey along the nephron is, in effect, an ultrafiltrate of plasma. The rest of the nephron is concerned with altering the composition and concentration of the filtrate until it leaves the renal pelvis as urine.

This process takes place along a tortuous route traversing the proximal and distal convoluted tubules (which lie in the cortex) and the loop of Henle, which is in the medulla and connects the two convoluted portions of the nephron to each other. Finally, the filtrate enters the renal pelvis. It then travels down the ureter to the bladder where it is stored, sometimes further concentrated, and then voided intermittently.

Most of the glomeruli lie superficially in the cortex and have short loops of Henle. Some, the juxtamedullary glomeruli, lie deep in the cortex and have long loops of Henle. It is these juxtamedullary glomeruli and their tubules which are preferentially concerned with sodium conservation and urine concentration.

Although only a litre or so of urine is voided in 24 hours, 100 or 200 times this volume may cross the glomerular basement membranes to form the ultrafiltrate. In other words, one-fifth of the plasma water is removed from the circulation at the glomerulus and most of this has to be rapidly returned to the blood if disastrous consequences are to be avoided. This return to the blood occurs mainly in the proximal tubule where active and passive reabsorption reclaims much of the plasma constituents lost upstream through the glomerulus. Failure of tubular function, therefore, can produce dramatic changes in fluid balance and nutritional status in a very short space of time.

The loop of Henle acts primarily as a concentrating mechanism through the effect of the countercurrent multiplication phenomenon. The distal convoluted tubule makes some additional fine adjustments to the tubular fluid composition and final concentration of the urine occurs as the collecting ducts traverse the medulla, passing as they do through an interstitium of steadily increasing solute concentration before the renal pelvis is reached.

One vital renal function is the ability of the tubules to act as secretors of additional substances into the ultrafiltrate. This function is invoked for a number of substances which do not readily pass the glomerular barrier, such as organic acids, and therefore many drugs and other toxic substances.

Disordered function

Many of the excretory, secretory and reabsorptive functions of the kidney are enzyme-dependent and therefore vulnerable to toxic substances. Failure of a complex substance to cross the glomerular membrane frequently results in its sequestration in the renal cortex. If it still has toxic properties, these will be concentrated in the cortical portions of the nephron, namely the glomerulus and the convoluted tubules.

Loss of nephrons occurs primarily through trauma, renal ischaemia or toxic damage. It is also a function of ageing. Whenever nephrons are lost or damaged, polyuria, thirst and nocturia will follow through loss of ability to concentrate urine.

Selective tubular damage may also induce partial resistance to antidiuretic hormone. Here the dysfunction is not the inability to maintain an osmotic pressure gradient between urine and plasma, but an inability to absorb more than a small fraction of the huge water volume contained in the glomerular filtrate.

Inability to concentrate urine is not, however, the most serious aspect of nephron loss or damage. The glomerulus may fail to filter enough solute and water, or, conversely, allow an excess amount to pass. Provided the proximal tubule is intact, the urine flow is more dependent on the tubular reabsorption than the glomerular filtration rate. This small segment of the nephron is responsible for reabsorbing most of the water from the glomerular filtrate, and it is also the main site of reabsorption of glucose and amino acids. As 150 g of glucose enters the glomerular filtrate each day, failure to reabsorb such vital materials can be nutritionally disastrous. Tubular failure will also affect calcium metabolism and acid-base balance, as well as the secretion of organic acids and bases.

Proteins in the urine are nearly always an early sign of glomerular disease. Renal disease rarely spares the glomeruli and rarely occurs without some degree of proteinuria. Not all proteins are prevented from entering the glomerular filtrate even in healthy kidneys. However, the quantities are small, as is the protein's molecular weight. The proximal tubule then mops up the filtered protein and none appears in the urine. It is therefore useful to establish which types of proteins are present in urine if a simple non-specific test for protein is positive. Small molecular weight proteinuria may be indicative of tubular, rather than glomerular, dysfunction.

Acute renal failure

Acute renal failure is a frequent sequel of massive toxic renal damage. The lesion is usually acute tubular necrosis but can be prerenal (massive haemorrhage, shock, hypovolaemia) or postrenal (obstruction due to casts or crystals). Acute renal failure is characterized by a fall in glomerular filtration rate. Oliguria is usual, but not invariable. Cortical necrosis, whilst a rare cause of acute renal failure, occurs more commonly with toxic nephropathies than other causes of kidney damage. Renal failure is also a common complication of advanced liver disease. Drug-induced renal failure can be a hypersensitivity reaction and the underlying pathology is an acute interstitial nephritis.

The nephrotic syndrome

This syndrome is characterized by massive proteinuria (more than 3.5 g/1.75 m^2 of body surface area per day). The cause of the heavy protein loss is damage to the renal glomerular capillaries so that there is an increase in glomerular permeability. Whereas most chemicals cited in the next section can cause acute oliguric renal failure with primarily a tubular defect, mercury, gold and bismuth are among the few 'occupational' causes of nephrotic syndrome.

Renal damage is not always sufficiently gross to be demonstrable on microscopic examination of kidney tissue. As many of the transport mechanisms need metabolic energy and appropriate enzyme activity, certain poisons can compromise or destroy these functions with little evidence of even microscopic anatomical disruption to the nephron. Nevertheless, the paralysis of tubular transport mechanisms can cause the appearance in the urine of glucose and amino acids, as well as the ions of sodium, potassium, phosphate and bicarbonate. The effects of such losses include osteomalacia, rickets, acidosis, renal calcinosis, renal stone, hypokalaemia and dehydration. Such syndromes can occur after occupational exposure to toxic materials.

Occupational renal disease

A list of the more common industrial toxins which can damage the kidney are listed in *Table 5.3*. Although an acute toxic insult may produce minor and transient renal effects, a more prolonged and heavy exposure can cause permanent renal damage. A severe acute poisoning can result in complete, if temporary, renal failure.

Metals

Acute poisoning from occupational exposure is rare. Most of the metals involve chronic low dose exposures; the most important nephrotoxic metals are cadmium and mercury.

Arsenic   and inorganic arsenical compounds can cause severe hypovolaemic shock which can give rise to acute oliguric renal failure.

Table 5.3. SOME CAUSES OF OCCUPATIONALLY INDUCED NEPHROTOXICITY

Inorganic	*Organic*	*'Biological'*
Arsenic	Aniline	Antimicrobials
Bismuth	Carbon tetrachloride	Cantharides
Cadmium	Chloroform	Chlorinated hydrocarbon
Gold	Ethylene glycol	insecticides
Iron salts (in overdose)	Ethylene diamine	
Lead	tetraacetic acid	Fungi
Mercury	Methoxy fluorane	Horse serum
Phosphorus	Methyl alcohol	X-ray contrast media
Potassium chlorate	Paraquat	
Thallium	Phenol	
Uranium	Toluene	

Prolonged exposures, in addition to causing the classic dermatological and neurological stigmata, can also lead to acute cortical or tubular necrosis due to enzyme poisoning effects. Renal disease is not, however, a primary manifestation of arsenical poisoning. Arsine gas, on the other hand, is extremely toxic, leading to catastrophic intravascular haemolysis. The effects of this on the kidney include haemoglobinuria, oliguria or anuria, uraemia and death.

Bismuth   is said to cause hepatic and renal disease, but rarely if ever from occupational exposures.

Cadmium   is a potent cause of proteinuria in workers chronically exposed to the metal and its salts. Cadmium is concentrated in the renal cortex. Low molecular weight proteins leak into the urine due to failure of the proximal tubular reabsorptive mechanism. Urinary concentrations of β2 microglobulin is used as a measure of cadmium induced nephropathy. Continued renal insult can lead to failure of other facets of the proximal tubular function, leading to glycosuria, aminoaciduria, phosphaturia, hypercalcuria and renal tubular acidosis. Advanced disease is rarely reversible though complete renal failure is virtually unknown.

Gold and iron   salts are reputed to cause renal disease, but this is almost always following excessive therapeutic use.

Lead   has a nephropathic effect and although the renal damage is less prominent than lead-induced effects on other organs, tubular damage can occur. Insidiously progressive chronic nephritis has been described in children exposed to lead paint. Lead seems to be more nephrotoxic in children but the authors have seen such renal damage in workers at secondary lead smelters.

Mercury,   like cadmium, is preferentially retained in the kidney, and excretion is reported to take place through the proximal tubular epithelium. Mercury can damage the tubular transport mechanism handling sodium and thereby induce a diuresis. Inhibition of sulphydryl-containing enzymes is part of the explanation for the tubular damage as proteinuria, aminoaciduria and glycosuria have been described. The renal effects of mercury can involve a vast outpouring of protein in the urine. Such heavy loss of protein must result from glomerular damage with or without renal ischaemia. The proteinuria of mercurialism is much greater than that seen in cadmium nephropathy.

Thallium and *uranium*   can be extremely toxic and are capable of causing severe renal damage. These effects are often, at least initially, overshadowed by their toxic action on other organ systems.

Organic compounds

Illness resulting from these compounds rarely, if ever, presents as renal disease. Carbon tetrachloride, chloroform, toluene, and phenol all tend to produce their primary effects elsewhere, with the kidney suffering secondarily. In the case of many of the organic solvents, the renal disease is not a direct result of a toxic effect of the solvents as much as it is a secondary effect of severe liver disease, leading to the hepatorenal syndrome. Ethylene glycol, however, by having the facility to produce large quantities of oxalic acid, may damage the kidney by tubular obstruction through the deposition of calcium oxalate crystals. Phenol can be a prerenal cause of oliguria due to intravascular haemolysis, shock and hypovolaemia.

Ethylene diamine tetra-acetic acid can have a direct nephrotoxic effect – a point worth remembering when using this chelating agent to cope with excessive heavy metal absorption.

Fluorinated hydrocarbons – particularly the fluoroalkenes–have been described as causing liver and kidney damage following chronic occupational exposure.

'Biological' compounds

Most of this group relate to antimicrobials which can cause renal damage by direct toxic effects (gentamycin, kanamycin, polymyxins) or hypersensitivity reactions (penicillin, sulphonamides, cephalothin). These are rarely important occupationally. However, certain persistent chlorinated hydrocarbon pesticides have been implicated in renal disease and the cantharides are well known for their adverse genitourinary effects.

Physical

Apart from direct trauma, the main physcial hazard associated with kidney damage is that resulting from ionizing radiation. Fibrosis, renal cell damage or death can lead to overt renal disease if the dose is high enough or prolonged enough.

Renal tract tumours

Occupationally related tumours of the renal tract are largely confined to the bladder which is the organ in greatest contact with the carcinogen.

The vast majority of urinary tract tumours are transitional cell epitheliomas. The tumours may be multiple or premalignant.

The link between urinary bladder tumours and occupational exposure to certain aromatic amines has been recognized for over 80 years. Such exposures occur in the manufacture of dyestuffs, chemicals, rubber and cables, as well as occupations using the carcinogens directly, such as chemistry laboratory workers, pest exterminators, pathologists and retort house operatives.

The aromatic amines which act as carcinogens are 1-naphthylamine, 2-naphthylamine, benzidine and 4-aminodiphenyl. They apparently act as carcinogens after metabolic transformation to orthoaminophenols which are restricted almost entirely to the bladder. The chemical similarity of 4-nitrodiphenyl, *o*-dianisidine and *o*-toluidine to these compounds, suggests that they should be suspected as carcinogens.

The prostate

Little is known of the normal metabolic processes of this organ but it seems to have a unique ability to concentrate (and excrete) heavy metals. Prostatic fluid, for example, has a naturally high concentration of zinc. Such an apparent affinity for heavy metals lends support to the epidemiological studies linking prostatic carcinoma with previous occupational exposure to cadmium.

Screening for occupational renal disease

There are two aspects of screening for renal disease.

1. The measurement of the toxic substance and/or its metabolites in body fluids.
2. The monitoring of renal function (*see* Chapter 11).

Urinary concentration of certain metals such as lead, cadmium and mercury, or the metabolites of organic compounds such as hippuric acid and phenolics, can be used to assess exposure and absorption of these substances. Such procedures can provide evidence of remarkably good correlation between exposure, absorption and excretion. However, if the substance under investigation is nephrotoxic, its concentration in urine may be increased, decreased, or remain constant, depending on the way the *damaged* kidney handles it compared with the normal kidney. Furthermore, some metals such as cadmium and mercury may be sequestered in the renal cortical tissue – a useful post mortem confirmation of exposure but hardly a practical everyday monitoring procedure. Specific sequelae of renal damage, such as β2 microglobulinuria in cadmium nephropathy, can also be a measure of exposure.

Prolonged exposure and marked nephropathic effects can make the monitoring of the offending substance somewhat academic. The management of the patient becomes less one of identifying the exposures as controlling the renal failure. At this stage, blood urea and uric acid, serum electrolytes, acid-base balance and creatinine clearances may be more appropriate tests (*see* Chapter 11).

THE SKIN

Structure and function

The skin consists of two basic elements. The outer layer or epidermis produces a protective coating of keratin as well as the pigment melanin. The inner layer, or dermis, consists of skin appendages and a fibrous covering.

Epidermis

The epidermis has a basal living layer which is slowly pushed closer to the surface as new cells are generated below and older cells are rubbed off at the surface. The mechanisms controlling basal cell division, and therefore the rate of epidermal cell production, are unknown, though differing rates exist in different parts of the body and these rates can be altered in response to external and internal stimuli. Friction, for example, increases the thickness of the flat keratinized dead cells of the horny layer. It is the keratin which gives the skin its toughness, and in combination with the cell membranes provides a remarkable degree of waterproofing.

Melanocytes are derived from the neural crest and lie on the basement membrane. These cells synthesize melanin, which gives the skin its colour. It is the rate of melanin production which determines whether the skin is black or brown or pink as there is no difference in the number of melanocytes in Negroid or Caucasian skin.

The epidermis is stuck to the dermis by a mucoprotein at the dermoepidermal junction.

Dermis

The dermis provides the inherent strength of the skin largely through its collagen content. There is great variation in the dermal thickness from, say, the skin of the back to the skin of the eyelid. The tethering of these collagen fibres also varies, depending on the need for mobility. For example, the structure is relatively immobile in the tips of the fingers but highly mobile over the elbow joint. Elastic properties are also present in the dermis to allow stretching to be followed by an orderly return to the prestretched state. Ageing diminishes the elastic and collagen content of the dermis.

Appendages

The skin has a large range of specialized appendages. Sensory nerves are vital in providing the organism as a whole with a clear sensory account of touch, pressure, pain and temperature in relation to the outside world.

Millions of hairs protrude from the skin. No new follicles appear after birth but the characteristics of the hairs can change dramatically with age or the onset of disease processes. Sebaceous glands secrete waxy, fat-soluble liquid into the hair follicle, which aids waterproofing as well as assisting thermal regulation. Sweat glands play a vital role in temperature regulation as well as exerting a subsidiary role to the kidney in maintaining salt-water balance. Both sebaceous and sweat glands are controlled by endocrine secretions. Nails are specialized plates of keratin which are important aids to the adapted use in man of the hands and feet.

Skin diseases may be characterized by discrete elevated lesions, such as pimples, but these do not necessarily imply a discrete stimulus. Patchy rashes can occur in response to a large area of skin being exposed to toxic influences, whereas a localized irritation from, say, the rubber of a fingerstall, will produce only a local response. The patient's reaction to skin disease is often seemingly irrational, with a feeling of concern or disgust out of all proportion to the severity of the disease. Furthermore, skin diseases in many countries are seldom contagious and rarely carry 'third party' risk, although many patients with skin disease, their relatives and fellow workers, think otherwise.

The localization of the rash is important. Even where the disease has a systemic origin, the type of rash is usually characteristic. However, as the skin varies in content and structure in different parts of the body, so its reaction, and therefore its appearance, can vary. This variation in visual effect is present when the same lesion occurs at different depths in the skin. Naevi are blue when deep and brown when superficial. Additionally, there is individual variation. Two people, equally sensitive to a given contact allergen, may not have identical rashes. Unlike the evaluation of most other organ dysfunctions, a detailed history is less important than close observation. Nevertheless, a history of the evolution of the rash, particularly in relation to drugs and occupational factors is important.

Occupational skin disease

While contact eczema accounts for most occupationally related skin disease, physical and biological agents may cause other types of disease. There are also occupational skin cancers which occur, either as the final phase of a long-standing skin disorder or as the first manifestation of prolonged occupational exposure. Finally, there are cutaneous manifestations of toxic substances which produce their more serious dramatic effects on other organ systems.

Primary irritant contact dermatitis

About 70 per cent of all occupational dermatoses are due to irritants which cause their effects by direct action on the normal skin at the site of contact. The cutaneous response depends on the strength of the irritant and the length of time it is in contact with the skin. Concentrated sulphuric acid or sodium hydroxide require only a few seconds to produce severe skin damage, whereas soap and water may take many days before a clinically noticeable change has occurred.

The mechanism of action of the primary irritants varies with the substance. Strong acids produce albuminates and the effect resembles a thermal injury. Strong alkalis combine with fats as well as having the ability to dissolve keratin. Lipid solvents remove the skin's protective oily secretions and disturb the waterproofing properties of the horny layer. Certain metallic salts of arsenic, mercury and chromium can combine with skin protein and initiate ulceration. Acne formation is accelerated by contact with chlorinated naphthylenes and cutting oils.

In addition to their direct effects, primary irritants can render the skin more vulnerable to a wide range of noxious influences and thereby facilitate injury from physical, biological and even sensitizing agents which would not otherwise have produced a cutaneous effect, or, if a response did occur, it would have been less severe.

Allergic contact dermatitis

Sensitizing eczemas account for about 15–20 per cent of all occupational dermatoses. The response is usually specific to one agent but the effect may be delayed for a week or more. Sensitization may require several hours of contact with the substance initially but subsequent exposure to the allergen usually produces a response in a matter of minutes.

As with primary irritants, the list of known sensitizers is long and increasing in length. Only the more important ones are mentioned.

The reaction producing the eczema is a delayed-type hypersensitivity which is produced by a lymphocyte mediated mechansism. The sensitizing agent combines with protein in the epidermis and this hapten-protein complex is engulfed by dermal macrophages. Following transportation to a regional lymph node, antibody to the complex is produced which circulates in the blood on small lymphocytes. A further exposure to the sensitizing hapten results in the release of antibody after the hapten-protein complex has been formed. The result may not be dissimilar in appearance to the effect of primary irritants – namely, erythema, papular eruption, vesiculation, oozing and desquamation. Chronic effects result in a thickened, fissured skin which may erupt into a more acute dermatitis on re-exposure to the antigen or following contact with an irritant substance.

The more common sensitizing agents are listed below:

1. Aniline derivatives; e.g. TNT, Tetryl, azodyes.
2. Antibiotics; e.g. penicillin, neomycin, tetracycline.
3. Dyes; e.g. paraphenylenediamine, ink, paints, cosmetics.

4. Metals; e.g. arsenicals, chromates, nickel, cobalt.
5. Resins; e.g. epoxy and formaldehyde resins, monomers of cellulose, vinyls and acrylics, colophony.
6. Rubber chemicals; e.g. accelerators, vulcanizers, antioxidants.
7. Plants; e.g. primula, buttercup, daffodil, chysanthemum, tulip.
8. Trees; e.g. West African mahogany, iroko, red cedar.
9. Pharmaceuticals; e.g. phenothiazines, procaines, tolbutamide, chlorothiazide (many of these are capable of photo sensitization).

The 'standard battery' of patch tests proposed by the International Contact Dermatitis Research Group contains 20 substances which represents a considerable number of groups of sensitizers listed above.

Physical agents

Occupational groups such as coal miners and construction workers are prone to cutaneous injury resulting from friction and laceration. Frictional dermatitis can present with an initial erythema, scaling, pigmentation and bullous formations which may resemble discoid or nummular eczema.

Any 'outdoor' worker may be exposed to wind, rain, frost and sun, which can lead to chapped and dessicated skin. Constant exposure to heat, or water can also cause purely physical damage to the skin. Sweat disturbances can result and conditions like dehydration oedema and prickly heat can be of occupational origin. Ultraviolet light causes premature ageing of the skin, pigmentation changes, acute burns, or carcinomata.

Biological agents

Rarely, biological agents may cause occupational skin diseases. Confectioners exposed to sugar, heat and moisture, are prone to monilial infection. Dockers, fishermen, millers. grocers, farmers, veterinarians and poultry men can contract infections such as erysipeloid, tuberculosis, tularaemia, dermatophytosis, anthrax, orf, cow pox, Rocky Mountain spotted fever and cutaneous larva migrans.

Cutaneous manifestations of systemic disease

In occupational health practice these can include the dermatitic effects of arsenic, chromium, lead, silver, aniline, and dinitrophenol, or the alopecia associated with arsenic, ionizing radiation or thallium. Furthermore, skin manifestations of drug therapy used to treat other, specifically dermatitic, conditions may confuse the clinical picture.

Skin tumours

Two hundred years ago Percivall Pott noted that chimney sweeps had an increased risk of acquiring scrotal cancer. Similar tumours were subsequently observed in shale-oil workers, mule spinners and other occupational groups who came into contact with mineral oils, soot tar and creosote. The tumours are frequently multiple, tend to occur on the skin surface most exposed to the carcinogen, and may be found as premalignant papillomata or frank squamous or basal cell carcinoma.

Ionizing radiations can cause skin cancer and those occupations most at risk include radiologists and other X-ray workers and radioisotope handlers. Ultraviolet radiation can cause squamous cell carcinoma and the risk is particularly high for the fair-skinned person engaged in outdoor work such as fishing or farming.

Arsenic and arsenicals are also capable of causing epitheliomatia in exposed workers.

The prognosis for skin cancer tends to be better than for many other sites because of earlier recognition and the practicability of local excision with or without radiotherapy.

Distinguishing features of contact dermatitis

Contact dermatitis has certain distinguishing features from other dermatoses.

History The lesions bear a temporal relationship with exposure to the suspected cause, with periods of remissions and exacerbations depending on exposure. The eruptions usually diminish at weekends. If the lesions fail to clear after a prolonged absence from the suspected agent, another agent or different aetiology should be considered.

Site The lesions due to direct exposure with the offending agent are usually on exposed areas of the skin such as the backs of the hand, ventral portion of the wrists, the forearms and the cubital fossae. Gases, vapours and fumes often affect a wider area such as face, eyelids, ears and the 'V' of the neck. Of contact ezcemas 80–90 per cent affect the hands.

Type Most occupational contact dermatoses tend to present as acute conditions with erythema, oedema, papules, vesicles, or bullae. Crusting, scaling and finally desquamation commonly occur. Chronic conditions more commonly associated with skin dehydrators or defatters may present as erythematous, scaly, dry lichenified or fissured skin.

Less common manifestations (which often imply a particular causal agent) include folliculitis or acne, neoplasia, pigmentation changes, chronic indolent granulomata or ulcers, and hair loss.

Screening for occupational dermatoses

Apart from the history and examination, special investigations may be useful in establishing the diagnosis. Skin biopsy, skin culture or microscopy are such special tests, but patch-testing for sensitizing agents warrants a specific account.

Special investigations   Having established (or suspected) that a skin condition is a contact eczema, patch-testing can be most useful in distinguishing irritants from allergens. False positive results, false negative results, failure to patch-test for the true underlying cause of the eczema, or failure to establish that a given allergen exists in the workplace will, however, prevent an accurate diagnosis and therefore effective treatment. A good knowledge of the workplace and the procedures and materials used are essential.

Patch testing   Strips of aluminium foil with lint attached to non-allergenic adhesive tape and the test substances are applied to the lint. The skin is marked next to the sites of the patches and after 48 hours the patches are removed, and skin reactions noted. A further reading at 96 hours is advisable for slow or weak sensitizers, and to avoid confusing the findings with an eczematous reaction to the plaster or even the enclosure of the skin for two days.

Patch-testing is not technically difficult, but misinterpretation of the results can easily occur. Widespread reactions may mask specific allergic responses, and primary irritant contact dermatitis may occur if the test solutions are too strong. False positive and false negative reactions are not uncommon. Retesting may therefore be necessary with modified solutions or new materials. Nevertheless, in experienced hands, patch-testing can distinguish allergic from irritant eczemas in over one-third of people who warrant testing, although one disadvantage of this test is the alleged risk of sensitizing any individual through the patch test itself.

The importance of establishing an accurate diagnosis cannot be overemphasized. Although occupational dermatoses are rarely, if ever, fatal they are by far the most frequent cause of time off work for occupationally related conditions. Relapse can occur after treatment or removal from exposure, but contact dermatitis is a preventable disease and should be controlled.

6

The Effects of Work Exposures on Organ Systems II – Blood, Lungs and Nervous System

H.A. Waldron and J.M. Harrington

THE BLOOD

Structure and function

The blood volume in a person of average build is about 70 ml/kg. All the blood cells are derived from the haemopoietic organs, which include the bone marrow, spleen, liver, thymus and the lymph nodes.

The cells of the blood have a common origin from reticulo-endothelial (RE) cells which give rise to the primitive haemocytoblast. From this cell, precursors of all the cell lines, red, white and platelets are formed, which, after varying stages of maturation, form the cells which circulate in the peripheral blood.

The red cell

The formation of the mature red cell is represented diagrammatically in *Figure 6.1*. Erythropoiesis is controlled through the mediation of erythropoietin, a hormone with a molecular weight of about 46 000, which is produced in the kidney in response to hypoxia.

The red cell has a life span of about 120 days and in health, senescent cells are removed by the phagocytic action of RE cells in the spleen and liver. The normal ranges for haemoglobin concentration, red cell numbers and red cell indices are shown in *Table 6.1*.

The white cell

The white cells are formed in the bone marrow and the mature forms are released into the blood where their life span is generally short. Following their release from the marrow, the lymphocytes undergo further modification in the lymphoid tissues, some in the thymus (the

115

Table 6.1. NORMAL RED CELL VALUES

Haemoglobin (g/dl)	13.5–18.0 (men)
	11.5–16.5 (women)
Red cells $(10^{12}/\ell)$	4.5–6.5 (men)
	3.9–5.6 (women)
MCV (femto litres)	76–96
MCH (picograms)	27–32
MCHC (g/dl)	30–35

MCV = mean corpuscular volume
MCH = mean corpuscular haemoglobin
MCHC = mean corpuscular haemoglobin concentration

Table 6.2. MAJOR FUNCTIONS OF THE WHITE CELLS

Neutrophil	Phagocytosis of small particles of foreign material; bacterial phagocytosis greatly facilitated by the presence of specific antibodies.
Eosinophil	Phagocytosis of antigen-antibody complexes, rheumatoid factor complexes and red cells (in the presence of immune sera).
Basophil	Release of histamine.
Monocyte	Production of tissue macrophages; phagocytosis (including that of large particles); pinocytosis of proteins and colloids.
Lymphocytes	Mediation of specific immune responses.

Table 6.3. NORMAL WHITE CELL VALUES

		%
Total white cells	$4.0–11.0 \times 10^9/\ell$	
Neutrophils	$2500–7500 \times 10^6/\ell$	(40–75)
Lymphocytes	1500–3500	(20–45)
Monocytes	200–800	(2–10)
Eosinophils	40–440	(1– 6)
Basophils	0–100	(0– 1)

T-cells) and some (the B-cells) elsewhere, probably in Peyer's patches in the small bowel. The T-cells and B-cells can be distinguished under the scanning electron microscope since the B-cells have a ragged appearance whereas the T-cells are smooth.

The prime function of the B-cells is to divide in response to contact with antigen to produce plasma cells which in turn produce specific antibodies. The T-cells help the B-cells in initiating primary immune responses and are themselves involved in cell-mediated responses, including rejection phenomena.

The most important functions of the white cells are outlined in *Table 6.2;* normal values appear in *Table 6.3.*

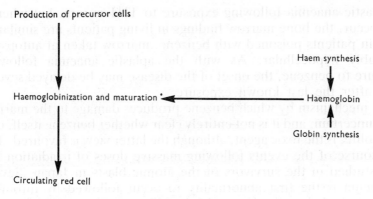

Figure 6.1 Stages of red cell maturation

Disordered function from occupational toxic agents

The red cell

The effect of damage to the red cell is the development of anaemia which may be trivial and asymptomatic, or so severe that the patient's life is put in jeopardy. Damage to the red cell may be sustained at any of the stages shown in *Figure 6.1;* some toxic substances act only at one stage, others, of which lead is the prime example, at more than one.

Production of cells

Bone marrow damage is rarely sustained in industry and the number of agents which are known to induce aplastic anaemia as the result of occupational exposure is small. The best known are benzene, trinitrotoluene (TNT) and irradiation, but a few cases have been reported following exposure to organic insecticides. Cadmium has also been reported to cause bone marrow changes. The changes which may be found in the peripheral blood of workers exposed to benzene depend on which cell line, or lines, suffer the brunt of the injury. Anaemia is generally the first effect noted, followed sequentially by leucopenia, thrombocytopenia and finally, if the damage to the marrow is severe, by pancytopenia.

These changes may follow from either acute or chronic exposure, and aplastic anaemia has been found in subjects whose last known exposure took place 10 years before the onset of symptoms. All manner of abnormalities may occur in the bone marrow, from acellularity to hypercellularity; the latter occurring even though there is no evidence of cell formation in the peripheral blood.

Aplastic anaemia following exposure to TNT is rare, but when it does occur, the bone marrow findings in living patients are similar to those in patients poisoned with benzene; marrow taken at autopsy is invariably hypocellular. As with the aplastic anaemia following exposure to benzene, the onset of the disease may be delayed several years after the last known exposure.

The mechanism by which benzene produces damage to the marrow is still uncertain, and it is not entirely clear whether benzene itself, or a metabolite, is the toxic agent, although the latter view is favoured. The time course of the events following massive doses of irradiation has been studied in the survivors of the atomic blasts in Japan. Severe leucopenia is the first abnormality to occur followed by thrombocytopenia and then by the production of anaemia; at this stage the bone marrow shows an almost complete loss of cells.

Haemoglobinization and maturation

Inorganic lead affects a number of the stages of haem synthesis by inhibiting the activity of the enzymes which control the progressive production of haem from glycine and succinyl co-enzyme-A (*Figure 6.2*). Organic lead compounds have no significant effect on haem synthesis except (slightly) to lower δ-aminolaevulinic acid dehydrase (ALAd) activity. Many of the enzymes in this pathway contain essential sulphydryl(-SH) groups for which lead has a high affinity; by combining with these -SH groups, lead renders the enzyme inactive. As haem synthesis becomes progressively depressed by the action of

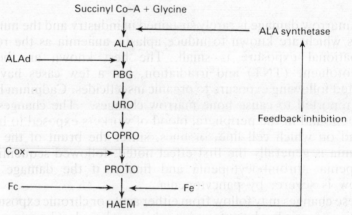

Figure 6.2.   Haem synthesis: the production of ALA by ALA synthetase is the rate-limiting step. Principal steps inhibited by lead are indicated by arrows on left. ALA = δ-aminolaevulinic acid; PBG = porphobilinogen; URO = uroporphyrinogen III; COPRO = coproporphyrinogen III; PROTO = protoporphyrin IX; ALAd = ALA dehydratase; C ox = coproporphyrinogen oxidase; Fc = ferrochelatase

lead, ALA synthesis is actually induced as the result of interference with the negative feedback loop shown in *Figure 6.2*. The complex impairment of haem synthesis produces characteristic changes in the levels of haem precursors in the blood and urine which should enable a definitive diagnosis of lead poisoning to be made (*Table 6.4*). Lead

Table 6.4. CHANGES IN HAEM METABOLISM IN LEAD POISONING

Blood	Urine
Erythrocyte protoporphyrin concentration elevated	Urinary ALA concentration raised
Red cell ALAd activity inhibited	Urinary coproporphyrin concentration raised
Serum ALA concentration raised	
Serum iron concentration raised	

also inhibits the synthesis of the globin moiety of the haemoglobin molecule, impairing the production of both α and β chains *in vitro*; the former to a greater degree than the latter. In lead workers the $\alpha:\beta$ ratio may increase, suggesting that there is a compensatory increase in α chain production *in vitro*. In other haem deficiency states, globin synthesis is impaired through the inhibition of the formation of a specific ribosomal complex and this is probably the mechanism in lead poisoning also.

The circulating red cell

Arsine is the most notorious of the industrial haemolysins and the intravascular haemolysis which it produces is often profound. The mechanism of action is not yet fully established but among the possibilities which have been discussed are the production of elemental arsenic, or arsenic dihydride, the inhibition of catalase with a resultant production of hydrogen peroxide, and the formation of complexes with -SH groups which impair the sodium-potassium pump. This multiplicity of explanations suggests that none is entirely correct. Likewise, the haemolytic actions of TNT, one of the few other industrial haemolysins, has not been adequately explained although it is interesting to note that acute haemolytic episodes have occurred in TNT workers who have a deficiency of glucose-6-phosphate de-hydrogenase (G6PD), which suggests that in these men the lytic action is due to the oxidizing properties of the compound.

Lead is the only other haemolytic agent which needs to be considered. It is only weakly haemolytic and it exerts this effect by producing abnormalities in the structure and function of the red cell membrane. Red cell Na^+ and K^+ ATPase is strongly inhibited by lead *in vivo* and *in vitro* and this produces a great loss of potassium from the cell. This loss is much greater than would be expected from the inhibition of ATPase alone, and the conformational changes induced in the membrane probably create a selective gate through which potassium (but not sodium) can leak. Changes in the shape of the red cell are also in themselves likely to accelerate their rate of removal from the circulation by RE cells and perhaps to increase their mechanical fragility. However, the haemolytic action of lead is of less importance in the production of anaemia than the alterations in haem synthesis and frank anaemia is a late manifestation of lead absorption.

Malignant change

The malignant potential of ionizing radiation has been exemplified by studies on patients with ankylosing spondylitis treated by X-irradiation and by the follow-up studies of the survivors of the atomic bomb blasts. The peak incidence of leukaemia occurs about four years following irradiation and the acute non-lymphocytic types predominate. The data from Japan indicated that there was a linear relationship between the radiation dose and the incidence of leukaemia above 100 rad, but below this dose the shape of the curve was uncertain. At Nagasaki no increase in incidence in leukaemia was observed when the dose was less than 100 rad, but at Hiroshima an increase was noted for doses down to 20 rad. The risk of leukaemia in the atomic industry is of a very low order. For persons exposed continuously to 10 milli-Sieverts per year (mSv/y) (the maximum dose usually experienced by workers in the British nuclear power industry) from the age of 20, it is between one in 100 000 and one in 500 000. The only other significant risk of developing occupationally related leukaemia appears to arise from exposure to benzene. There is a considerable number of case reports, mostly of the acute non-lymphocytic type, although an unusually large number of cases of erythroleukaemia have also been reported. Numerous chromosomal aberrations have been found in patients with leukaemia who have had chronic benzene exposure which is consistent with the impairment of DNA synthesis noted in experimental animals.

Epidemiological studies have tended, in general, to confirm that benzene may induce leukaemia, but not all authorities are convinced by the evidence.

Screening for occupational disorders of the blood

When all that is possible has been done in the way of substitution, segregation and environmental and personal protection, the control of hazards likely to cause disorders of the blood depends upon a knowledge of the mode and site of action of the materials under consideration. Lead workers can be supervised by monitoring the excretion of haem precursors in the urine or by estimating erythrocyte protophyrin concentrations; the measurements of erythrocyte ALAd is too troublesome a technique to be considered as a routine measure and is no more valuable an indication of metabolic effect than the other tests.

Anaemia is a late manifestation of lead poisoning, and haemoglobin estimations if used at all in routine screening programmes, must be supplemented by tests which will provide *early* warning of impaired haem synthesis. On the other hand, routine haemoglobin estimations and red cell counts will be helpful in the supervision of workers exposed to benzene or other haemolytic agents, since in their case anaemia is an early sign of potentially harmful effects. With the present radiation doses in industry, there is little point in undertaking any form of haematological monitoring except following accidental exposure to high doses.

The occurrence of leukaemia will be prevented only by limiting exposure and nothing is to be gained by looking for leukaemia or preleukaemia changes in the blood of those potentially at risk.

THE LUNGS

Structure

The lungs are composed of a number of topographical units called bronchopulmonary segments, which are roughly pyramidal in shape with their apices directed inwards and their bases lying on the surface of the lung. Each lung is composed of ten segments grouped together into lobes.

The airways

It is usual to describe the airways in terms of the number of divisions or generations which separate them from the main bronchus. Thus, the segmental bronchus is counted as the first generation, and its first branch as the second generation, and so on. There are about 15 generations of bronchi, the first five of which are 'large' bronchi which, by definition, have a plentiful supply of cartilage in their walls.

The fifth to fifteenth generations are 'small' bronchi whose walls contain only sparse amounts of cartilage. There are about ten generations of bronchi which ultimately open into the alveolar ducts from which the alveoli take their origin. The bronchiolus which opens into the alveolar duct is defined as a respiratory bronchiolus whilst the one immediately proximal to it is defined as a terminal bronchiolus.

The acinus and the lobule

The acinus is all that part of the lung distal to a terminal bronchiolus. It includes up to eight generations of respiratory bronchioli and their associated alveoli, and it is approximately 0.5–1 cm in diameter. The lobule is the term used to describe the three to five terminal bronchioli together with their acini, which cluster together at the end of any airway.

The alveolus

In the fully developed adult lung there are about 300 million alveoli which offer a surface area of some 70 m^2 for gas exchange. The alveolus is between 0.05–0.1 mm in diameter and its walls are formed from the cytoplasm of two types of epithelial cells, the type I and type II pneumocytes. The alveolar wall may be only 5–10 μm thick at its thinnest point, but the epithelial lining is continuous throughout. The type I pneumocyte is a large flat cell covering a much greater area of the alveolar wall than the type II cell, although it is less numerous than the latter.

The type II cell is a small cell containing characteristic lamellated inclusion bodies within its cytoplasm which are the origin of surfactant. The aveolus also contains macrophages which remove dust and other debris entering the alveolus, and then traverse up the airways aided by the cilia in the epithelial lining. On reaching the larynx, they and their contents are either passed outside in the sputum or swallowed. Foreign materials cleared from the lungs and substances may, of course, be subsequently absorbed from the gut.

The blood–gas barrier

The epithelial lining cells of the alveolar wall, together with the endothelial cells of the pulmonary capillaries (each with its own basement membrane) and the tissue fluid in the spaces between, make up what is known as the blood–gas barrier. This is the distance across which gas molecules or indeed any other material must pass into or out of the alveolus. The total distance involved is probably less than 0.001 mm.

The lining of the airways

The large airways are lined by a ciliated pseudostratified epithelium from which mucus is secreted by the goblet cells and the submucosal glands. The mucus traps particulate matter in the airways, and is constantly being moved up towards the larynx by the synchronous beating of the cilia. This so-called mucociliary escalator is one of the most important mechanisms by which the airways are cleared of particulate matter.

Ciliated cells are found down as far as the respiratory bronchioli so that this mechanism operates continuously from the bronchiolo-alveolar junction to the larynx.

Rate of flow in the airways

The consequence of the continual branching within the airways is greatly to increase the surface area whilst reducing the rate of flow as may be seen in *Table 6.5*.

Table 6.5. CROSS-SECTIONAL AREA OF THE AIRWAYS AND RATE OF AIRFLOW

	Area (cm^2)	*Flow rate* $(cm\,sec^{-1})$*
Trachea	2.0	50
Terminal bronchioli	80	1.25
Respiratory bronchioli	280	0.36
Alveoli	7×10^5	

* For a volume flow of 100 ml sec^{-1}

Lung function

The main function of the lungs is to supply the body with oxygen and to remove from it carbon dioxide. The processes involved in gaseous exchange can be subdivided into:

1. Ventilation.
2. Gas transfer.
3. Blood–gas transport.

These subdivisions can also be used to outline lung function tests in common use. None of these measures of lung function is diagnostic in itself and the degree of accuracy and sophistication used in measuring lung function depends on many factors including the facilities available, for example, whether the tests are undertaken by a health department in a factory surgery or by a cardiothoracic unit in a hospital; and the needs of the patient and the investigator. The comparatively simple tests used either to monitor an individual worker or as a measure of lung function in an epidemiological survey are described in Chapter 11.

Although tests of lung volume, ventilation, gas distribution and gas transfer may be useful in differing circumstances, measures of ventilatory capacity are indispensable in all investigations of occupational lung disease.

Gas transfer

The effective exchange of oxygen and carbon dioxide in the lungs depends on three processes:

1. The correct distribution of ventilated lung to blood-perfused lung;
2. The efficient diffusion of gases across the alveolar-capillary membrane;
3. The appropriate uptake or release of the gases by the red cell.

Measures of the ventilation/perfusion ratio are primarily research tools and frequently involve the inhalation of inert gases and the subsequent measurement of the exhaled concentrations. Alternatively, radioactive gases can be used in the lung. Assessment of the rate of diffusion of gases across the alveolar-capillary membrane is also a specialized laboratory technique, but the measurement of the gas transfer factor ($T\ell$) using carbon monoxide (CO) is widely used in assessing diffusion defects, as may occur in asbestosis and other fibrotic diseases of the lung (*see* Chapter 11).

Blood–gas transport

The final step in getting the oxygen to the tissues and transporting the carbon dioxide in the opposite direction involves the blood. The efficiency of blood–gas transport requires the measurement of the concentration of alveolar carbon dioxide as well as arterial oxygen, carbon dioxide and pH. Blood–gas tensions and pH are rarely measured in occupational health practice, though they are frequently vital in the measurement of respiratory failure in hospital practice.

Occupational lung disorders

Harmful effects to the lung produced by toxic agents can be grouped into six more or less distinct categories.

Acute inflammation

A number of gases and fumes produce acute inflammation in the airways, the most important of which are shown in *Table 6.6*. The resulting symptoms vary, depending upon which parts of the airways are affected, and this is in turn a function of the solubility of the substance. Thus highly soluble gases such as ammonia produce immediate effects on the upper respiratory tract (and the eyes) causing

Table 6.6. IRRITANT GASES OR FUMES

Gas or	Sources	Acute effects	Chronic effects
Ammonia	Production of fertilizers and explosives; refrigeration; oil refining; manufacture of plastics	Pain in eyes, and mouth and throat; oedema of mucous membranes; conjunctivitis; pulmonary oedema	Airways obstruction; usually clears in about a year
Chlorine	Manufacture of alkali bleaches and disinfectants	Chest pain; cough; pulmonary oedema	Usually none; occasionally causes airways obstruction
Nitrogen oxides	Silo filling; arc welding; combustion of nitrogen containing materials	Pulmonary oedema after lag of one or two hours; obliteration of bronchioli in severe cases after two to three weeks	Permanent lung damage with repeated exposure
Ozone	Argon-shielded welding	Cough; tightness in the chest; pulmonary oedema in severe cases	Usually none
Phosgene	Chemical industry; World War I gas	Pulmonary oedema after lag of several hours	Chronic bronchitis
Sulphur dioxide	Paper production; oil refining; atmospheric pollutant	As for ammonia	Chronic bronchitis
Mercury vapour	Chemical and metal industries	Cough and chest pain after lag of three to four hours; acute pneumonia	Usually none; pulmonary fibrosis rarely
Osmium	Chemical and metal industries laboratories	Tracheitis; bronchitis; conjunctivitis	None
Vanadium pentoxide	Ash and soot from fuel oil	Nasal irritation; chest pain; cough	Bronchitis; bronchopneumonia
Zinc chloride	Manufacture of dry cells; galvanizing	Tracheobronchitis	None

pain in the mouth, throat and eyes due to the swelling and ulceration of the mucous membranes. These symptoms are so intensely unpleasant that affected individuals will make every effort to remove themselves as quickly as possible from exposure. Following continuous exposure, or a single exposure to a very high concentration, the deeper airways are affected and pulmonary oedema results as a consequence of capillary damage. The subsequent impairment of gas exchange may cause serious symptoms if not adequately treated. A relatively insoluble gas such as phosgene produces no immediate effects on the upper respiratory tract but induces profound pulmonary oedema after a lag of several hours. Some of the pulmonary irritants also cause permanent lung damage if exposure is particularly high or frequently repeated, whereas others such as phosgene and sulphur dioxide are said to predispose to conditions such as pneumonia or chronic bronchitis.

Asthma

Occupational asthma develops as an abnormal immunological response to foreign materials which act as antigens. The inhalation and absorption of the antigen provokes the production of specific antibodies which in turn set in motion a series of events culminating in the release of histamine and other active materials which cause reversible bronchial constriction.

Asthmatic symptoms may follow immediately after exposure to the antigen but more commonly there is a delay of several hours and symptoms develop during the evening or at night. This often means that the relation of the symptoms to the patient's occupation is overlooked.

Some of the agents which are responsbile for occupational asthma are shown in *Table 6.7*. Others are continuously being identified, and the question whether or not this condition should be a prescribed disease in Great Britain is presently under review.

Byssinosis

This is a different type of occupational asthma which causes chronic obstructive airways disease. It can progress to irreversible impairment of pulmonary function. The disease is due to inhalation of cotton, flax, soft hemp or sisal dusts. While its pathogenesis is not clear, there is strong evidence to suggest that the chronic constriction is caused by a histamine releasing agent rather than an immune response as in other types of occupational asthma. Symptoms of chest tightness or

Table 6.7. SOME EXAMPLES OF OCCUPATIONAL ASTHMA AGENTS

Agent	Source
Isocyanates especially toluene di-isocyanate (TDI) and methylene bisphenyl di-isocyanate (MDI)	Paints and foams
Metals and their salts platinum, nickel, chromium, tungsten carbide	Refining and use in manufacturing processes
Natural resins colophony	Welding in the electronics industry
Proteolytic enzymes alkalase from *B.subtilis*	Biological washing powder
Rat and mouse urine	Laboratory workers
Wheat grain	Handling and moving grain
Wood dusts red cedar, iroko	Wood working

breathlessness are first noticed on Monday mornings when the cotton worker returns from a weekend away from exposure.

There may be an interval of several years from the time of first exposure to the onset of symptoms which, in the early stages disappear on the second day back at work. If exposure is continued, the symptoms are noted for longer periods during the week until finally they are continuously present throughout the whole of the week.

Pneumoconiosis

The term pneumoconiosis means 'dusty lungs' and for medical purposes it has been recommended that the term should be confined to mean permanent alteration of lung structure following the inhalation of mineral dust, and the tissue reactions of the lung to its presence. Despite this, clinicians still refer to 'benign' pneumoconiosis, which is an apparent contradiction in terms to the definition given above. The dusts which are most harmful to the lungs are silica (or quartz), coal dusts and asbestos. Silicosis, which follows exposure to fine free crystalline silicon dioxide or quartz has a claim to be the oldest of all the occupational diseases since there is no doubt that the palaeolithic flint tool makers would have suffered from it, though there is no objective evidence that this was the case.

During the last century, the disease was common in many industries including mining, quarrying, the pottery industry, in iron and steel foundries, and in sand blasting. In recent years, however, the number of new cases diagnosed has declined considerably as the result of improved working practices and the substitution of safer materials for silica wherever possible.

The presence of silica in the lung sets up a reaction which leads to the formation of small nodules of fibrotic tissue (*ca.* 1 mm in diameter) which increase in size and coalesce as the disease progresses. These nodules are seen radiologically as small, rounded opacities scattered throughout the lung fields but most prominent in the upper parts of the lung. Radiological changes are normally present before any symptoms of breathlessness appear. Early diagnosis of pneumoconiosis is essential as, if a sufficiently large quantity of dust has been inhaled, the symptoms may progress even after exposure has ceased. The progression of the disease is marked by increasing dyspnoea, and death is frequently caused by cor pulmonale. An unexpectedly large number of patients with silicosis also develop tuberculosis for reasons which have never been clear and this does nothing to help their dismal prognosis.

Pneumoconiosis in coalminers is conveniently subdivided in the simple forms and progressive massive fibrosis (PMF), although the latter always develops in the context of the former. The diagnosis of a simple pneumoconiosis is made when small rounded radiological opacities due to the presence of coal dust are found in the lung. These radiographic changes bear a close relationship to the total amount of dust inhaled.

Beyond a slight cough which produces blackish sputum, there are virtually no symptoms attributable to coalminers' simple pneumoconiosis and its importance is due to the fact that it is the precursor of PMF.

The reasons why simple pneumoconiosis progresses to PMF in a small proportion of miners are not clear, as attested by the numerous explanations which have been put forward. The most recent proposes that it is due to an antigen-antibody reaction in the lungs, but the mechanism underlying this reaction remains obscure. Radiologically, PMF is recognized by the presence of large, irregular, opacities usually confined to the upper lung fields.

In the lung itself these areas appear as hard, black masses often with a necrotic cavity filled with jet black fluid. Occasionally the lesion involves one of the larger airways. If the bronchial wall is eroded, the patient coughs up the contents of the lesion as inky black sputum. Moderate to severe forms of PMF produce sufficient lung damage to disable the victim and perhaps lead to premature death, but fortunately the number of cases of the severe form of the disease is declining rapidly with the better methods of dust suppression underground.

Exposure to asbestos is associated with pulmonary fibrosis known as asbestosis, with the development of bronchial carcinoma, and more rarely, with the development of pleural mesothelioma. In addition, asbestos bodies may be found in the sputum and calcified pleural plaques may be discovered incidentally on a chest X-ray.

Asbestosis manifests itself clinically by increasing dyspnoea, by a non-productive cough, finger clubbing and by weight loss. The radiographic changes are usually confined to the lower parts of the lung and show up as linear shadows which become larger and more irregular as the disease is clinically apparent. Although patients may die from uncomplicated asbestosis, perhaps as many as 50 per cent of those in whom the disease ends fatally die because of the development of carcinoma of the bronchus.

Smokers exposed to asbestos are at a very much greater risk of developing lung cancer than those who only have asbestos exposure, that is to say, smoking and asbestos dust act synergistically in the production of cancer in the lung. Pleural or peritoneal mesotheliomas are rare and seem to occur mostly in those who have been exposed to crocidolite or blue asbestos. It is of note that some of the tumours have developed in women whose only exposure to asbestos occurred when they washed their husband's dirty overalls. The tumour spreads from the pleura into the underlying lung tissue and is invariably fatal.

Asbestos bodies are commonly found in the sputum of persons exposed to asbestos. They are rod-like structures, 20–150 μm in length and often have a beaded appearance, and are coated with an iron-containing protein which can be stained so that they are visible under the microscope. The presence of asbestos bodies in sputum can be taken as evidence of exposure to asbestos but it does not imply disease.

Similarly, the calcified pleural plaques seen on a chest X-ray are a sign of exposure to asbestos, but are not themselves indicative of disease.

Benign pneumoconiosis The presence of dust particles in the lung will be detected radiologically if their atomic number is greater than that of calcium and the higher the atomic number of the dust, the denser the shadow produced on the X-ray film. The diagnosis of benign pneumoconiosis is entirely dependent on radiological findings since, by definition, the patient is symptom-free. Of the dusts which produce radiological change, the most important are iron, tin, barium and antimony. Although they are asymptomatic, the radiological changes due to the first three dusts are given specific names (siderosis, stannosis and baritosis respectively) which perpetuate the notion that they are diseases.

Extrinsic allergic alveolitis

Organic materials which are inhaled may have one of two distinct effects on the lungs: they may either induce asthma (as described above) or they cause alveolitis with a subsequent reduction in gas transfer. A wide range of organic materials may produce the disease but the most common are fungal spores (*Table 6.8*).

Many clinical varieties of the condition have been described, each due to one (or occasionally more than one) specific allergen. The symptoms in each variety are similar and the individual types of the disease are usually named after the occupational group in which it was first discovered, or in which it is most common.

The predominant type seen in this country is farmer's lung. Symptoms typically develop 4–8 hours after exposure to mouldy hay, after which time the patient presents with a fever, tiredness, chills and generalized aches and pains. Shortness of breath and an unproductive

Table 6.8. SOME TYPES AND CAUSES OF EXTRINSIC ALLERGIC ALVEOLITIS

Clinical condition	Due to exposure to	Allergen
Farmer's lung	Mouldy hay	*Micropolyspora faeni* *Thermoactinomyces vulgaris*
Bird fancier's lung	Bird droppings	Protein in the droppings
Bagassosis	Mouldy bagasse	*T. vulgaris*
Malt worker's lung	Mouldy malt or barley	*Aspergillus clavatus*
Suberosis	Mouldy cork dust	Cork dust
Maple bark stripper's lung	Infected maple dust	*Cryptostroma corticale*
Cheese washer's lung	Mouldy cheese	*Penicillium casei*
Wood-pulp worker's lung	Wood pulp	*Alternaria* species
Wheat weevil disease	Wheat flour	*Sitophilus granarius*
Mushroom worker's lung	Mushroom compost	*M. faeni* *T. vulgaris*

cough are then noted but there is none of the wheeziness so typical of the patient with asthma. After removal from exposure, the symptoms can be expected to clear up within 12 hours. Repeated exposure, however, may result in the development of pulmonary fibrosis with permanent impairment of lung function and X-ray changes most apparent in the upper part of the lung.

Malignant disease

Exposure to asbestos is the most important cause of occupational lung tumours, but there are other groups whose occupation predisposes them to an unduly high risk of this disease, foremost among whom are miners subjected to ionizing radiation. This group includes not only men mining uranium but also haematite miners in Britain and fluorspar miners in Canada. In each case the radioactive source is radon or its daughter products, polonium 218, 214 and 210. All these elements are gases which diffuse from the rock into the air in the mine shafts. They all emit alpha particles which have a penetration range in tissue cells of between 40–70 μm, just sufficient to enable them to damage the nuclear material in the basal cells of the bronchial epithelium and set in motion the train of events which culminates in the proliferation of a clone of malignant cells.

The excess number of cases of lung cancer in miners exposed to radon appears to be between 0.3–1.0/year, per rem per million exposed. In the past, an excess of lung tumours was noted in men engaged in the manufacture of arsenical sheep dips and in nickel refiners. Arsenical sheep dips have now been rendered obsolete by the development of other compounds, but arsenic is still encountered in other occupations and care must be taken to avoid both its acute and chronic effects. The risk of lung and nasal cancers in nickel refining disappeared when the process was altered in the 1920s and arsenic-free sulphuric acid was used to remove copper from the ore.

Concurrent with this change, the levels of nickel dust in the atmosphere were reduced and it has never been determined beyond all reasonable doubt whether arsenic or nickel (or indeed some other material) was the carcinogenic agent.

More recently it has been found that exposure to hexavalent chrome salts carries with it an increased likelihood of contracting lung cancer, although not so great as that following exposure to asbestos or radon. Coke oven workers, particularly those on the oven tops, have also been reported to suffer to excess from lung cancer, due to the presence of polycyclic aromatic hydrocarbons in the air in the vicinity of the ovens.

The only other tumour of the respiratory tract of note, apart from the pleural mesothelioma caused by crocidolite, is adenocarcinoma of the nasal sinuses which has been found in woodworkers in the furniture industry in High Wycombe, and in leather workers in the Northamptonshire shoe trade. This is an otherwise rare tumour, and came to light in the first place, as have a number of other occupationally induced tumours, when clinicians became aware that they were seeing an unusually large number of patients with uncommon diseases. Subsequent epidemiological studies were able to confirm the original clinical observations.

Screening for occupational lung disease

Exposure to potentially toxic dusts must be controlled by regular medical supervision (*see* Chapter 25). It is important to include lung function tests and radiography in any control programme (*see* Chapter 11) since the development of clinical disease is indicative of damage which may not be reversible. When interpreting the results of lung function tests due weight must be given to the smoking habits of the individual being examined since smoking may well impair his lung function to a greater degree than the dust to which he is exposed.

Susceptibility to allergens which cause asthma can be investigated by provocative tests; care is required both in the administration of the test and the interpretation of the results and both should be left to those with the necessary experience.

THE NERVOUS SYSTEM

Structure and function

The basic unit of the nervous system is the neurone, which may be divided into four principal areas: the cell body, the dendrites, the axon and the synaptic terminal. The dendrites are short, branching fibres which provide a large surface area for synaptic contact with other cells; impulses from the dendrites are transmitted towards the cell body. The cell body itself is covered with synaptic inputs from other cells (perhaps as many as 10^4 on each cell in the brain) and it is the region in which all the excitatory and inhibitory inputs are summated and from which the nerve impulse originates.

The term axon is strictly applied to a long single nerve fibre which transmits impulses *away* from the cell body, but in practice, any long nerve fibre is called an axon irrespective of the direction of conduction. At its distal end, the axon swells to form the presynaptic terminal. The synapse is formed of two membranes, presynaptic and postsynaptic, separated by a cleft of some 20–30 μm. Each presynaptic terminal is rich in mitochondria and synaptic vesicles which contain a neurotransmitter released into the synaptic cleft to act upon the postsynaptic membrane.

In its normal resting state, the membrane of the axon has a resting potential of about –85 m volts, that is, it is positively charged on the outside and negatively charged on the inside. This resting potential is maintained by means of the sodium pump, an energy-dependent mechanism which actively transports positively charged sodium ions from within the fibre outside. At the same time, potassium is actively pumped into the cell, although only about a third as much potassium is taken into the fibre as sodium is put out.

When the permeability of the membrane to sodium is increased, there is a rapid change in the membrane potential as sodium ions pour into the fibre along the concentration gradient. The influx of sodium ions reverses the polarity of the membrane so it becomes positively charged on the inside and negatively charged outside. Almost immediately after this depolarization takes place, the membrane reverts to being impermeable to sodium, but allows a rapid flow of potassium outward through the membrane to restore the original membrane potential. The normal balance of sodium and potassium ions is restored by the sodium pump.

Once the process of depolarization has been triggered off, it travels inexorably down the nerve fibres, followed by a wave of repolarization. The nerve impulse then, is nothing more than a train of electrical currents moving at a finite speed along the fibre. The speed of conduction of the impulses is greatly increased in myelinated fibres since the myelin does not allow ions to diffuse through it except at the nodes of Ranvier. Impulses in a myelinated nerve are conducted from node to node, rather than along the whole length of the fibre. This saltatory conduction not only increases the velocity of conduction but also conserves energy, for since only the nodes depolarize, comparatively little energy is required to re-establish the resting potential through the mechanism of the sodium pump.

Conduction along the axon is an all-or-nothing phenomenon, but in the dendrites it is decremental. That is to say, a considerable amount of the postsynaptic potential leaks away as it passes towards the soma, thus presynaptic fibrils which terminate near the soma of the postsynaptic neurone have a much better chance of causing the neurone to fire than those which terminate on distant points on the dendrites.

At the synapse, electrical energy is transformed into chemical energy by the release of neurotransmitters which may either be excitatory or inhibitory; the resultant effect of the excitator and inhibitory transmitters from the many inputs on to the postsynaptic membrane determines whether or not the postsynaptic neurone (or other target cell) becomes activated.

The signs and symptoms of toxic damage to the nervous system vary, depending on whether the site of action is peripheral or central.

Disordered function

Toxic peripheral neuropathy

In the peripheral nervous system, motor and sensory function may be affected independently, but it is more common in occupational practice to find that both are affected concomitantly. Loss of motor function is exhibited by symmetrical muscle weakness and wasting which is usually distal in distribution, producing foot or wrist drop in extreme cases.

Sensory disturbance is almost invariably distal in distribution and produces the well-known glove and stocking type of sensory loss and paraesthesiae. The tendon reflexes are frequently diminished or absent and disturbances of autonomic function, including anhydrosis, hypotension and sphincter disturbance may be present.

Most toxic substances which cause peripheral neuropathy affect the axons to a greater extent than the Schwann cell. In the former case, the distal ends of the nerve degenerate and this 'dying back' process continues along a considerable length of the fibre. The process occurs particularly in large diameter fibres and most of the damage is sustained by the intramuscular branches of the main nerve trunks which are not themselves greatly affected. The damage to the axons is thought to be secondary to some alteration in the metabolism of the nerve cell brought about by the toxic material, but the nature of this alteration is obscure.

If exposure to the toxic material ceases, the nerve fibres begin to regenerate, and if the dying back process has not progressed too far proximally, regeneration may be virtually complete. When damage has been severe, however, regeneration is slow and frequently ineffective, as the regenerating nerves do not always establish the right connections. If damage is directed mainly against the Schwann cell, segmental demyelination is produced. The small myelinated fibres are most severely affected, and the changes which are first seen at the node of Ranvier may spread to involve the whole internode. In the larger fibres, breakdown of the myelin is usually confined to the paranodal regions. The toxins which produce segmented demyelination inhibit the synthesis of proteins and lipids and the early changes seen at the nodes of Ranvier suggest that these are active sites for the production and repair of myelin.

Regeneration of the myelin sheath takes place either by extension of the internodal myelin back to the node of Ranvier, if the break in the sheath is small (less than about 15 μm) or by the migration of one or more Schwann cells into the gap. Repair is normally more or less complete if exposure to the toxic material is discontinued. In the peripheral neuropathy produced by n-hexane and methyl-butyl-ketone giant axonal swellings are found in the affected nerve fibres, most notably in their proximal parts together with the normal features of dying back. These swellings are seen with the electron microscope to contain densely packed, whorled masses of neurofilaments. The link between these two compounds is a common metabolite, 2,5 hexane-dione which is considered to be the agent responsible for the neurological damage.

Nerve conduction in peripheral neuropathy

In a degenerating nerve, conduction continues normally for a few days after the injury has been sustained and then stops with little intermediate slowing in conduction velocity.

If there is extensive axonal degeneration, the conduction may be reduced slightly owing to the loss of fast-conducting fibres of large diameter. When dying-back is predominant, the distal latency is increased because the distal ends of the nerve fibres are disintegrated.

Conduction velocities are slow during the early stages of axonal regeneration and may only return to only 60 per cent or 75 per cent of normal in two or three years.

Because the integrity of the myelin sheath is essential for the propagation of the nerve impulse, it follows that when it is damaged the conduction velocity will be impaired. In some of the more severe demyelinating neuropathies nerve conduction may be slowed by 30 per cent or more.

Disordered function of the central nervous system

Toxic organic psychosis

The clinical picture in this condition is due to the disruption of normal brain function and the symptoms are remarkably constant, whatever the nature of the agent responsible for their development. The primary change is impairment of consciousness ranging on a continuum from a barely perceptible dulling of mentation to profound coma. Characteristically this impairment fluctuates, being worse at night and when the patient is fatigued. These fluctuations, interspaced with lucid intervals, are of great value in separating organic from non-organic psychoses and acute from chronic reactions.

The condition is associated also with changes in motor behaviour, which decreases progressively *pari passu* with the impairment of consciousness, and with disordered thought.

The patient may develop ideas of reference, and delusions of persecution are especially common. Memory is disturbed and may be associated with disorientation for time, place and person, although the last occurs at a relatively late stage. Disorders of perception are common, and familiar objects may be misinterpreted and friends and relatives misidentified. Visual hallucinations are common, but tactile and auditory hallucinations may also occur.

In the early stages of the disease, the patient may exhibit mild depression or anxiety and irritability but later the most striking feature of the patient's affect is his apathy.

Occupational peripheral neuropathy

The most important of the substances encountered in industry which might produce peripheral neuropathy are listed in *Table 6.9*. In almost every case the neuropathy produces mild motor and sensory defects; lead is unusual in producing a pure motor neuropathy.

Organophosphates, for example, TOCP, di-isopropyl fluorophosphate (DFP) and mipafox are also potent anticholinesterase inhibitors

Table 6.9. SUBSTANCES CAUSING PERIPHERAL NEUROPATHY

Triorthocresyl phosphate (TOCP)

Acrylamide

Carbon disulphide

Mercury compounds (organic and inorganic)

Some organophosphate compounds

Diethyl thiocarbamate

n-hexane

Methyl-butyl-ketone

Arsenic

Lead (inorganic)

Thallium

Antimony

and on this account produce a number of symptoms including headache, abdominal pain, vomiting, sweating, miosis and muscular twitching. These can be controlled with atropine or cholinesterase reactivators such as 2-pyridine aldoxime methiodine (PAM). Such treatment, however, does not prevent the appearance of the other neurotoxic effects.

Subclinical neuropathy

Some workers exposed to potentially neurotoxic materials have been found to have electrophysiological abnormalities in the absence of overt signs or symptoms, and are described as having subclinical neuropathy. Lead workers are the group among whom this phenomenon has been most frequently reported. The changes reported include slowing of motor conduction velocity (including that in the slow conducting fibres) and prolongation of distal motor latency. These changes occur apparently in the absence of any marked histological change and are most probably due to minor defects in the nerve cell membrane. The significance of these changes, which do not correlate well with blood-lead concentration, is far from clear as it is not yet known whether they progress to frank neuropathy if exposure is continued, or how quickly they revert to normal after exposure ceases. There seems to be no relationship between slowing in motor conduction velocity and results in performance tests which suggests that affected workers are not put at any disadvantage.

Occupational organic psychosis

Substances which may produce a toxic organic psychosis include arsenic, lead, manganese, mercury and carbon disulphide (CS_2), although, fortunately none is likely to do so under modern conditions of work, in this country at least. During the last century, cases due to mercury were undoubtedly common, as may be judged by the fact that the classic description of the condition is the Mad Hatter in *Alice in Wonderland*. Exposure to CS_2, especially during the cold curing of rubber was also followed by the appearance of many cases of organic psychosis. At one factory, the workers took to throwing themselves from the windows, to which the kindly owners responded by fitting them with bars to make egress more difficult.

Poisoning with organic lead compounds is occasionally encountered when leaded petrol is inappropriately used as a solvent, as, for example, in the manufacture of cheap shoes or sandals. There have also been several reported cases of psychotic symptoms occurring in children in America who have been sniffing petrol for kicks.

Other central effects

Other central nervous system disorders which should be mentioned are epilepsy and parkinsonism.

Epilepsy

Epilepsy is seldom due to exposure to toxic materials but it may be a feature of lead encephalopathy in adults (which is itself rare, although the most common manner of presentation in children). It has also been reported to occur in men who have been accidentally over-exposed during the manufacture of chlorinated hydrocarbons (such as dieldrin, aldrin and endrin). Abnormalities in the EEG, in the absence of any clinical abnormalities have also been described in workers exposed to these compounds and also in workers exposed to carbon disulphide, methylene chloride, methyl bromide, benzene and styrene.

The significance of this subclinical encephalopathy, like subclinical neuropathy has yet to be established.

Parkinsonism

Parkinsonism, the syndrome characterized by hypokinesis, rigidity and tremor, has been known for many years to occur in workers exposed to manganese. Most cases nowadays come from Chile and from India. In the Chilean cases, parkinsonian features are almost always preceded by psychotic symptoms, including hallucinations, delusions and compulsive acts, and are referred to by the local population as 'manganic madness'. This phase lasts for up to three months, whether or not exposure is discontinued, at the end of which time the extrapyramidal symptoms make their appearance.

The lesion in manganese-induced parkinsonism is similar to that in the naturally occurring form; that is to say, the cells of the substantia nigra are deficient in melanin and the dopaminergic nigrostriatal pathways have a reduced content of dopamine which produces a reduction in the concentration of dopamine in the corpus striatum.

Behavioural changes

In the USSR, it has long been the practice to establish hygiene standards on the basis of changes in conditioned behaviour in animals. This approach, which has its roots in the work of Pavlov, has usually resulted in the proposal of standards (so called Maximum Allowable Concentrations) which are considerably lower than the Threshold Limit Values adopted by the USA and UK. Behavioural toxicology, as the subject has come to be known, has only recently begun to attract much attention in the West, but already some dozen or so solvents, gases or heavy metals have been reported as producing decrements in scores obtained in batteries of tests which measure speed of reaction, vigilance, dexterity and intelligence in the absence of clinically obvious signs of intoxication (*see Table 6.10; see also* Chapter 11).

Table 6.10 CHEMICALS INDUCING BEHAVIOURAL CHANGES

Carbon monoxide

Carbon disulphide

Inorganic mercury

Halothane

Methylene chloride

Trichloroethylene

Toluene

Methylchloroform

Styrene

White spirit

In all cases, the differences found between exposed populations and controls are small and it is often not possible to be certain that differences which are found are not due to pre-existing differences between groups studied, or to other factors such as learning, motivation or physical and mental fatigue. Nevertheless, with some refinements in technique, this seems to be an important area for future development.

Screening for occupational effects on the nervous system

The need for close medical and environmental control of men working with known neurotoxic materials, including regular biological monitoring, is obvious. Tests for measuring nerve conduction velocities and other screening procedures are discussed in Chapter 11. The occupational physician faced with a patient with peripheral neuropathy should remember that diabetes is a much more likely aetiological factor than any chemical likely to be encountered in the workplace, and that the most likely cause of toxic organic psychosis at work is alcohol abuse.

OTHER TARGET ORGANS

There are at least two other target organs, the cardiovascular and the endocrine systems, that deserve mention as there is growing evidence that these organs can be seriously affected by toxic exposures.

The cardiovascular system

Risk factors in ischaemic heart disease are now receiving greater attention (*see* Chapter 10). There is increasing evidence that occupational and environmental exposures can lead to arteriosclerotic heart disease (Rosenman, 1979).

The metals

Lead is known to cause chronic nephropathy which can lead to hypertension. It is possible, though not proven, that lead may have a direct toxic effect on the myocardium and thereby could contribute to a myopathy.

Arsenic has been described as causing peripheral vascular disease in populations drinking high concentrations in well water, and has been postulated as causing myocardial disease in young children. These studies have not been corroborated elsewhere but there is epidemiological evidence to suggest that arsenic smelter workers have a higher risk of cardiovascular disease than the general population (*see* Chapter 14).

The link between cadmium and heart disease is stronger. This may be secondary to the known renal effect of this metal, but as pulmonary emphysema is well recognized after chronic cadmium exposure, cor pulmonale is also a possibility.

Cobalt in high concentrations can cause cardiomyopathy in persons with normal hearts, but it is particularly potent if the cardiac muscle is already compromised by poor diet and high alcohol intake. Arsenic was similarly suspected of causing such an effect in the early part of this century. Other metals that have been linked with heart disease include: antimony, chromium, manganese, mercury, niobium, vanadium and zirconium.

Organic compounds

In contrast to the putative effects of metals on myocardial cells or secondarily through kidney or lung, organic chemicals seem to be associated with cardiac arrhythmias. This is particularly so for some of the aliphatic chlorinated hydrocarbons such as 1,1,1-trichloroethane, trichloroethylene, chloroform, carbon tetrachloride and halothane. The propellants in bronchodilator aerosols such as fluorocarbons may have been partly responsible for some of the unexplained deaths associated with excessive use by asthmatics. Vinyl chloride, which had a brief vogue as a propellant, has been reported as causing Raynaud's phenomenon among exposed workers.

Nitroglycerine (glyceryl trinitrate) and nitroglycol, used as explosives, cause vasodilatation (glyceryl trinitrate is used medicinally in the treatment of angina pectoris). Workers manufacturing these explosives may suffer from headaches, nausea and vomiting, and have lowered pulse pressures. They develop a tolerance, which may be lost over the weekend or on holiday. There is epidemiological evidence that they have a higher than expected mortality from acute myocardial infarction. Carbon disulphide is recognized as having an atherogenic effect and there are numerous studies linking carbon disulphide with an increased risk of coronary heart disease (*see* Chapter 14).

Gases

Asphyxiation causes considerable cardiovascular strain but carbon monoxide appears to have a direct myotoxic action as well. This may be one of the factors linking heart disease with tobacco smoking.

Physical agents

Environmental stress in the form of high or low ambient temperature is associated with heart disease and peripheral vascular changes. Extreme cold causes vasospasm and extreme heat causes high cardiac output strain. Both have been linked with overt cardiovascular morbidity and mortality.

Vibration is a well recognized cause of peripheral vasospastic disease which can be severe enough to lead to gangrene in the workers occupationally exposed, such as drillers and chain-saw operators. However, there is only tenuous evidence to suggest that noise or radiation can be a cause of cardiovascular disease.

Endocrine systems

The endocrine glands, and in particular the pituitary, control such a large range of bodily functions, and, in turn, respond to such a wide variety of intrinsic and extrinsic influences, that it is difficult to identify specific occupational or environmental causes for dysfunction. The only area where any knowledge exists is in relation to testicular and ovarian function.

The gonads

Recent evidence that certain chemicals, such as the nematocide dibromochloropropane, and the insecticide Kepone, can cause oligospermia or aspermia, has led to a reappraisal of the whole subject of the sensitivity of the gonads to chemicals. Animal work has shown that lead, organotins and cadmium are potent testicular toxins, and human research has studied the absorption of oestrogen derivatives such as diethylstilboestrol (DES) and its relationship to gonadal dysfunction, in both sexes.

DES, for example, seems to be more likely to produce organ specific teratogenesis in the offspring rather than act as a gonadal toxin in the parent. Childhood tumours may be a demonstration of such an effect or may follow transplacental absorption of the toxin. The organochlorine pesticides, chlordane and heptochlor, have been linked with childhood tumours though the evidence is disputed. Mercury has widespread cytotoxic effects and may have a gonadopathic effect.

The Minamata Bay disaster suggests that organomercurials can produce congenital psychoneurological disease, presumably by transplacental passage to the foetus.

Anaesthetic gases and vinyl chloride have been linked with various birth defects in the offspring of those exposed to the chemicals. Much of the evidence for such an association is based on epidemiological studies of poor or dubious design, but there is suspicion of a teratogenic effect which needs further study.

Ionizing radiations can produce gonadal cell death, lowered function or disordered function leading to parental tumours or congenital malformation and cancer. There is no clearcut evidence to link mutagenic, teratogenic or transparental carcinogenic effects with other physical agents such as noise, vibration or non-ionizing radiations.

Other endocrine glands

The thyroid gland is especially sensitive to bodily exposure to, and subsequent absorption of, the radionuclides of iodine. Thyroid function can be markedly altered by absorption of non-radioactive isotopes of iodine. Other goitrogens include cyanates, cobalt, and lithium.

The adrenal gland's secretion of corticosteroids is finely balanced, and exogenous absorption of such hormones can markedly disrupt this delicate feedback system. Therapeutic administration is a common cause of adrenal dysfunction but pharmaceutical workers manufacturing these steroids are also at risk. Cytotoxic drugs such as vincristine, cyclophosphamide, and methotrexate can have widespread endocrinopathic effects similar to ionizing radiations.

There exists a growing number of metals and organic chemicals which are being demonstrated as having effects on the endocrine system. Particular attention has been placed on gonadal responses which may manifest themselves as relative or absolute infertility, malformed offspring or childhood cancer.

Ionizing radiations and therapeutic cytotoxic drugs may have similar and widespread dysfunctional effects.

One group of workers who have received little attention to date, but are clearly at risk, are the pharmaceutical workers who manufacture the endocrinologically active drugs for therapeutic use.

FURTHER READING

Adams, R.M. (1969) *Occupational Contact Dermatitis*. Philadelphia: J.P. Lippincott

Bingham, E. (1977) *Proceedings of Conference on Women and the Workplace*. Washington DC: Society of Occupational and Environmental Health

Bodley Scott, R., ed. (1978) *Price's Textbook of the Practice of Medicine*, 12th ed. Oxford University Press

Boyland, E. and Goulding, R., eds (1974) *Modern Trends in Toxicology – 2*. London: Butterworths

Desoille, H. Sherrer, J. and Truhaut, R. (1975) *Précis de Medicine du Travail*. Paris: Masson et Cie

Campbell, E.J.M., Dickinson, C.J. and Slater, J.D.A., eds (1978) *Clinical Physiology*, 5th ed. Oxford: Blackwells

Daugaard, J. (1978) *Symptoms and Signs in Occupational Disease - a practical guide*. Copenhagen:Munksgaard

Hamilton, A. and Hardy, H.L. (1974) *Industrial Toxicology*. 3rd ed. Acton Mass: Publishing Science Group Inc.

Hunter, D. (1978) *The Diseases of Occupations*. 5th ed. London: English Universities Press

Medicine (1978 *et seq*) *The Monthly Add-on Journal*, 3rd Series. Oxford: Medical Education (International)

National Institute of Occupational Safety and Health (1977) *Occupational Diseases; a guide to their recognition*. Washington DC: DHEW (NIOSH) Publication No. 77–181

Parkes, W.R. (1981) *Occupational Lung Diseases*. 2nd. ed. London: Butterworths (In press)

Rosenman, K.D. (1979) 'Cardiovascular diseases and environmental exposure.' *British Journal of Industrial Medicine* **36**, 85–97

Waldron, H.A. (1979) *Lecture Notes on Occupational Medicine*, 2nd ed. Oxford: Blackwells

Zenz, C., ed. (1975) *Occupational Medicine, Principles and Practical Applications*. Chicago: Year Book Medical Publishers Inc.

7

Functions of an Occupational Health Service

R. Murray and R.S.F. Schilling

An occupational health service has to meet the special needs of the undertaking concerned and the people employed. With the enormous range and scope of industrial, commercial and agricultural activities, it is not possible to lay down detailed plans for a service which should be suitable for all undertakings in all circumstances. Important factors that have to be considered are the standard of medical care outside the workplace and the type and size of the organization to be served (Health and Safety Commission, 1977). Priorities are bound to vary. A service for a dam-building operation in a remote area has to provide primary health care and, in tropical countries, to control endemic disease which tends to proliferate where good standards of hygiene are difficult to secure. At the other extreme, a chemical works in a city in the western hemisphere has available to it all the resources of modern methods of medicine and surgery. There the main occupational health function will be the environmental and biological control of specific hazards; what might be called the treatment of the factory rather than the person.

In developing countries the maintenance of health in industrial communities presents special problems and priorities, which are discussed in Chapter 2. The functions of their occupational health services are described in detail in a recent publication of the African Medical and Research Foundation (1979).

A new challenge is how to deal with the psychosocial factors in the modern work environment that lead to mental health problems (WHO, 1980). This involves treating the organization rather than the individual worker, which may be beyond the scope of an occupational health service. However, physicians and nurses have an important role in identifying organizational causes of stress and taking the initiative to get them put right (*see* Chapters 1 and 23).

BASIC FUNCTIONS

The prime responsibility for health and safety rests with management. The function of an occupational health service is to promote and maintain the health of the enterprise as a whole by providing expert advice to help management achieve the highest possible standards of health and safety in the interests of that particular working community and the larger community of which it is a part. Advice may be given at top level, for example on general policy and planning. Day-to-day advice is given to line management by physicians and nurses and hygienists about particular problems concerning people and the work environment. It is against this background that the functions of a service must be considered.

Placing people in suitable work

The advice given to management on placing people in suitable work is based on the pre-employment medical or health examination discussed in Chapter 9. In the past its main objective was the selection of the fit and the rejection of the unfit. Nowadays it is increasingly used to detect particular physical or mental disabilities which would be a handicap in specific jobs, for example colour-blindness in drivers or chronic respiratory disease in dusty occupations. It should be a *preplacement* examination which assesses working capacity and, wherever possible, matches this capacity, no matter how limited, with a suitable job. This aim of preplacement and not rejection is ethically more acceptable to the physician and nurse, but it calls for intimate knowledge of the work and a higher level of professional judgement. Specific requirements can be laid down for some jobs, but in many instances the assessment of a person's fitness for a particular job has to be tailor-made. A worker who has the necessary professional or technical qualifications to do a job should not be denied the opportunity of fulfilment because of some condition which reduces the likelihood of reaching pensionable age. Provided the worker and the employer both know the score, they can make the necessary adjustments which might be in the best interests of both parties. An automatic matching of theoretical job requirements against the disabilities of an individual is not likely to be successful on its own.

In a large industry which is expanding rapidly or has a high labour turnover, preplacement medical examinations can take up so much time that the physician or nurse has little opportunity to do other important work. Moreover the routine superficial examination of large numbers of potentially healthy people can be intensely boring and lead to a lowering of professional standards. These disincentives can be overcome first, by using the examinations more selectively – for example by confining them to persons to be employed in occupations which are dangerous to themselves or others, such as the dusty trades, chemical and radiation work, transport and crane-driving, and food handling; and secondly by adopting screening procedures that can be undertaken by less highly trained staff, the physician seeing only those people who for one reason or another show a potential problem during initial screening.

While industry has over-exploited the medical examination for selection purposes, it has made much less use of vocational psychology. Psychological techniques for selecting and training skilled personnel in the armed forces developed rapidly in the Second World War. The hire-and-fire type of selection could not be tolerated at a time when the best possible use had to be made of all available manpower. The application of these techniques to civilian life has been limited by ignorance, indifference and sometimes by hostility to industrial psychology. Although aptitude testing is now sufficiently refined to be of practical value, the use of personality tests to identify potential managers is less precise and some firms that have tried them have yet to be convinced of their value.

Maintaining people in suitable work

It is also a function of the service to spot misfits and persons who, as a result of sickness or injury, are either no longer able to continue in the same job, or need to have it modified to reduce the work load. This may be done through consultations by workers at their own request or that of management or colleagues, routine screening or by examination after sickness absence (*see* Chapter 9).

Providing treatment

The provision of treatment used to be the main, and sometimes the only, function of a medical service at a workplace. In isolated groups such as construction workers in remote areas, a treatment service is still the first and foremost need. In countries which have comprehensive general practitioner and hospital services, there are those who do not attach much importance to treatment in an occupational health service and, indeed, see it as a diversion from the main preventive role. Many experienced occupational physicians and nurses, however, regard the provision of treatment as an essential part of the service. The reasons for this are discussed here, whereas the organization and functions of treatment services in the workplace are described in Chapter 8.

Efficient and speedy treatment of injuries, acute poisonings and minor ailments is important because it establishes confidence in the competence of the service, prevents complications and aids rehabilitation. In an industry where the risk of injury is high, it can boost morale by increasing the workers' sense of personal security. A good treatment service at the workplace can prevent unnecessary loss of working time and pay by eliminating travelling and waiting in crowded outpatient departments in hospitals and dispensaries. Many workers referred to outside agencies with relatively minor injuries and ailments are unnecessarily kept off work by physicians and nurses who do not know what the patient's work entails and thus tend to play safe. Physicians, nurses and auxiliaries trained in occupational health ask, as a routine, 'How was this worker injured?', and visit the department concerned to see how such injuries could be prevented. Any unusual clustering of cases, for example foreign bodies in eyes, is more readily noted by the occupational health service than by the general health services. In this way preventive action may be taken without delay.

A treatment service can also provide epidemiological evidence of occupational hazards. For respiratory or other symptoms, as well as common incidents such as minor injuries to eyes, the back and limbs, a simple method of recording and analysing the facts may give clues to the existence of unsuspected occupational risks and their causes. For example, at one factory the repeated attendance for treatment of back pain by men from one department revealed that they were using the wrong methods for loading trucks. Epidemiological methods can also be applied to the records of injuries and disease to determine more obscure causal factors in the working environment (*see* Chapter 13).

As well as safeguarding workers from physical hazards, the treatment service also provides evidence which leads to the identification of psychosocial factors causing ill health (WHO, 1973; Levi, 1980). It offers opportunities for personal counselling on social and emotional problems and for the health education of individual workers who attend for treatment and advice.

Controlling recognized hazards

One of the foremost functions of the service is to control recognized hazards of injury and disease. Equally important is the promotion of safety and the prevention of disability among those who have been injured (these functions are discussed in Chapters 8 and 22). While it may not be possible to eliminate accidents entirely, there is evidence to suggest that their prevalence and severity can be substantially reduced by an occupational health service (Schilling, 1963). Occupational disease on the other hand can often be eliminated completely, or at least successfully controlled (*see* Chapter 25).

The principle of finding a harmless substitute for a toxic material is the most successful of all the methods of prevention. Where substitute materials are impracticable, prevention of disease must depend on adopting other measures. As long as there is a potential risk both medical surveillance and environmental monitoring are usually essential. Both have become more effective with improved techniques for the early identification of health impairment (WHO, 1975) and for measuring environmental contaminants (*see* Chapters 17, 18 and 25).

Identifying unrecognized hazards

The identification of previously unrecognized hazards is an interesting, rewarding and often overlooked function of an occupational health service. Identifying new risks is sometimes considered quite wrongly to be the prerogative of research workers and labour inspectors. Much too often occupational physicians state 'There are no uncontrolled hazards in my factory'. This is because they are content to deal only with recognized hazards and because they are not trained to do this type of detective work. They have opportunities to pursue such investigations because of close and continuous contact with a working population. Occupational health services should be concerned with detecting all kinds of health hazards, from the apparently trivial, which do no more than interfere with comfort and efficiency, to those which endanger life and limb. The identification of previously unrecognized hazards depends on three distinct methods of enquiry: toxicity testing (*see* Chapter 24); clinical observation of individuals who seek treatment or advice (*see* Chapter 3); and epidemiological enquiry (*see* Chapters 13, 14 *and* 15). Toxicity testing is now required in many countries before new chemicals are used in industry.

The second method, clinical observation, is as old as occupational medicine itself. It was used by Ramazzini to identify asthmatic complaints in hemp workers; by Percivall Pott in the eighteenth century to suggest that soot caused scrotal cancer in chimney sweeps, and more recently by Creech and Johnson (1974) to highlight the risk of angiosarcoma of the liver among workers exposed to vinyl chloride monomer in the manufacture of polyvinyl chloride.

Epidemiology, the third method of enquiry, is also not new. Greenhow in the nineteenth century identified occupational risks of pulmonary disease in metal miners from their mortality rates (*see* Chapter 1). Epidemiology has been developed as a more precise method of identifying occupational hazards such as cancers of the respiratory tract and bladder, coronary heart disease and cataract (*see* Chapters 14 and 15).

Avoiding potential risks

Many potential risks of disease and injury can be avoided by planning the layout and design of a new plant or the modification of a new process. The occupational physician, hygienist and ergonomist can all make contributions to prevent risks if consulted at the beginning of the planning process. The health service should also be consulted on arrangements of hours of work and shifts. It may be possible to reorganize work to avoid unnecessary fatigue, inefficiency and adverse psychosocial factors (*see* Chapter 23).

Screening for early evidence of non-occupational disease

With better understanding of the influence that work has on health it becomes increasingly difficult to classify illness as occupational and non-occupational. Such nosological distinctions are often made solely to define areas of responsibility in medical care. There are many examples which illustrate that work has an important influence in the aetiology and management of chronic diseases as varied as ischaemic heart disease, peptic ulcer, rheumatic disorders and varicose veins. By studying the epidemiology of chronic diseases in industrial populations, the occupational physician may discover both occupational and other factors of aetiological importance.

The presymptomatic or early diagnosis of chronic disease by health screening has been adopted by many of the larger firms and applied particularly to executives. Its value as a function of an occupational health service is discussed in Chapter 10. Screening may help to control some diseases which are prevalent in certain industrial populations, including those of tropical countries. In this way occupational health services can play a part in helping to eradicate malaria and schistosomiasis and can contribute to the control of others such as coronary heart disease, hypertension and mental illness. In particular the occupational physician has almost unrivalled opportunities for identifying and dealing with mental illness in its early stages (*see* Chapter 23).

Supervision of vulnerable groups

In any working community there are specially vulnerable groups. The young, women of childbearing age, the aged, the disabled and those with prolonged or repeated absences from work, are examples of the groups of people who require special care (*see* Chapter 3). They can be supervised by routine examinations and other methods which are further discussed in Chapters 9, 10 and 11. They can be helped by counselling, by rehabilitation at the workplace and by modifying their work to remove any harmful influences. The service should also help to prevent the overloading of a worker whose skill or productivity has been reduced as a result of age, disease or injury.

Health education and safety training

The staff of an occupational health service can perform two different types of health education and safety training. First, by collaborating in the formal health and safety training programmes for management and workers (*see* Chapter 25); secondly, at attendances for treatment and other routine examinations, workers can be advised how to improve and maintain their health and safety and prevent the recurrence of injuries.

The physician and nurse also have opportunities for the education of employees towards healthier modes of living. As we have seen, this can be a day-to-day function directed towards those who attend for treatment, advice or routine examinations. While much of this education is aimed at maintaining health and safety on the job, there is no reason for limiting it so narrowly.

Apart from formal training programmes, opportunities also exist, or can readily be made, to educate management in its responsibilities for the health and safety of employees. The reputation of the service at all levels of management depends on giving sound advice about how to deal with environmental hazards and individual problems. To be effective, the advice must be based on an accurate and detailed knowledge of all the facts relevant to the particular problem. This must include a knowledge of the appropriate research work and how this can be interpreted and applied to the immediate local situation. The final analysis of all this material requires the exercise of good professional judgement. The physician and nurse are acting as advisers to others who have the executive power to act and, thus, to implement the advice. Many doctors and nurses, who, in clinical situations, have become accustomed to exercising the ultimate executive power and responsibility, may find it difficult to accept this role. Nevertheless, in industry, it is the management or other authority who will either accept and implement the advice given, or refuse it.

The primary objective of industry is to produce goods or services, and the strategy of achieving whatever targets may be laid down depends on many factors, including health and safety, which management cannot afford to ignore. Nevertheless, as cost factors must be taken into account, the ideal answer to a health and safety problem may not be technically or financially feasible. Compromises, sometimes involving additional responsibility on the occupational health service, may be necessary. Effective solutions to problems will ultimately depend on the quality, realism and authoritativeness of such advice. It must be presented clearly, concisely, persuasively and, wherever possible, must include concrete solutions to the problems posed. Inherent in this process is the professional ethical responsibility to care for the best interests of the individuals of the community affected and, equally, to ensure that management accepts its own responsibilities for actions which are an inevitable accompaniment of the power it wields.

Counselling

Counselling is increasing in occupational health practice and is undertaken especially by nurses. It should not be confused with health education. Essentially it consists of sympathetic listening and understanding which leads to self help. It is not merely giving advice. While it is mainly concerned with the worker's own health problems, it also deals with social and occupational problems, such as personal relationships. Sometimes the worker wants to discuss the health of other people, for example workmates or family (Williams, 1979).

The patient (though he is not necessarily sick) plays a major role in making decisions and taking action: while the counsellor has an ongoing supportive role. Counselling may include giving information or explanations about health problems or social service agencies, but above all, it entails careful listening and encouragement to the 'patient' to 'talk it out' and to take decisions about priorities where there are multiple problems. When the 'patient' has decided what to do, the counsellor helps with information, referral, if necessary, to more expert help, and encouragement to enable the necessary action to be taken.

The occupational health practitioner has to know the voluntary, statutory and other agencies which can help. There are some people in every organization who frequently need supportive care, but the counsellor will have to know when to terminate the relationship, especially where deep emotional problems are involved. An impartial professional attitude of friendliness, without familiarity, is most helpful to the 'patient', who must have confidence that no private matters will be divulged without permission. Thus the need for privacy is crucial. As counselling is most often sought during treatment, occupational health departments should have a separate room where a distressed person can confide or recover.

Surveillance of sanitary, catering and welfare amenities

The service generally has responsibility for advising management on requirements of sanitary installations such as toilets, wash places and facilities for storing and drying clothes. It should be responsible for the routine surveillance of these installations and other amenities such as kitchens, canteens, day nurseries and rest homes. Useful advice can be given on their design, construction and maintenance. In developing countries high standards for sanitary conveniences are particularly important where there is no public sewage disposal and where there are flyborne diseases. Kitchen and canteen staff in particular need routine surveillance to reduce the risk of infection of the food they prepare and serve (*see* Chapter 9). Advice may be given by the service on the nutritional content of meals and on diets for employees with special needs, such as diabetics.

Environmental control outside the workplace

It has been well said that every enterprise has three different audits, a financial audit, a personnel audit and an environmental audit. The staff of an occupational health service have opportunities for preventing the industries for which they work from adversely affecting the health of neighbouring communities through the discharge of toxic effluents from the workplace or other exposures which cause a nuisance. The service may extend its monitoring systems to communities outside the workplace, maintaining continuous observation of the pollutants emitted, and even evaluating, in collaboration with public health authorities, the effects of such pollutants on community health. The need for this is illustrated by the neighbourhood effects produced by local industry, such as the occurrence of mesothelial tumours, chronic beryllium disease, lead absorption and Minamata disease. This last was caused by inorganic mercury pollution in the effluent from a factory at Minamata Bay in Japan being taken up by fish and converted into methyl mercury. It led to the death or severe disablement of many people, caused by organic mercury poisoning from eating the fish caught in the Bay (Takeuchi, 1967). Industry can have less dramatic effects on the health of the community from noise produced by machinery or heavy traffic, and from the emission of inert dusts and unpleasant odours.

ROLE OF STAFF AND CONDITIONS OF SERVICE

A service in a large organization, with a team of physicians, nurses and hygienists trained in occupational health, should have no difficulty in carrying out these functions. Many services, however, in smaller workplaces are staffed by a nurse, usually with the support of a part-time physician with no training in occupational medicine. In such circumstances the extent to which these functions can be implemented depends on the nurse's training in occupational health. In principle there is no reason why these functions cannot be carried out competently by a fully qualified occupational health nurse, or by a trained medical aide in a developing country, who has the appropriate medical or other support for dealing with the more difficult problems concerned with fitness for work, identification of hazards and environmental control.

The physician or nurse who runs the service should be directly responsible to top management in the industry or to the governing body of the organization which it serves. In group services this body is usually a committee which includes representatives of both management and workers. While all occupational health service staff are professionally responsible to the head of the service, those in scattered units should be functionally responsible to line management.

To work effectively the staff must serve, and be seen to serve, the interests of the whole workplace, not any particular sections of it. One criticism of services provided and paid for by employers is that they serve the interests of management and therefore are not trusted by workpeople. The confidence of employees in the health service will depend on management having a clear understanding of the service's functions, and not abusing it, and on its staff being aware of their role of serving the community as a whole.

Staff need to have adequate safeguards in their terms and conditions of service to encourage impartiality and dispel any fear of victimization, however unfounded this may be. In Great Britain employers have a duty to consult with trade union representatives to encourage joint cooperation in promoting and developing health and safety measures. There seems to be no reason why this principle could not be extended to joint responsibility for the health service such as is found in group occupational health services.

ETHICS FOR OCCUPATIONAL HEALTH STAFF*

The ethical standards for occupational health physicians should be no different from those of other members of the medical profession. Although they have responsibilities to the organization which employs them, the interests of that organization are best served by a service which is scrupulously impartial and carefully observes professional ethics.

*Much of this section is based on the views of the Society of Occupational Medicine's Advisory Panel expressed in a letter (30 May 1979) to the Faculty of Occupational Medicine.

Medical records

Access to clinical records of individual workers must be limited to members of the health service staff. They, and others such as dentists and physiotherapists who may be privy to this information or a part of it, are therefore equally bound with physicians in the issue of confidence.

Medical records may be disclosed to other members of the medical profession if this is in the best interests of the worker. When an employee leaves an organization his or her records may be transmitted, with the knowledge of the employee, to the occupational health department of the new company, but to no one else.

In Britain safety representatives are legally entitled to have access to certain documents to enable them to carry out their duties under the Health and Safety at Work, etc. Act 1974. Such documents are those which the employer is required to keep under the provisions of the Act. *They do not include any employee's individual records.*

In litigation cases in England and Wales the court may direct disclosure of medical records to the applicant or his solicitors. Normally this would be only to a nominated medical adviser of the applicant. In other countries, for example Scotland and Northern Ireland, disclosure may be made direct to the applicant. Industrial tribunals also have power to compel disclosure of medical records. Thus, by law, an occupational physician in these circumstances is obliged to produce medical records.

Medical examinations

An important part of the daily work of a service is giving an opinion on the fitness or unfitness of individual workers. In most instances all that is required is a simple statement as to whether the person is fit or unfit. On some occasions the information deals specifically with the individual's functional ability and limitations. In other instances further details need to be given, but normally they should only be disclosed with the consent of the individual.

The physician and nurse have frequent contacts with representatives of the workers. These may be with workers' members of the safety committee, with departmental shop stewards or with full-time officers of the union concerned. Although it might be assumed that the trade union representative is acting in the worker's best interest, the same standards of conduct apply as in dealings with management. Only at the express wish of the worker involved should any details of his health status be discussed.

On some occasions giving clinical details about a worker to management may be to the worker's advantage, for example in persuading a manager to change or modify a person's job.

Where public safety, or the safety of fellow workers, is involved disclosure of details of unfitness may be necessary, even though the worker will not give consent. In practice, refusal in these circumstances is rare. Where consent is not given it is desirable first to obtain a second opinion. If this confirms the occupational physician's judgement, the position should be carefully explained to the worker. If in spite of earnest persuasion the worker remains resolute, the doctor has to abandon concern for the welfare of the patient in favour of the greater good of the community.

RELATIONSHIPS WITH GENERAL PRACTITIONERS AND OTHER DOCTORS

In spite of the growth of group practice, medical care in many countries is so organized that everyone is registered with a doctor who regards that person as his own patient. The occupational physician must recognize this situation. The general practitioner is likely to know more about the family circumstances, and is responsible for referring the patient to hospital or to a consultant. Therefore, it is no more than good manners that the works doctor should defer to the general practitioner.

The wise occupational physician will visit general practitioners in the area to acquaint them with the facilities that are available. Even better, he should invite local practitioners to the occupational health service to see for themselves. Good relations thus established are of the greatest help to all concerned. If the worker is aware of confidence between the general practitioner and the occupational physician this increases trust in both. The general practitioner can refer patients for ambulant treatment at the workplace. The works doctor can learn something of the family circumstances which will assist him in handling the patient's problems at the workplace.

Difficulties inevitably arise. Some general practitioners resent what they regard as interference with their patients, and although this type of objection is disappearing, it still occurs. Requests by workers to be treated by the occupational health service will be made frequently. Except for emergencies and minor ailments and injuries these should be resisted unless the consent of the patient's own doctor is obtained. Some general practitioners are willing to give general consent to treat their patients, relying on the occupational health staff to keep in touch on matters of importance. In this way mutual confidence is built up and professional relationships are not impaired.

One of the awkward problems in relation to general practitioners is sickness absence. The occupational physician is being unwise, if not unethical, in attempting to police sickness absence. Cases are bound to arise where the pattern of sickness absence raises doubts about the validity of the certification. These can usually be resolved by good relations between the doctors concerned, but there will always be difficult cases calling for resources of tact and discretion.

The code of ethics which governs the relationship with general practitioners applies with equal relevance to relations with other doctors, for example, consultants and house officers in hospitals, and physicians working for central and local government. The ideal is personal contact. The remoteness of correspondence, however well phrased, is often a barrier to effective professional liaison. Mutual respect is best fostered by free discussion of a problem so that all aspects can be aired.

ROLE OF GOVERNMENTS, INSTITUTES OF OCCUPATIONAL HEALTH AND PROFESSIONAL BODIES

While the basic functions of occupational health services may be similar in all countries, the method of providing them will differ. In some countries, for example in Eastern Europe, the government provides the service as an integral part of total medical care. In others, as in the United Kingdom, many workplaces provide their own occupational health services, with the government playing important advisory and supervisory roles while leaving many of the day-to-day duties to the in-plant services. Whatever the system may be, governments play a vital role in promoting occupational health by enforcing or encouraging higher standards of health, safety and welfare in workplaces, particularly in those which have no health services of their own. They can also provide information and advisory services to assist in the recognition and control of all types of occupational hazards.

Support for occupational health services is also provided by institutes, research units and university departments which have been established in countries all over the world, mostly during the last 30 years. An institute of occupational health includes experts in such fields as epidemiology and statistics, biochemistry, psychology, ergonomics as well as in medicine and hygiene. They act as a team to conduct research and to teach and provide information and advice.

Another recent development is the setting up of professional bodies like the new Faculty of Occupational Medicine, established in Great Britain in 1978. Such bodies provide an academic focus and maintain the necessary standards of training, competence and ethics. Through their various activities, governments, institutes and professional bodies are essential for raising and maintaining standards of service in workplaces.

REFERENCES

African Medical and Research Foundation (1979) *Occupational Health: a manual for health workers in developing countries,* ed. H. de Glanville, R.S.F. Schilling and C.H. Wood. Rural Health Series no. 11. AMREF, P.O. Box 30l25, Nairobi, Kenya

Creech, J.L. and Johnson, M.N. (1974) 'Angiosarcoma of the liver in the manufacture of polyvinyl chloride.' *Journal of Occupational Medicine* **16,** 150–151

Faculty of Occupational Medicine (1978) *Standing Orders Royal College of Physicians,* London

Health and Safety Commission (1977) *Prevention and Health: Occupational Health Services, The Way Ahead* (pp. 26). London: HM Stationery Office

Levi, L., ed. (1980) *Society Stress and Disease,* Vol. 4: *Working Life.* Report of a symposium sponsored by the University of Uppsala and the World Health Organization. London: Oxford University Press

Schilling, R.S.F. (1963) 'Developments in Occupational Health during the last thirty years.' *Journal of the Royal Society of Arts* **111,** 933–984

Takeuchi, T. (1967). *Pathology of Minamata Disease.* Japan: Kumamato University Press

Williams, M. Margaret (1979). A study of counselling by occupational health nurses in hospital and other occupational health services. MPhil.thesis, University of London

WHO: World Health Organization (1973). *Environmental and Health Monitoring in Occupational Health.* Technical Report Series No. 535. Geneva: WHO

WHO (1975). *Early Detection of Health Impairment in Occupational Exposure to Health Hazards.* Technical Report Series No. 571. Geneva: WHO

WHO (1980). *Health Aspects of Well-being in Working Places.* Euro Reports and Studies 31. WHO Regional Office for Europe, Copenhagen (in preparation)

8

Treatment and First Aid Services

P.J. Taylor and A. Ward Gardner

Whatever occupational physicians consider their main function should be, it is in the therapeutic role that they are most often judged by both employers and employees. It is a common dilemma of doctors working in industry that their own perception of duties may differ substantially from that of those among whom they work. There are many who believe that, except in emergencies, 'treatment' should not be included among the functions of an occupational health service. On the other hand, society usually expects a doctor to be a clinician, and it is as a clinician that he or she can most readily establish a reputation. Few actions of a doctor win the confidence and support of workers and management more than prompt and expert treatment of serious injury or illness occurring in the factory. The first medical director of the Slough Industrial Health Service maintained that the successful growth of the service was due to the considerable emphasis placed upon providing efficient first aid and follow-up treatment (Eagger, 1965). Nevertheless, it can be argued that this is an incorrect use of scarce medical resources which could be used to better advantage for larger populations by restricting the activities of occupational health services to preventive medicine. The whole issue of treatment provided as part of an occupational health service, including the question of cost-benefit studies of treatment services, has been discussed in the booklet *Occupational Health Services:The Way Ahead* (Health and Safety Commission, 1977).

In this chapter, the wide range of treatment facilities which can be found in occupational health services today is discussed and some guidelines for the services of the future suggested.

FACTORS INFLUENCING THE ESTABLISHMENT OF TREATMENT SERVICES

There are at least six important factors which influence the establishment of treatment services within an occupational health setting.

The level of medical care in the community

There are wide divergences both within and between countries in the availability of individual medical care for the sick and injured, and there are important differences in political philosophy that affect the administrative framework for the provision of medical care. Some countries have a comprehensive national health service, while in others, particularly developing countries, there may be in certain areas a virtual absence of any diagnostic or medical care. In developing countries it has become common practice for an organization wishing to establish a factory to be required to provide for the full medical care of all the employees, their dependants and sometimes for others in the local community (*see* Chapter 2). Some international companies have considerable experience in this field and the agreement to set up such a service often includes a provision for the ultimate transfer of medical responsibilities to the local government. At a symposium on the health problems of developing countries (Ross Institute of Tropical Hygiene and TUC Centenary Institute of Occupational Health, 1970) it became clear that the introduction of general medical services linked to occupational health services was an effective and economic method of raising the health status of a community.

Isolation from other medical services

Even in countries where medical care is generally available, circumstances arise where the place of work is relatively isolated and it is impracticable to rely upon the general medical services to provide treatment without a lengthy journey to a general practitioner or hospital. Even when isolation is not a serious problem, the provision of prompt medical care, including first aid and medical aid, is an essential function of any occupational health service. Breakdown in emergency care can lead to immediate emotional problems in a workforce, while a good performance in this field can earn a high reputation, confidence and gratitude.

Cost-benefits of treatment services

Very little has been done to measure the cost-benefit of occupational health services (*see* Chapter 1), but many managements believe that, apart from any of the less tangible effects on employee morale, a good case can be made out for the provision of a wide range of medical and ancillary services to save employees' time away from work. In our experience the provision of treatment facilities for minor conditions that can be treated at work eliminates the need for visits to the family doctor and enables people to stay at work, thus saving large amounts of time.

Minor sepsis, sprains and strains, minor trauma of all kinds and many other conditions can be treated successfully at work. For example, courses of injections, say for allergic desensitization, which have been prescribed by the family doctor, can more conveniently be given at work. The cumulative effect of such time saving in absences from work can be considerable. It also helps to 'sell' the medical services in general to the 'customer', and to make people aware of the facilities that are present for their use.

Specific hazards of the factory

Some factories have particular physical or chemical hazards, the effects of which can be mitigated by an efficient treatment service on the spot. For example, a large glass factory may have special arrangements for the definitive treatment of severe lacerations and tendon injuries, and chemical manufacturers will need to pay special regard to the immediate treatment of chemical burns both of the eye(s) and of the skin.

Medical ethics and rules

Another factor to consider is the effect that the attitude of the medical profession as a whole can exert upon the activities of an occupational health service. This can range from an absolute prohibition of any treatment other than first aid (as in France) to the permissive view of other countries. Most national medical associations have laid down sets of rules or advice to doctors working in occupational health (for example, British Medical Association, 1980) which try to ensure that the medical responsibilities of general practitioners are not usurped by doctors working in industry.

Attitudes of occupational physicians

The attitudes and interests of occupational physicians can and do differ widely. These are influenced, above all, by their training and experience and how they perceive their role in the organization. Not infrequently, those who transferred from general practice without a period of formal training tend to concentrate on the care of individuals rather than of the group. It is one of the attractions of occupational medicine that it allows the doctor to combine the roles of looking after people in groups, practising preventive medicine and administering a service with the more traditional diagnostic and therapeutic roles in relation to the individual. It is vital, however, that a correct balance is achieved.

MATCHING THE SERVICE TO THE NEED

It would be unrealistic to recommend one set pattern for all occupational health services. Each organization must be considered in relation to its needs taking the six general points already described into consideration. If treatment services are to be provided, their scale should be commensurate with the requirements of the organization objectively assessed in relation to the facilities within the community as a whole. To duplicate or compete with the therapeutic services already available in the area is not only wasteful of scarce medical and nursing manpower but can seldom be sustained by cost-benefit analysis. Undue concentration on treatment can, and often does, reduce the effectiveness with which the occupational health service can perform its unique task of protecting the health of all workers in the enterprise. A description of the treatment services provided in British Industry gives a general indication of their scope and their limitations and thus offers some guidance to industries both within and outside the British Isles.

TREATMENT SERVICES IN THE UNITED KINGDOM

In a country where family doctor, hospital and pharmaceutical services are available to all, one might imagine that there would be little need for elaborate treatment services in industry. Although no domiciliary services are provided – and a few organizations which did undertake these before 1948 have ceased to do so – there are a number of occupational health services that do provide what amounts to a comprehensive out-patient service. Even so, such organizations represent a very small minority of all factories. In the United Kingdom there is a wide divergence of treatment services from the statutory minimum of first aid to the large and prestigious medical centres providing care by nurses and doctors, drugs, laboratory and radio-graphic facilities, visiting specialists, dentistry, physiotherapy, and chiropody. Since the cost is wholly borne by the employer, it is inevitable that the most elaborate services are found only among larger companies. Others equally large often provide very much less.

Why do some firms spend so much? Which of these activities are essential and which are merely an extravagance? How much do they cost, and have any cost-benefit calculations been done? Clear answers are difficult or impossible to obtain, but the more important activities will be considered with these points in mind.

Treatment by first-aiders

Although there have long been legal requirements for first aid boxes to be available in most places where people are employed, and in many of them there must also be someone trained in first aid, the quality and effectiveness of first aid has varied very widely. British law in relation to occupational health and safety has always been designed on the principle of setting minimum standards to which all employers must adhere and thus many of the more progressive companies have provided first aid or other treatment to a much higher standard than the law required. The law assumes, for example, that a holder of a valid certificate in first aid (that is, a certificate which is less than three years old) is capable of providing effective first aid to a casualty. Sometimes this is certainly the case, but it is also true that the ability to answer questions about first aid theory and to apply a triangular bandage in an examination does not always indicate competence in a crisis to treat an injured colleague when the blood and pain is real. A good first-aider must have a sense of vocation for this work and this cannot be obtained by attendance at a course.

The objectives of first aid at work have traditionally been: first, to preserve life and to minimize the consequences of serious injury or major illness until professional medical or nursing help can be obtained; and second, to treat minor injuries which might otherwise go untreated. There are two further objectives at least as far as occupational first aid is concerned: the provision of redressing or other simple continuation treatment and the treatment of minor illnesses arising at work. Although these extra responsibilities have been undertaken by some experienced first-aiders for many years, this policy has its critics since the last proposal, treatment of minor illnesses, in particular implies that the occupational first-aider must be trained to make some sort of diagnosis, even if it is merely the exclusion of conditions for which the services of a doctor would be mandatory.

At the time this chapter was being revised (1979) the law relating to first aid at work still derived from four Acts (Factories; Offices, Shops and Railway Premises; Mines and Quarries; and Agriculture) and no less than 29 sets of statutory regulations. It is expected, however, that a complete revision of the law is likely within the near future.

Proposals for this change are being widely discussed and they will unify the laws relating to first aid at work under the Health and Safety at Work Act of 1974. The intention is to permit considerable flexibility so that the extent of first aid cover can be adjusted to meet the needs of each place of work, both in terms of risk and in the size and nature of the population. All workplaces would be required to hold an appropriate amount of useful first aid materials which, depending on circumstances, could range from a small box of dressings to a well-equipped treatment room. Similarly, it is proposed that there should be four levels of training and competence for first-aiders. At the most simple level there would be a short (four-hour) course on emergency life-saving first aid which would cover resuscitation, care of the unconscious patient and the control of bleeding. The second level would be to the standard currently required for the ordinary first aid certificate, that is first-aiders as legally defined at present. The third would be first-aiders who had taken additional training in 'occupational first aid' and these would be in charge of treatment rooms. This training covers topics such as the keeping of records, the diagnosis of minor illnesses and the administration of simple treatments, the application of dressings with a no-touch technique and so on. Finally the most elaborate training would be required for the so called 'rig medics' who work alone on oil rigs, often several hours away from a doctor.

Managers sometimes have problems in providing adequate numbers of trained first aid staff, even under existing regulations. They require, for example, that in factories where there are 50 persons employed at one time there shall be a first aid box and a trained first-aider 'always readily available during working hours', and for each additional 150 there must be another box and trained first-aider. The basic training for the official certificate in first aid is either a four-day full-time course or attendance at 10 out of 12 two-hour training sessions. As many employees are reluctant to give up their free time, some factories arrange first aid courses during working hours and others offer overtime pay for attendance. As a further incentive to become trained first-aiders, some companies pay an allowance or an annual lump sum to those who hold a valid certificate; others give extra holidays with pay. Difficulties can arise when first-aiders go on holiday or when shift work is done, since the law requires a first-aider to be 'readily available'. Despite all these problems there are still a few companies with a strong tradition in first aid where there are more than enough volunteers even when no monetary incentive is offered. Many of the problems for management arise because factory first-aiders almost always have a regular job and only undertake first aid as and when it is required. It is partly for this reason that many larger companies employ a nurse to run a treatment room.

Treatment by nurses

The provision of emergency treatment to employees taken ill or injured at work is only one of the accepted functions of the occupational nurse or doctor, although for some nurses employed in industry in the past this has been their only function. In many smaller factories the nurse may actually have no formal qualification at all, but in others the nurse may be state enrolled or state registered. The employment of a highly trained registered nurse with an occupational health nursing certificate (OHNC) solely to provide treatment is a gross misuse of scarce resources, and in practice such a nurse would probably not wish to remain for long in this restricted type of job. On the other hand an enrolled or registered nurse without occupational health training can provide professional and competent treatment for a wide range of injuries and ailments, thus saving employees taking time off work to see their doctor or to attend the local hospital. They can also provide continuation treatment prescribed by a doctor including such things as courses of injections.

There are, however, some medicolegal problems if a nurse has to work entirely alone without a doctor available, on a part-time or 'on call' basis. The recent Medicines (Prescription Only) Regulations 1978 have greatly restricted the range of drugs, and in particular injections, that may be obtained and administered to patients without a doctor's prescription and written instructions. The Royal College of Nursing, Society of Occupational Health Nursing (1979) has issued a useful guidance note for occupational health nurses which should be consulted for further information.

For the large factory where there is justification for the provision of a nurse-staffed treatment service because of the hazardous nature of the work, the best solution might be to have a senior nurse with an OHNC to provide the full range of occupational health care, and a treatment room staffed by enrolled nurses who are professionally responsible to the senior nurse. In such a situation there would probably be an occupational physician, either full-time or part-time, who would also be able to provide professional support and guidance in the day-to-day problems that can arise in the treatment room. Such elaborate arrangements and staffing, however, will be found only in a minority of factories and could never be justified in terms of need or cost-benefit in the majority of workplaces. This is particularly true in the United Kingdom where the freely available facilities of the National Health Service exist.

The cost of employing a nurse is largely determined by the salary scales laid down for guidance by the Royal College of Nursing. To this must be added the usual overheads for all employees, and a moderate revenue expenditure for drugs and dressings. Some capital outlay will also be required to provide a surgery or treatment room with basic furnishings, equipment and adequate washing facilities.

Treatment by a doctor

While it is unnecessary to specify what treatment can be provided by an occupational physician, it may be of value to indicate some of the limitations. There are important differences in practice between services where the doctor is full-time or part-time, visiting the workplace only once or twice a week.

An occupational health physician must not come between an employee and his or her own family doctor. Any treatment provided by the factory doctor should only be in the nature of first aid or, where it is not, the patient's own doctor should be consulted or informed. Since occupational physicians do not provide domiciliary services, any condition that may require medical attention at home should not be definitively treated by them. While a close and friendly liaison with local family doctors is necessary for most activities, in none is this more important than in the field of treatment. Patients should only be referred direct to hospitals or specialists in an emergency, and then the family doctor must be informed; at any other time such referrals should be only after consultation.

Occupational physicians not infrequently find themselves being used as 'second opinions' by employees. This, although understandable on the part of the patient, can be a pitfall for the unwary. Some patients succeed in playing one doctor off against another, with unfortunate consequences which can affect all three. In most cases, however, the employee only wants an explanation of what may be a standard medical procedure.

The provision of an 'open door' consultation service by the factory doctor undoubtedly limits the time available for activities more strictly related to the practice of occupational health. It is difficult to lay down guidelines but it seems that in areas where general practitioners' lists are large this may on average amount to between half and one hour a day for each thousand employees. Where a nurse and a doctor work together, it is usually possible to reduce this by suggesting that patients first see the nurse, who will then refer a selected number to the doctor.

The cost of employing a doctor can be considerable; with the salary range laid down by the British Medical Association, a full-time experienced doctor in a large factory may be paid as much as the senior members of management. As with the nurse, an appreciable capital outlay is also required. A part-time doctor may share the nurse's accommodation, but a full-time doctor will need his own examination room. One must not presume, however, that these expenses are all incurred under the heading of 'treatment' since most of the activities requiring extra expenditure, including clerical assistance, will be incurred from medical examinations, environmental control, and advice to management on many problems unconnected with treatment.

The ability to provide an open consultation service without jeopardizing the doctor's ability to undertake other occupational health responsibilities must clearly depend on the size of the population. For a full-time doctor in an average factory this may be somewhere between 3000 and 5000 employees, depending on the routine examination load and other factors. Where possible, some degree of personal consultation service is of the greatest value, since it enables the doctor to get to know some employees quite well and to obtain the confidence of the workforce in the easiest and most traditional way. It also provides the doctor with the possibility of recognizing previously unknown hazards of the physical or psychosocial environment (*see* Chapter 13).

This activity is one of the more stimulating aspects of the routine work in an occupational health service. In some cases, problems may first be raised by the employees themselves, but in others astute observation by the nurse or doctor may suggest that a problem exists. Two new cases of dermatitis of the hands from one workshop may be an obvious clue that there is a hazard. The recognition that several patients with assorted minor complaints all come from the same working group may suggest a problem in morale of employees or management. If one is to verify a hunch of this sort, it is essential that adequate records of treatment be kept. These will be discussed in a later section of this chapter. Both consultations and treatment records provide invaluable data for epidemiological studies, described in Chapter 13.

Medical supplies

A glance at the drug cupboard and medical stores can be an effective way of assessing the scale of treatment services provided by the medical and nursing staff. These can range from basic supplies of analgesics and antacids, and mixtures for coughs and diarrhoea, through to a range of drugs that would not disgrace a pharmacist's shop. Except for underground mines and ships at sea, for which special provisions are made by law, dangerous drugs such as morphine are only held in departments with medically qualified staff.

Although expenditure on drugs is a relatively small item in the budget compared to salaries and overheads, this is one of the areas in which savings can often be made. As in all other medical establishments, the contents of the drug cupboard bear witness both to fashions in prescribing and to the preferences of previous doctors or nurses. A stock check for valuation might prove a salutary exercise in those departments where it is not normally practised. The use of standard preparations instead of proprietary equivalents can usually result in appreciable savings, while realistic forecasts of consumption can allow bulk purchasing, particularly for those organizations having several medical centres, and this can usually be arranged at an appreciable discount.

Similar arguments may also be applied to dressings, but the benefits of pre-packed sterile dressings in terms of a lower rate of secondary sepsis have probably saved more money by reducing time off work than the increased cost of materials involved. In some larger medical centres or groups of centres, it can be more economic to use an autoclave and pack department dressings during slack periods, such as on the night shift, rather than to buy from commercial sources.

Radiography

Many large occupational health centres have their own radiographic apparatus. Although there can be little justification for elaborate equipment of hospital standard which is rarely, if ever, used to its full potential, most organizations with radiographic apparatus are convinced of its value. The capital outlay is considerable, not only for the apparatus but also for its screening, the space it requires, and the accompanying darkroom. Few departments employ a full-time radiographer, but they may have one who also undertakes other duties, or more frequently a nurse has been taught to take the relatively simple films usually required.

There are two reasons for having radiography in a medical centre: to take films of injuries which might be fractures, and to take chest and other films for routine pre-employment or periodic medical examinations. The former is most relevant in the context of this chapter, but the latter can provide a substantial reduction in expenditure at the local chest clinic, if, as in many companies, a normal chest radiograph is required before accepting a new employee, and labour turnover is high.

In a country such as Britain with a National Health Service, there is no reason why any employee with a suspected fracture should not attend the local hospital for a radiograph to be taken, and this is routine procedure in most workplaces throughout the country. But where hospital facilities are distant much time and therefore money can be saved by taking films on the spot.

Were such films requested only when there was clinical evidence of bone injury, the argument for radiography in a factory medical department could not be sustained. Unfortunately, and largely for medicolegal reasons, it is now the practice to obtain a radiograph of all injuries which might conceivably show an abnormality. This applies particularly to occupational injuries, but also to a lesser extent to other types of injury. If between 20 and 40 per cent. of all occupational injuries reported to the medical department (many of them minor) come under the diagnostic group of 'fractures, sprains or bruises', the number requiring a radiograph can be appreciable.

The loss of working time involved in sending an employee to the local hospital, and the cost of transportation there and back, can be substantial. When the majority of such patients will prove to have no bone injury and should return to work, the unnecessary loss of working time will also be high, particularly if, as is often the case, a number of them then take a day or more off work. With a population of several thousand manual workers in a high risk industry, there would be little difficulty in demonstrating the cost-benefit of simple radiographic facilities, even taking into account the depreciation on the capital investment.

Specialist treatments

Physiotherapy

The main justification for providing physiotherapy is also one of saving in time lost from work. In contrast to the provision of radiography, however, the capital outlay is not so high and the amount of time provided can more easily be adjusted to meet the needs of the population at risk.

As with all treatment services other than first aid, it can be argued that physiotherapy is freely available at the local hospital. Here, however, the demand is considerably greater than the supply, and treatment twice or three times a week is the rule, not because it is ideal but because of the pressure of work. It is also true that for the majority of patients regular attendance at a hospital physiotherapy department involves continued sick absence until the course of treatment is complete.

Cost-benefit studies in a number of organizations have indicated a real saving by the provision of physiotherapy. The numbers of hours per week required will depend on the size of the workforce and the degree of manual labour involved. It is most important, however, that close liaison be maintained with the orthopaedic and physical medicine consultants in the area so that treatment prescribed by them can be undertaken by the occupational health service. One solution which works well is a joint appointment for the physiotherapist, shared between the service and the local hospital.

Dentistry

Relatively few dentists are employed in occupational health centres because of the shortage of dental surgeons and the heavy capital outlay involved in equipping a dental surgery.

The economic justification depends once again upon a saving in lost working time. It has become almost impossible in the United Kingdom to obtain a dental appointment outside normal working hours and for many people each visit involves the loss of half a day from work. Even though only a small proportion of the working population attends regularly twice a year, the numbers doing so are rising. The independent contractor status of dental surgeons and their payment on a fee per item of service by the National Health Service allows a dentist who works in industry to claim the usual fees from Health Service funds. In some cases the employer pays the dental surgeon a salary and recoups fees from the state, and in others the dentist is paid a retainer and collects the fees himself.

The size of the capital investment, however, usually restricts dental services in industry to organizations employing several thousands, or to group health services covering similar numbers.

Ophthalmic services

Some of the larger occupational health services have arranged regular visits by an ophthalmologist to undertake the testing of vision, refraction and the prescription of spectacles. This is relatively inexpensive since the only requirement is for a suitable room to be made available for the sessions. The system of payment under the National Health Service is, like dentistry, based on a fee per item of service and the consultant obtains his fees direct. It is usual for a dispensing optician to attend so that employees can be fitted for the spectacle frames. The standard health service charges are made and so the employer is involved in no regular financial responsibility. The provision and fitting of safety glasses, both with plain and prescription lenses is often made easier if ophthalmic services are available.

The rationale for such services is twofold: not only is it important that all employees should be able to see clearly for their work, but, as in many of the other treatment services, this arrangement can also save a substantial loss of working time since without them such appointments usually involve half a day away from work.

The use of visiting specialists

Some occupational health services have arrangements with local consultants to do occasional outpatient clinics at the workplace.

Although this is not widely done, the specialties usually provided are orthopaedics, physical medicine, dermatology and psychiatry. The main reason for this development in some parts of the United Kingdom has been the excessive waiting time for first appointments with such specialists within the National Health Service. Many of the treatment services provided by industry could not be justified if the National Health Service facilities were close to the factory and prompt treatment and rehabilitation were readily available. Unfortunately the waiting lists of some specialists can amount to two months or more for a case that is not medically urgent. Sometimes, however, the patient may remain certified unfit for work (or unfit for his usual work) during this waiting period. The real necessity for this may be questionable, but general practitioners and occupational physicians may find it difficult to insist that the patient is fit for work while requiring him to see a specialist for an opinion, for example, on the advisability of a meniscectomy, or treatment for recurrent episodes of back pain.

Chiropody

In the past decade several occupational health services have introduced chiropody services for employees. Although they started in organizations such as department stores and factories with a predominantly female population who had to stand for most of their work, they are now to be found in other industries. Many people have minor foot ailments and there is little doubt that efficient chiropody can alleviate and sometimes cure the condition. Even so, some chiropody services in industry have closed because, after initial enthusiasm, the number of employees wishing to attend have dropped below the level to justify them.

Capital outlay is small since a part-time chiropodist can use a treatment room chair. In some cases the chiropodist is paid by the employer and the service is offered free, but in others a charge (often subsidized by the employer) may be made. This is because chiropody is only provided free by the National Health Service to a limited number of patients referred by doctors, such as those attending diabetic clinics.

CONCENTRATION OF TREATMENT SERVICES

It is the experience of all who have been involved in the provision of treatment services in industry and elsewhere that the most efficient and the least expensive are those which concentrate trained medical, nursing and other resources as far as possible. Much better work will be done by a few highly skilled people in purpose-designed premises than can be done by more numerous but less efficient first-aiders in workshops or offices. The rates of secondary sepsis, return treatments, and the cost of dressings will all be lower, with a much greater chance of consumer satisfaction.

This will require both efficient and rapid means of transport for the injured to the treatment centre, whether by trolley or wheelchair in the workshop or by a motor vehicle in more dispersed sites – and also the training of as many employees as possible in emergency life-saving first aid, but not necessarily in the whole rigmarole of an officially approved first aid course.

MEDICAL RECORDS

Treatment records

The need for the maintenance of adequate personal treatment records is accepted in all types of medical, nursing and ancillary medical services, but the particular needs and functions of an occupational health service may require more than is customary elsewhere. The records must be available, not only to ensure efficient care of the individual, but also for epidemiological investigations for the identification of health hazards and the evaluation of services (*see* Chapter 13). In the United Kingdom only one record is required by law in every place of employment, namely the Accident Book. This must be kept in a prominent place and be available for the recording of all occupational injuries.

Records kept by the medical department for internal use can be considered in two categories: individual record cards (*Figure 8.1*) or files, and those of a chronological nature, listing all attendances for treatment, usually called 'day sheets' (*Figure 8.2*).

Personal records

A centralized system of filing in which all medical records about an individual employee are kept in the same folder or envelope is the obvious and most efficient arrangement, but it is surprising how few occupational health services actually do this. Since a central registry administered by a clerk/receptionist is the usual arrangement in most hospitals and health centres of the National Health Service, it is of

INDIVIDUAL TREATMENT RECORD

NAME

PERSONAL NO. | DEPARTMENT

ADDRESS

SUPERVISOR/FOREMAN | OCCUPATION

DATE OF BIRTH | PATIENT'S OWN DOCTOR

DATE	TREATMENT		Occup.	Non occup.	DISPOSAL

Figure 8.1 An example of an individual treatment record card

DAILY ATTENDANCE RECORD

TIME	NAME & INITIALS	PERSONAL NUMBER	DAY OR SHIFT	SUPERVISOR/ FOREMAN	AREA OF INJURY	CAUSE (IF AN ACCIDENT)	DEPARTMENT

Figure 8.2 An example of a 'day sheet'

interest to consider why this is not so widely found in industry. Individual treatment record cards (*Figure 8.1*) are often held in a filing box or in drawers or, sometimes, in rotating drums in or close to the treatment room itself. The other personal records of routine examinations, doctors' letters, reports of investigations and so on, are often kept in another place. While the disadvantages are obvious, the reason stated is usually that waiting time is of the greatest importance for

SUMMARY OF ATTENDANCES

Sheet no..............

Day...................   Date...................

OCCUPATIONAL								NON-OCCUPATIONAL											OTHER				REFER		DISPOSAL				
CUTS, BRUISES, ABRASIONS OF SKIN	BURNS AND SCALDS	SPRAINS, STRAINS, SUSPECTED FRACTURES	OCCUPATIONAL DERMATITIS	EYE CASES — FOREIGN BODIES	EYE CASES — OTHERS	CHEMICAL INJURIES	OTHER INDUSTRIAL ACCIDENTS CONDITIONS	RESPIRATORY DISORDERS	DIGESTIVE DISORDERS	RHEUMATISM GROUP	FUNCTIONAL NERVOUS DISORDERS	NON WORKS ACCIDENTS	BOILS AND CARBUNCLES	NON OCC SEPSIS	NON-OCC SKIN CONDITIONS	CONDITIONS OF EYE AND VISUAL DEFECTS	CONDITIONS OF EAR	OTHER NON INDUSTRIAL CONDITIONS	RETURN FROM SICKNESS	MEDICAL EXAMINATION	INOCULATION	OTHER REASON	RE ATTENDANCE	DOCTOR	PHYSIOTHERAPY	WORK	HOME	HOSPITAL	REFER TO OWN DOCTOR	OVERTIME	INITIALS
A	B	C	D	E	F	G	H	I	J	K	L	M	N	O	P	Q	R	S	T	U	V	W	X	1		2	3				

Figure 8.3 An example of a form summarizing attendances. This summary is completed during, or at the end of, the day or shift by placing a tick in the appropriate box. Totals for the day (or shift) or longer periods, may be summed without difficulty for statistical returns

visits to the treatment room and further, that the records of these visits are usually maintained by nurses, medical orderlies or first-aiders.

It is important to insist that all attendances and all treatments are noted on the records. Some departments note only the first attendance on the individual's record and put reattendances down on the day sheet. Since there are strong medicolegal reasons why all attendances for occupational injuries should be recorded, however trivial they may appear on first attendance, it is advisable to maintain scrupulous records for all attendances.

Daily records

All treatment rooms should keep a record in summarized form of every attendance (*Figure 8.2*). Although this may be done in a ledger, it is more usual for them to be entered on large 'day sheets' with printed columns, which greatly facilitate the subsequent preparation of statistical returns (*Figure 8.3*). This system should be accompanied by a separate entry on the employee's personal file – a duplication of effort that is frequently omitted by busy treatment room staff. Alternatively, an individual record may be made in duplicate at the treatment centre (*Figure 8.4*). One copy is kept at the centre as an

DOCKYARD MEDICAL CENTRE
INJURY/ILLNESS REPORT Serial No. 1 030717 6 Name

7 Date | Yd D | Yard Number 20 | Centre | Age | Sex | Employer | Follow-up No. 37

38 Occupation 57 | Date Injured/Illness | Illness | Time of Injury 68

69 Part(s) of body injured 84 | 85 Cause of Injury 104

For foot, head and eye injuries ONLY, was:
a. Relevant protection issued? 105
b. Relevant protection worn? 106

Medical notes

109 Treated by

Direct return to work

112 No. attendances

Signature ....

Figure 8.4 An example of an individual record of treatment made in duplicate. One copy for the individual's file and the other for statistical (computer) analysis

individual record. The duplicate copies are available for statistical analysis at the end of each month or a longer period (Grant Macmillan, personal communication).

It is the responsibility of the person in charge of the treatment room to ensure that an entry is made for every attendance. The basic information required consists of the date and time, the name, identification number and department of the patient and, for first attendances, a brief description of the condition, its treatment, its cause if an injury, and a note of disposal. Where more than one person works in the treatment room, the identity of the therapist should also be included. This information can subsequently be used for statistical analysis of treatments given, the number and nature of occupational injuries, and the departments in which these occurred. Not infrequently they prove to be important if, at a much later date, any enquiry is raised about the attendance of an employee with a specific injury. These records should therefore be kept for not less than five years, and longer if space is available.

Records of treatment or attendance at other parts of the medical centre such as radiography, physiotherapy and so on are also required. It is important to ensure that the entire set of one person's medical documents are kept in one place and can be rapidly collected together. Clearly, the more separate records there are, the more advisable it is to have a central filing system.

REHABILITATION AND RESETTLEMENT AT WORK

Rehabilitation is the process of restoring as much function as is possible in the person or in the disabled limb or other part of the body. Resettlement means placing the person in suitable temporary or permanent work.

The processes of rehabilitation and resettlement following injury or illness may be the responsibility of an occupational health service but they will require close and extensive cooperation with hospital specialists, family doctors and social workers (*see* Chapter 3).

The doctor in industry should be aware of the schemes in the community – both in the health and social services and those organized by employment ministries – which can help the disabled, and of the assistance which can be given by other groups concerned first with special disabilities such as blindness, deafness and limblessness, and secondly with such diseases as diabetes, alcoholism and multiple sclerosis.

Within industry there are many different kinds of schemes for rehabilitation ranging from the elaborate formalized to *ad hoc*

CONFIDENTIAL

NOTIFICATION OF RESTRICTED DUTY OR OF CHANGE OF MEDICAL CATEGORY

1. OHS records
2. Supervisor
3. Personnel records

From: OHS

To:

No:

Date:

Restricted duty for

1. No climbing of vertical ladders ⎱ See
2. No work at unguarded heights ⎰ remarks
3. To work at ground level only
4. No heavy lifting (40 lb. +)
5. No stooping or bending
6. No heavy work
7. No field work
8. Sedentary work only
9. No work near moving machinery
10. No driving of cranes/automotive equipment
11. No arc welding
12. No work involving the use of right/left hand/arm/leg
13. To avoid contact with ....
14. No overtime
15. Should not work more than .... hours overtime per week
16. Should work on days only
17. Should do no work involving rapid action or decision taking

The doctor would be grateful if this employee could be given work with the restriction(s) indicated for the next .... days. If this should prove difficult or impossible, please refer the employee back to the Medical Centre

Signed .... OHS

This employee's medical category has been changed to ....

Remarks:

Figure 8.5 An example of a certificate of notification of restricted duty or change of medical category

arrangements. In general any scheme is only as good as the interest of those who supervise it. There is no substitute for personal concern on the part of doctors, nurses, and supervisors for the restoration to the affected person of full physiological function and a satisfying job - meaningless labour is not good enough. The medical and work aspects must be given equal attention if success is to be assured.

Doctors working in industry are in a privileged position with regard to rehabilitation and job placement: the doctor has - or should have access to all relevant facts about the medical condition and the job requirements. He can also discuss the problems with supervisors, management and union and thus stimulate interest and catalyse activity in rehabilitation and resettlement.

The aim should always be to get the person back to his normal job as soon as possible. It is usually best to modify normal work to suit the worker's medical condition rather than to uproot him from his regular work and social environment. Although the work task may be easier in specially created jobs, the social content is often much less good and usually less helpful in achieving forward-looking attitudes or attitude-change on the part of ill or injured people. The man who is among his colleagues will generally be motivated to pull his weight as soon as he can and will more often have a sympathetic group around him. Transferring a man to modified work in another department can have disastrous results since his new colleagues may resent an impaired 'outsider' whereas his own friends would have helped him.

Should it be necessary to carry out rehabilitation away from the workplace, the occupational physician can still show interest and help to guide his medical colleagues at the place of treatment by giving the sort of information which will help them to apply their efforts in the best direction and to enable them to feel happy about sending patients back at an early date to suitable work.

Job placement

In communicating with managers and supervisors about job place-ment, the doctor should try to avoid all such vague phrases as 'light work'. Many different interpretations of this phrase are possible, and can lead to people being less well placed than is desirable or to supervisors creating 'jobs' which consist of sitting down and doing nothing. It is best to try to spell out exactly what a person can and cannot do. An example of this approach is shown in *Figure 8.5*.

Another useful approach to supervisors is to remind them not only of what is not working–for example, 'He has a bad leg'–but of what is working–for example, 'Although he has some trouble with his left leg and cannot climb stairs, he has no other disability'. The focus on *ability* should be at least as sharp as that on *dis*ability. In this way, useful productive and creative work can be arranged which is within the capacities of the disabled person. Should there be a large number of problems, job analysis as well as disability and ability analysis may be formalized to aid administration. This sort of approach will only be required in very large organizations. Special rehabilitation workshops are occasionally found in very large organizations. They either use modified machinery to manufacture items for part of the factory's production schedules, or are set up to make things such as safety gloves. Only when the total employed population on one site is in the order of 20 000 or more can such special workshops be justified on economic grounds. For the smaller factory a great deal can be achieved by enthusiastic medical staff, an understanding management and a sympathetic work force.

REFERENCES

British Medical Association (1980). *The Occupational Physician*. London: BMA
Eagger, A. Austin (1965). *Venture in Industry. The Slough Industrial Health Service, 1947 - 1963*. London: Lloyd-Luke
Health and Safety Commission (1977). *Occupational Health Services. The Way Ahead*. London: HM Stationery Office
Ross Institute of Tropical Hygiene and TUC Centenary Institute of Ocupational Health (1970). *Proceedings of the Symposium on the Health Problems of Industrial Progress in Developing Countries*. London School of Hygiene and Tropical Medicine
Royal College of Nursing. Society of Occupational Health Nursing (1979). Information Leaflet No. 11 (third issue). London: RCN

9

Preliminary, Periodic and Other Routine Medical Examinations

P.J. Taylor and P.A.B. Raffle

The performance of large numbers of routine medical or health examinations can be one of the more tedious aspects of the work of an occupational physician and nurse. This chapter describes the reasons for such examinations, discusses their value and the alternative methods by which they can be performed, and considers the extent to which they can be justified on economic as well as medical grounds in the modern practice of occupational health.

Routine examinations are undertaken on a considerable scale by occupational health services throughout the world. A substantial proportion are done on grounds that would not withstand a critical analysis in terms of costs and benefits. A large and influential section of the lay public, however, appears to place great store upon such procedures and this is one reason why it is a great deal easier to institute regular examinations than it is to dispense with those already established. Nevertheless, all such examinations should be reviewed regularly both for their purpose and cost effectiveness. This chapter is not concerned with two types of examination which are considered elsewhere, namely, the examination of persons exposed to specific toxicological hazards (*see* Chapters 11, 12 and 25), and screening well people (*see* Chapter 10).

THE CONDUCT OF ROUTINE MEDICAL EXAMINATIONS

The performance of routine medical examinations differs fundamentally from the more usual activities of the doctor or nurse, since they do not normally begin with any therapeutic intent. In general practice and in hospital, the contact between doctor and patient arises because the latter wishes to consult the doctor, and thus the traditional doctor–patient relationship, with all that this implies, is established at the outset. With routine medical examinations, however, although the subject is a patient the full relationship may not be established.

In some medical examinations the subject may have to take part either because the examination is statutory or because it forms part of his terms and conditions of employment; others may be completely voluntary. In the first category the doctor may be obliged to make known the salient results of his examination to a third party, and in this respect the examination is similar to one undertaken on behalf of a life insurance company. In the voluntary type, the doctor operates under the usual medicolegal and ethical rules about unauthorized disclosure of confidential information. Nevertheless, for neither type of examination is the usual doctor–patient relationship fully developed.

This can, and often does, colour the attitudes of both subject and physician, particularly in the pre-employment situation where there can be intentional concealment of significant aspects in the previous medical history. Managers seldom appreciate just how much a physician depends on the subject's medical history, particularly when many important conditions, including epilepsy, seasonal asthma and so on, cannot be detected by a physical examination alone. In one internal study of 516 epileptics in the Post Office, for example, it was found that over one-third of cases had not volunteered the information at the pre-employment medical examination. Carefully designed questionnaires can reduce this, since subjects are often reluctant to reply incorrectly to a specific question, even though they may not offer information without prompting. Not all omissions are intentional; many people frequently forget about quite important episodes in their own medical histories. Investigations have shown that when it comes to minor illnesses, reliability of recall falls off rapidly after two weeks and it is for this reason that the General Household Survey in the United Kingdom only enquires about health in the preceding two weeks.

The problems, as well as the atypical doctor–patient relationship, can make the physician's adjustment to the role of 'medical examiner' a difficult one. Additional problems may arise when the doctor detects unrecognized and untreated conditions. In most countries occupational physicians are forbidden by the code of medical ethics to treat patients of other practitioners except in an emergency. This too may add to the doctor's sense of frustration with routine medical examinations. The place of treatment services is discussed elsewhere (*see* Chapter 8), but in most circumstances therapy does not play a part in routine examinations. It is, however, the duty of the occupational physician to refer subjects found to have abnormalities which would benefit from treatment, to their general practitioner.

The procedure and organization of medical examinations should also be assessed. All too often not enough thought has been given to the real objectives of the examination and thus a 'full standard medical' may be undertaken without modification for pre-employment, job placement, routine health checks, post-sickness absences and medical retirement purposes. This procedure is often aided by a standard examination *pro forma*, which, although of value in reminding the physician of measurements or tests that can be made, tends to make the whole procedure mechanical. When combined with pressure to perform many such routine examinations in a day on a basically healthy population, this can sometimes result in the exercise becoming a charade in which clinical acumen is blunted and abnormal physical signs missed or ignored.

There is a need to assess the objectives of any routine examination, and wherever possible to adjust the procedure involved so as to obtain the most meaningful results. Whereas in an ideal world it might be useful in a non-specific manner for the physician to be able to spend an hour or more with each subject and use the medium of the examination to establish rapport, take a full medical history, conduct a complete general physical examination, supplementing it with assorted special tests, and impart some essentials of health education and perhaps some treatment as well, this is an uneconomic and inefficient use of the scarce medical resources in occupational health (Norman, 1960).

If routine examinations should be planned in both objective and content with these points clearly in mind, it is necessary first to consider the main reasons why such examinations are undertaken. Most are required for either or both of the following reasons:

1. To ensure that the person is fit for the work.
2. To detect whether the work has had any adverse effect upon the worker's health; this latter, when in relation to specific occupational hazards, is considered in Chapters 23 *and* 25.

STATUTORY EXAMINATIONS

One of the reasons why routine medical examinations are undertaken is that some of them are required by law. Although national factory or health and safety laws often include requirements for such examinations, there are great differences between countries in the extent to which these are obligatory. In some they are restricted to include certain 'at risk' or vulnerable groups, including, for example, young persons, new employees or those in defined occupations such as heavy goods vehicle drivers and so on. In other countries, such as France, there are requirements for *all* workers to be medically examined once a year. The European Economic Community has not yet produced directives on this aspect of occupational health and although the Treaty of Rome requires harmonization of such requirements between member states, it may well take time before an acceptable compromise policy can be achieved. Meanwhile the value of routine examinations has been seriously questioned since there is no evidence that those who have them enjoy better health and longevity than those who do not (de Souza, 1978), and in 1978 the Trades Union Congress decided to change its long-established policy urging routine annual examinations for all workers.

Young persons

Laws requiring routine medical examinations can be repealed if the historical reasons for undertaking them are no longer applicable. The history of the statutory examination of all young persons under 18 years employed in factories in Great Britain illustrates this point. The social evils of the early factory system, particularly in cotton mills, which included the employment of small children in appalling conditions, resulted in a requirement, included in the Factory Act of 1833, that a doctor must certify that a child to be employed had the strength and appearance of one not less than nine years old. The subsequent development of compulsory education, the progressive rises in the school leaving age, and even more the rise in standards of living and community health, steadily changed the needs for, and thus the nature of, the statutory examination of young people working in factories. Finally it became clear that there was no longer a medical justification to require over one-third of a million examinations every year, since the detection rate of previously unrecognized abnormalities was negligible. The law was changed again in 1972 and now occupational health care in this respect is concentrated upon young persons known to have health problems which have been detected by general practitioners and the School Health Service.

Enabling legislation

Because of the considerable inertia involved in any legal system several countries now incline towards a greater use of 'enabling legislation' which empowers a minister to make regulations or orders as and when they are required. Although this has the merit of greater flexibility in implementation and withdrawal, there is also a risk that evanescent opinions or fashions can influence the practice of occupational health without adequate consultation or a consensus of approval. The medical content of the examination is not usually spelt out in any great detail; it is often left to the discretion of the individual doctor. In recent years, however, there has been a tendency for some aspects of certain examinations to be specified (as in the heavy goods vehicle driving licence examination), and others may require supplementary haematological or other tests.

Drivers

In the United Kingdom, the medical requirements for holders of licences to drive heavy goods vehicles (h.g.vs) and public service vehicles (p.s.vs) are that a report signed by a medical practitioner must be submitted with the first application for a licence. Thereafter a medical report is required, after reaching the age of 60, by an *h.g.v.* driver each time he seeks a further licence and by a *p.s.v.* driver each time he seeks a further licence after reaching the ages of 50, 56, 59 and 62, and at age 65 and annually thereafter. The statutory requirements for examinations often represent minimum standards, and simple adherence to the law may not meet the best standards of occupational health practice. For example, holders of h.g.v. driving licences in Britain are not required to be re-examined medically until they reach the age of 60. Thus the gap in time between the first and the next examination may be more than 30 years. A health declaration is required at each renewal of licence, and the licensing authority (Chairman of Traffic Commissioners for the Traffic Area in which the driver lives) may, in special circumstances, call for a certificate at other times. The omission of routine examinations until age 60 or after can partly be attributed to reluctance to add further to the workload of already scarce national resources of medical manpower, for a very small yield of conditions likely to affect road safety (*see* below, section on examinations after sickness absence), but also because of the provisions under the Road Traffic Act 1974 for all holders of ordinary driving licences (and hence h.g.v. drivers) to notify 'relevant' and 'prospective' disabilities to the Driver and Vehicle Licensing Centre as soon as they become aware of them. 'Relevant disabilities' are described in detail in the Appendix to *Medical Aspects of Fitness to Drive* (Medical Commission on Accident Prevention, 1976) but are, broadly: epilepsy, severe mental subnormality, liability to sudden attacks of disabling giddiness or fainting, inability to meet the prescribed eyesight standards or any other disability likely to cause driving of a vehicle by the licence holder to be a source of danger to the public. 'Prospective disabilities' are disabilities which may become 'relevant disabilities' and include epilepsy, provided the licence holder has not had an epileptic attack either awake or asleep with or without treatment for three years, and the use of a cardiac pacemaker. The law does not allow the holding of an ordinary licence by those with 'relevant disabilities' and a licence valid for one, two or three years may be granted on the advice of the Medical Advisory Branch of the Driver and Vehicle Licensing Centre to those with 'prospective disabilities'. The only other bar to holding a h.g.v. licence is suffering from a epileptic attack after the age of three. The licensing authorities for both ordinary and h.g.v. licences are guided by the recommendations in the appropriate sections of *Medical Aspects of Fitness to Drive* and the Medical Advisory Branch of the Driver and Vehicle Licensing Centre is helpful in advising on difficult individual cases.

Some firms, such as transport undertakings, go beyond the requirements of the law with respect to medical examinations for h.g.v. and p.s.v. drivers. Other firms subject all employees who drive professionally, including chauffeurs and drivers of commercial vehicles and forklift trucks, to pre-employment, periodic and after sickness examinations.

Recommendations on standards of fitness required by drivers are not immutable, but must evolve in the light of new knowledge about the relationship between ill-health and fitness to drive (Raffle, 1974a). There are also provisions to harmonize licensing requirements within the United Nations. A UN Agreement on Minimum Requirements for the Issue and Validity of Driving Permits, commonly referred to as the APC Agreement (*Accord des Permis de Conduire*), is now open for signature. The United Kingdom will have to decide whether to subscribe to the Agreement and, consequently, to meet the medical standards it lays down. This would mean that for h.g.v. drivers epilepsy at any time of life, insulin-treated diabetes and diplopia would become bar disabilities; visual standards would be higher and would also effectively make monocularity and treated cataracts bar disabilities.

VOLUNTARY EXAMINATIONS

The use of the word 'voluntary' is solely to distinguish these routine examinations from those required by law. Many organizations have their own requirements, and these are often part of the terms and conditions of employment, and thus for practical purposes are compulsory for those employees covered by them. These may be required by the management or their personnel or pension fund advisers, by the trade unions, or by occupational physicians.

The commonest variety is undoubtedly the pre-employment medical examination, although in effect most of these examinations are more accurately described as 'preplacement examinations'. The distinction may appear semantic but in modern occupational health practice very few applicants are totally rejected for any type of employment, even though some with disabilities may not be fit for certain jobs. In these circumstances the occupational physician would be well advised to state that the applicant was found unfit for employment as a driver, food handler, or whatever, rather than unfit for any employment.

Other types of voluntary examination include those on employees returning to work after periods of sickness absence, the periodic examination of employees in certain occupations such as crane or vehicle drivers, firemen and so on, and in recent years the offer of 'check ups' for older workers or senior managers. Although not required by law, some of these regular examinations have become almost inviolate within an organization because of 'custom and practice' and it may be very difficult, without a good deal of active re-education and persuasion of both managers and employees, to abolish or reduce in frequency those that are of little value. For this reason an occupational physician should always consider most carefully the needs and likely benefits before introducing a new scheme of regular medical examinations.

Pre-employment examinations

The objectives of pre-employment examinations are:

1. To ensure that the subject is fit to undertake the job without risk to himself or to his colleagues, or in certain circumstances to the general public.
2. In those organizations requiring wide job flexibility under productivity agreements, that the subject is also fit to undertake any other jobs which may be required.
3. Where applicable, that the subject is an acceptable life or disablement risk to the pension fund.

These apparently simple objectives conceal many real difficulties few of which are appreciated either by management or the subjects themselves. Fitness to undertake a specific job should imply that the examining physician be fully conversant with the physical, mental and psychological demands of the job on the one hand, and be able to match them with an individual's state of health on the other. General medicine is still an imprecise science and the facilities for physiological and psychological testing on ergonomic principles are not normally available to assist the occupational physician. Fortunately, in the majority of cases such elaborate tests are neither necessary nor desirable. Even when they can be used, their artificial nature restricts their value. The usual type of medical examination done for this purpose is able to do no more than assess that the applicant is *not clinically unfit* at that time for the job.

Usually the physician is quite unable to predict the person's future state of health, and it is certainly true that rarely, if ever, can he predict the subsequent sickness absence of an employee from evidence usually available at this stage. This latter point seems ill understood by many members of line management or personnel departments. Recent work, however, shows that the main reliable predictor of future sickness absence is the subject's own experience in previous occupations (Froggatt, 1970); this may only be revealed when references are taken. Employees with chronic disabling conditions, for example, have been shown to have a wide range of sickness absence experience ranging from none at all to excessive amounts, and objective medical evidence seems to play a small part in determining what pattern an employee will adopt (Thompson, 1974).

The use of pre-employment examinations to determine whether or not an applicant can be allowed to join the company's pension scheme occasionally complicates the situation. Most companies do not lay down specific instructions, and their pension schemes are designed to accept the normal range of life risks in any working population; this applies whether the company runs its own pension scheme or uses a life insurance company to provide a scheme to meet its requirements. Some organizations have a waiting period or trial of up to one year or so before a new employee is admitted to the pension scheme; most occupational physicians are not directly involved since the decision is taken according to actuarial rules. Some physicians seem to take it upon themselves to act on behalf of the company pension fund even when they have not been instructed to do so. This is most unwise. Such a step might lead them into difficulties if they decide to reject applicants with for example, well controlled diabetes as unfit for employment in clerical jobs. If one were to take the cynical view, the 'ideal employee' would be one who worked without disability until reaching retirement age, and then died suddenly, leaving no dependants. Chronic non-fatal disabling conditions which cause an employee to be prematurely retired on pension are most expensive to the fund. Nevertheless, all funds should be designed to cater for the small proportion of employees who will fall into this category, and despite protestations to the contrary, private pension fund rules can be changed without difficulty if they are unduly restrictive, provided the company is prepared to increase its contribution. In our opinion, the practising occupational physician might well be advised that when a subject is found fit for employment he should also be considered fit for superannuation unless there are incontrovertible grounds for an adverse opinion. As an example, experience in London Transport has shown that while 11 per cent of applicants for bus driver positions are rejected for that occupation (60 per cent due to defective eyesight) less than 4/1000 applicants for clerical jobs are rejected.

Examinations after sickness absence

Examinations after absence through sickness are one of the most useful types of examination, but some occupational physicians rarely undertake them. The main and most obvious reason for these examinations is to ensure that an employee who has been absent with a medically significant condition is fit to undertake his usual occupation when he returns to work. While it is neither practicable nor necessary for the physician to see all employees who have been away sick for short spells of absence, it is usually advisable to see all those in occupations involving theoretical or real risks to the employee or to others after absences of a month or so, or after certain specified illness, such as vertigo, syncope and heart disease.

TO	FROM
.... 19	

CERTIFICATE OF MEDICAL EXAMINATION

Name .... Normal grade .... Number ....

Reason for examination .... Location ....

I certify that the above named has been medically examined and is:-

	Signature of Medical Officer
Medically unfit for employment in normal grade .... Fit for alternative employment as .... To be re-examined if found alternative employment 19	

REMARKS

....
....
....
....
....
....
....
....
....
....
....
....
....
....
....

MEDICAL OFFICER

Figure 9.1 An example of a certificate of medical examination

The physician may arrange the screening of post-sickness cases to be done by an occupational health nurse who seeks a medical opinion only where this seems necessary. The decision as to what length of absence or type of illness should be followed by examination must be left to the occupational physician who will be guided by information on sickness absence patterns in the organization and the risks involved in the various occupations (*see* Chapter 16).

The other main objective of these examinations is to assist those who are not fit to return to their usual occupation to be resettled temporarily or permanently in other jobs. Properly applied, with adequate sickness records and foresight, this procedure can also be an effective means by which lengthy or unduly frequent sickness absence can be monitored. The 'light work' certificate so often used by some general practitioners causes a great deal of difficulty to employers and to the patient's immediate supervisor. The phrase itself is meaningless – to one person it may mean that he can do virtually nothing, to another that he can lift only 20 instead of 40 kg.

Wherever possible, employees with such certificates from their own doctor should be seen by the occupational physician so that they can be assessed in a way more meaningful to the management. The critical factor is the physician's intimate knowledge of the physical and mental demands of all the occupations in the organization. An example of a certificate of medical examination (in this case that of London Transport) intended for management is given in *Figure 9.1*. In the certificate, space is provided for recommendations for alternative employment in specified occupations. Open ended limitations are a constant source of irritation to management and unions, as well as to the employee.

Periodic examinations

Periodic examinations are practised extensively in some occupational health services, scarcely at all in others. The reasons usually given for their non-performance are either that there are no occupations in the organization which require them, or that there are so many pre-employment examinations that there is no time available to do them. The former explanation may be defensible, the latter is not. Seldom should time be devoted to routine periodic examinations at the expense of examinations of those returning from sickness or accident, except those required for the surveillance of specific toxic hazards. The yield from periodic examinations in terms of significant pathology found or advice on lifestyle which is heeded is very small. For instance, Raffle (1974b) showed that 85 per cent of the disabilities rendering bus drivers aged

50–54 unfit to drive a bus, either temporarily or permanently, were picked up at postsickness or postaccident examinations and only 15 per cent at routine age examinations. Since this chapter is not concerned with specific toxicological hazards, a good example of an occupation requiring this type of periodic medical is that of vehicle driver. The frequency of such examinations varies greatly throughout the world. For bus drivers, data from the Union Internationale des Transport Publics (Raffle, 1979; personal communication), indicates frequencies varying from annually at all ages to the statutory requirements in Great Britain of examinations at the licence renewal following the 50th, 56th, 59th, 62nd and 64th birthdays and annually thereafter. The points to watch for and the methods of this examination have been described in the publication *Medical Aspects of Fitness to Drive* (Medical Commission on Accident Prevention, 1976) which was circulated in 1978 to all medical practitioners in the National Health Service as a guide to them in advising their driver/patients of all kinds. The recommendations therein, though not identical, are similar to those published in other parts of the world. In addition the British Medical Association (1975) publishes, notes for the guidance of doctors completing medical certificate forms in respect of applicants for public service vehicle driving licences and heavy goods vehicle driving licences.

Serious difficulties can arise for the physician who is responsible for such periodic examinations if there is no agreed and acceptable procedure within the company for cushioning the effect on an employee of a decision that he or she is no longer fit for that occupation. Maintenance of earnings for a minimum period and some degree of security of employment for those fit to do other jobs are essential if the physician and the employee are to co-operate in such a programme of periodic examinations. In the United Kingdom it is important for the physician to be fully aware of the provisions of recent industrial relations legislation, especially the unfair dismissal provisions of the Trade Union and Labour Relations Act 1974, in order that employees who become unfit to continue in their normal work are treated not only humanely, but also 'fairly' within the meaning of the legislation.

.The periodic examination of senior or other key staff should be briefly considered. It has become fashionable in some quarters to organize these regular medical examinations and the initiative has often come from the senior executive themselves. Viewed as yet another example of periodic examination of fitness for employment in specified occupations, this procedure may have some merit. The main hazards of such jobs to the individual, and to others, are said to be in the realm of stress and its psychological or psychosomatic consequences. This type of examination is dealt with in Chapter 10, 'Screening Well People'.

EXAMINATION OF FOOD HANDLERS

All doctors and nurses are trained to be conscious of the risks to health if those who prepare or handle food are, like Typhoid Mary, carriers of enteric disease. Occupational health services have always taken particular care of this aspect of their responsibilities in food factories and also in other industrial canteens or restaurants. Recent expansion of the food industry and the development of new products have made this aspect of occupational health even more important. The increased popularity of delicatessen foods and the new rapid freezing techniques involving the preparation of precooked meals and snacks mean that the potential for large-scale outbreaks of food poisoning is greater than ever before. A group of experienced physicians in the British food industry has prepared guidelines on the medical aspects of food handler supervision and the points made below follow these recommendations.

A food handler is defined as one who handles food itself, the implements used in its preparation and the containers with which it may be in direct contact. In each situation the critical factor is whether contaminated food will be able to support bacterial growth and whether it is to be eaten without further *thorough* cooking. If these conditions apply, even greater care is required in medical selection and surveillance. The guidelines for the pre-employment selection of food handlers suggest that the conditions indicated below would normally disqualify the applicants:

1. A history of typhoid or paratyphoid disease (with such a history the applicant should be permanently excluded from food handling).
2. A history of intestinal disease, unless freedom from the carrier state can be established by bacteriological tests.
3. Septic inflammatory disease of the eyes, ears, nose, mouth and throat as well as persistent or recurrent staphylococcal inflammation of the skin.
4. Persistent chest disease with sputum production during working hours.

All food handlers should be carefully screened by occupational health nurses (or physicians) before employment by use of an appropriate health questionnaire and physical assessment, and stool tests arranged for appropriate cases. Health assessments should also be made on any employed food handler who returns to work after sickness absence and if the illness has been gastrointestinal further stool tests may be required. Staff returning from holiday should also be asked if they have been ill and, if so, clearance will be required by the occupational health nurse or physician (*Figure 9.2*).

FORM FOR FOOD HANDLERS ALREADY IN EMPLOYMENT

For use by Supervisors, Managers &.Personnel Officers

This form must be completed by all food handlers on return to work after absence due to illness, injury or holidays.

NAME...	FOR HOLIDAYS ETC COUNTRIES VISITED	LENGTH OF STAY
ADDRESS..	....	....
....	....	....
....	....	....

1 (a) Since you have been away have you suffered from sickness, diarrhoea or any stomach disorder?

YES/NO

1 (b) Have you had any 'flu like' symptoms?

YES/NO

2 Have you been in contact with anyone with typhoid, typhus, paratyphoid, cholera, gastro-enteritis, or any other of the symptoms in 1 (a) and 1 (b)?

YES/NO

3 (a) Are you suffering from any infectious conditions of the skin, nose, throat, eyes or ears?

YES/NO

(b) Have you suffered from any of these conditions since you have been away?

YES/NO

This form must be completed under the supervision of the Departmental Supervisor and signed by the employee

Signature of employee...

Date......................................

Job title...

NOTES FOR GUIDANCE FOR MANAGERS AND SUPERVISORS

A. If the answer to any of these questions is 'YES' the person must not be allowed to return to work until he or she:

1 has been seen by the Company Nurse or Doctor, or
2 has been cleared by the Local Health Authority.

C. If at any time there are any queries on the use of this Health Declaration Form, you are advised to contact your Company Nurse or Doctor.

Figure 9.2 An example of a health declaration form for food handlers already in employment

Routine pre-employment chest X-rays are not recommended unless there is a clinical indication because, contrary to popular belief, human tuberculosis is not normally a foodborne disease and in many countries it has become relatively uncommon. Also, it is now universally accepted that no one should be unnecessarily exposed to X-rays. The indications for such radiographs must either be on individual clinical grounds decided by the doctor or, in any location where the district community physician (or chest physician) advises that such routine films should be taken for public health reasons. Any food handler found to have open (sputum positive) disease should of course be kept away from work until effective treatment has started and the chest physician gives clearence to resume food handling. Routine serology, such as the Widal reaction, is *not* recommended at pre-employment as the results can be difficult to interpret in a healthy person. Decisions on the need for stool-testing (normally three consecutive daily specimens) must be taken by the company's occupational physician who may choose to discuss individual cases with the Medical Officer of Environmental Health or the Public Health Laboratory Service. As a guideline, such tests are often required of applicants who have a history of intestinal disease, but it is the policy of some food companies to require them routinely on employment for handlers of fresh cream or other prepared foods which are not to be cooked after they have been handled. It is as well for the physician to be aware of the possibility that stool specimens submitted by one applicant for employment may not invariably come from that individual, but there is no easy solution for this dilemma.

It cannot be emphasized too strongly that the safety of all food products depends in the final analysis on the strictest application of the highest standards of food hygiene both by the managers and by the food handlers themselves. The careful medical selection of employees alone will be insufficient, and in Britain the Food Hygiene Regulations of 1970 require that any person engaged in the handling of food who becomes aware that he or she is suffering from any salmonella infection, amoebic or bacillary dysentery or any staphylococcal infection likely to cause food poisoning shall inform the manager, who in turn shall immediately notify the Medical Officer of Environmental Health.

PROCEDURES FOR MEDICAL EXAMINATIONS

An efficient and effective occupational health service must always be aware of its costs. The 'full standard medical examination' as described above can be most costly to the organization in terms of the physician's time, and if done on a large scale in a normal working population, it can also be inefficient in terms of the detection of ill health.

Teamwork

In recent years, the development of professional and highly trained occupational health nurses and other medical auxiliaries has allowed the methods and procedures of medical examinations to be improved. If the objectives of each examination are kept constantly in mind, it is clearly possible to design procedures to achieve them at least cost.

Trained nurses can be taught to administer a standard health questionnaire and follow it up with a semistructured interview designed to clarify matters of relevance such as fitness for specific occupations. They, or other ancillaries, also undertake the straightforward measurements of height, weight, vision, hearing, blood pressure, urinanalysis, and ventilatory function.

In the pre-employment situation, many occupational physicians now encourage occupational health nurses to undertake such procedures for all those applicants from whom medical screening is necessary except those who are being assessed for fitness for the more hazardous occupations. The trained occupational health nurse is usually delegated the authority to pass a subject as fit, but not to reject one. The nurse refers to the occupational physician those about whom there is uncertainty. Such an arrangement usually means about 10 per cent of applicants need to be seen by the physician, and these come with a medical history and basic investigations already recorded on the notes. For others whose occupation requires that they be seen by the doctor the same preliminaries ensure that he can allocate his time fully to the use of clinical skills (Harte, 1974).

Pro formas

Virtually every occupational health service throughout the world has designed its own particular form for routine medical examinations. Despite the essential similarity of problems throughout occupational health practice, it appears that there must be a wide divergence of interest in symptomatology since seldom are two forms alike. Few, if any, questionnaires used in occupational health have been validated in a satisfactory manner, and although this may not seem of much moment, the internationally accepted value of the British Medical Research Council's *Questionnaire on Respiratory Symptoms* shows how much can be achieved by the use of a well designed and tested questionnaire.

Any standard questionnaire must be designed not only to provide information of relevance but also to be easily understood by the subjects upon whom it is to be used. The vocabulary of different strata of society within one country can differ considerably – and usage of

IN CONFIDENCE

HEALTH DECLARATION

A To be completed by the Recruitment Office

Applicant for employment as .... at ....

Name in full: Mr/Mrs/Miss ....
(Surname in block capitals)   (Other names)

Address: ....

Date and Place of Birth ....

B Applicants should read the following carefully:

1 The questionnaire below should be completed as fully as possible. All questions must be answered. The information will be treated in confidence.

2 WARNING In completing the questionnaire you are responsible for the accuracy of your statements. If information is withheld, suppressed, is deliberately misleading or false you may be liable, if employed, to dismissal.

C Please give the following details:

1 What is the name and address of your family doctor? ....

....

2 What is your height (without shoes)? ....

3 What is your weight (in indoor clothes)? ....

4 Please list below all absences from work/school for health reasons during the past 12 months.

Length of Absence (Days)	Cause	Length of Absence (Days)	Cause

D		Yes	No
5 Do you wear glasses/contact lenses?			
6 Is your sight in each eye separately good for all usual activities with glasses/contact lenses if necessary?	Right		
	Left		
7 Is your hearing in each ear separately good enough for all normal purposes including telephoning (with a hearing aid if necessary)?	Right		
	Left		
8 Do you have a discharge (running) from either ear from time to time?	Right		
	Left		

Figure 9.3. Health declaration form

E   Please complete the following questions by ticking the appropriate box. If the answer is 'Yes' give details in the remarks column.

9   Have you ever in your life, including childhood, had any of the following? If the answer to any of the questions is 'Yes' please give full details including:

(i)   date

(ii)   amount of time lost from work/school

(iii) whether you require or are awaiting treatment

	YES	NO	If Yes Please give date and details
a. Fainting Attacks and Giddiness			
b. Tuberculosis			
c. Sinusitis			
d. Bronchitis, Asthma or Pneumonia			
e. Recurring Headaches or Migraine			
f. Dermatitis or other Skin Disorders			
g. Foot or Knee Trouble			
h. Varicose Veins Causing Trouble			
i. Rupture/Hernia			
j. Recurrent Indigestion/ Dyspepsia			
k. Kidney or Bladder Disease			
l. Blackouts, Epilepsy or Fits			
m. Heart Trouble, Heart Attack or Angina			
n. Raised Blood Pressure			
o. Diabetes			
p. Severe Shortness of Breath			
q. Nervous Disorders. 'Nerves' or Breakdown			
r. Back or Neck Trouble. Sciatica/Arthritis			
s. Serious Injury or Accident			

Figure 9.3 continued

	YES	NO	If Yes Please give date and details
10 Have you ever had any operations or been admitted to hospital for any reason?			
11 Are you currently attending any hospital, clinic or out-patients department?			
12 Are you currently attending a doctor?			
13 Are you at present having any medication or any other treatment prescribed by a doctor?			
14 Do you have any symptoms which frequently prevent you from going to work, school, etc. for a day or two longer?			
15 Do you have, or have you had any defect, disorder or other condition, mental or physical not already mentioned in any of your answers?			
16 Have you ever had a health examination for PO employment?			
17 Have you left a job (or been discharged from HM Forces) because of ill-health?			
18 Are you in receipt of a war pension or any other disability benefit?			
19 Are you or have you ever been Registered Disabled?			
20 Have you had a chest X-ray within the past 12 months? If so state place, date and result.			

Figure 9.3 continued

F DECLARATION

I declare that to the best of my knowledge all the foregoing statements are correct. I fully understand
the warning on Page 1 of this form and appreciate that a health interview or examination may be required.

Signature .... Date ....

G STATEMENT OF APPLICANT

I agree that if required a medical report may be obtained from my doctor or a specialist by the Occupational
Health Service acting for the Post Office. I understand that the report will be treated in confidence by the
Occupational Health Service but that advice based upon it may be given to Post Office Management.

Signature .... Date ....

FOR OFFICIAL USE ONLY

To OHS for recommendation in view of answer(s)

to question(s)

....

Department .... Signature .... Date ....

Tel   Name in Capitals ....

Figure 9.3 continued

some words can and does vary across regional as well as national boundaries. Some questionnaires seem to have been designed at the time when the infectious diseases were predominant, others place undue weight upon trivial conditions which have little or no relevance to fitness for work (Alexander *et al.*, 1975).

An example of a self-administered questionnaire (or Health Declaration) is shown in *Figure 9.3*. It was derived from a number of sources and has now been used by the Post Office on thousands of applicants for employment. The form asks for the information usually required at a pre-employment examination stage and includes a

---

HEALTH ASSESSMENT FORM

IN CONFIDENCE

A   (To be completed by Post Office)

NAME .... Date of birth ....

Applicant for ....

Office of Employment ....

DELETE
PART B
IF NO
EXAMINATION
IS REQUIRED

B   (To be completed by OHS staff, GP or LMO)

1   Height .... Weight (without coat or shoes) ....

2   Vision   Right | Left

   a. Distant (Snellen Chart)

      Without glasses   6/....... | 6/.......

      With glasses (if worn)   6/....... | 6/.......

   b. Near - J or N Type (with glasses if worn)   .... | ....

   c. Visual fields by hand test   .... | ....

3   Cardiovascular

   a. Pulse   Rate ..../ min   REGULAR/IRREGULAR

   b. Blood pressure ..../ ....

4   Urine

   a. Albumen   YES/NO

   b. Sugar   YES/NO

5   Other positive findings or observations

PART C — See overleaf

Figure 9.4 Health assessment form

warning, repeated on the last page, about false statements. It also enquires about recent sickness absence experience, since it has been found that many previous employers are unable, or unwilling, to provide this information. The final statement includes consent for the occupational health service to obtain a report, should one be required, from the applicant's own doctor. There are no questions about previous jobs as this information is already available on the initial job application form. Depending on the job for which the applicant is to be recruited and also on his or her answers to the questions about health, the applicant may then be accepted for clerical or office work,

---

C   6 Comments and opinion on points mentioned in the Health Declaration Form.

7 Any other comments on health of applicant, e.g. Skin disease, asthma, bronchitis, recurrent back pain, psychiatric illness, and for food handlers any hygiene risk.

8 CONCLUSION

  FIT/UNFIT OR DOUBTFUL (Please specify below)

Signature .... Date ....
Name in block letters ....
Occupation ....
Address ....

  ....
  ....

Figure 9.4 continued

or examined by an occupational health nurse with or without reference to a physician, before a final decision on fitness or rejection for employment is taken. The examination form used is shown in *Figure 9.4* and although this is quite simple, the most important questions are numbers 6 and 7. Any additional investigations such as haemoglobin, X-ray, lung function tests, audiometry and so on, are only done where there is a clearcut need and then at the discretion of the physician. Some occupations such as teachers and staff working in hospitals are currently required to have a chest X-ray, but this is primarily to protect the health of others. Our general view is that such tests should only be required in cases where they are justified; they should not be done merely for the sake of leaving no stone unturned.

Much less elaborate *pro formas* would be sufficient for subsequent or periodic examinations, but it is essential that the initial set is readily available to the nurse or doctor. It is also necessary for the employee's personal record of sickness absence to be available at the same time; a sample of such a form can be seen in Chapter 16.

In conclusion we would reiterate that routine examinations should only be done where there is a clear need for them; they are costly in time and easier to introduce than to abandon. Whatever procedure is adopted for their conduct, whether they are done by a nurse, by a doctor, or partly by both, and whatever *pro formas* may be used to assist in their performance, the crucial points to consider are why they are to be done, who will benefit from them and how much detail is necessary. Only then will it be possible to determine the methods to be used in order to provide the most useful answers at the lowest cost.

REFERENCES

Alexander, R.W., Maida, A.S. and Walker, R.J. (1975) 'The validity of pre-employment medical evaluations.' *Journal of Occupational Medicine* **17**, 687–692

British Medical Association (1975) *Notes for the Guidance of Doctors Completing Medical Certificate Forms in Respect of Applicants for Public Service Vehicle Driving Licences and Heavy Goods Vehicle Driving Licences.* London

de Souza, M.F. (1978) 'The value of screening and health surveillance to employment.' *Journal of the Royal College of Physicians, London,* **12**, 230–239

Froggatt, P. (1970) 'Short-term absence from industry.' *British Journal of Industrial Medicine,* **27**, 199–224

Harte, J.D. (1974) 'Why a medical examination?' *Proceedings of the Royal Society of Medicine,* **67**, 177–180

Medical Commission on Accident Prevention (1976). *Medical Aspects of Fitness to Drive.* London

Norman, L.G. (1960) *The Value of Routine Medical Examinations in Modern Trends in Occupational Health,* ed. R.S.F Schilling. London: Butterworths

Raffle, P.A.B. (1974a) 'Fitness to drive.' *Medical Society of London,* **90**, 197–205

Raffle, P.A.B. (1974b) 'Disability rates of bus drivers.' *British Journal of Industrial Medicine,* **31**, 152–158

Thompson, D. (1974) 'Civil Service experience of pre-employment examinations.' *Proceedings of the Royal Society of Medicine,* **67**, 182–184

10

Screening Well People*

H. Beric Wright and Alan Bailey

It is now accepted that good health is largely a manifestation of individual and community lifestyle. Health education, which at long last is becoming both accepted and more effective, aims at establishing more sensible living standards and habits, working through schools, families and the community in general. It is also known that in relation to many of the significant diseases like coronary thrombosis, breast cancer and perhaps long-term effects of the contraceptive pill, it is possible to identify potentially vulnerable individuals by screening the 'at risk' population in relevant ways. Those identified as being vulnerable can then be offered intervention and/or treatment to reverse the trend and improve their health. This chapter outlines current thinking about the value of screening with particular reference to management and industrial populations.

The occupational physician is in a unique position to influence the total health as well as the occupational hazards of his community. Screening allows the physician to identify those who are vulnerable, and with the support of management the company can improve individual, community and family health by adopting sensible lifestyles and by furthering the development of health education.

Health maintenance is an important activity which opens the door to an active health education and promotion programme, supported by management and carried out by the health and personnel departments. A positive attitude within the organization on fitness, smoking, alcohol and so on, could be productive over the years.

*Since this chapter was written we have completed and analysed the results of a pilot screening survey of over 500 members of the Electrical Trades Union. As far as we know this is the first time that a significant number of shop floor workers has been submitted to an executive-type screening programme.

The average age of the group was 35 which made them 12 years younger than the managers attending our centre. They were 'risk rated' for a number of coronary risk factors and had as high or in some cases a higher rating than the older managers. In addition, 10 per cent had a major correctable abnormality and 16 per cent had a minor one.

In relation to the known higher mortality and morbidity rates in social classes III, IV and V, these findings appear significant. The screening and supporting insurance cover were employer paid and this footnote has been added at the proof stage because of the importance of the findings in relation to the likely spread of health screening at shop floor level over the next few years. As in the USA, employers and their medical departments will increasingly be pressed to provide this facility.

WHAT IS SCREENING?

The screening process is the identification of clinical vulnerability. Vulnerable people fall into three groups: first, those who have established disease but are unaware of it; secondly, those who do not appear to have disease but whose physiological make up puts them in groups that experience a high disease rate; and thirdly, those whose physical condition makes them vulnerable to certain environments, especially their occupation.

An example of the first group may be someone who harbours an early cervical cancer; this is usually symptomless and can be cured by a simple operation. When it becomes symptomatic it may have spread outside the cervix and require far more extensive treatment with less chance of a cure.

The second group is exemplified by coronary risk factors. It is known from prospective studies that a man with a high blood pressure and high blood lipid levels, and who smokes, places himself in a group whose incidence of ischaemic heart disease is far higher than average.

A simple example of the third group is the man who has occasional fits, being vulnerable if he has to work with complicated plant or climb ladders.

Screening services can be simply designed to search for one group of vulnerable people such as the nationwide cervical screening service, or it can be a comprehensive or multiphasic system examining all areas of vulnerability where it is likely that something positive can be done for the vulnerable group. Comprehensive screening depends for its justification on the fact that it can be carried out easily on large numbers of people rapidly, relatively cheaply and without discomfort. The precision and accuracy of the tests must be known, together with the prevalence of the condition to be tested. It is most satisfactory when the prevalence is relatively high. It is seldom worth doing when it is low.

There are two important aspects of screening about which there is much discussion. The first is whether presymptomatic diagnosis actually alters the natural history of the disease, or merely makes the patient aware of the disease for a longer period. Screeners believe that the natural history is altered and there is good evidence that in the case of breast cancer for instance, this is true. Other studies set up to determine the truth of this belief are discussed later. If it is true in the case of ischaemic heart disease which affects one in five people before retirement age, then the potential saving of life is enormous.

The second and more technical issue is that of the 'normal' or reference range for various indices. Clinical medicine traditionally works to relatively low tolerance levels; for example, the diagnosis of diabetic ketosis or malignant hypertension requires much less precise tests than the identification of a diabetic vulnerability or a potentially harmful blood pressure level. It is in these grey areas that screeners need to set ranges which are different from traditional 'ill-health' ranges and which take account of factors such as age, sex, race, diet, alcohol and medication, smoking and level of physical fitness. Moreover trends over time, within the reference range, may be important for detecting the vulnerable, so the accuracy of the test must be high.

In addition, there is a valuable spin-off from a comprehensive or multiphasic screening facility. It provides a way into health care for 'the worried well' described by Sydney Garfield (1978), the pioneer of the Kaiser Permanente Service.

TYPES OF SCREENING PROCEDURES

Screening started in Britain as single-disease or 'one-test' screening. It gained much of its popularity through the development of mass radiography for pulmonary tuberculosis. In 1948 the mortality from this disease was still considerable and the anti-tuberculosis drugs were being successfully developed, so that during and immediately after the war it was expedient to develop and offer chest radiology which cheaply and rapidly 'screened' large numbers of people for early and asymptomatic chest lesions. Because of the knowledge about the disease and the availability of effective treatment, the public lined up to have their chests X-rayed in factories, offices and on street corners. Along with screening, improved hygiene and effective treatment led to a significant reduction in the occurrence of tuberculosis, to the extent that mass screening of this type is no longer necessary.

Single-disease, one-test screening, is still carried out on neonates for phenylketonuria and is being developed for genetic abnormalities in older mothers. It is likely that over the next decade estimation of alpha-fetoprotein will become a routine antenatal procedure for spinal abnormalities. Similarly there is a single-disease screening of communities for common diseases like diabetes and glaucoma and the routine checking of school children for auditory and visual defects. Industry has routine blood or urine tests for workers known or likely to be exposed to toxic hazards. These are done in conjunction with environmental monitoring, to detect the earliest signs of changes which might arise from exposure to lead compounds or complex organic chemicals (*see* Chapter 25).

Probably the best known single test in developed countries is the cervical smear for cancer of the cervix. Although this is a relatively uncommon condition compared with breast cancer, women have been persuaded and facilities have been made available to monitor cellular changes in the cervix to detect precancerous changes. These procedures were largely demanded by women themselves. There is evidence which has taken 20 years to gather that the death rate of the disease has been reduced in communities which have been vigorously screened. A well-known screening survey is that in Framingham, which is a stable community suburb outside Boston in the USA. Starting in 1948, the adult citizens were offered periodic and ongoing screening for cardiovascular disease. The survey was conceived as a monitoring epidemiological exercise, with no treatment for high risk cases being offered. Much of our knowledge of coronary risk factors and the natural history of this multifactorial disease is based on the long-term follow-up and continued screening of this population, now in its third generation.

The Kaiser Permanente medical organization in California has pioneered development of prepaid medical care on a group basis for local residents, operating out patient and hospital facilities to a high standard and at economic rates, in an area where medical charges were high. Although Kaiser itself is a large industrial corporation, its first medical officer, Dr Sydney Garfield, was a pioneer in occupational medicine. The Permanente medical facility evolved from its industrial medical service and under Dr Morris Collen developed the technique of automated multiphasic screening. By using computer technology it proved possible to offer a battery of tests, including a self-recording clinical questionnaire, to members of the Permanente group.

This was an extension of the periodic medical or executive health examination that developed in America before and rapidly after the war. Although it became something of a status symbol in American management, it established the concept of routine health maintenance on a defined working group; the idea being to screen individuals for early signs of disease, particularly cardiovascular and cancerous conditions, and to instigate treatment. Just as Garfield and Collen hoped by early diagnosis to reduce the cost and the toll of illness, industry hoped to keep its management active by providing good health maintenance. Many of the larger organizations in the USA now offer this facility, ranging from the janitor to the president. Unions, in the USA and now in this country, are increasingly interested in obtaining screening for their members.

What Kaiser and Collen demonstrated was that modern computer-linked technology lent itself naturally to the screening process. The advent of the biochemical autoanalyser made it no more expensive to perform a battery of tests, rather than a single test. Using a computer as a data gatherer and digester, not only could the information about an individual be presented in relation to medical priorities and deviation from a predetermined reference range, but also the information from large numbers of individuals could be co-ordinated and analysed in various ways, with the minimum of effort. Working on similar lines, the British United Provident Association (BUPA)

Medical Centre was set up in London in 1970. It evolved from and absorbed the original Insitute of Directors Executive Health Clinic, which opened in 1964. It showed that by using a relatively simple computer as a data gatherer, it was possible to deal, in a fairly sophisticated and rapid way, with more than 100 patients a day. Not only were individual records produced which could be used as a clinical data base, but research statistics were generated which facilitated the establishment of reference ranges and allowed longer-term follow-up of trends and changes in the various indices, as well as making it easier to monitor individuals. This Centre has continued to expand its range of activities. It currently sees about 30 000 men and women per year and runs executive health supervisory schemes for over 1000 companies. A far more flexible computer system is now installed. It has an automated personal and medical history question-naire which is a valuable method of establishing a data base and identifying problem areas. When these are linked with other measure-ments or clinical assessments, such as blood pressure, the biochemical profile, electrocardiographic function and X-rays, the doctor can be presented with a medical profile to which it is quick and easy to relate his physical findings and overall clinical assessment. Although a medical judgement is essential as the culmination of the exercise, the data-gathering and measurement can be done by machines and specially trained paramedical personnel.

Similar centres are rare in Europe, but flourish in Australia and Japan. In the USA such automated centres exist, but usually in a more specialized form related to hospitals, group practices or industrial complexes. This type of automation lends itself to mass screening in developing countries. By setting different standards it is relatively easy to identify people with florid disease rather than early deviations from normal. It is also easy to monitor the effects of treatment.

Automated well-person screening, properly directed and aimed at identifying vulnerability, holds out great hopes of establishing, for the first time, a personal medical data base or log book which could be of immense value to the study of the natural history of disease.

ATTITUDES TO SCREENING

Those of us who are involved in the practice of either occupational or community medicine, need no convincing as to the value or import-ance of preventive medicine, but primary care physicians and hospital based doctors still need to be convinced that it has much to offer. Their doubt to some degree influences their view of screening, presymp-tomatic diagnosis and the value of early treatment of certain common conditions. The climate is, however, changing with more general acceptance of the need to give higher priority to and perhaps to derive greater satisfaction from, preventive practices. These, of course, hinge round more effective methods of health education and a willingness to treat early.

Our experience is that the public demand for screening has consistently been ahead of the medical profession's willingness to supply it. As patients' expectation of the medical profession continues to increase, demand will certainly grow. As has been said, some group practices have successfully instigated simple screening programmes with encouragingly high yield rates; it is then easy to carry out the necessary treatment.

As a matter of contemporary interest and justification for this statement, one can add that in August 1979 two groups of workers, one large and one much smaller, negotiated the provision of private health insurance. The larger group started by wanting work site health checks, realized that a need for treatment might follow and went on to cover this by insurance. Because this group is in the construction industry the screening was to be provided from mobile units. This has now been abandoned. The screening will be done in static units.

This marks a breakthrough for this country in two respects: first, that a fringe benefit rather than wages was asked for, and secondly that this was for health screening. If the growth of this follows the American pattern, larger companies may well and quickly be forced to provide screening as part of their inhouse facilities. Obviously the occupational physician will be responsible for this. On this basis the demand for screening may grow very rapidly over the next five years.

Much of the criticism of screening hinges round the lack of statistics as to the benefit of early diagnosis. Statistical evaluation of screening is not an easy task and may prove impossible in comprehensive screening. Nevertheless, it has been subjected to more evaluation than any other medical technique. There is evidence that a true benefit accrues from screening, provided any necessary treatment takes place as a result; for example, for phenylketonuria (Raine, 1974); breast cancer (Shapiro, 1977); cervical cancer (Macgregor and Teper, 1978) and blood pressure (Veterans Administration, 1970). In one study on the value of screening (South East London Screening Group, 1977), although a considerable amount of pathology was detected, it seems that little positive action was taken, thus it was hardly surprising that there was little difference in the results between the screened and unscreened groups. The medical care system needs to be 'preventive orientated' for screening to be effective. Preventive medicine is to some doctors a boring option, and certainly little can be achieved in the standard six-minute GP consultation. The study emphasizes the need for more intervention as a result of screening.

Dales, Friedman and Collen (1979) in an eleven-year controlled trial of screening found a difference in mortality between the unscreened and screened groups; although lower in the latter, the difference did not reach significant levels; in contrast to the South East London study, there was no increase in the use of hospital services, and self-reported disability in the 45 – 54 year olds was reduced in the screened group. Finally, in some realistic economic calculations a substantial cost benefit was demonstrated in the screened group.

Collen (1978) has reviewed the work and development of the Kaiser centre with encouraging results. His centre being an integral part of an ongoing medical service makes intervention much easier. About five years ago, Kaiser set up a health education unit to supplement the more traditional treatment services.

Intervention trials in several countries (WHO, 1974) are under way but not yet completed, to test whether the risk of coronary heart disease in industrial populations can be reduced by persuading people to change their lifestyle (*see* Chapter 14) in order to reduce coronary risk factors (*see* pages 210–215). A most interesting outcome of these trials will be to find out how much lifestyles can be changed. But to be effective the medical system involved must also be willing to treat patients with early deviations from reference values as in blood pressure and lipids.

RECENT DEVELOPMENTS IN SCREENING

For various but largely economic reasons BUPA has developed screening in this country to provide health checks for managers and business executives – the so-called executive health examination (EHE). Although originally controversial it is now widely accepted, at least by management, as being of value, and there are few large concerns in this country that do not make EHE available to senior staff.

Some organizations, notably Marks and Spencer Ltd have provided a more specialized facility for screening their women staff at all levels for breast and pelvic disease. This facility is much appreciated, reveals enough pathology to justify itself and is spreading to other organizations, some of which use the BUPA facility, often including the wives of their staff.

Screening is very much the 'art of the possible' and no protagonist of it would claim to provide a facility which hopes to diagnose or pick up early signs of almost every disease. To look, for instance, for stomach or bowel cancer, in the absence of symptoms, would be prohibitive in terms of time and money, as well as being uncomfortable; (procedures for the former are routine in Japan where stomach cancer is relatively common).

The screening package must be designed to focus on conditions known to be common in the group. In BUPA this is ischaemic heart disease (IHD) in men and breast cancer in women. But a properly run multiphasic screening, as distinct from a single disease screening programme, does provide an opportunity to look at and measure some aspects of *the whole person* who may not have any overt complaints or appreciable anxieties. In our view, there are two distinct aspects of the screening process. The first is the measuring of various indices and the establishment of a medical data base. The second is clinical consultation and lifestyle assessment, followed if necessary by a counselling session or appropriate referral. Adequate use must be made of the data generated, and the process repeated at an appropriate interval. The first examination charts a baseline against which future change can be measured.

Screening is increasingly providing a way into health care for 'the worried well'. They come for screening because of anxiety or dissatisfaction, or because they want reassurance. This provides the screener with an additional challenge to be generally and clinically alert, which is essential for competent well-person screening. Recent experience in BUPA has highlighted some of the problems or conditions in which screening can be valuable.

CORONARY RISK FACTORS

Although the benefits of modifying or improving coronary risk factors, particularly the lipids, is still a controversial subject, there is general agreement, based largely on the results of the Framingham Study (1970) that groups who have certain physical and behavioural characteristics have a measurably increased risk of suffering from ischaemic heart disease (IHD). The risk factors are now well known, but merit brief mention.

Age, sex and race   IHD is the commonest cause of death in the 45–65 age group – far more common in men and almost unknown in underdeveloped countries. In most western countries the incidence among men is levelling out, but rising alarmingly in women, perhaps because of increased smoking and prolonged use of the contraceptive pill.

Cigarette smoking   Heavy smokers are three times more prone to IHD than non-smokers.

Blood pressure and blood cholesterol levels   The incidence of IHD increases with an increase in these factors, gently at first and more steeply later on. The further separation of heavy and light lipoprotein fractions gives added information on risk. Minor electrocardiographic changes of a non-specific nature are associated with an increased risk of IHD.

Personality, physical activity, 'stress' Various factors associated with personality, stress at home and work, and the sedentary way of life have been implicated in the aetiology of IHD. Because of the difficulty in measuring these factors research has not produced clearcut conclusions

Genetic factors Some genetic factor is undoubtedly present and this may modify other factors. Some people think this is the most important factor and are sceptical of our ability to prevent IHD because 'we cannot choose our parents'.

Other factors There is an association between the oral contraceptive and IHD (Shapiro *et al.*, 1979) and some evidence that the hardness of the drinking water supply is related (Stitt *et al.*, 1973).

(Thus many factors need to be measured by special techniques in order to determine whether the patient falls into a high, medium or low risk group and what preventive action should be taken.)

This is not the place to discuss the details of modifying or assessing these risk factors, but it is obvious that many of them are behavioural and relate directly to lifestyle at work and at home. Drugs can be used to reduce blood pressure and, to some degree, lipids, but these have their dangers (Committee of Principle Investigators, 1978). The main action required is behavioural. The necessity to motivate behavioural changes puts the screener into close liaison with health educators. Because of the close-knit group at work and the fact that the climate within a company can and must influence the workers, the occupational physician is in a key position, backed by screening results, to influence this climate, particularly so with support from management. Companies can discourage smoking, take an interest in canteen diets, encourage physical fitness and discourage excessive alcohol consumption.

One company has for several years discouraged smoking in its head office until after 5 pm, so much so that notices in the reception area more or less say 'abandon your tobacco here'. Shops increasingly forbid smoking, and staff away from the counters are also discouraged from smoking in their offices.

A military unit with a hard-drinking commanding officer will consume more alcohol per head than one with a more abstemious commander. For the same reason some organizations drink more than others. With increasing concern about alcoholism and liver damage due to alcohol, companies need to develop a more positive policy about the amount drunk as part of the organization's 'pesonality', and to take positive steps to identify and deal with those who drink too much. The Alcohol Education Centre can give detailed information about what is being done and what help is available.

The BUPA Medical Centre has always had a policy, as far as is ethically possible, of telling an individual about his state of health and what steps he should take to alter his lifestyle. A health promotion centre was opened in 1979 to give advice and information on fitness, smoking, alcohol consumption, diet and mental health. Advice on how to manage stress and relaxation techniques are also available. Discussion groups and personal counselling can be set up to meet problem areas thrown up by a screening programme. This type of activity is likely to grow and can usefully be encouraged by an occupational health service.

It is now relatively simple to compile a coronary risk rating score on an individual, in relation to a range of risk factors. In our experience, to confront an individual and his wife with such a 'score' can be highly motivational in promoting change and improvement. Based on five factors, a score for each is given and these are added up to calculate a risk score. *Table 10.1* shows the probability of a man developing IHD according to his cumulative risk score. As can be seen, there is a ninefold difference between the low and the high scorer. These figures are based on eight years of Framingham experience. In the population examined at the BUPA Medical Centre about 20 per cent of men have a higher than average risk of developing IHD.

Lipid measurement and control

Elevation of serum cholesterol has long since been established as a major coronary risk factor, although there have been frequent, and until now unexplained, examples of individuals with both high and low cholesterol levels developing IHD at an early age. Over the past decade or so more has been learnt about the relatively rare genetically determined group of hyperlipidaemias to the extent of being able to differentiate five groups or types of this condition.

Lipid typing into density fractions has become a routine procedure and it is now relatively simple and cheap to break down the traditional cholesterol package into high and low density fractions. Using these techniques Miller and Miller (1975), Gordon *et al.* (1977) and

Table 10.1 RISK OF DEVELOPING ISCHAEMIC HEART DISEASE (IHD: SEE TEXT FOR EXPLANATION) ACCORDING TO SIMPLE RISK RATING BASED ON CHOLESTROL LEVEL, PHYSICAL ACTIVITY, CIGARETTE SMOKING, DIASTOLIC BLOOD PRESSURE (BP) AND RELATIVE WEIGHT

Simple risk rating	*IHD rate/1000*
0	18
1	18
2	32
3	34
4	48
5	51
6	55
7	95
8	115
9	168

Scoring system

Diastolic BP		*Physical activity* (special questionnaire)	
<90 mmHg	score 0	vigorous exercise	score 0
90–109 mmHg	score 1	moderate exercise	score 1
>110 mmHg	score 2	quite inactive	score 2

Relative weight (%)		*Cigarette smoking*	
<110 of predicted	score 0	non or ex-smoker	score 0
110–119 of predicted	score 1	<20 per day	score 1
>120 of predicted	score 2	>20 per day	score 2

Cholesterol	
<260 mg/100ml (<6.7 mmol/ℓ)	score 0
260–300 mg/100ml (6.7–7.7 mmol/ℓ)	score 1
>300 mg/100ml (>7.7 mmol/ℓ)	score 2

Williams, Robinson and Bailey (1979) have published extensive data relating high and low density lipoprotein levels to IHD and other risk factors.

High density lipoproteins (HDL) appear to be protective and physiologically related to normal fat utilization. It is encouraging in health education terms, that exercise and moderation of alcohol consumption increase HDL, and cigarette smoking lowers it. We believe that, if carefully interpreted, the HDL to total cholesterol ratio is an essential measurement in a screening profile. The Framingham studies have produced a risk rating for men and women based on the

Table 10.2 VARIOUS LEVELS OF HIGH DENSITY LIPO-
PROTEINS/TOTAL CHOLESTEROL RATIO RELATED TO
CORONARY RISK.

Coronary risk based on Framingham Study	HDL/Total cholesterol (%)	
	Men	Women
Half average	>29	>31
Average	20	23
Twice × average	10	14
>three × average	< 5	< 9

ratio of HDL to total cholesterol at the time of their examination and
calculated from the incidence of IHD in that population (*Table 10.2*).
The ratio has also been shown to correlate well with our simple risk
rating as shown in *Figure 10.1* (Williams, Robinson and Bailey, 1979).

Smoking and carbon monoxide

One of the toxic products of inhaled cigarette smoke is carbon
monoxide (CO). As haemoglobin has a greater affinity for CO than
oxygen, a proportion of the circulatory haemoglobin is, in a smoker, in
the form of carboxyhaemoglobin (COHb). Astrup (1972) suggests that
the presence of COHb in a smoker's blood is one of the factors in
atherogenesis. The level of COHb attained by a smoker is determined
by the type of cigarette, depth of inhalation, state of the lungs and the
amount of physical activity being taken. The BUPA Centre is
conducting prospective studies of the link between smoking and IHD
and is following up a group of more than 12 000 men whose smoking
habits and COHb levels have been carefully studied. Early findings
indicate that the general switchover from plain to filter cigarettes that
occurred in the late 1950s and early 1960s may in fact increase
individual COHb levels (Wald *et al.*, 1975). It seems that the weaker
the cigarette in terms of tar and nicotine yield, the deeper the smoker
inhales to obtain whatever effect it is that makes him smoke (Wald,
personal communication).

   Carboxyhaemoglobin is simple to estimate and has been used to
check on the truth of people's smoking history, and as a motivator in
the health education field.

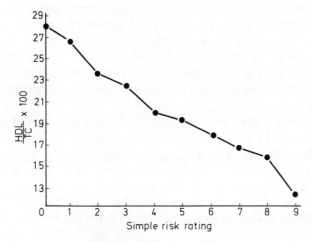

Figure 10.1 High density lipoproteins/total cholesterol ratio in relation to simple risk rating (see Table 10.1)

SCREENING WOMEN FOR BREAST AND PELVIC DISEASE

Women have formed 20 per cent of the people attending BUPA screening centres. Some are wives, sent by their husbands (some send their husbands), others come of their own volition, and others are working women who attend as part of company schemes or on their own.

There is thus a significant demand for screening by women and interestingly it is by no means confined to the breast and pelvic regime. Women play a key role in family management, and as is generally recognized, have major anxieties and problems, particularly in middle age. It is relatively difficult for many of them to get advice from their general practitioner and so they come to screening centres as 'worried well'.

BUPA aimed its first series of Health Promotion Seminars at various aspects of health awareness in women. There is a significant demand for sensible advice on premenstrual tension, menopause control and minor gynaecology. A screening service which looks at the whole woman can be very much a 'way in' to care and could be operated in occupational health services' departments. Breast cancer is a very emotive anxiety for women, responsible for 13 000 deaths a year in England and Wales. But it is also, at least until recently, an area of medical defect in that the results of various forms of treatment have shown little improvement.

Shapiro (1977), using a healthy insurance population, showed that mortality could be reduced and survival increased by early diagnosis. Much effort has since been devoted in this country to developing breast screening facilities and techniques and promoting regular self-examination. The Department of Health and Social Security is now sponsoring various trials, and more figures for confirmed cancer findings through screening are available.

By using clinical tests and specialized X-ray techniques, it is possible to detect small lumps in the breast, some of which may not be palpable. Needle biopsy (a simple procedure needing some skill and sterile facilities) will often confirm a diagnosis. When the lesion is not palpable, but is visible on the X-ray, biopsy under X-ray control is necessary to establish the diagnosis. This occurs in one quarter of BUPA patients with positive findings. The prognosis after treatment seems excellent.

Results suggest that there is merit in early diagnosis and treatment and that the cost is not prohibitive (Bailey *et al.*, 1976). Cost – benefit appraisal is difficult to apply realistically to preventive medicine because long-term surveillance and follow-up will always be more expensive than short-term treatment at the acute stage followed by early death.

If women can be persuaded to come for breast screening, as they have for cervical cytology, and to undertake regular self-examination, their attitude to the possiblity of having cancer will be changed for the better. There is an urgent need to allay the fear of cancer, to get people to realize that many cancers are successfully treated and that many of those that are fatal occur in old age, when death is to be expected.

WELL-PERSON SCREENING AND OCCUPATIONAL HEALTH

Occupational medicine is a speciality which aims at keeping people fit and monitoring health, rather than treating established disease. To this extent it is relevant for all those involved to understand about the scope and possibilities of screening in general.

Occupational health practice provides one of the few opportunities to look at men and women in relation to their environment, rather than to deal with the symptoms they choose to present. Over the next few years, the demand for screening in industry will almost certainly grow so that occupational physicians and others involved will benefit from understanding its aims and problems. General screening is a logical extension of what is now done to monitor vulnerable groups within industry, either for specific toxic exposures or for public health and safety as in the food industry.

The range is wide, from complicated biochemistry, detailed history questionnaires and specialized X-ray mammography, to the measurement of height, weight, blood pressure and a urine test. But whatever is done, two things seem certain. The first is that there will be a significant yield of treatable unsuspected disease. The second is that the activity will be appreciated and provide a lead into a situation where health promotion and lifestyle advice will be welcomed. Well-person screening is the starting point for a more positive and productive approach to health education and lifestyle alteration.

Executive health supervision

When we started in this field in 1958, Executive Health Examinations (EHEs) were a controversial topic, partly because they raised all the issues about screening – who advises what and to whose patient. They also touched on the sensitive spot of what right has management to confidential medical information about individuals. This largely reflects the relationships and the climate within the organization, and the way in which they trust or distrust each other. Most of this is now ancient history, and the great majority of firms encourage their more senior staff to have periodic medical supervision.

There are several problem areas in this type of well-person screening:

1. The extent and periodicity of the examination.
2. Who should do it?
3. What use should be made of the findings of the examination?
4. Several less important problems arise, such as how far down the staff hierarchy the examination should be offered, and whether it should be voluntary or obligatory.

The extent of the examination

The critical investigations probably are the biochemical profile, with lipid and liver function estimations, blood pressure, height, weight, lung function and cardiograph. The others are desirable extras. The chest X-ray sanctified by tradition is expected by patients but produces a low yield. Abdominal X-rays often reveal unsuspected stones and other occasional abnormalities.

There are two parts to an EHE, both critically important and often neglected. These are the screening profile and the clinical personality assessment. A clinical examination of the simple life insurance type is not adequate. If the diseases one gets are related to the life that is led –

and 'life' consists of work, home and leisure – it is essential to look at the individual as a totality and to assess his or her performance, personality and problems against this background. In psychosomatic terms the 'why' of disease is just as important as the 'what'.

A consultation along these lines can seldom be done in less than an hour, and may well take longer. It should establish a good rapport with the patient and lead to counselling in relation to the screening findings, clinical problems and lifestyle. If this approach is accepted, it is desirable that the same doctor should see all the people from a particular working group in order to give some insight into their problems and relationships. There should also be continuity over the years. This makes the assessment of improvement and deterioration easier and more accurate. An ongoing rapport and understanding is both essential and satisfying. Staff doctors thus build up a relationship with the company as well as the individual and are available to advise them.

The frequency of the examination

Periodicity is a difficult problem on which to be precise. If an EHE is regarded as a health maintenance procedure, it should be done regularly and as has been said as far as possible by the same doctor. Annual examinations may be unnecessary for younger staff and those under less stress. Senior people, those who travel a lot and the over fifties should be seen yearly, and the rest every two years. Two years is still quite a long time in the natural history of some diseases – perhaps too long in, say, hypertension. Thus for budget purposes and in terms of actually working through the examinations on any large group, periodicity works out at about every eighteen months.

Who does the examination?

What is done and how the examination is conducted is more important than who does it. All the investigations should be done at one centre and properly coordinated. Moreover, busy executives dislike travelling around different centres to get their consultation and investigations. Very few occupational health services have all the facilities required but some have most of them or have a good working relationship with a local hospital. Given adequate facilities, and here the specialized centre has enormous advantages, the four main options are:

1. The company's occupational physician.
2. A nominated specialist.

3. A doctor of the individual's choice.
4. An independent specialized unit such as the BUPA Centre.

In our view, option 3 is a non-starter and a waste of money. With say 20 executives going to 20 different doctors, there is no uniformity and a wide range of standards. Option 2 can and does work well provided the specialist is interested and experienced in company life. The danger is that he is clinically rather than lifestyle orientated. Being usually a traditional clinician he is likely to be unwilling to communicate with either patient or company, and may be disinterested in the early treatment of presymptomatic disease. Nevertheless, much of this work is done by consultants and the companies seem happy.

The company's own doctor would seem to be the obvious person and often does this well. He has one advantage and two disadvantages. The advantage is that being the company doctor he knows the scene, the people and the problems. But the obverse of this is that being part of the scene managerially, he has to relate administratively to the people. Thus to get involved in their personal problems may not be sensible, particularly with the senior staff.

Another possible disadvantage is that he may be too busy to give the requisite time and may not have the full range of facilities. Time being a matter of allocating priority, this implies that some occupational physicians do not regard these examinations as either important or in the main stream of what they see as their expertise, which is the assessment of fitness for work and the control of occupational hazards.

A unit such as the BUPA Centre (option 4) has specialized expertise, appropriate equipment and complete independence to express a view about the way in which management should treat people. The Centre nominates one or more doctors to be responsible for a group, makes arrangements for holidays and emergencies and manages to achieve a high degree of continuity. From experience and statistical back-up, such centres can express an informed view and call on a range of nominated consultants.

Two compromises are possible with this system. First, that a screening centre carries out the technical profile and sends the results to the company or nominated part-time doctor. Second, that the company doctor sees his people on the neutral territory of the screening centre using their records and working to their standards.

Reporting the results

Problems of confidentiality may arise concerning an individual patient, employee or staff member. Doctors working in industry should be prepared to give professional advice to management which is often based on 'confidential' information about situations or individuals. The doctor who 'won't tell anything' is not serving the best interests of the patient or the organization.

The one person who has the right to know the findings of an EHE is the patient. He or she must then decide who else should be told. It is in practice essential to determine this in advance, as part of the agreement under which the examination is carried out. Some companies are happy not to know the results, but merely to pay for the screening. This we feel is shortsighted and prevents the company and the patient from reaping the full benefits of the examinations. We recommend that a report be sent to the company, but only with the full knowledge and agreement of the individual. If there is a difficult situation involving promotional prospects or the immediate future of the patient, this must be discussed. It is seldom difficult to persuade the individual that the recommendation is for his or her long-term good.

There are theoretical difficulties but they seldom appear in practice, except in unhappy and competitive companies where there is insecurity and bad relationships. Doctors involved in these conditions are well advised to have a direct relationship only with the individual. In practice such companies are seldom willing to pay for the examinations.

A compromise often used, and one which works well, is for the screening centre to report in detail to the company doctor, who decides what use is made of the information and what advice is given and to whom. If he is a competent occupational physician this works well. Where there is no such person, we persuade the individual to take action with the company.

Who should be screened?

Because industry's resources are limited and because some categories of staff are less replaceable and more responsible than others, the EHE tends to be confined to senior management. The probability is that this will extend downwards over the next few years. In our view, it is important to include the 'high flyers', young though they may be, because a great deal is being invested in them and they are under stress. Some companies include senior wives and senior women staff. On the whole this is a much appreciated facility.

It is unwise to make screening obligatory unless it is written into the original conditions of service. The main purpose of the activity is defeated if it is perceived with suspicion. It must be seen as being for the mutual benefit of both the individual and the organization, as they share responsibility to maintain health and effectiveness.

In America EHEs are often regarded as a condition of service and made obligatory. This is seldom the practice in Britain, except in American companies, which often insist on a report going to the corporation medical adviser. When this obligation is written in to their contract of employment, there is no difficulty.

When a firm decides to start a scheme for existing staff, whose conditions of service have already been agreed, it is wise to make the examination voluntary and unwise to make it obligatory. In the great majority of cases the facility is appreciated and there is no difficulty in persuading managers to attend.

Pre-employment medical examinations, discussed in Chapter 9, are particularly important in the choice of new executives. These examinations should also be made prior to promotion. Time and a great deal of money is invested in executives, and given a choice, which may not always be the case with scarce technical experience, it is better to take on or promote a fit man than one with, for instance, a high coronary risk rating. If it is essential to employ someone who has serious medical disability, the fact that both sides know that there is a problem and what it is makes it easier to deal with should there be a breakdown in health.

REFERENCES

Astrup, P. (1972) 'Some physiological and pathological effects of moderate carbon monoxide exposure.' *British Medical Journal*, **4**, 447–452

Bailey, A., Davey, J., Pentney, H., Tucker, A. and Wright, H.B. (1976) 'Screening for breast cancer – a report of 11654 examinations.' *Clinical Oncology*, **2**, 317–322

Collen, M.F. (1978) *Multiphasic Health Testing Services*. New York: Wiley

Committee of Principle Investigators (1978) 'A co-operative trial in the prevention of ischaemic heart disease using clofibrate.' *British Heart Journal*, **40**, 1069–1118

Dales, L.G., Friedman, G.D. and Collen, M.F. (1979) 'Evaluating periodic multiphasic health check-ups: A controlled trial.' *Journal of Chronic Diseases*, **32**, 385–404

Framingham Study, The, ed. W.B. Kannel and T. Gordon. (1970) US Government Printing Office.

Garfield, S. (1978) In *Multiphasic Health Testing Services*, ed. M.F. Collen, pp, 454 *et seq*. New York: Wiley

Gordon, T., Castelli, W.P., Hjortland, M.C., Kannel, W.B. and Dawber, T.R. (1977) 'HDL as a protective factor against coronary heart disease.' *American Journal of Medicine*, **62**, 707–714

MacGregor, J.E. and Teper, S. (1978) 'Mortality from carcinoma of the cervix uteri in Britain'. *Lancet*, ii, 774–776

Miller, G.J. and Miller, N.E. (1975) 'Plasma HDL concentration and development of IHD.' *Lancet*, i, 16–19

Raine, D.N. (1974) 'Inherited metabolic disease.' *Lancet*, ii, 996–998

Shapiro, S. (1977) 'Evidence on screening for breast cancer from a randomised trial.' *Cancer*, **39** (6), 2772–2782

Shapiro, S., Sloane, D., Rosenburg, L., Kaufman, D.W., Stolley, P.D. and Miettinen, O.S. (1979) 'Oral-contraceptive use in relation to myocardial infarction.' *Lancet*, i, 743–747

South East London Screening Group (1977) 'A controlled trial of multiphasic screening in midde-age.' *International Journal of Epidemiology*, **6** (4), 357–363

Stitt, F.W., Clayton, D.G., Crawford, M.D. and Morris, J.N. (1973) 'Clinical and biochemical indicators of cardiovascular disease among men living in hard and soft water areas.' *Lancet*, i, 122–126

Veterans Administration Co-Operative Study Group (1970) 'Effects of treatment on morbidity in hypertension.' *Journal of the American Medical Association,* **213** (7), 1143 *et seq.*

Wald, N., Howard, S., Smith, P.G., Bailey, A. (1975) 'Use of carboxyhaemoglobin levels to predict the development of diseases associated with cigarette smoking.' *Thorax,* **30** (2), 133–140

Williams, P., Robinson, D. and Bailey, A. (1979) 'HDL and coronary risk factors in normal men.' *Lancet* i, 72–75

World Health Organization Collaborative Group (1974) 'An international controlled trial in the multifactorial prevention of coronary heart disease.' *International Journal of Epidemiology,* **3,** 219–224

11

Screening Organ Systems I—Cardiovascular System, Blood and Blood-forming Organs, Lung, Liver, Kidney and Bladder, and Central and Peripheral Nervous Systems

Alan Bailey

In this chapter and the next the screening of organ systems is reviewed with particular emphasis on what can be done by an occupational health service concerned with prevention and the early detection of disease. They describe the more commonly used 'tests' that are available for the following:

1. Detecting early disease.
2. Assessing systems which may for some reason be more vulnerable than average.
3. Detecting occupationally induced disease at an early stage when it is hoped that intervention will lead to a reduction in mortality and morbidity.

Thus the screening of organ systems is considered in three different ways: for well-person screening, for the pre-employment examination and for detecting effects of occupational exposures.

This chapter deals with the cardiovascular system, blood and blood-forming organs, the lung, liver, kidney and bladder, and central and peripheral nervous systems. Audiometry is discussed in Chapter 12. Some examples are given of occupations where screening may be relevant in the identification and control of occupational hazards. Such occupations are more fully covered in Chapters 5 and 6.

THE TECHNIQUE OF SCREENING

In looking for early disease and high risk subjects the combination of a battery of tests gives more information about individual risk than does a single test with its limitations on sensitivity and specificity. The screening for each system should start with a list of key questions and the testing may be modified in the light of the answers. Common to almost all systems, and necessary to interpret results, are the patient's age, sex, smoking and alcohol habits, type of occupation, past history and present drug therapy. For each system there is a small set of questions which must be answered, for example cough and sputum production in lung disease, and physical signs, such as finger-clubbing or basal crepitations. Many of the procedures described here can be carried out by suitably trained paramedical staff and historical information can be gathered by a number of questionnaire techniques.

Certain of the physiological measurements and biochemical tests act as markers for disease, for example, the blood cholesterol level in coronary heart disease. These markers may or may not have a direct aetiological role to play in the natural history of the disease. Nor is it clear that modification or removal of the marker will necessarily alter the course of the disease.

CARDIOVASCULAR DISEASE

Some degree of screening for cardiovascular disease is included in most, if not all, pre-employment medical examinations. Its extent will depend on the subject's occupation and whether or not the occupational health service is pursuing a well-person screening programme (*see* Chapter 10).

Certain occupations require complete regular cardiovascular screening because a sudden loss of consciousness may produce a major disaster. In the United Kingdom, for example, an airline pilot after a certain age requires six-monthly examinations, and a London taxi driver's licence may not be renewed if he gives a history of coronary thrombosis (*see* Chapter 9). Regular screening should also be considered for certain toxic exposures (*see Table 11.1*).

A complete examination of the cardiovascular system requires a listing of the major risk factors, namely smoking and drinking habits, positive family history, oral contraceptive pill usage, physical activity index and presence of angina. Examination must include blood pressure, assessment of heart rate, rhythm and size and auscultation to exclude valvular disease (*see Table 11.1*). Chest X-ray gives an objective measurement of heart size, electrocardiograph (12 lead) will give evidence of pre-ischaemia or established ischaemic heart disease (IHD), and measurement of lipids, particularly total cholesterol and high density lipoprotein (HDL) will allow a predictive index to be calculated for each subject (*see* Chapter 10).

For younger people, where there are no occupational indications for the examination, chest X-ray and electrocardiography have a low yield of useful indicators for IHD and can be omitted. It is argued that there is some theoretical use in having chest X-rays and electrocardiographs (ECGs) in normal people as a baseline for future comparisons, but the practical advantages have not so far been convincingly demonstrated. The ECG is of low sensitivity for incipient coronary disease, changes usually occurring after infarction. Its sensitivity for coronary disease is increased if carried out before and after an exercise test (Sheffield and Roitmann, 1976). However, about 5 per cent of symptomless patients at the British United Provident Association (BUPA) Medical Centre

Table 11.1 CARDIOVASCULAR DISEASE SCREENING OF INDUSTRIAL POPULATIONS

	*Routine**	*Pre-employment*
Smoking history	+ +	+ +
Blood pressure	+ +	+ +
Physical examination	+ +	+ +
Lipids	+ +	+ +
Electrocardiograph	±	+ (sometimes exercise ECG)
Chest X-ray	±	+

+ + = essential; + = of value; ± = of doubtful value
* should be done for toxic exposures such as carbon disulfide (NIOSH, 1979) and nitrates for making dynamite

have non-specific ST and T wave changes reported in two or more leads, and this group have a sixfold increase in coronary incidents over a five-year period compared with people who have normal ECGs.

Although it is possible to identify groups of people with varying risks for ischaemic heart disease (*see* Chapter 10, page 210), it is still impossible to predict for an individual with any certainty, emphasizing the multifactorial nature of the disease. Even an apparently completely healthy person still has some risk of IHD. For this reason those whose occupation is such that they might endanger other people's lives were they to have a disabling or fatal attack must work where possible with others who can take over in such an emergency. This is the case for commercial airline pilots; partly for this reason IHD is not a common cause of airline disasters.

RESPIRATORY SYSTEM

The two measurements of lung function which are most useful in occupational health practice are ventilatory capacity and gas transfer.

Ventilatory capacity

Ventilatory capacity of the lungs may be measured by either a peak flow meter or a spirometer. Peak flow readings are not equivalent to spirometer readings, nor are they as valuable. If a single test of lung function is to be used, most authorities would advocate the forced vital capacity (FVC) and forced expiratory volume in one second (FEV_1).

All such tests should be accompanied by information on the subject's age, height, weight, sex, ethnic group and smoking habit. Prediction formulae incorporating these variables are available and the results are frequently expressed as a percentage of the predicted value, or as a residual value which is the difference between the observed and predicted values (*see* Chapter 15). Total lung capacity at full inspiration is governed to a large extent by body size, which in normal subjects correlates well with height. Chest deformities, inspiratory muscle weakness and increased lung stiffness (loss of compliance) will all reduce this volume. During maximal expiration, some air is left in the lungs and this is known as the residual volume. The physiological event which limits expiration is closure of the intrathoracic airways. Their patency depends not only on the elastic tissues of the lungs holding them open, but also on the strength and integrity of their walls and the presence or absence of fluid or muscle spasm which may narrow the airways. In asthma and bronchitis, for example, the airways tend to close prematurely because of disease in the walls of the airways.

In normal subjects 75 per cent or more of the vital capacity (VC) can be expired in one second and the remaining 25 per cent takes a further two to three seconds. Diffuse airways obstruction, as in asthma, bronchitis or emphysema, causes irregular and premature airways closing during expiration. Therefore, not only do these patients frequently have a reduced vital capacity but also they may be unable to expire a large portion of this air in the first second. FEV_1 is diminished in all lung diseases which reduce the VC, but the FEV_1/FVC ratio is reduced only in airways obstruction. This ratio is useful for measuring deterioration over a period of time. It is a valuable baseline to have in the pre-employment examination and is an essential test of persons exposed to dusts or other aerosols which may cause pulmonary disease.

When the subject has restrictive lung disease, as for example in diffuse pulmonary fibrosis from asbestos exposure, the VC and FEV_1 are reduced, but the ratio is normal. Variable airways obstruction is a characteristic feature of asthma and can be assessed by measuring the FEV_1/FVC ratio before and after the administration of bronchodilator drugs. Measurement of FEV_1 before and after work is useful in assessing the acute effects of dust or other exposures. During the last few years, increased attention has been paid to 'small airways' disease which seems to be an early sign of impaired ventilatory capacity. It can be estimated from a spirometer tracing of the FEV_1 by measuring the rate of expiratory flow during the middle of expiration. Thus the forced midexpiratory flow rate (FMF) may reveal early airways disease. This measure, sometimes referred to as the FEF 25–75, is of value in the investigation and detection of occupational asthma.

Maximum expiratory flow volume curves are being increasingly used to detect early obstructive disease. These curves show more inter-individual variation than the volume measurements (FVC, FEV_1) but in certain circumstances may be better than the FEV_1/FVC in discriminating between groups of normal and abnormal subjects as well as detecting abnormal individuals (Bouhuys, Schoenberg and Beck, 1977).

For the three types of screening in occupational health practice – well-person, pre-employment examinations and detecting effects of occupational exposures – FEV_1 and FVC provide the simplest and most useful measurements of ventilatory capacity (*see Table 11.2*).

In the United Kingdom the McDermott and the Vitalograph dry spirometers have been recommended for standardized tests of FEV_1 and FVC. Standard wet spirometers (e.g. the Stead-Wells spirometer) are also acceptable but are less suitable for field surveys. Many electronic devices with direct read-out of FVC and FEV_1 are sold commercially; some of these are inaccurate. If more than one instrument is used in a survey they should be compared to ensure that they give the same results. Large (2-litre) syringes offer a simple means to check volume calibrations. A calibration device which delivers a mechanically determined forced expiratory blow with a fixed volume and flow volume pattern offers a more sophisticated method of standardizing spirometric and flow volume curve tests. For large-scale surveys and routine medical surveillance programmes, data collection with a computer helps to avoid errors in recorded data and eases the labour of tabulations and data analysis.

To perform the measurement of FEV_1 and FVC, the subject should be seated upright in front of the measuring instrument. Normally, only simple instructions are needed. The subject should make a maximum inspiration, taking in as much air as possible before blowing into the instrument. He should then be instructed to blow out as hard and as deeply as possible. If necessary the observer should demonstrate using an unattached mouthpiece. The observer should watch for air leakage between the lips and the mouthpiece, and explain that the mouthpiece should be inserted between the lips. Spuriously high values may be recorded if the subject purses the lips, like a trumpet player, against the mouthpiece.

The subject needs to be encouraged so that the forced expiration is continued to its very end. Otherwise, low values of FVC may be recorded. However, persons with severe ventilatory limitations may require considerable time for the delivery of a full vital capacity. It is recommended that in such cases the forced expiration should not be continued beyond eight seconds. In most examinations a single test consists of five separate forced expirations. The final results from the data obtained from these five blows may be expressed as the average value of the last three blows, or the average of the two highest out of the five.

The person who supervises the test may be a physician, nurse or technician, and needs to be trained to make the subject blow in the prescribed manner.

FEV_1 and FVC values should be corrected to body temperature, pressure (prevailing atmospheric), and saturation (water vapour) (BTPS). Standard tables can be used for tests with wet spirometers. With dry spirometers and electronic devices, one usually needs calibration against wet spirometers.

Gas transfer

The diffusion of gases across the alveolar capillary membrane is assessed by measuring the gas transfer factor ($T\ell$) using carbon monoxide (CO). The quantity of pure CO which crosses the alveolar capillary membrane in one minute when the difference between the partial pressure of CO in the lung and blood is 1 mmHg tension, is known as the $T\ell_{CO}$. Three measurements are required for the estimation of $T\ell_{CO}$; the volume of CO taken up by the pulmonary capillary blood, the partial pressure of CO in the alveolar air, and the mean partial pressure of CO in the pulmonary capillary blood. The procedure involves breathing a known concentration of CO and air, or oxygen (O_2), usually with helium, in order to assess the dilution effect of the residual volume.

The $T\ell_{CO}$ is not solely governed by the thickness of the alveolar capillary membrane. Many other factors are involved, of which ventilation/perfusion inequalities and the haemoglobin concentration in the blood are among the more important. Nevertheless, the $T\ell_{CO}$

Table 11.2 RESPIRATORY DISEASE SCREENING PROCEDURES

	Pre-employment	*Routine*	*High risk* occupations*
Clinical examination and questionnaire	+	+	+ +
FEV₁/FVC	+	+	+ +
Chest X-ray	For certain occupations and immigrant workers	±	+ +

+ + = essential; + = of value; ± = of doubtful value; *see* Chapter 6

remains the most useful, convenient and most widely quoted measure of diffusing capacity.

Gas transfer is not measured in routine screening. It is used to detect diffusion defects in workers exposed to asbestos and other dusts which cause fibrotic disease of the lung. Screening procedures for the respiratory system are summarized in *Table 11.2*.

The chest X-ray

The chest X-ray has a revered status in health screening, mostly because of the success of mass miniature radiology (MMR) for the detection of tuberculosis. Its routine use in young symptomless people in Britain today, not exposed to occupational risks, is of doubtful value. The exception to this is in immigrant workers from countries where tuberculosis is either still common or is so rare that host resistance may be low (e.g. Southern Ireland). The chest X-ray used as a screening tool for lung cancer in older people is questionable. There is no evidence to show that patients whose lesions have been picked up by routine X-ray have a better prognosis than others. Chest X-ray must be available for investigating symptoms. It is necessary for screening people in dusty occupations in which there is a risk of pneumoconiosis. Although mass miniature radiography has been used to advantage in many countries to identify pulmonary tuberculosis, the pneumoconioses require X-ray films of greater definition to establish accurate classification of the stage of the disease (International Labour Office, 1971).

BIOCHEMICAL AND HAEMATOLOGICAL DATA

The reference range

Increasingly organ systems are being tested by studying a pattern of biochemical and haematological estimations. For this procedure to be meaningful, it is necessary to know the frequency distribution of a particular test and what is 'normal'. As normal has a statistical meaning which is nothing to do with being healthy, it is better to abandon the term and describe for each biochemical test a 'reference range'.

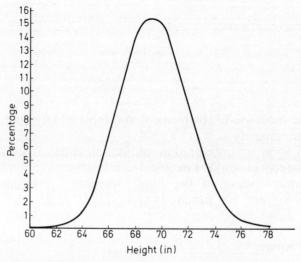

Figure 11.1 Frequency distribution of height in a population of males to demonstrate a Gaussian distribution

In statistical terms, the word normal is used to describe a variable that is distributed in a Gaussian manner. *Figure 11.1* shows the frequency distribution of height which approximates closely to a Gaussian distribution. To indicate the spread of heights in a population the reference range might be defined as the mean ± 1.96 standard deviation, which covers 95 per cent of the population (2.5 per cent at either end falling outside this range).

Most biochemical variables are not distributed in a Gaussian way. *Figure 11.2* gives an example of the distribution of gamma-glutamyltransferase (gamma-GT), which is not symmetrical but skewed to the right. A log transformation converts this to a more Gaussian appearance and the original distribution is often referred to as a 'log normal'. For a log normal distribution the application of a reference range of mean ± 1.96 standard deviation is not satisfactory

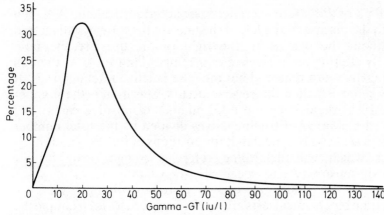

Figure 11.2 Frequency distribution of gamma - GT in males to illustrate a log-normal distribution

because the distribution is not symmetrical. In these types of distribution it is often more useful to divide into percentiles and set reference ranges accordingly.

What reference range does one apply?

The distribution of a test in a population in which some people have the disease that the test detects may be as in *Figure 11.3*. Here, normal has been divided from the abnormal but there is an area of overlap. If

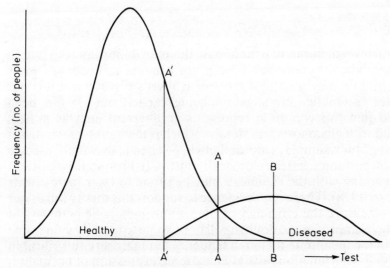

Figure 11.3 Determining the reference range

the upper limit of the reference range is set at the vertical line AA then people with the disease to the left of the line are the false negatives and people without the disease to the right of the line are the false positives. By altering the reference range upper limit to A'A' then all the false negatives are removed but the false positive rate increases. If the limit is set at BB then the reverse occurs. Sensitivity (the number of true positives divided by the total number of positive cases) and specificity (the number of true negatives divided by the total number of negative cases) can be calculated for positions of line AA and this is one factor which will determine reference range. (For further discussion of sensitivity and specificity *see* Chapter 15.)

Furthermore, the results of prospective studies will alter reference ranges. Certain biochemical tests are markers of disease, for example the higher the blood cholesterol level (or lower in the case of HDL or HDL/serum cholesterol ratio), the greater the probability of getting heart disease. Knowledge of this will affect the reference range. Finally, such factors as age, sex, race, diet and drugs all influence particular tests (Wilding and Bailey, 1978).

Bearing these points in mind, two systems where biochemistry tests are most often used are examined.

LIVER FUNCTION

Bilirubin

In hospital investigations of liver disease this is an important test, but, in an apparently healthy non-jaundiced group, it may be misleading. Levels over 25 μmol/ℓ may lead to a label of Gilbert's disease and further, sometimes traumatic, investigation being carried out. It has been suggested that this condition represents an aberrant enzyme system concerned with gluconoryl transferase, which is the metabolic route for detoxifying, for example, chlordiphenyls — chemicals widely used in fungicides, perfumes and the electrical industry. If this system is deficient people working with the chemicals may be prone to their toxic effects (Calabrese, 1978). There is no evidence to support this theory but as our ability to measure the efficiency of the various metabolic pathways at enzyme level improves, no doubt it will be possible to identify vulnerable people. In the meantime there is a tendency to regard a raised bilirubin with normal liver function tests as an extreme expression of normality, not indicative of disease (Bailey, Dawson and Robinson, 1977).

Gamma-glutamyl transferase

Increasing gamma-GT activity is specific for disease of the liver, biliary tract and pancreas, but in apparently healthy people elevated levels are associated with excessive alcohol consumption. However, there are some limitations to its efficiency as a screening test and accurate interpretation is not easy.

Although the BUPA Centre initially reported that higher levels of gamma-GT were seen in groups of people admitting to heavy drinking (Rollason, Pincherle and Robinson, 1972) a more recent study (Robinson, Monk and Bailey, 1979) shows no correlation between quantity of alcohol consumed and gamma-GT levels. It is possible that a gamma-GT response to alcohol is a marker for future disease. Where gamma-GT is elevated and alcohol ingestion stopped, the levels of gamma-GT fall to below 30 units/ℓ within six to eight weeks provided there is no demonstrable or irreversible liver damage. Hospitals tend to give a reference range of 0–30 units/ℓ. In the business population screened by BUPA, consumption of alcohol is above average. More than one-third of them would fall outside this range but it is estimated that not more than 10 per cent are heavy drinkers (i.e. greater than 360 g alcohol per week). For this population we set a more flexible range (0–50 units/ℓ for women, 0–80 units/ℓ for men) but individual change within that range may be meaningful.

Elevated levels may be seen in patients taking enzyme-inducing drugs such as the antiepileptic drugs and possibly some antidepressants, thus complete history of drug-taking is necessary when interpreting a result.

In practice, gamma-GT is a useful test for detecting the covert alcoholic (someone who minimizes his reported drinking but has elevated levels), and for monitoring heavy drinkers who are reducing their alcohol intake. In suitable patients it is a good motivator for reducing alcohol intake.

Elevated levels that do not appear to be related to drug induction or alcohol should be further investigated. In a small series we have found chronic liver disease in two such patients (confirmed by liver biopsy).

Other liver function tests

For routine purposes, the serum aspartate aminotransferase (SGOT) and alkaline phosphatase usually complete the liver profile. Neither are specific for the liver, but high values in a screened population are usually due to liver disease. If this is the case SGOT indicates hepatocellular damage and alkaline phosphatase biliary tract disease. SGOT may be affected by heavy alcohol ingestion occurring a few hours before the test is carried out (Wilding and Bailey, 1978). In the BUPA population, a raised alkaline phosphatase is more often associated with benign bone disease (e.g. Paget's disease). It is important that patients with high results are fully investigated, as many other system disorders produce elevated levels of these two tests.

In screening for occupational exposures which damage the liver, such as vinyl chloride monomer, these biochemical tests are insensitive, only altering when considerable damage has occurred. In this area more sensitive tests of liver function may be necessary. There are

Table 11.3 LIVER FUNCTION SCREENING TESTS

	Pre-employment	*Routine*	*High risk occupations**
Bilirubin	−	−	+ or ± (doubtful value)
SGOT	+	+	+
Alkaline phosphatase	+	+	+
Gamma-GT	+	+	+ +

+ + = essential; + = of value; ± = of doubtful value; − = of no value; *High risk occupations: industries manufacturing and using hepatotoxic chemicals (*see* Chapter 5), brewers and distillers, professions, actors and publicans

a number that measure specific functions of the liver. These include the clearance tests: bromsulphthalein (BSP), bile salts and aminopurine, and the estimation of galactose elimination and the measurement of hepatic blood flow. As our knowledge of the cellular chemistry of the liver increases, so will the opportunities to screen for vulnerable people. The use of liver function tests for pre-employment, routine screening and occupational health practice is summarized in *Table 11.3*.

THE RENAL SYSTEM

Three routine blood tests affected by renal function are urea, creatinine and urate (uric acid). If renal function is so poor as to move these outside the reference range, other signs of disease will probably be present. Raised levels do not help with bladder tumours unless an obstructive nephropathy has occurred. The most likely disease to cause obstruction is benign prostatic enlargement (hypertrophic prostatism). The tests come as part of a battery and, if abnormal, must be investigated further. Urea and creatinine are moderately specific for the kidney (although the former rises in gastrointestinal haemorrhage and after a high protein meal in some instances). Uric acid on the other hand may be raised in gout and by other metabolic processes outside the kidney.

Other renal function tests

Urine mirrors the functions of the kidneys and bladder. It provides more sensitive measures of these functions than blood and is easily obtainable. It can be tested for toxic substances or their metabolites, organisms and cells (*see Table 11.4*). The measurement of abnormal chemicals in the urine resulting from occupational exposures is discussed in Chapter 5. Screening for diabetes by testing the urine for sugar is important in the industrial context because of the importance of its long-term control particularly in certain jobs such as driving of public service and heavy goods vehicles, and shift working. In our experience at BUPA about half the cases of glycosuria are due to a low renal threshold and not to pancreatic disease. All cases with glycosuria must be followed up with a glucose tolerance test.

Heavy proteinuria is probably the first sign of a major renal problem and certainly more reliable than serum urea, creatinine or urate.

There is considerable discussion about the value of screening for organisms, particularly in asymptomatic women where the detection rate is moderately high but the significance of the result in question. In patients with symptoms, it is a necessary test and can be done simply using a dipslide.

Table 11.4 URINE TESTS USED IN RENAL SYSTEM SCREENING

	Pre-employment	*Routine*	*High risk occupations**
Protein	±	+	+
Sugar	+	+	+
Organisms	−	−	−
Abnormal cells	−	−	++
Chemicals	−	−	++

++ = essential; + = of value; ± = of dubious value; − = of no value; *see* Chapter 5

Workers exposed to the risk of bladder tumours (*see* Chapter 5) need to be monitored by screening the urine for various chemicals involved (e.g. aromatic amines). The exfoliative cytology or Papanicolaou test is of value. It may warn of the development of a tumour months or years before symptoms appear and is believed to improve the prognosis. Tumours occur many years after the original exposure. The Papanicolaou test is not useful after treatment of a bladder tumour; regular cytoscopy is required (Gadian, 1977).

Many countries have not yet controlled this risk and in the United Kingdom there is a hangover from lack of control in the past.

BLOOD AND BLOOD-FORMING ORGANS

The production of red cells, white cells and platelets may be disturbed by disease and by occupational hazards. Red and white cell counts and the red cell indices can be measured simply by automated machinery. Microscopy is required for red and white cell morphology and platelet counts.

For the early detection of anaemias and leukaemias due to natural causes we have found screening need consist only of haemoglobin and erythrocyte sedimentation rate (ESR) measurement. If either test is outside the reference range then a complete examination of the blood film is made. This action will detect anaemias and acute leukaemias where anaemia is an early sign. It will not detect a chronic leukaemia which is symptomless, but there is no evidence that the patient is better off if it were detected. If an alcohol problem is suspected red cell indices need to be measured because certain people develop macrocytosis due to excessive alcohol that may lead to anaemia which responds to folic acid supplements (and withdrawal of alcohol).

Full blood counts must be carried out on people returning from the tropics or areas where parasitic disease is endemic. In non-Caucasian races a simple sickle-cell screening test is available and people from thalassaemia areas will need full indices and haemoglobin electrophoresis if this is suspected.

In industries where there are known hazards to the blood-forming organs (e.g. from exposures to lead, benzene and other haemolytic agents, *see* Chapter 6), pre-employment examinations should include a full blood count and haemoglobin estimations. These will need to be repeated at intervals according to the amount of exposure. Early depression of any of the elements is reversible if detected early enough and the cause removed but such early changes may well occur within the so-called reference range. Thus a baseline measurement before exposure has occurred is useful. Fluctuations occurring in the peripheral blood

Table 11.5 SCREENING TESTS FOR BLOOD AND BLOOD-FORMING ORGANS

	Pre-employment	Routine	High risk occupations* and travellers to and from certain areas
Haemoglobin	+	+	+
ESR	+	+	+
Full peripheral count and indices	−	±	+
Marrow examination	−	−	+
Special tests	−	−	+

+ = of value; ± = of dubious value; − = of no value *see* Chapter 6

where there is suspected occupational risk should be followed up with marrow examinations if there is any doubt as to the cause and severity. Screening tests for blood-forming organs are summarized in *Table 11.5.*

ALLERGY

People with a personal or family history of allergy present a particular problem because they may be susceptible to otherwise harmless environmental substances (allergens) which may cause respiratory or dermatological symptoms. The allergic reaction may be an immediate (anaphylactic) response (Type I) which causes bronchospasm, urticaria, and angioneurotic oedema if severe, or it may be the Type III response when antigen/antibody complexes are formed which may cause lung disease similar to an alveolitis and also affect other organs such as the kidney. People who react in such a way are called atopic, and atopy in industry can be detected by several skin tests. A suitable battery of tests would be a control, grass pollen, *Aspergillus fumigatus,* house dust and *Dermatophagoides pteronyssinus.* Small amounts are administered in a skin-prick test and the skin-weal size correlates with levels of specific IgE antibody responsible for the reactions (Hendrick *et al.,* 1975). This battery of tests will identify the majority of atopic people. Their value in predicting susceptibility to occupational allergens is limited, and is discussed in Chapter 25.

Skin testing for sensitivity to particular substances to identify causes of disease must be done carefully with specially made up solutions. A

severe atopic may have a serious reaction so adrenalin and hydrocorti-
sone must be available for combating such a reaction. It is possible for
some skin-testing to sensitize a potentially allergic person, but this is
not likely if the battery mentioned above is used (Juniper *et al.*, 1977).

THE NERVOUS SYSTEM

Visual tests

A standard lens system can allow quick tests of phoria, visual acuity,
depth perception and colour vision to be made, and this is particularly
important in the transport industries. Measuring visual fields should be
done on people over 55 with a positive family history of glaucoma.
Although glaucoma is a common disease of the aged, screening by
estimating intraocular pressure is not reliable. Intraocular pressure
varies with the time of day and it is not possible to set a satisfactory
reference range for acceptable sensitivity and specificity levels.
Although the air puff technique is non-invasive, it is not helpful in
patients with corneal disease, in whom glaucoma is most common.

Normal colour vision is required by workers dependent on colour
signals and codes, as in transport services or in some colour-matching
tasks.

Several compact lens systems are available for testing visual acuity,
phoria and colour vision. These range from the most sophisticated
(and expensive) Bausch and Lomb Orthorater to the robust but
cheaper Keystone instrument.

Other nervous system tests

Hearing is dealt with in Chapter 12. During the course of the physical
examination testing of the other cranial nerves and the motor and sensory
peripheral nervous system is quickly performed.

As industrial processes become more complicated, there may be a
need for more regular biological monitoring of nerve transmission.
This is particularly important when dealing with organophosphorus
compounds, lead and other heavy metals. Plasma and erythrocyte
cholinesterase can be estimated at fortnightly or monthly intervals and
compared with an individual's own baseline. A depression in the
region of 30 per cent is considered to be an indication of exposure.
Regular electromyography will show changes in neuromuscular
transmission; if this occurs the subject should be removed from the

toxic environment until the pattern returns to normal. Finally, velocity of nerve conduction can be measured at six-monthly or yearly intervals, and early changes may be important in detecting and preventing peripheral neuropathies (Burgess, 1979).

Mental function

Some assessment is often needed of intelligence, personality and psychotic tendencies. Many questionnaires are available for each aspect of mental function, ranging from the Goldberg simple 12 questions for covert depression (Goldberg, 1972) to the Minnesota Multiphasic Personality Inventory which consists of 566 statements to which the patient responds 'true' or 'false'.

On the simple questionnaire used at the BUPA Medical Centre, 10–15 per cent of patients score high, suggesting an underlying depression or disequilibrium with their environment. An experienced physician is likely to identify such conditions during the interview and examination. One advantage of the questionnaire is that it can be quickly and cheaply administered by non-medical personnel.

CONCLUSION

The present state of the art of screening individual organ systems has been outlined. Health screening is still a controversial subject although its benefit to people working in certain occupations is beyond doubt. Health screening may be a most powerful tool for disease prevention and health maintenance on a much broader front. It is only by evaluation of screening procedures that it will be possible to tell whether these expectations are fulfilled. Occupational health practice offers excellent opportunities to do this.

REFERENCES

Bailey, A., Dawson, A.M. and Robinson, D. (1977) 'Does Gilbert's Disease exist?' *Lancet*, i, 931–933

Bouhuys, A., Schoenberg, J. and Beck, G. (1977) 'Discriminating power of MEFV curves.' *Federation Proceedings*, **36**, 468

Burgess, J.E. (1979) Personal communication. Occupational Health and Hygiene Services, Kings Lynn, Norfolk

Calabrese, E.J. (1978) In *Pollutants and High Risk Groups* p.2 *et seq.*, New York: Wiley

Gadian, T. (1977) 'Prevention and early diagnosis of occupational bladder tumours.' *Proceedings of Symposium on Health Screening*, pp. 25–29. London: Society of Occupational Medicine.

Goldberg, D. (1972) *The Detection of Psychiatric Illness by Questionnaire*. London: Maudsley Monograph

Hendrick, D.J., Davies, R.J., D'Souza, M.F. and Pepys, J. (1975) 'An analysis of skin prick test reactions in 656 asthmatic patients.' *Thorax,* **30,** 2–8

International Labour Office (1971) *Classification of Radiograph of Pneumoconiosis.* Geneva: 1972

Juniper, C.P., How, M.J., Goodwin, B.F.J. and Kinshott, A.K. (1977) 'Bacillus subtilis enzyme: a 7-year study.' *Journal of the Society of Occupational Medicine,* **27,** 3–12

NIOSH: National Institute for Occupational Safety and Health (1979) *A Recommended Standard for Exposure to Carbon Disulfide.* US Government Printing Office

Robinson, D., Monk, C. and Bailey, A. (1979) 'The relationship between serum gamma-glutamyl transpeptidase level and reported alcohol consumption in healthy males. *Journal of Studies on Alcohol,* **40,** 896–901

Rollason, J.G., Pincherle, G. and Robinson, D. (1972) 'Serum gamma-glutamyl transpeptidase in relation to alcohol consumption.' *Clinical Chimica Acta,* **39,** 75–80

Sheffield, L.T. and Roitman, D. (1976) *Progress in Cardiovascular Diseases* **19,** No.1, 33–49

Wilding, P. and Bailey, A. (1978) 'The normal range.' In *Scientific Foundations of Clinical Biochemistry,* pp. 451–459. London: Heinemann

12

Screening Organ Systems II – Audiometry

T. J. Wilmot

Screening the hearing of a section of the population is a relatively recent concept. It has been used particularly in children where the presence of deafness has profound educational importance. The problem in small children has been to devise methods of obtaining reliable and repeatable results and to minimize subjective errors. Special techniques have been developed to do this. In older children clinical tests of hearing and pure-tone audiometry are more reliable. Special problems related to unilateral deafness or to hearing loss confined to the higher frequencies may, however, arise, and the pitfalls of audiometry are now understood and avoided.

Most of the work in relation to children's hearing loss has been done by co-operation between special units within the National Health Service (NHS) (usually a hospital based otological unit), the educational authorities (through the educational psychologists and trained teachers of the deaf) and those responsible for the actual screening of the children (trained health visitors, district medical officers etc.). Expertise has developed, with the result that the majority of children can be screened effectively without the need for very specialized testing, which is reserved for the genuine problem case or the child with multiple handicaps. When screening hearing there are problems concerned with the detection of deafness, and with the causation and treatment of the deafness so discovered.

In adults the same problems exist, although the emphasis tends to be on different aspects of these problems. The main effects of hearing loss are on work suitability and social adaptability in two main groups affected.

GROUPS AFFECTED BY HEARING LOSS

The first group is composed of adult working men and women whose hearing may be diminished as a result of their employment, with consequent effects on their efficiency and ability to do their particular job, as well as the profound social handicap it entails.

The second group is the elderly, especially those in homes or institutions or with coincident mental handicap. Many of these may have lost hearing from previous occupational exposures to noise, from senile changes or from other causes. In this group the social disabilities of deafness predominate and the purpose of detection of hearing loss is to organize rehabilitation.

Thus early detection of hearing loss may be of profound importance in relation to education, to the working environment and to social problems among the elderly. Screening techniques, initially developed for children, can now be applied to adult groups. The screening process, in relation to the working population, will be considered as a part of occupational health practice primarily concerned with people who are, or have been, exposed to noise levels which are a hazard to hearing. Noise-induced hearing loss should be detected at an early stage and all personnel working in harmful noise levels should be protected.

Environmental pollution is of topical interest and noise is one of the main culprits (Glorig, 1958; Bell, 1966; Burns, 1968; Burns and Robinson, 1973, 1976). The fact that compensation is now available for those with severe hearing loss from occupational causes has emphasized the need for accurate measurement of hearing on a scale not previously contemplated. The assessment of hearing loss from noise damage is an admission of failure, and ideally the emphasis should be on prevention rather than on compensation.

Many individuals with considerable social and working hearing handicap caused by their occupation are at present neither entitled to compensation (Pelmear, 1978) nor helped to any significant degree by the standard NHS hearing aids. In the United Kingdom compensation is currently awarded only to those with severe losses in certain occupations (i.e. amounting to at least 50 decibels in the better ear), and the presence of recruitment (a paradoxical excessive sensitivity to noise amplification, discussed below, p. 253) in this type of hearing loss, makes satisfactory hearing aid amplification unsatisfactory. Recruitment reduces the normal wide range of tolerance to sounds so that the individuals concerned frequently find that once amplified sounds become audible, they often become uncomfortable. If social sound levels were at a constant level this would not be a problem, but they are not, and the degree of automatic attenuation necessary has not yet proved technically possible.

As criteria for compensation in the United Kingdom have become less restrictive it is to be hoped that technical improvements in hearing aids will enable more individuals with this type of deafness to cope more satisfactorily with their disability. The question of protection must assume a more dominant role and ultimately will depend not only on better protection for the ears but also on the reduction of sound levels in working areas. Economic factors are of prime importance but the fact that severe noise-induced hearing loss is now widely recognized, is becoming less acceptable socially, and qualifies for compensation, will induce design engineers to modify existing machinery to create new machines with lower sound levels.

AUDIOMETRIC SCREENING

Objectives

The objectives of an audiometric screening programme are:

1. To establish quantitatively the hearing status of an individual.
2. To monitor hearing loss during the period of employment and to control the risk of occupational hearing loss.
3. To regulate a hearing conservation programme.
4. To demonstrate the benefit of using personal ear protection as part of a health education programme.
5. To identify those people who are especially noise susceptible who may be affected even when using protection, or when working unprotected in a noise level not harmful to the majority of people.

Screening methods

Screening of adults with normal hearing or with noise-induced hearing loss, although simpler than in children, has not commanded the same attention. Where it has occurred it has been done largely outside the NHS, not in close collaboration with it. It is only since noise-induced hearing loss became a prescribed disease that the necessity for accurate measurement of hearing and skilled assessment of the cause of the hearing loss has been appreciated and action taken to remedy this.

The principles involved are:

1. The initial screening test should be relatively simple, quick and reliable.

2. Those who fail such a test must be examined clinically to exclude wax or ear disease, and must have a repeat screening test done under good conditions.
3. Those who fail the second test must be seen by an occupational physician who, in turn, may refer them to a consultant otologist with full technical back-up facilities.

In some cases a further examination may be desirable where tests for non-organic deafness or other unusual hearing patterns are available. Such highly skilled examinations would normally only be provided at special regional centres. Thus the initial screening test and the simple clinical evaluation of those who fail the test will normally be performed by the staff of the occupational health service of the industrial unit or group.

Where necessary the occupational physician refers patients to the local consultant otologist of the NHS for an opinion. If the otologist is doubtful of the nature or true extent of the hearing loss he should refer the case to a regional neuro-otological centre which can give the highest level of professional advice.

Aspects of audiometry in industry have been described by Atherley and others (1973, 1977), by Bryan and Tempest (1976) and in the Health and Safety Executive discussion document, *Audiometry in Industry* (1978).

Provision of screening

It is not only the principles involved in screening that have to be considered, but also the problems of examining large numbers of individuals at work. This entails assessment of noise levels, considerable expenditure on testing and the organization required to do the testing.

In the United Kingdon some nineteen million people work in factories of one sort or another. Some will have no noise problem, some will have a definite noise problem and others will be working where there may be exposure to high or variable noise levels, which require monitoring.

According to Faiers (1978) there are half a million workers in the United Kingdom exposed to risk of noise-induced deafness with over two million employed in industries where noise exceeds accepted levels.

The first priority is to assess the noise level and the number of workers who must be screened. In assessing noise levels, with rare exceptions, a single measurement is of limited value. The technicalities are of considerable importance — the sound pressure level, the duration of exposure, the frequency spectrum, the area affected, the quality of the instrumentation, the location of the measuring device and whether the noise is of sufficient duration to measure (*see* Chapter 18). Noise sources (e.g. vehicles and aircraft) may be mobile and this may further complicate matters. In a factory machine layout may undergo frequent alteration, requiring repeated surveys. To undertake noise measurement it is essential to have an occupational hygienist or other scientist available either on the staff or available on a contract basis.

If the number to be screened is large and the business a large one, arrangements will best be made for screening within the firm's premises, provided a suitable quiet testing area can be found or constructed. This will minimize the time off work for each individual screened. It will, however, impose a considerable work load on the occupational physician and his staff.

Small firms with small numbers of workers exposed to harmful noise levels may find such measures uneconomical and contract with larger firms, or with private screening units, for such services. Mobile audiometric units exist but both capital costs and running costs are high.

Screening itself can be done by trained technicians or nurses testing each individual, or by self-testing techniques using special audiometers (Hinchcliffe, 1959). Self-recording audiometry has a number of advantages. These and the recommended characteristics of suitable self-recording audiometers together with details of calibration and maintenance are described in *Audiometry in Industry* (1978). It is likely that industry will adopt the latter method generally while reserving screening by trained staff for those who fail the initial test in companies where large numbers are affected. Expenditure on audiometers and sound-treated testing booths will be necessary and extra personnel may be required. Organization will also be needed to provide and maintain the necessary equipment, to find a suitable testing time for workers (especially shift workers), to explain the nature of the test and how to operate the equipment, train those involved in the screening and record the findings.

Further, the whole question of establishing a system within the factory or industry contemplating such screening must be considered. First a noise survey is essential in order to designate the areas with harmful noise levels – these are sometimes referred to as 'red areas' and suitable warning notices must be placed which define such areas. All personnel working within them must be provided with personal ear protection. Such surveys should be done at regular intervals not exceeding a period of four years. It may also be necessary to undertake extra surveys after the installation of new machinery.

Many industries have a large turnover of workers. If a noise problem exists it is advisable to institute pre-employment screening which establishes a norm for the hearing of each new employee. This will be in addition to screening the hearing of all those working in, or part of whose duties lie within, the red areas. It will also be advisable to screen those leaving the factory even after a relatively short period of employment, or on retirement. A system of notification will therefore be required by which all starters and leavers are notified to the occupational physician. Estimates of the screening case load will then have to be made and the necessary arrangements established. The question of whether one or more hearing booths are necessary and what personnel are required must be decided by the occupational physician concerned and recommendations made to management accordingly. Considerable documentation is required and it is likely that at least a full-time nurse or technician will be needed to cope with all aspects of this work. One occupational health nurse responsible for one automatic (Rudmose) audiometer can provide such a service for up to 15 workers/day giving an annual turnover of perhaps 1000/year.

Screening problems

Many problems will occur:

Instruction and language

Instruction for workers in testing techniques may require extra time or the use of an interpreter. Simple written instructions as to how to perform the test should be given, followed if necessary by verbal instructions.

Documentation

This process is time-consuming and requires a separate filing system within the medical office. *Audiometry in Industry* (1978) suggests data for the keeping of records as well as depicting suitable data cards, audiogram charts and medical record sheets.

Interference with work

This is to some extent unavoidable. Appointments should be arranged through the appropriate foreman. In large factories covering a lot of ground it may be necessary for an employee to be absent from work for more than a minimal period. It is obviously essential to obtain the co-operation of management and employees.

Avoidance of temporary threshold shift (TTS) of hearing

Harmful noise levels may cause a TTS which disappears after the noise exposure ceases. Ideally all personnel working in red areas should have their hearing tested in the morning before starting work. A minimum period of 12 hours without noise exposure is recommended. If this is not possible and workers are coming direct from work in a 'red area' to have their hearing screened they must have worn ear protection from the commencement of their work. It may not be possible in a busy factory to implement fully such a policy but retesting under proper conditions will be required in those who fail their initial screening test.

Technical problems

Periodic calibration of the audiometer(s) will be required. The use of a sound-treated room or booth is obligatory.

Practical problems

The presence of wax plugs may invalidate a test and syringing of the ears affected will be necessary. While this can legitimately be done by an occupational health nurse it would not be part of the duties of a technician audiometrician. It may be necessary to train the nurse in methods of examining and cleaning ears. Correct fitting of the testing earphones is also important.

Assuming that a self-recording audiometer in a special sound-treated booth has been provided, with suitable low ambient noise levels inside it (*Audiometry in Industry,* 1978), and that an audiometrician is available, there are important decisions to make before screening begins with regard to the choice of frequencies for testing, technique used for testing, categorization of audiograms, the referral system necessary for further diagnosis and opinion and personnel to be screened.

Choice of frequencies

Screening of hearing can be used for several purposes: to detect a hearing loss, to diagnose the type of hearing loss present and to ascertain whether a hearing loss is connected with noise damage to the hearing. So far as occupational health is concerned, while physicians may be concerned with all three indications, management is interested only in the last.

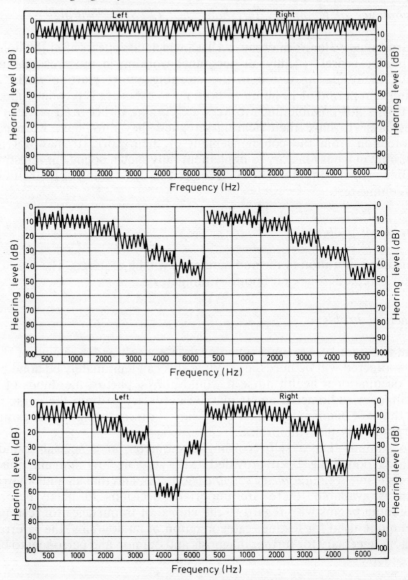

Figure 12.1 Normal young adult. Early presbycousis. Noise-induced deafness (bottom graph)

The vast majority of cases of noise-induced hearing loss commence at the frequency of 4kHz (*see Figure 12.1*). With continued exposure damage to the cochlea becomes more extensive and tends to spread into the neighbouring frequencies involving 3 and 2kHz as well as the higher frequencies (Schneider *et al.*, 1970). This is of considerable clinical significance as these lower frequencies are of importance in the understanding of speech.

The major complicating factor is the involvement of many of these frequencies in the normal ageing process. This process is variable but a loss in hearing can be expected at frequencies above 3kHz in all normal individuals over the age of 60 and in a proportion of people in the age group 50–60. Standard corrections of audiograms for ageing may be applied in British subjects (Burns and Robinson, 1976).

In a case of pure noise-induced hearing loss the hearing chart can be expected to show a typical notch with the loss maximal at 4kHz and less severe on each side of this. When noise-induced loss is superimposed on presbycousis the loss will naturally be greater, but it may be difficult, or indeed impossible, to allocate exact reponsibility to each of these causes. In both types of hearing loss both ears are usually involved in a symmetrical manner (*Figure 12.1*).

In industrial hearing screening there has to be a compromise. If the single frequency of 4kHz was tested, those with normal hearing could immediately be excluded from the noise-induced hearing loss group. On the other hand it might not be economic to screen a single frequency using an automatic (self-operated) audiometer when each worker has to be instructed in its use and when no hearing chart (audiogram) appeared as an end result.

It seems logical, therefore, to screen six frequencies 0.5 kHz, 1 kHz, 2 kHz, 3 kHz, 4 kHz and 6 kHz and to do this for both ears (*Audiometry in Industry*, 1978).

The technique of testing is set out for both manually operated and self-recording audiometers together with much helpful advice on necessary preliminaries in *Audiometry in Industry* (1978).

Categorization of audiograms

The recommendations of *Audiometry in Industry* (1978) may be summarized as follows:

Category 1   If either the sum of the low frequencies 0.5, 1 and 2 kHz or the high frequencies 3, 4 and 6 kHz show an increase of 30 decibels (dB) or more compared with a recent preceding audiogram, or 45 dB when the preceding examination exceeds three years the case should be categorized 1.

Category 2   Includes those cases where the difference of the sums of hearing between the two ears exceeds 45 dB in the low frequencies or 60 dB in the high frequencies.

Table 12.1 CHART FOR CATEGORIZATION OF HEARING LEVELS

Age in years	Sum of hearing levels (dB) in each ear			
	0.5, 1 and 2 kHz		3, 4 and 6 kHz	
	Warning level	Referral level	Warning level	Referral level
20–24	45	60	45	75
25–29	45	66	45	87
30–34	45	72	45	99
35–39	48	78	54	111
40–44	51	84	60	123
45–49	54	90	66	135
50–54	57	90	75	144
55–59	60	90	87	144
60–64	65	90	100	144
65–	70	90	115	144

Source: Audiometry in Industry (1978)

Category 3 Cases in which the sum for either ear exceeds the 'referral' level, set out in *Table 12.1,* either for low or high frequencies or both.

Category 4 Cases in which the sum for either ear exceeds the 'warning' level either for low or high frequencies, or both, but does not exceed the 'referral' level.

Category 5 Cases which do not fall into any of the above categories.

The occupational physician will take further action with categories 1, 2 and 3. This includes a clinical examination and referral where necessary to a consultant otologist (*see* page 251).

The action to be taken for persons with evidence of occupational deafness, may be summarized as follows: Those in category 4 should be warned to take precautionary measures to preserve their hearing. Those in categories 1, 2 and 3 must be informed that their hearing loss is considered to be noise induced. Except in persons with 'tender ears', or on other medical grounds, noise-induced hearing loss is not in itself a good reason for removing someone from further noise exposure, provided he or she can have adequate ear protection. Medical reasons for removal are diseases of the external and middle ear which preclude sufferers from wearing ear defenders, and other ear diseases which may be exacerbated by further exposure. Those who show evidence of noise-induced hearing loss for the first time should have audiograms repeated after six months. If this shows evidence of susceptibility to noise, removal should be recommended.

Other simpler methods of categorization are possible and can work quite effectively in practice. For example:

1. Where there is no more than a 25 dB loss in any frequency.
2. Where there is a 25 dB loss in one frequency, plus a loss greater than this in the higher frequencies (4 kHz and above).
3. Where there is a loss of 25 dB or more affecting the speech frequencies (0.5, 1, 2 and 3 kHz).

In this system categories 1 and 2 will not pose working or social problems and can be dealt with by the nurse (audiometrician) at local level. Wax removal may be necessary. Category 3 will necessitate referral to the occupational physician who will take a clinical history, examine the individual concerned and possibly repeat the audiogram. In the event of doubtful findings, or unilateral hearing loss or the presence of ear suppuration he will refer the case to a consultant otologist either directly or through the general practitioner.

Referral system for further diagnosis and opinion

The main responsibility of an occupational physician is to safeguard the hearing of those workers at risk. He is not primarily concerned with the diagnosis of obscure ear conditions nor in the treatment of an ear condition unless this interferes with the safety or efficiency of the worker.

He does, however, need to know how severe a hearing loss is, whether it will increase with further noise exposure, whether it is connected with organic ear disease and whether it could lead to danger to the worker or to the machinery for which he is responsible. In short he requires advice from a consultant otologist who is aware of, and interested in, the problems involved and who is not solely concerned with the clinical aspects of the case. It is obvious, therefore, that there are many advantages in direct liaison between the occupational physician and an individual consultant familiar with his problems. Thus a choice of consultant is desirable and direct communication over problem cases is sometimes essential. In all cases the general practitioner should be notified and his permission for referral obtained.

Personnel to be screened

Ideally perhaps all factory workers where there is any noise hazard should be screened but this at the moment is quite impractical.

It is desirable to screen all starters and leavers, including those who are retiring. In the long term perhaps newcomers have the greatest priority as a baseline of hearing can then be established but a loss of hearing subsequent to this need not necessarily be due to occupational noise exposure. It can be caused by ear disease, by meningitis, by primary cochlear degeneration, by head or skull trauma, by presbycousis, or by noisy hobbies. Pre-employment audiometry furthermore is not much of a problem as it can readily be fitted into the pre-employment examination schedule.

Apart from these groups, 'red area' workers, and those regularly visiting 'red areas', should be screened at regular intervals, usually of one year or more frequently where indicated. Management, however, may well feel that screening is a luxury which can be dispensed with. The argument that it is the responsibility of management to define 'red areas' and to provide adequate protection for all those working in such areas is a valid one. Any individual claiming that his hearing has been damaged by noise exposure must, therefore, have been negligent in using his protection and the responsibility is his and not management's. In practice, however, the exact definition of 'red areas' is not easy and may be impossible — the susceptibility of some individuals to noise is much greater than others and the floating population in a factory who may, for one reason or another, visit a 'red area' at some time without protection is very difficult to control. It seems likely, therefore, that there will be increasing use of screening in the future and that enlightened management will come to appreciate that this is the lesser of two evils and will pay for itself in the long run.

Audiometry in Industry (1978) gives excellent advice on this subject and also includes a glossary and definition of terms which will be of considerable value to an occupational physician unfamiliar with this subject.

DIFFERENTIAL DIAGNOSIS

The occupational physician utilizing screening will need to know some basic facts about other common types of deafness and how these affect the audiogram.

Conductive deafness

The audiogram is almost invariably asymmetrical (*Figure 12.2*). The Weber tuning fork test (tuning fork at vertex) is lateralized to the deafer ear and the Rinne test shows bone conduction (mastoid placement of fork) hearing to be more acute than air conduction (fork at opening of external auditory meatus). Tuning forks with a frequency of 256 Hz or 512 Hz should be used.

Common causes of conductive deafness are wax, Eustachian tube obstruction, previous or present suppurative otitis media, tympano-sclerosis and otosclerosis. The history and a simple otological examination can often decide which of these is present.

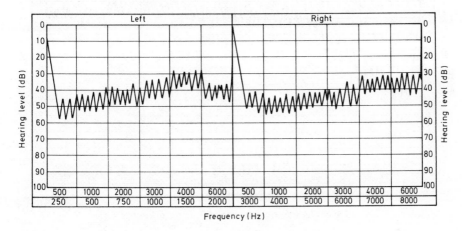

Figure 12.2 Bilateral middle-ear deafness

Bilateral conductive deafness will tend to protect the ears from noise damage and workers so affected should not necessarily be excluded from red area work.

Workers with known otosclerosis which has been operated on for the relief of deafness, however, are at risk and should be excluded from such areas.

Sensorineural deafness

This type of deafness can affect the end-organ (cochlea) or the eighth cranial nerve and its central connections. True central deafness is rare, is often characterized by better pure tone findings than the clinical picture suggests, and will normally be diagnosed only at consultant otologist or consultant neuro-otologist level.

Cochlear deafness is characterized by the presence of recruitment. This is the phenomenon of excessive sensitivity once threshold of hearing has been reached. It is often characterized clinically by intolerance to noise above a certain level and it makes the use of a hearing aid more difficult. Noise-induced deafness occurs in the cochlea and some recruitment is always present. Some types of presbycousis and Ménière's disease also produce cochlear loss and recruitment (*Figure 12.3*).

Retrocochlear deafness is less common and is characterized by tone decay. This is the tendency for a given note at constant amplitude to

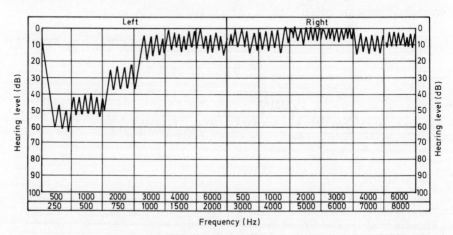

Figure 12.3 Early unilateral Ménière's disease

fade in the affected ear. Retrocochlear deafness is commonly unilateral and may be associated with an acoustic nerve neuroma. Any unilateral sensorineural deafness should be referred for consultant opinion.

Clinically, recruitment is suspected in people with obvious hearing loss who are still tolerant to noise above a certain level or who complain of distortion of sounds.

Non-organic deafness

Such deafness should be suspected when conversational levels of hearing in an individual do not appear to tally with the pure-tone threshold audiogram. While hysterical deafness in one or both ears can occur – usually in children or young adults – it is very uncommon in the middle-aged or elderly. In most adults non-organic deafness is more likely to be due to malingering or to a desire for compensation associated with head or ear trauma or the effects of noise upon the ears.

Whereas some individuals with non-organic deafness have difficulty in repeating a particular pattern of audiogram, others become very adept at this.

Most of these cases will need to be referred to a consultant otologist and some of them will require specialized testing methods. There are, however, several well recognized ways in which their simulated hearing loss can be disproved and there has been considerable advance in objective tests which do this.

LEGAL ASPECTS

Screening at factory or industry level may have little or no legal significance as a measure of hearing loss. For legal purposes audiometry must be performed by a trained audiometrician in a sound-treated room who is responsible to a consultant otologist who is himself responsible for assessing both the patient and the audiogram. In some cases, particularly those involving non-organic deafness, referral to a neuro-otologist or to a unit where special tests can be performed, may be necessary.

Hearing loss may be multifactorial in origin and the assessment of such cases may be controversial and difficult to evaluate.

For an industrial injury to qualify as a good claim in common law it must be shown that the employer has been negligent or that the employer is in breach of the relevant Act of Parliament or Regulations. Since 1963 employers (in Great Britain) should have been aware of their liabilities.

If an employer is responsible for a noise hazard situation he should institute a hearing conservation programme, remedial and palliative, noise surveys, warning signs, hearing protection education, the provision of individual hearing protection, the provision of noise refuges, an audiometric service, noise reduction projects (and the measurement of neighbourhood noise).

The law requires the employer to warn, not merely by posters and exhortations but also by instruction and education, to provide adequate and efficient ear protection, to fit such protection properly and instruct employees how to use it, and to ensure that the protection is used. Non-users should be warned and such warnings logged. The wearing of ear protection in certain areas should be made a condition of employment. The employer must further ensure that the exposure to noise is reduced.

REFERENCES

Atherley, G.R.C., Duncan, J.A. and Williamson, K.S. (1973) *Journal of the Society of Occupational Medicine,* **23,** 19–121
Atherley, G.R.C., Merriman, R.J. and Phillips, M.R. (1977) 'Occupational audiometry. Its place, functions and evaluation.' *Health Screening.* (pamphlet). London: Society of Occupational Medicine, pp. 32–52

Audiometry in Industry (Discussion document) (1978) *Report of Health and Safety Executive Working Group,* London: HM Stationery Office

Bell, A. (1966) *Noise, an Occupational Hazard and Public Nuisance,* Public Health Papers, No. 30. Geneva: WHO

Bryan, M.E. and Tempest, W. (1976) *Industrial Audiometry.* Salford: University of Salford

Burns, W. (1968) *Noise and Man.* London: Murray

Burns, W. and Robinson, D.W. (1973) *Journal of the Society of Occupational Medicine,* **23,** 86–91

Burns, W. and Robinson, D.W. (1976) *Hearing and Noise in Industry.* London: HM Stationery Office

Faiers, M.C. (1978) 'Noise deafness and the employer's liability.' *Journal of the Society of Occupational Medicine,* **28,** 20–24

Glorig, A. (1958) *Noise and Your Ear.* New York: Grune and Stratton

Hinchcliffe, R. (1959) 'Self-testing automatic recording audiometry – an appraisal.' *Journal of Laryngology and Otology,* **73,** 795–812

Pelmear, P.L. (1978) *Practitioner,* **221,** 668–674

Schneider, E.J., Mutchler, J.E., Hoyle, H.R., Ode, E.H. and Holder, B.B. (1970) 'The progression of hearing loss from industrial noise exposure.' *American Industrial Hygiene Association Journal,* **31,** 368–376

Further official sources

Department of Employment (1972) *Code of Practice for Reducing the Exposure of Employed Persons to Noise.* London. HMSO

Industrial Health Advisory Sub-Committee on Noise (1974) *Framing Noise Legislation.* London: HMSO

Industrial Injuries Advisory Council (1973) *Occupational Deafness,* Cmnd 5461. London: HMSO

Industrial Noise and its Effect on Hearing (1969) White Paper. Cmnd 4145. London: HMSO

Noise and the Worker (1963) Health and Safety at Work, Booklet No. 25. London: HMSO

Wilson Committee (1963) *Final Report on the Problem of Noise.* Cmnd 2056. London: HMSO

13

Epidemiology I — Uses and Descriptive Methods

R.S.F. Schilling and Joan Walford

Epidemiology is the study of the distribution and determinants of disease and other indices of health in populations. In occupational health, epidemiology is needed in research and daily practice to study environmental and personal factors that determine the incidence of disease and promote health and well-being, and also to evaluate health services.

USES OF EPIDEMIOLOGY

Epidemiology has at least nine uses in occupational health:

Identifying occupational disease and work related illness

Epidemiology can identify diseases of occupational origin which may escape detection when they occur commonly in the community and it can find out how the work environment influences the frequency and severity of disease. A study of an industrial population may be the only way of identifying an occupational risk of lung cancer, bladder cancer or chronic obstructive pulmonary disease; or of determining the influence of work factors in the aetiology of coronary heart disease, varicose veins or rheumatic disorders.

Finding causes

A particular environmental contaminant such as a dust or vapour may be identified as the cause of a disease by investigating exposed groups. Epidemiological studies of workers in Lancashire cotton mills, together with measurements of their dust exposures, made it clear that cotton dust was the cause of chronic obstructive pulmonary disease suffered by card room operatives at a time when air pollution outside the factory was considered to be the major cause of their disease (Schilling, 1956).

Controlling hazards

The surveillance of a group of workers exposed to a recognized hazard makes it possible to detect susceptible subjects, to remove them from exposure and to determine the value of environmental control measures such as exhaust ventilation and personal protective equipment. Data derived from routine periodic health examinations designed to protect the individual should be used epidemiologically to study groups of workers. This demands that the reliability of the methods of examination should be of the same high standard as those required in field surveys (*see* Chapter 15). Special epidemiological experiments may also be made to test new control measures (*see* Chapter 14).

Establishing and revising permissible levels of exposure

A more detailed examination of the relationship between prevalence and severity of disease and levels of exposure to a causal agent makes it possible to determine dose or uptake/response relationships* in the form of a curve, from which permissible levels are derived (*see* Chapter 25). Many of the existing permissible levels have been based on inadequate data from animal experiments and routine human observations. Reliable epidemiological studies of workpeople and their exposure to contaminants provide the most valid data on which to base permissible levels. Good examples are those for cotton dust (US Department of Labor, 1978), styrene vapour and hexavalent chromium (Volkova, 1975).

In order to draw reliable conclusions on cause/effect and exposure/response relationships, it is essential to relate the prevalence and severity of disease, or its early manifestations, to exposure levels. Measurements of exposure to a harmful agent should include its concentration, characteristics (such as particle size in the case of dusts), duration and other concomitant exposures. Current exposures

---

*Uptake may be used in preference to dose: it is the amount of the agent taken up by the body per unit time over a period of time (WHO, 1977). A dose or uptake/response relationship indicates the relationship between the level of exposure to or uptake of a contaminant and the proportion of individuals affected.

can be measured at the time of the investigation, but data on past exposures may not be available. From existing information and the subjective opinions of management and workers a crude classification into light, moderate and severe exposures may be possible, thus enabling the extent of the risk to be assessed, as was done for asbestos workers by Newhouse and Berry (1979) (described in Chapter 14). Tentative hygiene standards can be established until more accurate data are available (Roach and Schilling, 1960).

Establishing priorities

Epidemiology measures the prevalence and severity of disease. In a factory, office, school or military establishment there are likely to be several health problems to be investigated. Knowing how many people are affected, and to what extent, enables priorities to be set so that preventive action can be taken where it is most needed. For this purpose, and to evaluate its own effectiveness, an occupational health service has to keep continuous records of health examinations, accidents, treatments, sickness and other absences, and environmental measurements, These are the basic data of descriptive epidemiology, which is often the starting point for more sophisticated studies.

Determining 'expected' rates and normal values

In studies of disease incidence or prevalence, epidemiology is used to establish expected rates based on the mortality or morbidity experienced in specially selected control groups and to determine normal physiological values, such as lung function, that may be adversely affected by work exposures.

Expected rates

In mortality studies, the expected numbers of deaths (or expected death rates) may be derived from national or local mortality data. Such comparisons are of value as a preliminary reconnaissance and in identifying an excessive risk in a common disease such as lung cancer or myocardial infarction. Because of the influence of self-selection and medical selection in a particular occupational group, comparison with the general population can be biased and interpretation of results may be unreliable. If possible, comparisons should be made between exposed and unexposed groups, living and working in the same area and similar in all relevant factors such as race, age, sex and length of employment. Where this is not possible, heavily and slightly exposed groups may be compared.

Deaths have the advantage over morbidity indices in that they are easy to count, but they are only of value in identifying risks of fatal diseases such as most cancers and cardiovascular disease. Morbidity indices like sickness absence, attendances for treatment, or even prevalence of symptoms, are less reliable than mortality because they can be influenced by a variety of factors other than disease. Considerable care has to be taken in assessing an occupational risk by such indices. It is only by applying sound epidemiological methods that reliable baselines can be set in order to make meaningful assessments of risk.

Normal values

In studies in which the aim is to discover early adverse effects of work exposure it is necessary to know the normal range of function of the organ system being investigated, for example, lung in the case of respiratory disease, liver for exposures to hepatotoxins and hearing for noise exposures. Normal values or reference ranges (*see* Chapter 11) can only be derived from epidemiological studies of representative, but healthy, populations. Assessment of risk of occupational respiratory disease depends increasingly on lung function measurements and their comparison with predicted normal values (*see* Chapter 15).

Identifying work factors that promote health

The promotion of health and well-being is one of the aims of occupational health defined as long ago as 1950 by a joint committee of the World Health Organization (WHO) and International Labour Organization (ILO). It has been neglected in favour of the prevention of sickness and injury, largely because positive health and well-being are difficult to define and measure, and health personnel are not trained to do this. Attempts to quantify physical well-being have been made by measuring physiological function and the absence of the precursors of disease. Mental well-being, or the absence of it, which is one of the most important human problems in modern industry, is generally evaluated negatively in terms of mental and psychosomatic disorders, absenteeism, alcohol abuse, labour turnover, dissatisfaction and social unrest (Levi, 1978). Psychosocial and physical factors related to these disorders may be identified by epidemiological methods. In this way the effects of shift work on well-being have been investigated (Torsvall and Akerstedt, 1978).

The next step forward in health surveillance is likely to be a change in emphasis from negative to positive aspects in attempts to identify sources of people's good health (Morris, 1975).

Evaluating health and safety care at the workplace

In addition to its use in evaluating measures for controlling occupational hazards, epidemiology can help to assess the quality of health care (Morris, 1975) by finding out how services are used, their success in reaching certain standards, and by investigating the opinion of users of the services (Phillips and Hughes, 1974).

In many countries studies are now being made to find out how to make medical practice more relevant to health needs, and to provide a sound basis for health care planning (Hartgerink, 1976; White *et al.*, 1977). Little seems to have been done for occupational health apart from surveys that reveal the uneven distribution of occupational health care in Western countries (Health and Safety Commission, 1977; Ashford, 1976), and a few studies of cost-effectiveness (*see* Chapter 1).

Providing information on community health problems not caused by work

The analysis of the daily experience of an occupational health service permits the early recognition of epidemic outbreaks of diseases such as influenza, and can give information on general community health problems, particularly in developing countries where an occupational health service in a large enterprise may be an important source of national health data.

Where occupational health is an integral part of total medical care, there are even better opportunities for investigating community health problems. The South Eastern Indian Railway, which runs from Calcutta to the southern tip of India, provides a comprehensive medical service for more than a million employees and their dependants. As the railway runs through regions with different ethnic groups and climates, the central medical service has invaluable data from patients' records for investigating environmental causes of disease (Malhotra, 1970). Occupational health services may also participate in randomized controlled trials for new prophylactic measures such as vitamin C for the common cold and more ambitiously, in intervention trials for heart disease (*see* Chapter 14, section on experimental epidemiology).

TYPES OF EPIDEMIOLOGICAL STUDIES

Epidemiological investigations may be carried out by surveys (studies) of records or populations, or by designed experiments. A survey may be purely *descriptive,* aiming only to describe the situation in the population; or it may be *analytical,* seeking explanations and testing hypotheses about the aetiology of disease by case-control or cohort studies (*see* Chapter 14). Often it is a mixture of both descriptive and analytical methods. If the survey provides information concerning the situation at a given point in time, it is called *cross-sectional,* while a study that collects information at different points of time over a period, is called *longitudinal.* In this chapter the methods and uses of descriptive epidemiology are described. Analytical and experimental epidemiology are discussed in the next chapter.

DESCRIPTIVE STUDIES

Descriptive studies can be based on three quite distinct sources of information. First, there are the official statistics and records, routinely collected and kept by corporate bodies, which may be useful in epidemiological studies, but are not collected primarily for that purpose. It is therefore essential to know how they were collected, what they mean and whether they are accurate enough to be used (Case and Davies, 1964). Examples are:

1. National and local population census records of births, marriages and deaths.
2. Social security data on sickness incapacity, accidents and occupational disease.
3. Hospital and autopsy records.
4. Industrial pension and funeral benefit schemes and records kept by employers and trade union organizations.

Secondly, there are the records of employment, sickness, injury, treatment, health and environmental surveillance routinely kept by occupational health services or personnel departments.

Thirdly, a study of working groups may be designed solely to discover the magnitude of a problem (e.g. an estimate of disease prevalence or concentrations of air contaminants).

The variables commonly examined in the above studies are descriptive of time, place and persons. Secular or time changes in prevalence may be associated with changes in industrial processes or the introduction of new materials or preventive measures. There may be differences in the distribution of disease by place (e.g. geographical areas, different factories within an industry and different workplaces within a factory). There may be differences related to personal characteristics such as age, sex, ethnic group, socioeconomic status, occupation and personal habits such as smoking and alcohol consumption. Simple analyses of such data often provide the clues for formulating hypotheses to explain the causes of disease which may then be tested by specially designed analytical studies.

Official mortality data

Every ten years in England and Wales the mortality rates of the more populous trades and professions are published by the Office of Population Censuses and Surveys in the Decennial Supplement on Occupational Mortality (the most recent is for 1970–72: Registrar General, 1978). Crude rates which make no allowance for differences in age distribution are of limited value as indices of risk. For this reason, age-standardized indices of mortality are used. There are two main methods of age standardization: direct and indirect. Both require the choice of a reference or standard population, which is usually the total male or female population of England and Wales. The population being investigated, for example an occupational group, may be known as the study population.

Direct standardization

Age-specific death rates from the study population are applied to corresponding age groups in the standard population to give the number of deaths expected to occur if the standard population had experienced the same death rates as the study population. The 'expected' deaths in each age group are then totalled and compared with the total observed deaths in the standard population.

Two mortality indices can be calculated; the comparative mortality figure (CMF) and the direct standardized death rate (SDR).

$$\text{CMF} = \frac{\text{Total expected deaths in standard population}}{\text{Total observed deaths in standard population}} \times 100$$

$$\text{Direct SDR} = \frac{\text{CMF}}{100} \times \text{Crude death rate in standard population}$$

Indirect standardization

Age-specific death rates from the standard population are applied to corresponding age groups in the study population. The 'expected' deaths in each age group are then totalled to show how many deaths would have occurred in the study population had it experienced the death rates of the standard population. This total is then compared with the observed deaths in the study population. An indirect standardized death rate can be calculated, but the better known index of occupational mortality is the standardized mortality ratio (SMR).

$$\text{SMR} = \frac{\text{Total observed deaths in study population}}{\text{Total expected deaths in study population}} \times 100$$

$$\text{Indirect SDR} = \frac{\text{SMR}}{100} \times \text{Crude death rate in standard population}$$

Table 13.1 shows the calculation of the SMR for Social Class I males for England and Wales , 1970–72. The deaths are those of men aged 15–64 years, registered in the three years 1970–72, while the population numbers are estimated from a 10 per cent sample of the 1971 census. The standard population is that of all males in England and Wales. The three-year death rate was calculated by dividing the number of deaths in 1970–72 by the estimated population in 1971. There were 8586 observed deaths of Social Class I males, aged 15–64, in 1970–72; and the total of the expected deaths is 11 222.4. Therefore:

$$\text{SMR} = \frac{8586}{11\ 222.4} \times 100 = 77$$

This indicates that, after standardizing for age, mortality in Social Class I males was 77 per cent of that of all males in England and Wales.

The crude death rate per 100 000 living *per year* in the standard population is:

$$\frac{0.017912}{3} \times 100\ 000 = 597.06$$

$$\text{Indirect SDR} = \frac{77}{100} \times 597.06 = 460 \text{ deaths per 100 000 living per year}$$

Table 13.1 INDIRECT AGE-STANDARDIZATION OF MORTALITY OF SOCIAL CLASS I MALES, 1970–72

Age group	All males England and Wales 3-year death rate	Social class I males		
		Population 1971	Expected deaths based on all males rates	Observed deaths 1970–71
	(1)	*(2)*	*(1)* × *(2)*	
15–24	0.002772	95 190	263.9	193
25–34	0.003014	214 680	647.0	431
35–44	0.006923	171 060	1184.2	854
45–54	0.021594	137 080	2960.1	2079
55–64	0.061672	100 000	6167.2	5029
All ages	0.017912	718 010	11 222.4	8586

$$\text{SMR} = \frac{8586}{11\ 222.4} \times 100 = 77$$

Source: Registrar General (1978)

Standardized rates are fictitious and will vary according to the standard population chosen. This does not invalidate comparison between occupational groups as the relative positions of the groups, based on rates using one standard population, remain unchanged when another population is used. They provide a convenient summary measure, but are not as informative as the age-specific death rates themselves.

The validity of age-standardized indices depends on the assumption that the relative increase in mortality with age is the same in both study and standard populations. When this is so the CMF and SMR give similar values. In the choice of index, the CMF has the disadvantage that when the study population is small the age-specific death rates can be unreliable and, if applied to the standard population, too much weight can be given to too few deaths. The SMR has the advantage that it can be calculated when the number of deaths in individual age groups in the study population is unknown, and is also less affected by sampling fluctuations. Traditionally, the SMR is used in occupational mortality studies. Apart from official mortality statistics, these methods of standardization are commonly used in epidemiological studies of morbidity and mortality where comparison may be made between several occupational groups after standardizing for age and other factors, such as smoking habits. In these special studies, national data may not be available to provide a standard population and standardization may be carried out by the indirect method, using age-specific rates calculated from the combined data of the occupational groups as standard rates.

Proportional mortality rates and ratios

In some situations it would be inappropriate to calculate the CMF or the SMR because of inaccuracies in the enumeration of the populations. Many people aged 65 years or over state their occupation on the census as 'retired'. In occupations where there is a tendency for early retirement, such as firemen, policemen, aircraft pilots and the armed forces, there is under-representation of these particular occupations at the census. In mortality studies of women classified by their own occupation, discrepancies arise because a woman who is married or widowed will record her own occupation at the census, but unless she was employed at the time of death, or had been in paid employment for most of her life, her husband's occupation is entered on the death certificate.

Without an accurate estimate of the population at risk it is not possible to calculate mortality rates, and analysis has to be restricted to mortality from specific causes considered on a proportional basis.

$$\text{Proportional mortality rate} = \frac{\text{Deaths from specific cause}}{\text{Deaths from all causes}}$$

The official mortality index used for persons aged 65–74 years is the proportional mortality ratio (PMR)

$$\text{PMR} = \frac{\text{Proportional mortality rate in study population}}{\text{Proportional mortality rate in standard population}} \times 100$$

In 1970–72 there were 277 168 deaths from all causes and 8249 deaths from cancer of the stomach in the male population of England and Wales aged 65–74 years, giving a proportional mortality rate of $8249 \div 277\ 168 = 0.02976$. In Social Class I males in the same age range there were 7206 deaths from all causes and 130 deaths from cancer of the stomach, giving a proportional mortality rate of 0.01804. The PMR for stomach cancer in Social Class I males is, therefore, $0.01804 \times 100 \div 0.02976 = 61$. For mortality analysis of persons aged 15–64, where the CMF or SMR is inappropriate, an age-standardized PMR can be derived by indirect standardization (*see* Registrar General's Decennial Supplement for 1970–72, page 11).

The disadvantage of the PMR is that, since it makes no use of the population 'at risk', it does not give a mortality rate, but simply an indication of whether the proportion of deaths from a particular cause was high or low. The proportion may be high because the rate for that cause is high, or because the rate for some other major cause is low. The PMR has the advantage that it is not influenced by differences between information given at the census and at death. It has value as an indicator for a more detailed study of an occupational group, but should be considered in conjunction with the corresponding cause-specific SMR. It should not be used in place of the SMR or CMF when reliable populations 'at risk' are available.

Limitations of national mortality rates

National mortality rates have serious limitations and may be inaccurate. They are now accepted as possible indicators of an occupational risk, but this can be misleading, as they may be based on incorrect or incomplete information about the cause of death and the occupation. First, the information that the individual provides about his or her occupation at the National Census is likely to be correct. Information given by next of kin about a deceased person's occupation is not infrequently wrong, because the deceased is afforded a higher status than that in fact attained. Also people live longer nowadays and change their jobs more often, thus their occupation at death may be quite different from their main occupation. Workers disabled with pneumoconiosis may die in sedentary 'end occupations' as storekeepers, watchmen or clerks (*see* Chapter 3).

Secondly, national occupational mortality rates may be given for a whole industry and not for its separate trades. The mortality and morbidity rates for chronic respiratory disease among United States textile workers were no greater than those for all males. This was used as evidence against there being a risk of byssinosis in US cotton mills. The number of workers employed in dusty departments of cotton spinning mills comprised a relatively small proportion of all textile workers who mostly work under good conditions. The national figures obscured high rates in the small group of dust exposed workers. Field surveys in US textile mills have since shown relatively high prevalences of byssinosis in card rooms and other dusty processes (US Department of Labor, 1978).

Thirdly, there may be unsuspected anomalies in the method of collecting deaths. In England and Wales, historically the SMRs for occupational accidents among deep-sea fishermen grossly underestimated the risk because no account was taken of deaths at sea. The inclusion of the latter raised the fishermen's SMR for accidents in

1959–63 from 466 to 1726; that means the risk was seventeenfold, not about fivefold as indicated by the Registrar General's national statistics (Schilling, 1971).

Other mortality data

Records of death are available to the epidemiologist from local sources such as company and trade union pension schemes, local death registers and hospitals. They are used more frequently in analytical, rather than in descriptive studies. The following three studies, although testing hypotheses, are given as examples of the use of data routinely collected for other purposes.

A study of the pension records of a group of companies making lead accumulators revealed a significant mortality excess from cerebro-vascular disease among workers who had been heavily exposed to lead (Dingwall-Fordyce and Lane, 1963).

Tiller, Schilling and Morris (1968) used data from local death registers in an area where there were three viscose rayon factories to make a preliminary study of cardiovascular mortality. The initial data comprised 223 male viscose workers and 174 male non-process workers who had died between 1933 and 1962. As the size and age structure of the population at risk was unknown, the data were analysed by calculating the age-standardized proportional mortality rates using indirect standardization. The deaths were allocated to three age groups in each of six quinquennial periods, and the number of expected deaths calculated using the age-specific proportional mortality rate derived from the corresponding national mortality records. Over the whole period 94 viscose rayon workers died from coronary heart disease against 41.5 expected (*Table 13.2*). The corresponding figures for the non-process workers were 41 observed and 30.9 expected. Such a large excess of deaths from coronary heart disease was likely to be real. It led to an hypothesis which was tested and confirmed in a cohort study of viscose rayon workers.

The third example of making the best use of incomplete data comes from a small factory making arsenical sheep dip. The works manager thought that an unduly high proportion of his workers were dying of cancer. Over the previous 40 years the factory had kept lists of the deaths of employees but the cause of death was often absent. As there were no population figures to relate to these deaths it was not possible to calculate ordinary death rates. Hill and Faning (1948) searched through the local death registers of the town in which the factory was situated and showed, by using proportional mortality rates for different age groups, that a significantly high proportion of sheep dip workers, 29.3 per cent, had died of cancer compared with 12.9 per cent

Table 13.2 CALCULATION OF OBSERVED AND EXPECTED DEATHS FROM CORONARY HEART DISEASE IN VISCOSE RAYON WORKERS, USING, PROPORTIONAL MORTALITY RATES

Period	35–44 years		45–54 years		55–64 years		35–64 years	
	Ob-served	Ex-pected	O	E	O	E	O	E
1933–37	0	0.1	0	–	0	0.2	0	0.3
1938–42	0	0.2	2	0.6	0	0.4	2	1.2
1943–47	1	0.2	7	0.9	1	0.6	9	1.7
1948–52	3	0.5	9	2.9	8	2.8	20	6.2
1953–57	2	0.3	10	3.3	16	8.0	28	11.6
1958–62	0	0.2	8	4.9	27	15.4	35	20.5
1933–62	6	1.5	36	12.6	52	27.4	94	41.5

Source: Tiller, Schilling and Morris (1968)

Table 13.3 PROPORTION OF CANCER DEATHS OF MALES BY OCCUPATION IN SHEEP-DIP FACTORY

	Total deaths	Cancer deaths	Percentage due to cancer
Chemical workers	41	16	39*
Engineers and packers	10	3	30
Others†	24	3	12.5*

†includes non process workers such as printers, watchmen, boxmakers and carters
*$\chi^2 = 3.95$; $P = 0.047$
Source: Hill and Faning (1948)

in other occupational groups living in the same town. A further and illuminating analysis of the sheep-dip workers' deaths was possible by classifying the deaths according to actual job descriptions based on available factory records and memory. It revealed a significant excess of cancer deaths among the chemical or process workers, who would be more exposed to the sheep-dip powder than the other groups (*Table 13.3*).

The above study illustrates how descriptive data may be crucial to the actual identification of a risk. Without factory records it would have been difficult, if not impossible, to confirm the works manager's hunch that the sheep-dip workers had an occupational risk of cancer.

National morbidity data

National records of morbidity concerning the health and safety of people at work are available in most countries, but are even less reliable than mortality records because denominators (populations at risk) are inaccurate or not available, and seldom measure up to the strict requirements of the epidemiologist.

In Britain information about the number of people injured or killed each year as a result of accidents at work is available from the two sources:

1. The Department of Health and Social Security (DHSS) records of people receiving industrial injury benefit;
2. The Health and Safety Executive (HSE) notifications of industrial accidents.

DHSS figures provide the more reliable information since the number of injuries reported to the HSE is known to be incomplete (Case and Davies, 1964). A nationwide survey of personal injuries indicated that the DHSS records are incomplete as far as women and the self-employed are concerned (Royal Commission on Civil Liability, 1978).

Compensation data

The number of workers receiving compensation for occupational disease are sometimes used as an index of risk, with secular trends in these numbers indicating successes or failures in prevention (Newhouse, 1976). These numbers can only provide a crude index of secular trends as they are influenced by various factors such as changes in the populations at risk, the numbers eligible for compensation, the increasing knowledge of the compensatable diseases and changes in the criteria for diagnosis.

In most countries fairly strict criteria must be fulfilled before an occupational disease becomes compensatable. In the United Kingdom the cause and incidence of the disease must be shown to be a risk of a certain occupation and the attribution of particular cases to the nature of employment must be established with reasonable certainty. This is the reason why there are delays in officially recognizing occupational diseases. It also means that some of the less obvious ones are never made compensatable.

In a national survey in the United Kingdom in 1973, it was found that for every person compensated for occupational disease there were at least five others who believed that work was the cause of their illness or had made a major contribution to its development (Royal Commission on Civil Liability, 1978). This finding may exaggerate the risk of occupational illness since it is based on subjective judgements, but it suggests that the official statistics may underestimate the true incidence of occupational disease.

Notification of occupational disease

The notification of occupational diseases is not a reliable measure of their extent and severity, as they may not be correctly diagnosed and physicians seldom fulfil their statutory obligation to notify. Of 130 patients with squamous epithelioma, 81 were likely to have had work exposures to known carcinogens and thus be notifiable. Only three of these were notified (Murray, 1958).

National sickness absence data

Difficulties arise in the interpretation of national data on sickness absence as diagnosis may be inaccurate and some of the absence is known to be due to causes other than illness. Nevertheless they do give some indication of the health status of the working population and provide a basis for more detailed investigations.

National figures for sickness absence in 1972–73 showed substantially different rates for regions in the United Kingdom. They were particularly high in Wales, Northern Ireland and the North East and low in London and the South East, and in East Anglia. Possible causes for these differences could be occupational, economic, cultural or even genetic factors. The figures led to a study of sickness absence rates in Post Office workers, whose records are not included in the national figures (Taylor, 1976). As their rates are measured in calendar days, while the national figures are based on a six-day week, the ratios of days lost per man in a region to days lost per man in the United Kingdom were used to make comparisons between the Post Office and the Social Security data (*Table 13.4*). For Wales and the North Eastern Region, the Department of Health and Social Security (DHSS) ratios are well above the corresponding Post Office regions. For London and the South East the Post Office ratio is higher than the Social Security ratio. In all other regions the ratios are similar. The high ratios for

Table 13.4 CERTIFIED SICKNESS ABSENCE RATIOS FOR MEN IN STANDARD REGIONS OF UNITED KINGDOM. SOCIAL SECURITY 1972–73, POST OFFICE 1974–75

Standard Regions	Social Security	Post Office
Wales	191	121
Northern Ireland	160	161
North East	139	108
North West	124	122
Scotland	118	117
South West	95	85
East and West Midlands	91	83
East Anglia	76	87
London and South East	62	82
United Kingdom	100	100

Ratios are calculated as follows: $\dfrac{\text{Age standardized days lost per man in region}}{\text{Days lost per man in United Kingdom}} \times 100$

Source: Taylor (1976)

Wales and the North East could be due to a preponderance of men in heavy industry (e.g. mining), while the low ratio for London and the South East may be caused by the great numbers of workers in light industry and offices. The study of Post Office workers showed that variation between regions cannot be explained entirely by differences in types of industry, since the same trend is shown in the Post Office figures which are effectively standardized for occupation.

Health records at the workplace

Occupational health services keep details of medical examinations and treatment of workpeople as well as of mortality and morbidity. These records help the medical staff with the management and treatment of patients and the investigation of compensation claims and provide essential data for annual reports. They have a further value as a source of descriptive epidemiological data.

Medical examinations

Routine surveillance can provide data which are useful in maintaining control of recognized hazards. For example records may reveal that a particular group of workers have raised levels of absorption of a toxic contaminant due to some fault in control measures. A periodic review of the results of routine examinations helps to determine their value, and when it is necessary to modify or stop using them.

Treatment records

Attendances for treatment indicate dangerous, or potentially dangerous, practices and identify persons who are prone to accidents. Epidemiological studies of treatment records may reveal hazards that cannot be recognized from a perusal of individual occurrences. Although such records do not give a true indication of minor injury and illness rates because many factors influence attendance for treatment, they provide useful pointers to more detailed studies.

Table 13.5 ATTENDANCE FOR CHEST COMPLAINTS AT A DISPENSARY IN A KENYAN SISAL FACTORY

	No. of workers	Attendances for chest complaints	Attendance rate (%)
Card room	130	70	54
Other departments	908	232	26

Source: Stott (1958)

In an African sisal factory records of treatment at the dispensaries showed that the attendance rate for chest complaints was much higher for workers in the card room where dust exposures were excessive, than for other departments (*Table 13.5*). Further analysis of hospital treatments showed that card-room workers, who represented one-eighth of the factory population, comprised about one-quarter of all workers admitted for acute chest conditions. This led to a full investigation of card-room workers, some of whom were found to be suffering from occupational pulmonary disease (Stott, 1958).

Data on treatment for serious injuries are more reliable in detecting causes as the variables affecting attendance virtually cease to operate. In collieries with thin seams miners have to move along in the prone position and use their elbows for propulsion. Treatments in one coalfield with thin seams showed a high incidence of 'beat elbow'. Elbow pads were developed and treatment improved with the result that beat elbow was virtually eliminated (Archibald and Kay, 1966).

The physician and nurse, as well as using their records to reveal unidentified hazards, should think epidemiologically in their daily task of treating individual workers. Inspired observations by clinicians on small groups of patients have led to the discovery of hazards such as nasal cancer in wood workers (Macbeth, 1965), and acro-osteolysis from heavy exposure to vinyl chloride monomer (Bastenier, Cordier and LeFevre, 1971).

Table 13.6 FALL IN PERCENTAGE OF WORKERS REFERRED TO HOSPITAL FOR EYE INJURIES AFTER APPOINTMENT, IN 1949, OF A SPECIALLY TRAINED NURSE

Year	Eye injuries treated	Eye injuries referred to hospital	
		Number	*Percentage*
1948	663	103	16
1949	698	32	5
1950	1250	26	2
1951	1466	62	4

Source: Roberts and Schilling (1952)

Treatment records can also provide valuable information on the work of a health service: its deficiences, how much it is used and how much time it saves the employer and the worker — all of which can help to raise standards. In an engineering factory where eye injuries were prevalent, treatment records revealed hazards which could be corrected, and this justified the employment of a more highly skilled nurse with ophthalmic training. After her appointment the number of workers seeking treatment increased, but the proportion sent to hospital was significantly reduced (*Table 13.6*).

Sickness absence at the workplace

Sickness absence recording and analysis are described in Chapter 16. They are helpful in the management of individual health problems but can also be used epidemiologically to assess sickness and behavioural characteristics of working groups, and to study effects of working conditions such as shift systems and toxic exposure.

Routinely collected data on sickness absence and lateness, and other forms of absence, in oil refinery workers revealed that shift workers had consistently and significantly lower rates of sickness than day workers (Taylor, 1967). The age-standardized inception rate (spells per 100 men), in the period 1962–63, was 182 per cent for day workers and 108 per cent for shift workers. Lateness and absenteeism showed similar wide differences between the groups. The findings led to a series of detailed studies of mortality and morbidity in day shift workers (Taylor and Pocock, 1972; Taylor, Pocock and Sergean, 1972a), and of absenteeism in different types of shift system (Taylor, Pocock and Sergean, 1972b).

CONCLUSION

Descriptive epidemiology is an essential tool of an occupational health service and of government departments responsible for the health and safety of people at work. It can provide clues for forming hypotheses about the aetiology of disease which may be tested by analytical and experimental methods (*see* Chapter 14). It provides data which are required for the control of recognized occupational hazards. It helps to establish priorities, to indicate needs in health care and provide information on community health problems. Descriptive epidemiology depends to a great extent on data routinely collected at workplaces. In the United Kingdom, and most other industrialized countries, the employer is not obliged to keep health records of employees. There is a need for two levels of recording. For persons exposed to recognized hazards, there should be detailed information on the occupational history, health examinations, illnesses and environmental exposure. For those not so exposed less detailed information may be recorded, but it should be sufficient to help in the identification of unrecognized risks.

REFERENCES

Archibald, R. and Kay, D.G. (1966) 'A study of beat elbow and related conditions at the colliery.' *Occupational Health*, **18**, 118–124
Ashford, N.A. (1976) *Crisis in the Workplace: Occupational Disease and Injury*. Cambridge, Mass: The MIT Press
Bastenier, H., Cordier, J.M. and LeFevre, M. (1971) 'Occupational acro-osteolytis.' *Occupational Health and Safety*, Vol. A-K, pp.33–34. Geneva: International Labour Organization
Case, R.A.M. and Davies, Joan M. (1964) 'On the use of official statistics in medical research.' *The Statistician*, **14**, 89–119
Dingwall-Fordyce, I. and Lane, R.E. (1963) 'A follow-up study of lead workers'. *British Journal of Industrial Medicine*, **20**, 313–315
Hartgerink, M.J. (1976) 'Health surveillance and planning for health care in the Netherlands.' *International Journal of Epidemiology*, **5**, 87–91
Health and Safety Commission (1977) *Occupational Health Services. The Way Ahead*, pp. 26. London: HMSO
Hill, A. Bradford and Faning, E. Lewis (1948) 'Studies in the incidence of cancer in a factory handling inorganic compounds of arsenic.' *British Journal of Industrial Medicine*, **5**, 1–6

Levi, L. (1978) *Quality of the Working Environment: protection and promotion of occupational mental health*. Laboratory for Clinical Stress Research, Report No. 88, 1–25. Stockholm: Karolinska Institute

Macbeth, R. (1965) 'Malignant disease of the paranasal sinuses.' *Journal of Laryngology*, **79**, 592–612

Malhotra, S.L. (1970) 'The use of epidemiological studies.' *Journal of Tropical Medicine and Hygiene*, **73**, 275–280

Morris, J.N. (1975) *Uses of Epidemiology*, 3rd edn, pp. 71–97. Edinburgh: Churchill Livingstone

Murray, R. (1958) 'Occupational cancer of the skin.' In *Cancer* (ed. R.W. Raven), **3**, 334–342. London: Butterworths

Newhouse, M.L. (1976) 'The prevalence of occupational disease.' *Annals of Occupational Hygiene*, **19**, 285–292

Newhouse, M.L. and Berry, G. (1979) 'Patterns of mortality in asbestos factory workers.' *Annals of the New York Academy of Science*, **330**, 53–60

Phillips, R.M. and Hughes, J.P. (1974) 'Cost benefit analysis of occupational health programme.' *Journal of Occupational Medicine*, **16**, 158–161

Registrar General (1978) *Occupational Mortality Decennial Supplement for England and Wales 1970–72*. Series D S, No. 1. London: HM Stationery Office

Roach, S.A. and Schilling, R.S.F. (1960) 'A clinical and environmental study of byssinosis in the Lancashire cotton industry.' *British Journal of Industrial Medicine*, **17**, 1–9

Roberts, Mary and Schilling, R.S.F. (1952) 'Assessing our value.' *Journal for Industrial Nurses*, **4**, 69–74

Royal Commission on Civil Liability and Compensation for Personal Injury (1978) *Statistics and Costings*, Vol. 2 Cmnd 7054–II London: HM Stationery Office

Schilling, R.S.F. (1956) 'Byssinosis in cotton and other textile workers.' *Lancet* ii 261–265 and 319–325

Schilling, R.S.F. (1971) 'Hazards of deep sea fishing.' *British Journal of Industrial Medicine*, **28**, 27–35

Stott, H. (1958) 'Pulmonary disease among sisal workers.' *British Journal of Industrial Medicine*, **15**, 23–37

Taylor, P.J. (1967) 'Shift and day work. A comparison of sickness absence, lateness and other absence behaviour at an oil refinery from 1962–1965.' *British Journal of Industrial Medicine*, **24**, 93–102

Taylor, P.J. (1976) 'Occupational and regional associations of death, disablement and sickness absence among Post Office staff 1972–75.' *British Journal of Industrial Medicine*, **33**, 230–235

Taylor, P.J. and Pocock, S.J. (1972) 'Mortality of shift and day workers 1956–68.' *British Journal of Industrial Medicine*, **29**, 201–207

Taylor, P.J., Pocock, S.J. and Sergean, R. (1972a) 'Shift and day workers absence: Relationship with some terms and conditions of service.' *British Journal of Industrial Medicine*, **29**, 221–224

Taylor, P.J., Pocock, S.J. and Sergean, R. (1972b) 'Absenteeism of shift and day workers. A study of six types of shift system in 29 organizations.' *British Journal of Industrial Medicine*, **29**, 208–213

Tiller, J., Schilling, R.S.F. and Morris, J.N. (1968) 'Occupational toxic factor in mortality from coronary heart disease.' *British Medical Journal*, **4**, 407–411

Torsvall, L. and Akerstedt, T. (1978) *Summary of a Longitudinal Study of Shift-work Effects on Well-being*. Report No. 92B from the Laboratory for Clinical Stress Research, Stockholm: Karolinska Institute

United States Department of Labor (1978) 'Occupational exposure to cotton dust. Final mandatory Occupational Safety and Health Standards.' *Federal Register*, **43**, No. 122. Washington DC: US Govt Printing Office

Volkova, Z.A. (1975) 'Methods used in the USSR for establishing biologically safe levels of toxic substances', pp. 160–168. Geneva: WHO

White, K.L., Anderson, D.O., Kalimo, E., Kleczkowski, B.M., Rurola, T. and Vukmanovic, C. (1977) *Health Services: Concepts and information for national planning and management*. Public Health Papers No 67. Geneva: WHO

World Health Organization (1977) *Methods Used in Establishing Permissible Levels in Occupational Exposure to Harmful Agents*. Technical Report Series, No. 601. Geneva: WHO

14

Epidemiology II — Analytical and Experimental Methods

M.L. Newhouse and R.S.F. Schilling

Methods of analytical and experimental epidemiology are used to test hypotheses on the aetiology of disease, often following the findings of descriptive surveys. They are also used to evaluate preventive measures and to establish and revise permissible levels of exposure.

ANALYTICAL EPIDEMIOLOGY

Analytical surveys, like descriptive surveys, can be *cross-sectional* providing information about a situation at a given point in time, or *longitudinal,* providing data relating to more than one point in time. Cross-sectional surveys are described separately in Chapter 15.

There are two methods of analytical epidemiology: the case-control study and the cohort study. Case-control studies are backward-looking in that causes are sought from the past history by comparing patients who have disease with controls, and for this reason are called retrospective studies. In cohort studies, which are essentially forward-looking, the relationship between causes (for example occupational exposures) and the incidence of disease is investigated in a defined population. A cohort study can be *historically prospective* — that is one in which the population entered employment many years ago and is followed up from past and future records of mortality or morbidity, or it can be *truly prospective (see* page 283).

THE CASE-CONTROL STUDY

In the case-control study a comparison of past exposure to possible causal factors is made between a group of persons with the disease being investigated (the index group) and an otherwise similar group without the disease (control group), to see if a suspected factor occurs more frequently, or to a greater degree, among the cases than the controls. It is a quick and comparatively cheap method of investigation which is particularly useful when the disease is too rare for a population survey to be practical.

Methods

The disease being studied must be adequately defined and it is helpful if supporting objective evidence such as radiological findings, autopsy reports or histological material are available for the cases. The study need not be limited to those cases occurring in a narrow time limit or in one particular area. The choice of an appropriate control group is more difficult. It must be similar in age and sex structure to the index group, come from a similar socioeconomic group and be derived from similar sources. Cases drawn from a hospital's records should be matched with controls drawn from that hospital, or another that is similar. Those drawn from a general practitioner's records should have controls from the same or similar sets of records. This helps to avoid differences in personal characteristics or socioeconomic circumstances that would occur if the index group was drawn from a general hospital and the control group from a fee-paying nursing home. Control groups which minimize these factors can be selected from among the patients' relatives or neighbours (Morgan and Shettigara, 1976). In some studies each case is individually matched with a control, and analysis of the data is based on the matched pairs; this is the approach to be preferred although overmatching can be a fault in the design of a study. For example, suppose the effect of occupation is under scrutiny and cases and controls are strictly matched by place of residence: since people usually live near their work, both groups may be influenced by environmental factors due to emissions from the factory where the index cases were employed.

Details of past history relevant to the study may be obtained from medical or other records, or, more commonly, by interview with the patient or, if deceased, with a surviving relative. The same method of obtaining information must be used for both cases and controls, using one or more investigators, who should have similar training. In order to avoid bias in interviewing, the investigation wherever possible should be done blind, that is without knowledge of whether the subject is a 'case' or a 'control'. This was possible in a hospital-based investigation into the association of asbestos exposure with laryngeal cancer. Occupational histories were taken from patients the day before laryngoscopy which established the diagnosis (Newhouse, Gregory and Shannon, 1980). The data should also be interpreted and coded blind.

A structured questionnaire should be used to record the information, first to assist in covering adequately the area of inquiry, and secondly to avoid the use of leading questions or of overemphasis in any particular area. The smoking history should be recorded, particularly where it is known to be relevant as in respiratory, cardiovascular and digestive tract diseases. It is important to obtain a high trace-rate of cases and of controls, since a high proportion of untraced individuals casts doubt on any conclusions drawn.

Examples of case-control studies

Newhouse and Thompson (1965) used the case-control method to obtain information about the association of asbestos exposure and subsequent development of mesothelial tumours. At the London Hospital histological specimens are kept and filed under diagnosis. All specimens with the diagnosis of mesothelioma were reviewed. Eighty-three were accepted as mesothelial tumours and 43 rejected (Hourihane, 1964). All but four of these patients were deceased, so a surviving relative was identified and interviewed and an occupational

Tables 14.1 TYPES OF EXPOSURE TO ASBESTOS OF 76 PATIENTS WITH MESOTHELIOMA AND 76 CONTROLS

Type of exposure	Mesothelioma series	Control series
Employed at one asbestos factory Delivered goods to factory	18 (25.0%) 1	1 (1.3%)
Employed at other asbestos factories	5 (6.6%)	1 (1.3%)
Insulators and laggers	7 (9.2%)	4 (5.3%)
Relative worked with asbestos	9 (11.8%)	1 (1.3%)
Dockers handling asbestos cargo	0	2 (2.6%)
No history of exposure to asbestos	36 (47.4%)	67 (88.2%)

Difference between positive occupational and domestic exposures to asbestos in mesothelioma patients and control patients: $\chi^2 = 27.11$; $P < .001$

history was taken. Histories were obtained for 76 of the 83 mesothelioma cases. Two control series were chosen.

For the first, each case was matched with a control selected from among patients in the London Hospital and a neighbouring geriatric hospital. The cases and controls were matched for date of birth (within five years), sex and approximate residential area. The design of the study was not ideal as the interviewer knew whether she was dealing with a case or control since all but four of the cases were deceased and all the controls were alive. The second control series consisted of the patients who had originally been diagnosed as dying of mesothelial tumours, but in whom the diagnosis had been rejected. It was difficult to locate relatives of those who had died before 1950, therefore only the 17 who had died after that date were included in the series. Again this series was not ideal as the possibility of bias on the part of the interviewer could not be excluded.

Forty (52.6 per cent) of the 76 mesothelial patients had a history of occupational or domestic exposure (living in the same house as an asbestos worker), compared with nine (11.8 per cent) of the patients in the first control series (*Table 14.1*). The investigation also suggested that neighbourhood exposures were important. Among 36 mesothelioma patients with no occupational or domestic exposure, 11 (30.6 per cent)

Table 14.2 RESIDENCE OF PATIENTS WITH NO OCCUPATIONAL OR DOMESTIC EXPOSURE TO ASBESTOS

Category	Lived within 0.5 mile of asbestos factory	Lived more than 0.5 mile from asbestos factory	Total
Mesothelioma series	11 (30.6%)	25 (69.4%)	36
Control series	5 (7.5%)	62 (92.5%)	67

$\chi^2 = 7.85$, $P < 0.01$
Source: Newhouse and Thompson (1965)

compared with five (7.5 per cent) of the 67 control patients lived within half a mile of an asbestos factory (*Table 14.2*).

Comparing the 72 confirmed mesothelioma cases who died after 1 January 1950 with the 17 patients in the second control series in whom the diagnosis was rejected on histological grounds, 50 of the former and none of the latter had a history of asbestos exposure.

A different approach was taken by a Swedish group (Axelson *et al.*, 1978) to study the association between arsenic exposure and mortality from lung cancer and other diseases. The investigation took place in a semirural area where a copper smelter was the principal industry. It was based on a local register of deaths which contained the diagnosis given on the death certificate. Three hundred and twenty-five subjects were available for the study after excluding unsuitable subjects such as those with vague diagnoses. Medical files were studied to ascertain smoking habits and histological types of cancer. The cases were those who had died from lung cancer or cardiovascular diseases or cirrhosis of liver.

Table 14.3 CALCULATION OF CRUDE RISK RATIO (INDIRECT RELATIVE RISK) IN UNPAIRED CASES AND CONTROLS

	Number of deaths from lung cancer	
	Exposed	*Not exposed*
Cases	(*a*) 18	(*c*) 11
Controls	(*b*) 18	(*d*) 56

$$\text{Indirect estimate of relative risk} = \frac{a\ d}{b\ c} = \frac{18 \times 56}{18 \times 11} = 5.1$$

The controls were those dying of all other diseases. Those who had worked in the smelter (the exposed) were identified. The crude risk ratio (indirect relative risk) based on the total data for all ages was calculated, together with a separated component estimating the risk associated with exposure when the age effect is eliminated. For both the crude risk ratio and the age-standardized ratio, the exposure group compared with the controls had a fivefold excess in lung cancer (*Table 14.3*) and about a twofold excess in cardiovascular disease mortality. The calculation of the crude risk ratio for deaths from lung cancer is shown in *Table 14.3*. It is used when cases and controls are not individually matched (i.e. not paired). It is assumed that cases and controls are reasonably representative of people with and without diseases in the population from which they came, and that disease prevalence is small. It was difficult to determine the interaction of other agents, particularly sulphur dioxide, copper, nickel, lead and other metals. A possible source of bias in this study was selected emigration out of the local population of those with lung cancer to seek treatment elsewhere. This would lead to underestimation of the risk. On the other hand, practitioners who were aware of the association between arsenic exposure and lung cancer might have over-diagnosed this disease in people who were known to be employed at the smelter. Such sources of bias can be eliminated, or at least lessened, by obtaining information on local inhabitants dying outside the area and by eliminating from the study all cases in which the diagnosis has not been confirmed by clinical investigation or histology.

Limitations

As case-control studies are retrospective, the quality of information about the disease and the particular factor under investigation may be incomplete and not readily validated, and the possibility of unrecognized confounding factors remains. Other disadvantages are that the selection of controls requires skill and judgement; care has to be taken to avoid bias not only in the selection of controls, but also in history taking and data interpretation and coding (*British Medical Journal*, 1979).

Case-control studies permit only indirect approximate estimates of relative risk, unlike cohort studies where a direct measure may be made by taking the ratio of the incidence of disease in the exposed group to that in the unexposed. Case-control studies generally need confirmation.

COHORT STUDIES

The word 'cohort', which originally meant the tenth part of a Roman legion, has been adopted by epidemiologists to describe a group of people defined by their date of birth, or some other common characteristic. In occupational medicine, a cohort study follows a group of workers and measures their mortality and/or morbidity, which may be compared with corresponding rates in a specially selected group or in the general population. Comparisons may also be made between groups in the same population with different levels of exposure (examples of mortality studies are given later in this chapter). The cohort study is used to test precisely formulated hypotheses about the aetiology of disease (Rose and Barker, 1978), to provide a direct measure of risk and to obtain dose response relationships from which permissible levels of exposure may be derived. The cohort study is seldom used as an exploratory method of investigation because it is usually time consuming and costly. The cohorts are defined in terms of characteristics manifest before the appearance of the disease under investigation. For example, they may be employees of a particular factory (Newhouse, 1969) or retired workers in a large industry (Enterline and Henderson, 1973).

Types of study

The cohorts are observed over a period of time to determine the frequency of disease. The investigation may be *historically prospective* with the relevant events (causes and effects) having already occurred when the study begins, or *truly prospective*. In the latter type, the relevant causes may or may not have occurred at the time the study is begun, but the cases of disease will not have occurred and the investigators have to wait for the disease to appear among the members of the cohort.

In the historically prospective cohort study information is obtained from existing records, without undue delay and at lower cost. Its shortcomings are that exposure data from the past may be unreliable or unobtainable and that there may be difficulties in tracing workers who have left employment. Historical studies are only possible where employment records of good quality have been kept and preserved. They combine the economy of the case-control study with the advantages of the cohort study, in providing direct estimates of relative risk from the ratio of incidence of disease in exposed and unexposed groups. This gives a more accurate measure of risk than the indirect method used in case-control studies.

The truly prospective cohort study has several advantages. The groups to be studied can be accurately defined. It is easier to continue observation of those who leave employment. Exposure can be monitored throughout the period of the follow-up. Its most serious disadvantage is that a long time must elapse before data are available. Sometimes conflicts between scientific and ethical principles occur. For example in the prospective study of viscose rayon workers (Tolonen *et al.*, 1975), it became evident after five years of investigation that exposure to carbon disulphide (CS_2) substantially increased coronary heart disease mortality. Steps had therefore to be taken to reduce concentrations of CS_2 in the air and to transfer workers with symptoms of coronary disease to departments with no exposures. The measures taken for ethical reasons, disrupted the original study design and the continued follow-up was less informative (Hernberg, 1974).

Cohort studies should preferably be reserved for diseases which have a relatively common outcome of exposure. Otherwise both the size of the cohort and the length of the observation time required may make the study impracticable.

Problems of follow-up

Occupational cancers become manifest only after a considerable period from first exposure, indeed, quite commonly after the worker has either left the industry in question or retired altogether. For occupational bladder cancer among dyestuff workers the *mean* interval between first exposure and development of tumour (i.e. the *mean latent* period) was 18 years with a range of 5 to 45 years (Case *et al.*, 1954). For mesothelial tumours in asbestos workers the *mean latent* period was 30 years with a range of 20–50 years (Newhouse and Thompson, 1965). Therefore it is necessary in many investigations to be able to follow the experience of workers for years after they have left the particular job where the exposure under investigation had occurred.

Studies of mortality data in the United Kingdom are made easier by the fact that information on individuals can be obtained by recognized research workers from the Office of Population Censuses and Surveys (OPCS). The Central Register of the Department of Health and Social Security (DHSS) will provide a similar service. Although both these sources are able to trace individuals, given name and date of birth, it helps if the National Health number can be provided for the OPCS, and the National Insurance number for the DHSS. In the United States similar investigations can be carried out, using the Social Security number given to each employed person, though confidentiality laws which protect the names of deceased persons make them far less complete. However, in the US good follow-up rates have been achieved by supplementary searches, through other governmental agencies.

If morbidity is being investigated, for example, the occurrence of coronary thrombosis, follow-up is possible with the help of local hospitals and physicians. It does require the cooperation of many individuals and is likely to be incomplete as persons leave the district and women change their names on marriage and cannot be traced.

Methods used in mortality studies

In a mortality study the cohort to be investigated is defined and the number and causes of death in the cohort over a certain period are ascertained. For example the cohort may consist of all persons starting employment in a particular factory, say between the beginning of 1930 and the end of 1950. An historically prospective study started in 1975 can be based on deaths occurring between 1 January 1951 and 31 December 1974; and it may be continued by collecting deaths occurring in a future period (*see Figure 14.1*).

In cohort studies, as in case-control studies, a comparison has to be made internally or with a control population (*see* Chapters 13 and 15). For mortality studies the national statistics of the country are commonly used as control data. In the United Kingdom the death rates for England and Wales, which are published annually, are available in the Registrar General's Statistical Reviews up to 1973, and from then on in OPCS publications (series DH). These are not ideal because comparison of a specific occupational group with the general population may be unreliable where the occupational group is self-selected or medically selected. Regional statistics are also published and in some instances may be preferable as control data.

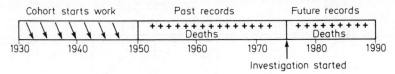

Figure 14.1 Example of a historically prospective cohort study

The occupational risk is estimated by comparing the observed number of deaths in the cohort with the number which would be expected if their mortality experience were the same as that of unexposed individuals of the same age and sex, dying in the same period in the general population. The man-years method of calculating the number of deaths to be expected was first used by Case and Lea (1955). The number of years at risk each individual has spent during the period of study is computed with each individual contributing from his date of entry to the cohort to his death or to the end of the follow-up period. A table is then constructed for all individuals, giving the total number of man years at risk in each five-year period. The death rates in the general population for all causes of death and for particular causes are then obtained for each five-year age group and each five-year secular period. Those for cancer in England and Wales are now published by the Institute of Cancer Research (1976). Each cell in the table of man years is multiplied by the appropriate death rate(s) and summed over rows and columns to give the expected number of deaths. There are computer programmes for calculating these figures. *Figure 14.2* shows how individuals contribute to the total man years. As they age they pass from cell to cell. The death rates for lung cancer are indicated as rising with increasing age and rising steadily during the period 1930–70.

The mortality of the group under investigation must be examined according to different characteristics of the group. In addition to sex and date of birth, the type of occupation, length of employment and degree of exposure are important. Much of this information is available in the personnel records of the factory, and some firms retain complete files on their past and current employees which give the basic data. Information on smoking habits has to be sought elsewhere than in personnel records.

There are certain difficulties in historically prospective cohort studies as the mortality relates to the conditions of many years ago. Methods of measuring environmental contaminants may change or measurements may be unavailable or unreliable. Length of employment may be taken as a guide to the degree of exposure, or the various jobs may be categorized

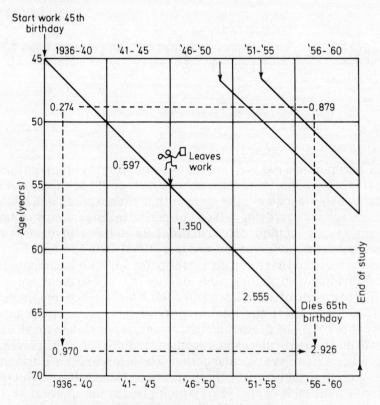

Figure 14.2 National death rates per 1000 man years for cancer of lung and pleura shown by broken line and on diagonal. Diagonal lines indicate individuals' total number of man years in the cohort, e.g. the man starting work on 45th birthday at beginning of 1936 contributes five man years in each quinquennium until he dies.

according to past experience and sporadic measurements of airborne concentrations (Newhouse and Berry, 1979). If sufficient environmental data are available and a detailed occupational history of each individual can be obtained, it may be possible to calculate the cumulative dose of, for example, asbestos dust to which a worker was exposed during his time in the industry. By these methods internal comparisons, as well as comparisons with control groups, give valuable information.

Examples of mortality studies

Three examples are given of cohort studies in which mortality rates are related to occupational exposures, estimated in different ways. Lee and Fraumeni (1969) used the length of employment as an index of the degree of exposure in a study of the effect of arsenic exposure on the mortality of workers who had been employed at a copper smelter. Mortality risk increases with intensity of exposure, whatever the length or period of employment (*Table 14.4*).

Table 14.4 STANDARDIZED MORTALITY RATIOS OF COPPER SMELTER WORK-ERS WITH MORE THAN ONE YEAR'S EXPOSURE TO ARSENIC AT DIFFERENT TIMES AND FOR DIFFERENT LENGTHS AND DEGREES OF EXPOSURE

Description of cohort	*Degree of exposure and no. of workers*		
	Light	*Medium*	*Heavy*
	3257	*1526*	*402*
15 years or more exposure before 1938	250	667	800
15 years or more exposure after 1938	310	545	667
Up to 15 years exposure after 1938	214	263	444

Newhouse and Berry (1979) made a historically prospective study of a cohort of workers in an asbestos factory who were first employed between 1 April 1933 and 31 December 1964. They included all those who had been employed for more than 30 days. First they divided them into those employed for up to two years and those with longer than two years. They then separated them according to their degree of exposure as judged by their occupational history. The cohort was finally classified into 'low to moderate' and 'severe' exposure categories (*see Table 14.5*). Expected deaths were based on the mortality experience of the male population of England and Wales. There were statistically highly significant excesses of deaths from all causes and from cancers of the lung and pleura among all workers with severe exposures. Groups with low to moderate exposures showed a significant excess of deaths from cancers of lung and pleura, but only among those with more than two years employment.

Table 14.5 MORTALITY EXPERIENCE OF MALE ASBESTOS FACTORY WORKERS FROM 1933 TO END OF 1975

	Exposure category							
	Low to moderate				Severe			
	< 2 years (884)		> 2 years (554)		< 2 years (937)		> 2 years (512)	
Cause of death	Observed	Expected	Observed	Expected	Observed	Expected	Observed	Expected
All causes	118 (4)	118.0	89 (7)	95.3	162*** (16)	122.2	176*** (19)	102.5
Cancers of lung and pleura (ICD 162, 163)	17 (3)	11.01	16* (1)	9.0	31*** (6)	12.8	56*** (7)	10.4
Gastrointestinal cancer (ICD 150–158)	10	9.0	9 (4)	7.3	20** (6)	9.5	19** (8)	8.2
Other cancers	6	7.4	8 (1)	5.8	16** (3)	7.9	16*** (4)	6.3
Chronic respiratory disease	19	17.5	16	14.7	20 (1)	17.6	28**	15.9

***p <0.001
**p <0.01
*p <0.05   Figures in brackets indicate the number of mesothelial tumours in each category.

Source: Newhouse and Berry (1979).

It was not possible to calculate expected figures for mesothelial tumours as they are so rare among the general population that they are not itemized in national statistics. In this study, validation of death certificates by examination of autopsy reports and histological material revealed a higher number of these tumours than was recorded on the death certificates (Newhouse and Wagner, 1969). Death rates for mesothelial tumours were calculated, using, as the denominator, the number of subject years at risk in each subdivision of the cohort (*Table 14.6*). A definite dose-response relationship is shown between the lowest and the highest exposure category.

Table 14.6 MESOTHELIOMA DEATH RATES AMONG ASBESTOS WORKERS

Exposure and duration (years)	Pleura	Peritoneum	Subject years	Rate per 100 000* man years
Males				
Low to moderate				
<2 years	3	1	12 031	33
>2 years	3	4	7500	93
Severe				
<2 years	6	10	15 428	104
>2 years	7	12	7827	243
Laggers				
<2 years	3	2	7893	63
>2 years	1	4	2690	186
Females				
Low to moderate	1	0	2066	48
Severe				
<2 years	8	5	9538	136
>2 years	4	3	4388	360

*Total number of mesothelial tumours × 100 000

number of subject years

Source: Newhouse and Wagner (1969)

In another historically prospective cohort study, McDonald, *et al.* (1980) analysed cancer mortality of chrysotile miners and mill workers in relation to accumulated dust exposure. This was based on dust measurements and length of employment. Their cohort consisted of 11 000 males and 440 females, born between the beginning of 1891 and the end of 1920, who had worked in the Quebec mines or mills for at least a month. Over 90 per cent of the cohort was traced and death certificates obtained for the deceased. Serial dust counts measured by the midget impinger for the period 1949 to 1966 were available. Figures for earlier years were estimated. Nearly 6000 jobs were identified from detailed job histories and for each one for each year an average dust concentration was estimated. Thus the accumulated dust exposure for each subject over the period of his employment, and expressed as million particles per cubic foot years (mpcf·y) could be calculated. Two methods of analysis were used. The first was to calculate standardized mortality ratios (SMRs) from the ratio of observed deaths to the expected deaths derived from the male mortality of the Province of Quebec, according to man

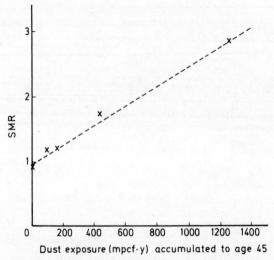

Figure 14.3 Lung cancer standard mortality rates in relation to dust exposure in million particles per cubic foot years accumulated to age 45 years. The broken line has been fitted by a modified least-squares technique. Source: McDonald et al. (1980)

years of exposure. The results for cancer of the lung (*Figure 14.3*) show a linear exposure effect relationship. Deaths from pneumoconiosis follow a similar pattern. In the second method internal controls were used to calculate relative risk or risk ratios of dying of certain diseases such as lung cancer (McDonald *et al.*, 1980). Three controls for each lung cancer death were selected at random from among men born in the same year as the case and known to have survived at least into the year following that in which the case died. This method avoids the use of an external group which may not be strictly comparable. It also facilitates the examination of the interaction between variables such as smoking, dust concentration

Table 14.7 RELATIVE RISK OF DEATH FROM LUNG CANCER IN CHYRYSOTILE ASBESTOS MINERS AND MILL-WORKERS

	Dust exposure mpcf.y*				
	<30	*30<300*	*300<1000*	*≥1000*	*Total*
Deaths	89	73	56	27	245
Controls	333	243	127	32	735
Relative risk†	1	1.1	1.7	3.2	

*Million particles per cubic foot years
†*See Table 14.3* for method of calculation
Source: McDonald *et al.* (1980)

and length of occupation. *Table 14.7* shows the relative risk of dying from lung cancer in groups with dust exposures greater than 30 mpcf·y. Their risk is compared with the risk for those with a cumulative exposure of less than 30 mpcf·y (*see Table 14.3* for method of calculation).

Cross-sectional surveys of respiratory symptoms and function and radiological status of these millers and miners in Quebec were also made and demonstrated a similar pattern of exposure response (McDonald *et al.*, 1974).

Example of a truly prospective cohort study

A prospective cohort study in occupational medicine is described to illustrate the methods used and the problems that may arise. Historically prospective studies in England (Tiller, Schilling and Morris, 1968) and in Norway (Mowé, 1971) had shown an increased mortality from coronary heart disease (CHD) among viscose rayon workers exposed to carbon disulphide (CS_2). A prospective study to examine the risk of fatal and non-fatal CHD was undertaken in Finland to determine whether CS_2 exposure accelerated the risk of developing CHD and whether it caused a more severe prognosis in persons who already had atherosclerotic changes in the coronary arteries (Tolonen *et al.*, 1975). In 1967 two cohorts, each with 343 men, were defined. One comprised viscose rayon workers with at least five years exposure to CS_2 between 1942 and 1967. The other (the control group) consisted of paper mill workers with no exposure to CS_2. The men in the two cohorts were individually matched for age, district of birth and type of work. At the first survey in 1967, relevant coronary risk factors (e.g. smoking history, physical activity and blood lipids) were equally distributed in both groups. Blood pressure was slightly higher in the CS_2 group and was interpreted as an effect of exposure rather than an independent risk factor. At the end of the five-year follow-up the only change in these risk factors was a slightly higher blood cholesterol in the exposed group.

During the five years, 14 of the exposed group and 3 of the controls died of coronary heart disease. Other causes of deaths were evenly distributed. There were 11 non-fatal first myocardial infarctions in the exposed group and 4 in the controls. At the end of the survey 25 per cent of the exposed group and 13 per cent of the controls gave a history of angina. The prevalence of electrocardiograph (ECG) abnormalities was slightly higher in the exposed group.

Table 14.8 RELATIVE RISK* OF CORONARY HEART DISEASE IN VISCOSE RAYON WORKERS ACCORDING TO THE SEVERITY OF MANIFESTATION

Manifestation	Relative risk (%)
Fatal infarction	4.8
Non-fatal infarction	2.8
Angina history	2.2
ECG changes	1.4

*Relative risk is the ratio of the incidence of disease in the exposed group to that in the unexposed group

Source: Tolonen *et al.* (1975)

Table 14.8 shows that the relative risk is highest in the fatal infarctions and lowest for ECG findings, indicating that exposure to CS_2 seems to worsen the prognosis of CHD as well as increasing its incidence. These findings, which indicated a serious occupational risk, led to a substantial reduction in CS_2 exposure levels. The Finnish authorities lowered their threshold limit value (TLV) for CS_2 from 20 to 10 parts per million. This truly prospective cohort study made it possible, for a limited period, to examine an occupational risk in more detail and more accurately than would have been possible with a historically prospective study. When the study had revealed the extent of the risk, preventive measures had to be taken, thus impairing the value of continued follow-up (*see* page 283).

With the increasing use of ongoing records of biological and environmental data, the truly prospective study should become more common in occupational health research.

EXPERIMENTAL EPIDEMIOLOGY

In analytical surveys groups are studied as they are found in their natural setting. In experimental epidemiology the investigator manipulates the factor being studied. Its uses in occupational medicine are to identify a suspected aetiological agent, or to test a preventive measure. In this way it may be possible to get quicker and more reliable answers to questions about aetiology and prevention. Industrial populations, because they are accessible and easily defined groups, may be used for field experiments such as clinical trials of drugs and vaccines or large-scale trials of health education measures.

Methods

The randomized controlled trial

'A controlled clinical trial is one in which the two series, under simultaneous investigation, are as alike as possible in every respect except that in one series the patients receive the control drug or procedure. The phrase "as alike as possible" means that these conditions have to be maintained both at the outset and during the course of investigation (Colton, 1974).'

Mistakes have been made in allocating people to the treatment and control series. A frequent source of error is comparison of volunteers for treatment with those who do not volunteer (Hill, 1977). The difference between them can be considerable. The only reliable method of assigning individuals or population groups to experimental and control groups is by a process of randomization (the randomized controlled trial). Such assignments can be made by using tables of random numbers or random permutations. It is an assurance against selection bias, but it does not guarantee that experimental and control groups will be identical.

The subjects may be arranged in pairs, matched for age, sex and other characteristics. The two members of the pair are assigned at random, one to the experimental and the other to the control group. Another method is to stratify subjects into age and sex groups and then make random assignments to the experimental or control categories within each age/sex group.

Pairing can be difficult and time-consuming. Sometimes the relevant factors for pairing or stratification are not known. Often there may be individuals who cannot be suitably matched. In such circumstances comparison between unmatched experimental and control groups can be made, using statistical techniques to allow for differences between the groups, with respect to age structure and other factors.

Bias in the handling and evaluation of the treatment and control groups also has to be avoided. Blind techniques and placebos make it possible to eliminate factors unintentionally associated with treatment. In a single blind trial patients do not know whether they are in the treatment or control group. In the double blind trial, both patients and investigators are unaware of who is assigned to each group.

Ethical issues Controlled experiments have ethical problems as control subjects may be deliberately deprived of some supposedly useful treatment or preventive measure. This can be justified if the potential value of the treatment or preventive measure is not clear and not strongly supported by suggestive evidence. Experiments are made easier when life and death, or serious disability, are not at issue (Hill, 1962).

Difficulties in design Randomized controlled trials, as well as presenting ethical problems, are difficult to achieve, and there are fundamental objections to their experimental design (McDonald, 1980). They are so highly controlled that it becomes unreal and difficult to generalize from their results. A randomized controlled trial demands a degree of rigidity which may prevent the application of lessons learned until the trial is over. It is expensive both in time and effort.

Other methods

Other methods include the use of subjects as their own controls and the 'non-experimental' before and after study. The main problem is to collect data free from subject and observer bias. It may be difficult or impossible to disguise the intervention and much then depends on objective measurements by observers and attempting as far as possible to avoid bias in the collection and interpretation of data (McDonald, 1980).

Aetiology of occupational disease

In the first clinical environmental survey of cotton workers undertaken in England (Schilling and Roach, 1961) there was a close correlation between the prevalence of byssinosis and concentrations of total dust. Unexpectedly there was no correlation between prevalence of disease and concentrations of respirable dust particles (i.e. less than 7μ

equivalent diameter). This indicated that the larger dust particles were more likely to be the cause of byssinosis. The acute effects of total and of respirable dust on lung function and chest symptoms were tested in an experiment in which 12 volunteers (including 10 cotton workers) were exposed, in a plastic tent, to total dust and to respirable dust for periods of three or four hours. The subjects acted as their own controls, the forced expiratory volume ($FEV_{0.75}$) and chest symptoms

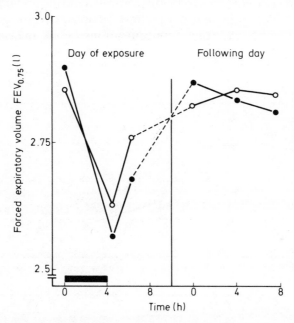

Figure 14.4 Mean change in $FEV_{0.75}$ caused by exposure to total cotton dust (●) and fine dust, less than 7μ (○) in 12 men. Period of exposure is indicated by ——
Source: Schilling (1979)

being recorded before and after exposure to each type of dust (*Figure 14.4*). For lung function there was little difference in response to the two types of exposure and symptoms of chest tightness showed no obvious differences (McKerrow *et al.*, 1962). This result indicated that the respirable dust particles could cause the disease.

Prevention of occupational disease

In laboratory trials the steaming of cotton as a measure to prevent byssinosis reduced dust levels and its biological activity, without impairing its spinning qualities (Merchant *et al.*, 1973). An intervention trial was subsequently set up in a cotton mill to test the effect of steaming, on dust levels, decrement in FEV_1 during the shift, and chest symptoms on a selected panel of 62 workers with byssinosis. While steaming reduced the dust levels and the acute effects on workers in the preparatory processes, it made conditions worse in later processes by binding the fine dust to the cotton fibre (Merchant *et al.*, 1974).

A before-and-after study in another cotton mill took advantage of the fact that steaming was to be discontinued during installation of new equipment. The prevalence of chest symptoms, changes in FEV_1 during the shift and dust concentrations were measured during

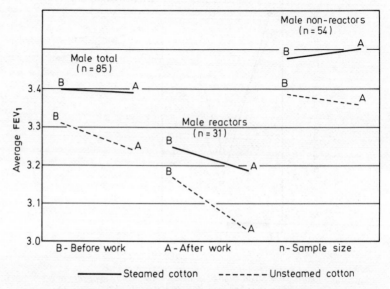

Figure 14.5 Average FEV_1 before and after work for males, based on average of two highest readings. This shows beneficial effects of steaming: (1) Lower drop in FEV_1 during shift and (2) higher pre-shift FEV_1 during steaming.
Source: Imbus and Suh (1974)

steaming and again in the period of no steaming. Among the 85 male workers tested there were 31 reactors to cotton dust, comprising those with byssinosis symptoms and those who experienced a definite drop in FEV_1 during the work shift. Steaming significantly reduced the acute changes in FEV_1 during the shift for both the reactors and non-reactors (*Figure 14.5*). While steaming was in operation the level of FEV_1 was higher by 60–100 ml in all groups (*see Figure 14.5*). It also reduced average respirable and total dust levels by about 30 per cent. However, steaming did not reduce symptoms (Imbus and Suh, 1974). These experiments showed that steaming was, at best, an adjunct to exhaust ventilation, but not an alternative. Both are examples of the use of subjects as their own controls.

Where adverse effects take a long time to develop, as from exposure to asbestos, the efficacy of new preventive measures may be evaluated by airborne dust sampling. The effects on the people at risk can be assessed by a before-and-after study of mortality. It is the ultimate measure of success or failure. The improvement in dust control in asbestos factories started with the implementation of new regulations in 1933. A follow-up study in one factory by Peto *et al.* (1977) to the end of 1974 was made of three cohorts: those employed before 1933, those first employed between 1933 and 1950, and those first employed in 1951 and later. It reveals a substantial lessening of the risk of lung cancer following improvements in dust control in 1933 (*Table 14.9*).

Table 14.9 NUMBER OF OBSERVED AND EXPECTED DEATHS IN AN ASBESTOS TEXTILE FACTORY BY DATE OF FIRST EXPOSURE

Employment	Number of workers	Lung cancer deaths*		
		Observed	Expected	Ratio observed/expected
Before 1933	143	25(5)	4.5	5.5
Between 1933–1950†	616	30(5)	16.1	1.9
Since 1951	347	6(0)	3.2	1.9

*Deaths due to pleural mesothelioma are included in observed number of lung cancer deaths and also given separately in brackets
†Implementation of new regulations for dust control in 1933
Source: Peto *et al.* (1977)

Because of the long delay of 15 years or more between first exposure and any resulting cancer, these figures do not reflect the effects of more recent improvements in dust control, which have led to a threefold reduction in dust concentrations.

Testing other preventive measures

Occupational health services in large organizations participate in epidemiological experiments or trials to test a vaccine, drug or a particular health education programme. They can offer a large potentially healthy population which is accessible, and from whose members health and sickness data can be collected without much difficulty. Since the experimental population chosen for such trials should be similar to the population to which the measure, if successful, will be applied, industrial groups are generally preferable to hospital patients. The British Medical Research Council's influenza and other vaccine trials have been carried out in industrial populations with the help of occupational health service staff (Medical Research Council, 1953, 1964).

Drug trials

A randomized controlled trial was designed to study the effect of ascorbic acid taken for 2.5 days from the start of the first symptoms of the common cold (Tyrrell *et al.*, 1977). Volunteers (1524) were recruited from seven different working groups in different parts of the country. Each volunteer was given a tube of 10 effervescent tables. All 10 tablets in the tubes given to those receiving active treatment contained ascorbic acid 1 g; all those in the placebo group contained inert substances of identical appearance and taste. Allocation to active or placebo treatment was random and double-blind. A small subsidiary trial confirmed that volunteers could not detect the difference between the two preparations by taste. Each volunteer was given a card on which to record the presence or absence of specified symptoms in the event of a cold. Twenty-three volunteers were excluded because of discontinuation of treatment or because their cards were spoiled, and 482 had one or more colds. There was no evidence in those who developed colds that either the upper respiratory or the general constitutional symptoms were alleviated by active treatment. The only finding of possible interest was that among the men who developed colds significantly fewer of those on active treatment than those on placebo treatment had multiple colds.

Heart disease prevention project

There are good predictors of risk of coronary heart disease which can be measured without much difficulty. They are raised blood pressure, cigarette smoking, lack of physical activity, obesity and raised blood cholesterol levels. The important question is whether by diminishing these risk factors in patients who have them, it is possible to reduce the risk of coronary heart disease. In an attempt to answer this question an international controlled trial in the prevention of coronary heart disease has been organized by the World Health Organization European Collaborative Group (1980). As part of this project Rose and others (1980) set up a randomized controlled trial in collaboration with occupational physicians in 24 factories or other occupational groups in the United Kingdom with a total of 18 210 male participants, aged 40–59 years. These 24 factories were paired as well as possible for size, region and nature of industry. Within the pairs they were randomized to either:

1. An intervention factory where participants were screened and given advice on smoking, diet and exercise, and treated for any hypertension detected; or
2. A control factory in which there was screening only for a 10 per cent sample of its members, but no intervention.

Examinations in the control group were limited in this way because a heart examination might alter the men's attitudes and behaviour and itself be an intervention measure. In the intervention factories the following treatment or advice was given: cholesterol-lowering diet (all subjects); cessation of cigarette smoking (smokers); daily physical exercise (sedentary workers); weight reduction (men 15 per cent or more overweight) and hypotensive drug therapy (men with raised systolic blood pressure).

First, the level of success in the control of each of the risk factors has been measured. There were small, but significant, reductions mainly in the high risk groups. (They comprised the 12–15 per cent of the examined men in each factory with the highest coronary risk scores.) These men were recalled for consultation with the occupational physician and advised or treated according to which of the five intervention factors were appropriate. In these groups there was a reported fall in cigarette consumption of 30 per cent at the end of the trial. Average blood pressure levels were lowered by a few milli-metres. Dietary changes and weight loss were not sustained once encouragement was relaxed.

At a later date the outcome of the trial will be assessed on the basis of total mortality and of clinically diagnosed events such as fatal acute myocardial infarction, sudden death (presumed coronary heart disease), non-fatal acute myocardial infarction and angina pectoris. After five years there was a final examination of all survivors to estimate differences between the intervention and control groups in risk factor levels and prevalence of disease.

In trials in which communities are allocated randomly to intervention and control groups, the same epidemiological principles apply as in trials with allocation of individuals. Once the two groups have been assembled, it is essential to ensure that there are no important differences occurring between them, by chance, that might affect the incidence of disease being studied. During the study the management of the two groups must be similar in every respect, except for the intervention measures being tested. One important source of bias is that defaulters from these prevention trials are more likely to be those who have not complied with the advice or treatment given (Rose and Barker, 1978).

ACTION ON RESULTS OF EXPERIMENTS

There are often problems in deciding what action has to be taken once the experiment is over. Such difficulties are succinctly summarized by Rose and Barker (1978).

'A positive outcome of a trial, in terms of decreased mortality or morbidity in the experimental group, will require balancing of the magnitude of the decrease against the costs of mounting a definitive service for the reference population. The statistical significance of a difference between the experimental and control groups will depend not only on the magnitude of the difference but on the number of participants. In a large trial a statistically significant result may fall short of what is of practical importance. Conversely and more frequently, a difference large enough to be of practical importance may fail to achieve statistical significance in a small trial. A trial may have a negative outcome either because the preventive measures are inherently ineffective or because compliance with them is poor. Either way, that particular preventive programme has failed – but it is important that there should be provision in the trial for measuring compliance so that potentially effective methods are not prematurely discarded.'

REFERENCES

Axelson, O., Dahlgren, E., Jansson, C-D. and Rehnlund, S.O. (1978) 'Arsenic exposure and mortality, – a case reference study from a Swedish copper smelter.' *British Journal of Industrial Medicine*, **35**, 8–15

British Medical Journal (1979) *The Case Control Study*, **2**, 884–885

Case, R.A.M., Hosker, M.E., Drever, B. McD., and Pearson, J.T. (1954) 'Tumours of the urinary bladder in workmen engaged in the manufacture and use of certain dyestuffs intermediates in the British chemical industry.' *British Journal of Industrial Medicine*, **11**, 75–104

Case, R.A.M. and Lea, A.J. (1955) 'Mustard gas poisoning, chronic bronchitis and lung cancer.' *British Journal of Social Medicine*, **9**, 62–72

Colton, Theodore (1974) *Statistics in Medicine*. Boston: Little Brown and Co.

Enterline, P.E. and Henderson, V. (1973) 'Asbestos and respiratory cancer.' *Archives of Environmental Health*, **27**, 312–317

Hernberg, Sven (1974) 'Epidemiological methods in occupational health research.' *Scandinavian Journal of Work Environment and Health*, **2**, 59–68

Hill, A. Bradford (1962) 'The Clinical Trial.' In *Statistical Methods in Clinical and Preventive Medicine*, p. 18. London: Livingstone

Hill, A. Bradford (1977) *A Short Textbook of Medical Statistics*, pp. 294. London: Hodder and Stoughton

Hourihane, D.O.B. (1964) 'The pathology of mesotheliomata and an analysis of their association with asbestos exposure.' *Thorax*, **19**, 268–278

Imbus, H.R. and Suh, M.W. (1974) 'Steaming of cotton to prevent byssinosis – a plant study.' *British Journal of Industrial Medicine*, **31**, 209–219

Institute of Cancer Research (1976) 'Division of epidemiology serial mortality tables, neoplastic diseases.' *Deaths and Death Rates by Sex, Age, Site and Calendar Period* (4 volumes). England and Wales 1911–1970

Lee, A.M. and Fraumeni, J.F. (1969) 'Arsenic and respiratory cancer in man: an occupational study.' *Journal of the National Cancer Institute*, **42**, 1045–1052

McDonald, J.C. (1980) 'Epidemiology.' *Occupational Lung Diseases: research approach and methods* Eds H. Weill and M. Turner Warwick. New York: Marcel Dekker Inc.

McDonald, J.C., Becklake, M.R., Gibbs, G.W., McDonald, A.D. and Rossiter, C.E. (1974) 'The health of chrysotile, asbestos mine and mill workers of Quebec.' *Archives of Environmental Health,* **28,** 61–68

McDonald, J.C., Liddell, F.D.K., Gibbs, G.W., Eyssen, G.E. and McDonald, A.D. (1980) 'Dust exposure and mortality in chrysotile mining 1910–1975.' *British Journal of Industrial Medicine,* **37,** 11–24

McKerrow, C.B., Roach, S.A., Gilson, J.C. and Schilling, R.S.F. (1962) 'The size of cotton dust particles causing byssinosis: An environmental and physiological study.' *British Journal of Industrial Medicine,* **19,** 1–8

Medical Research Council (1953) and (1964) Reports of Committee on Influenza and Respiratory Virus Vaccine Trials. *British Medical Journal* **2,** 267–271 and *British Medical Journal* **2,** 1173–1177

Merchant, J.A., Lumsden, J.C., Kilburn, K.H., Germino, V.H., Hamilton, J.D., Lynn, W.S., Byrd, H. and Baucom, D. (1973) 'Preprocessing cotton to prevent byssinosis.' *British Journal of Industrial Medicine,* **30,** 237–247

Merchant, J.A., Lumsden, J.C., Kilborn, K.H., O'Fallon, W.M.O., Copeland, K., Germino, V.H., McKenzie, W.N., Baucom, D., Currin, P. and Stilman, J. (1974) 'Intervention studies of cotton steaming to reduce biological effects of cotton dust.' *British Journal of Industrial Medicine,* **31,** 261–274

Morgan, R.W. and Shettigara, P.T. (1976) 'Occupational asbestos exposure, smoking and laryngeal cancer.' *Annals of the New York Academy of Science,* **271,** 308–316

Mowé, G. (1971) 'Coronary heart disease and occupational exposure to carbon disulphide.' *International Symposium on Toxicology of Carbon Disulphide, Yugoslavia 1971,* Abstract II 25–28, ed. Dusan Djuric *et al.* Belgrade Institute of Occupational and Radiological Health

Newhouse, M.L. (1969) 'A study of the mortality of workers in an asbestos factory.' *British Journal of Industrial Medicine,* **26,** 294–301

Newhouse, M.L. and Thompson, H. (1965) 'Mesothelioma of pleura and peritoneum following exposure to asbestos in the London area.' *British Journal of Industrial Medicine,* **22,** 261–269

Newhouse, M.L. and Berry, G. (1979) 'Patterns of mortality in asbestos factory workers in London.' *Annals of the New York Academy of Science* **330,** 53–60

Newhouse, M.L. and Wagner, J.C. (1969) 'Validation of death certificates in asbestos workers.' *British Journal of Industrial Medicine,* **26,** 302–307

Newhouse, M.L., Gregory, M.M. and Shannon, H.S. (1980) 'Aetiological factors in carcinoma of the larynx.' In *Biological Effect of Mineral Fibres.* Lyon: IARC, Scientific Publications No.30

Peto, J., Doll, R., Howard, S.V., Kinlen, L.J. and Lewinsohn, H. (1977) 'A mortality study among workers in an English asbestos factory.' *British Journal of Industrial Medicine,* **34,** 169–173

Registrar General *Statistical Reviews of England and Wales, up to 1973, and since. Office of Population Censuses and Surveys* (Series D H). London: HM Stationery Office

Rose, Geoffrey and Barker, D.J.P. (1978) 'Epidemiology for the uninitiated – experimental studies' *British Medical Journal* ii, 1687–1688

Rose, Geoffrey, Heller, R.F., Tundstall Pedoe, H.D. and Christie, D.G.S. (1980) 'The heart disease prevention project: a randomized controlled trial in industry. Changes in coronary risk factors.' *British Medical Journal,* **280,** 747–751

Schilling, R.S.F. (1979) 'Focus on grain dust and health.' In *Occupational Pulmonary Disease.* New York: Academic Press

Schilling, R.S.F. and Roach, S.A. (1961) 'Safe levels of dustiness in cotton spinning mills' *Pure and Applied Chemistry,* **3,** 69–75

Tiller, J.R., Schilling, R.S.F. and Morris, J.N. (1968) 'Occupational toxic factor in mortality from coronary heart disease.' *British Medical Journal,* **4,** 407–411

Tolonen, M., Hernberg, S., Numinen, M. and Tiitola K. (1975) 'A follow-up study of coronary heart disease in viscose rayon workers.' *British Journal of Industrial Medicine,* **32,** 1–10

Tyrrell, D.A.J., Craig, J.W., Meade, T.W. and White, T. (1977) 'A trial of ascorbic acid in the treatment of the common cold.' *British Journal of Preventive and Social Medicine,* **31,** 189–191

World Health Organization European Collaborative Group (1980) 'Multifactorial trial in the prevention of coronary heart disease, I. Recruitment and initial findings.' *European Journal of Cardiology, (in press)*

15

Field Surveys and Sampling Methods

Joan Walford

Routine records of mortality and morbidity are often used in studies of the health of industrial populations and, in particular, to measure the occupational risk of certain diseases. These methods of investigation are useful in studying diseases of low prevalence, and in providing some indication of the presence of an occupational hazard, but they have several limitations. Studies using routine data are often bedevilled by changes in nomenclature from one period to another, and by the inaccurate recording of occupation and cause of death. Information relevant to the aetiology of the disease is not usually available. For a more accurate and penetrating study of the prevalence and severity of disease and its relation to the working environment it is necessary to carry out a field survey with planned observations made at first hand.

A field survey may be cross-sectional or longitudinal. It may be designed as a purely descriptive study to assess the magnitude of a problem and estimate disease prevalence or incidence, but more often it is used in an analytical study to test aetiological hypotheses or measure a dose-response relationship. It may also be used to determine 'normal' values of physiological and biological measurements which can subsequently be used to assess the health risk in suspect occupations. The collection of first-hand data is necessary for the accurate evaluation of a medical service or health education programme, or to test the validity of routine statistics. The disadvantage of the field survey as a method of investigation is that special skills and techniques are usually required and much time, effort and cost can be involved.

PLANNING A SURVEY

The objectives of the survey should be clearly defined: why it is being carried out, what questions it seeks to answer, and what its limitations are likely to be. Before planning the survey it is important to be familiar with the work of previous investigators in the field of study.

A survey of the working environment is usually required at the same time as that of workers. Close collaboration should be maintained between the medical observers and the occupational hygienists, both in the planning and the execution of the survey.

Coverage

The coverage of a medical survey can range from a few case studies to examination of the entire population at risk. If the population is very large complete coverage would be far too costly and time consuming, and it is usual to select a representative sample to provide estimates of population characteristics. This has the advantage of permitting a more detailed and careful investigation to be carried out than would be possible in a survey of the entire population. The disadvantage is that the sample results will be inaccurate to the extent by which they differ from those that would have been obtained had the whole population been examined. Statistical methods can provide an estimate of the magnitude of this inaccuracy, and it is sometimes possible to calculate the sample size that is needed to give results of a required precision. Even if a whole population can be examined it may still have to be regarded as a sample since it is a sample in time.

In some studies of industrial populations a sample has been selected from the community in which the industry is situated instead of from the factory population itself (Cochrane *et al.*, 1956; Higgins *et al.*, 1959; Bouhuys *et al.*, 1969). The advantage of this method is that it provides control groups of workers from industries other than that being specially investigated, but exposed to the same environment outside the workplace. Also, in a community survey retired men will be included in the sample as well as working men, thus eliminating the element of bias that is often present in surveys of factories due to the factory population being a survivor population. Community surveys are, however, very time consuming since they necessitate some form of preliminary census of the whole community. It is necessary to select a town large enough to provide suffcent numbers of workers employed in the industry that is under investigation. Factory surveys are generally a more practical proposition.

The individual elements that make up a population are called sampling units and a list of all units is called a sampling frame. In surveys of factory populations the sampling units are likely to be the individual workers and the sampling frame might be the payroll, or some card index used for administrative purposes.

Definition of population

The population to be covered must be defined precisely. It may consist of men and women of all ages, employed in all types of occupations throughout the industry, whose names are on the payroll on a specified date, or it may be restricted to men and women from certain selected age groups and occupations, or to men or women only. Definition of the population will depend on the purpose of the survey. In studying a disease that is only likely to affect the middle-aged or elderly there may be no point in including the younger workers. In a study of lead absorption in workers exposed to lead it may be desirable to restrict the population to those whose exposure has been fairly constant.

When the population has been defined it must be decided whether it should be fully or partially covered. If a sample is to be taken it must be decided what type of sample, and how large it should be to give the degree of precision required in the results. These decisions are governed by what is practicable as well as what is desirable. The sampling frame must cover the defined population completely. It should not be subject to duplication and its accuracy should be checked.

In the investigation of a suspected occupational health hazard it is usually necessary to examine a group of workers for comparison with the group at risk. Ideally, they should be as similiar as possible to the workers at risk in all characteristics relevant to the investigation except that they are not exposed to the suspected health hazard. Factors that commonly have to be considered are age, sex, ethnic group, socioeconomic status, and area of residence (*see* Chapter 13).

To prevent the comparison between the control and the study group from being biased by extraneous factors the controls are sometimes matched to members of the exposed group. Matching may be done by pairing individuals from the control and study groups, or by ensuring that the composition of the groups as a whole is the same with respect to the matching criteria. Theoretically, it is better to use matched pair samples, but experimentally the two methods give similar results (Billewicz, 1964). Matching may not be practicable when several extraneous factors are involved, and it is often preferable to use unmatched samples and employ special statistical techniques which allow for the effect of differences between the samples in the distribution of the factors. Some of these techniques enable the importance of a factor to be assessed in relation to the indices being studied. Statistical tests used in the comparison of matched pairs differ from those used for unmatched samples and, as long as the matching is efficient, they are the more sensitive tests. In efficient matching the two sets of paired observations should show a strong positive relation.

Sometimes it is possible to choose as controls non-exposed workers from the factory that is being surveyed or from a neighbouring factory. Care must be taken that the controls are not exposed to some other health hazard that would invalidate comparisons. A factory population is likely to be a survivor population since workers susceptible to disease caused or exacerbated by exposure to an industrial agent, will often leave the industry. This occurs particularly among older workers in industries where there is a high prevalence of chronic disease causing increasing disability with age. Where stringent pre-employment medical examinations are used to exclude the less fit, or those likely to be susceptible to a particular exposure, a factory population may also be a selected population. The bias that could arise from sampling from these populations must be borne in mind when inferences are made from the survey results.

Time of survey

The time of year chosen for the survey may be important. Seasonal variation has been found in various physiological and biological measurements including serum cholesterol, blood pressure and ventilatory capacity. In a survey to measure the prevalence of chronic respiratory disease the results may differ according to whether the survey is carried out in the summer or the winter. In a study by Crooke *et al.* (1964) coal miners were interviewed first in June and again in the following January. In the winter month, the prevalence of respiratory symptoms was found to be higher and the average ventilatory capacity lower than in the summer month. The decrease in ventilatory capacity remained after the men were subdivided according to smoking habits and the presence or absence of respiratory symptoms. 'Lapse' rates are likely to be higher in some months of the year due to increased sickness absence in the winter and annual holidays in the summer months. Some factories close down for their annual holiday, and a survey should not start immediately on return to work if a period free from exposure is likely to bias the results.

Circadian rhythm is known to occur in several physiological functions such as body temperature, blood pressure, haemoglobin levels and ventilatory capacity. To avoid bias in comparison between occupational groups the time of day that tests are carried out should be similar in the groups. When measurements are taken before and after a shift to assess the acute effects of exposure to toxic substances an unexposed control group should be tested at the same time of day for comparison.

Collection of data

It must be decided exactly what information is required and what are the best methods of obtaining it. The initial aims of the survey should be kept strictly in mind to avoid the temptation to elaborate and collect too much information. The methods of processing and analysing the data should be discussed at this stage as these will influence the recording of information and questionnaire design.

In addition to measuring the prevalence of a disease, medical surveys are usually designed to throw light on its aetiology. In a survey of industrial populations, in addition to any suspected occupational factors likely to influence the onset, severity, and progression of disease, the relevance of factors outside the working environment must be considered. Statistical methods enable the effect of these factors on the condition or disease to be assessed and also make possible the comparison between prevalence rates or results from physiological or biological tests in occupational groups after correcting for the effects of non-occupational factors. A competent investigator would not now carry out a survey of industrial pulmonary disease without obtaining information on the smoking habits of the workers.

In surveys involving tests of ventilatory capacity the age and height of the workers are always recorded, since ventilatory capacity is negatively associated with age and positively associated with height. Studies of men and women with no known pulmonary disability have produced linear equations relating ventilatory capacity to age and height, which can be used to predict expected normal values of various indices of lung function for a man or woman of any given age and height. These predicted values can then be compared with the observed values to see if there is any evidence of pulmonary dysfunction (Cotes, 1979). More sophisticated prediction equations which include the effect of weight have been produced by Schoenberg, Beck and Bouhuys (1978). Often the observed value is expressed as a percentage of the predicted normal value, and in some surveys of working populations the mean values of this index have been compared between occupational groups. However, the use of the index in this way ignores the lack of precision in the predicted value due to the omission of pertinent but unknown factors from the prediction equation. A different method of comparison uses a statistical technique called an analysis of covariance. This enables the mean ventilatory capacity in groups of workers to be compared after adjusting for differences in factors such as age and height using equations calculated from the data of the survey itself. Assuming a simple linear relationship, the adjustment for age and height can be made by calculating independent estimates of the average change in ventilatory capacity that would be expected for an increase in age of one year and the average change expected for an increase in height of one centimetre, or one inch, depending on the unit of measurement. Using these estimates, the observed mean values of ventilatory capacity in the groups are adjusted to a chosen standard age and height and compared by statistical methods. The same technique can be used to adjust for additional factors such as weight or even smoking habits. The disadvantage of this method is that it gives only an internal comparison and ignores the question of whether or not the mean ventilatory capacity in the groups is lower than the expected normal.

In a study of the prevalence of varicose veins in women cotton workers, apart from information on the type of job and whether they stood or sat at their work, data were also recorded on age, height, weight, parity, method of stocking support, constipation and family history (Mekky, Schilling and Walford, 1969). Comparisons of prevalence could then be made between women who sat and women who stood at their work after correcting for differences in the other factors.

Questionnaire design

Field surveys require specially designed forms for recording clinical and other information. In some cases a simple questionnaire can be completed by the worker himself. The use of self-administered questionnaires is a convenient and cheaper alternative to the employment of trained observers and avoids observer bias. It is a useful method of collecting information when questions require a considered answer rather than an immediate response, but it does require motivation on the part of the respondent to complete and return the form. Questions must be self-explanatory, which may limit their scope, and it is not a suitable method for questions seeking depth. Clear instructions must accompany a self-administered questionnaire, together with an explanation of the purpose of the survey and the importance of full response, its sponsorship, and a guarantee of confidentiality. Usually it is preferable for questions to be asked by a trained medical observer at a personal interview, as long as standardized techniques and questions are used so that bias arising from different methods of questioning is avoided.

In constructing a questionnaire the aim of the survey must be defined so that questions can be formulated to cover the area of enquiry adequately. The questions should follow some sort of logical order, but it is advisable to keep questions that might cause embarrassment to the end. The order of the questions can sometimes influence the answers. A lower prevalence of smoking may be found when questions on health are asked first. Preferably questions should be 'closed ended' with answers to be chosen from a number of fixed alternatives rather than the respondent being allowed to answer in his own words. The alternatives should be mutually exclusive and cover all contingencies, although some investigators think the inclusion of a 'Don't know' category may encourage indecision. Where possible, questions should lead to simple 'yes' or 'no' answers. The questions should be simple and unambiguous, and kept to a minimum, resisting the temptation to collect more information than is needed. Questions that depend on the memory of the respondent or on his inadequate knowledge are unlikely to be worth asking. It is important to express the questions in everyday language to suit the population being studied, avoiding jargon and technical terms. In different ethnic groups with different cultures the questions may not be interpreted in the same way.

In opinion studies special scaling methods have been devised to measure the strength of an attitude or opinion rather than its mere presence or absence. The simplest form can consist of a straight line marked in intervals representing a range of attitudes from one extreme to another, and the respondent marks his position on the scale. Bortner (1969) used this method to construct a series of rating scales to measure type A behaviour, characterized by an enhanced sense of urgency, competitiveness, impatience, excessive drive and ambition. This behaviour pattern has been found to be associated with manifestations of coronary heart disease, and the Bortner rating scale was used by Heller (1979) in a study of the relationship between coronary heart disease and type A behaviour. More sophisticated and sensitive methods are described by Moser and Kalton (1971).

The questionnaire should be designed with subsequent processing and analysis of the data borne in mind. The layout should be as clear and simple as possible to facilitate editing and coding of the completed forms. Coding is a means of turning answers into numbers so they can be transferred to punch cards for processing by electronic computer. The conventional punch card has 80 columns and each column can be punched in 12 positions (*Figure 15.1*). The ten lowest positions are numbered 0–9 and are used for the main information. The two upper positions, 11 and 12, sometimes called X and Y, tend to be used as reserves. In general, the answer to a question is represented by one or more specific columns, each column having a hole punched in one position only. Numerical replies can be punched straight on to the cards matching each digit to the numbered positions in the columns, but non-numerical data must be coded. A system of coding involving the use of more than one position in a column is used for the alphabet

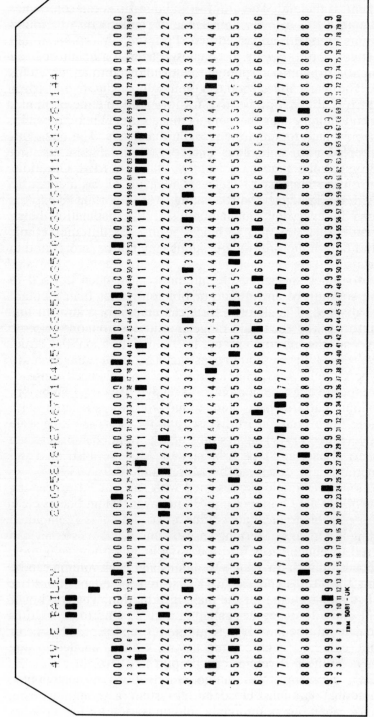

Figure 15.1 Example of 80-column punch card showing serial number, name, sex, marital status and other survey information

and special characters. An operator feeds the cards into a card punch machine where the holes are automatically punched by using a keyboard.

It is an advantage to design the questionnaire so that information can be punched straight on to the cards without the time-consuming chore of first copying the data on to special forms with the risk of introducing errors. It may be possible for information to be pre-coded by printing code numbers on the questionnaire so that the observer has only to ring the appropriate number or mark a coded box. One method of pre-coding is to have a set of boxes representing the possible replies to the question printed on the right hand side of the questionnaire, together with the relevant punch card column number or numbers. Each box bears the code number representing its position in the column on the card. Column numbers can be distinguished from code numbers by printing them in bold type. In some instances it may be difficult to enter information directly into pre-coded boxes but they can still be printed on the questionnaire and the replies coded and transferred to the boxes after the interview.

An example of pre-coded boxes is shown below with code numbers printed in ordinary type and punch card column numbers in bold type. The punch card would show that the respondent is a married woman by having holes punched in position 2 in columns 21 and 22 of *Figure 15.1.*

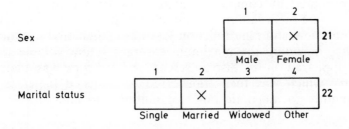

A self-administered questionnaire can have column and code numbers on the right-hand side, which the respondent is told to ignore. On completion, the replies are coded by the investigator who rings the appropriate number. An example is shown below with the column number printed in bold type on the left and the code number on the right.

If experienced interviewers are used and the questions are closed ended, mainly requiring a 'yes' or 'no' response, an alternative is to construct a coding schedule before the survey, so that the interviewer can enter the code number directly into a box.

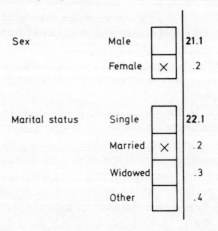

The principles of medical questionnaire design, with brief notes on some of the documented questionnaires that are relevant to medical research, are given by Bennett and Ritchie (1975).

The most sophisticated method of recording data was used in community studies of lung disease in the United States by Mitchell, Schilling and Bouhuys (1976). Two interview stations and a computer were housed in a trailer that could be moved to various sites in the communities. Each station contained a graphic display terminal and pneumotachograph–transducer–amplifier system linked to the computer. Each terminal was also connected to a hard copy unit. Questions were prompted by computer and read by trained interviewers, the answers being recorded in the computer memory. When the questionnaire interview and lung function test were completed a summary of the data, including the maximum expiratory flow-volume (MEFV) curve, appeared on the display screen. A hard copy was made of the summary which was then transferred to punched paper tape and subsequently to magnetic tape for editing and analysis.

Reliability and validity

It is not easy to test the reliability and validity of a questionnaire. The questionnaire may be administered twice to the same group of individuals in an attempt to test its reproducibility, but if the lapse of time between the two interviews is small the replies at the second interview could be influenced by memory of the first, whereas a longer time gap may allow a genuine change in the condition of the respondent. In a study of attitudes and opinions the first interview may cause more thought to be given to the topic, and lead to a change of opinion at the second interview.

A test of validity requires an independent criterion which is not always easy to find. If an established diagnostic test exists, the validity of questions on disease can be assessed on a group of people by comparing the test results with the answers to the questions. The number of false negatives and false positives produced by the questionnaire can then be counted. Questions with high sensitivity will yield few false negatives and those with high specificity will yield few false positives. The calculation of indices of sensitivity and specificity is described in the next chapter section, on diagnostic criteria. A perfect diagnostic test may not exist, and other imperfect measures have to be used instead. Sputum samples might be used to test the validity of questions on cough and phlegm, and electrocardiograms to test the validity of questions on chest pain designed to identify people with a high risk of coronary heart disease.

The London School of Hygiene questionnaire on chest pain and intermittent claudication was originally validated by comparison with diagnosis by physicians, and found to have high specificity and good sensitivity (Rose, 1962). It has since been widely used and evaluated. Results from a self-administered version of this questionnaire have been compared with those obtained by interviewer administration in two groups of men randomly allocated to the two methods (Rose, McCartney and Reid, 1976). For both 'angina' and 'possible infarction' the prevalence of positive results was about twice as high in the self-administration group, although comparison with electrocardiogram results and assessment of predictive power by a study of subsequent mortality from myocardial infarction showed that self-administration did not appear to lead to any major loss of specificity. However, this experience demonstrated that results obtained from two different modes of administration cannot be directly compared.

Diagnostic criteria

Diagnostic criteria must be agreed on and precisely defined, including grading of severity of disease where appropriate. It may be advisable to use the same criteria and grading used by other investigators so that the survey results are comparable.

The ability of diagnostic tests and criteria to detect disease or measure severity accurately should be assessed if it is not already known. In practice, an assessment of validity is often difficult to achieve, since it requires an independent measure of disease. Sometimes an investigator may wish to save time and costs by substituting a cheaper or less time-consuming test for a well-established one. In this case both tests could be used to examine a small group of people before the main survey, and the results compared. Among the criteria in the evaluation of a screening test, discussed by Cochrane and Holland (1971), are those of sensitivity and specificity where sensitivity is defined as the ability of the test to give a positive finding when the person screened has the disease under investigation, and specificity is defined as the ability of the test to give a negative finding when the person does not have the disease. High sensitivity means few false negatives, and high specificity means few false positives. The calculation of indices of sensitivity and specificity is shown in *Table 15.1*.

Table 15.1 CALCULATION OF INDICES OF SENSITIVITY AND SPECIFICITY

Results from screening test	Results from established test	
	With disease	Without disease
Positive	a True positives	b False positives
Negative	c False negatives	d True negatives
Total	$a + c$	$b + d$

$$\text{False negative rate} = \frac{c}{a + c} \times 100$$

$$\text{False positive rate} = \frac{b}{b + d} \times 100$$

$$\text{Sensitivity} = \frac{a}{a + c} \times 100 = 100 - \text{False negative rate}$$

$$\text{Specificity} = \frac{d}{b + d} \times 100 = 100 - \text{False positive rate}$$

Sensitivity is inversely related to specificity, and the value of the indices can be varied by changing the level of the criterion used to determine a positive result. A stricter diagnostic criterion will improve specificity but reduce sensitivity. Since a reduction in the number of false positives is only achieved at the expense of an increase in the number of false negatives and vice versa the relative importance of each type of error has to be considered and a compromise made. In comparative prevalence surveys the need for high specificity tends to take precedence over high sensitivity.

It is also important that results from diagnostic tests are consistent and should be reproducible. An index of reproducibility can be devised by examining the same group of people twice in similar circumstances and calculating the percentage deemed to be positive on both occasions out of those positive on one or both occasions. However, it is difficult to assess the accuracy of this index since reproducibility will be influenced by the biological variation within the person examined as well as instrument and observer variation. The efficacy of a test in selecting diseased persons from a population can be assessed statistically (Thorner and Romein, 1961).

Biological tests and measurements of physiological functions should be simple enough to be easily carried out under field conditions, and in some cases capable of being used by paramedical, or even non-medical personnel. An example of a simple but useful test of ventilatory capacity used in many surveys of respiratory disease is the forced expiratory volume usually now measured over one second (FEV_1). It is sometimes expressed as a percentage of the forced vital capacity (FVC), which can be measured at the same time. More recently, maximum expiratory flow volume curves (MEFV) have been measured in the field and recorded on-line in a computer equipped mobile laboratory (Schoenberg, Beck and Bouhuys, 1978). These curves have been found to be more sensitive in detecting obstruction in the small airways of the lung.

Since a high rate of cooperation is important all tests should be acceptable to the workers. The determination of lead in blood is commonly considered the most useful test of lead absorption, but it suffers from the disadvantage that to monitor workers in the lead industry regular venepuncture is required. However, it is known that lead intoxication leads to elevated zinc protoporphyrin (ZPP) levels in the blood, which can now be measured quickly, inexpensively and conveniently by a portable haematofluorometer using a drop of undiluted blood obtained by finger prick. The usefulness of ZPP as a biological indicator of chronic lead intoxication was assessed by Eisinger *et al.* (1978). They concluded that for chronically exposed individuals the ZPP level is a better guide to the evaluation of several adverse health effects of lead exposure than is the blood-lead level, and offers a preferred primary screening test for lead exposed populations.

All diagnostic tools, including questionnaires, should give results that are reproducible. Measurements taken on the same person on different occasions by the same observer, or by different observers, should vary as little as possible when the actual condition of the person is unchanged. Reproducibility can be improved by using standard methods in physiological and biological tests. Where several instruments of the same type are used, they should be repeatedly calibrated to check that they do not differ from each other. When using questionnaires, there should be set rules on how to ask the questions and how to record the answers. The study of the problem of reproducibility in questionnaire response led to the production of the Medical Research Council's standardized questionnaire on respiratory symptoms, which has become the basic instrument for surveys on chronic respiratory disease in many parts of the world (Medical Research Council, 1976).

In surveys of industrial populations it is very important to obtain an adequate and accurate occupational history (*see* Chapter 3). It may be found that the control group includes workers who have been transferred from the factory or department being surveyed because their health had been affected by exposure to an injurious working environment. Ideally, occupation should be recorded at the end of the interview so that knowledge of exposure to a particular agent cannot either consciously or unconsciously influence the observer. This precaution may be ineffective when the observer is the factory medical officer, since he is likely to know the workers who are at risk.

Sources of error

Errors in the survey results can originate from several different sources. They should be anticipated and ways of minimizing them considered.

In a sample survey there will always be some error because only part of the population has been examined, but apart from errors due to sampling there are likely to be non-sampling errors. These may be divided into random and systematic errors.

Random errors

Random errors are those which occur when a measurement is overestimated one time and underestimated the next. They will contribute to the overall variability of the measurement and reduce the precision of means and other measures calculated from the sample data, but they should on the average cancel out, and their effects can be minimized by increasing the sample size.

Systematic errors

Systematic errors are those which would occur if an observer consistently recorded readings from an instrument that were too high or too low, or if there were a consistent error in the instrument itself. Similarly, an observer may systematically over-record or under-record clinical signs and symptoms. Systematic errors occur in biological measurements if the sample is contaminated, and careful instructions should be given to workers if they are required to take samples themselves, for example in estimating lead in urine in workers exposed to lead, where contamination could arise from clothing. An important source of systematic error can arise from a low response rate, or an inadequate sampling frame. In a factory where there is a high turnover of labour the sampling frame may be inaccurate. It is important to avoid non-sampling errors by careful training of observers, checking of instruments, and standardization of techniques. Every attempt should be made to achieve a high response rate.

Systematic errors do not cancel out and they will always cause bias in sample estimates of population averages and prevalence rates.

Care to avoid error must also be taken in editing, coding, processing and analysing data. Computer programs are designed to examine the data for inconsistencies and gross errors before commencing the analysis, and computers are likely to play an even larger part in the quality control of survey data.

Observer variation

In a survey of any appreciable size it is usually necessary for the field work to be carried out by more than one observer. The variation in results caused by differences between observers in clinical judgement and in the reading of instruments is one of the main problems of medical surveys. A certain amount of observer variation may be unavoidable, not only between observers but also in assessments made by the same observer on different occasions, but steps should be taken to keep it to a minimum.

In recording signs and symptoms inaccuracies may occur through poor technique of the interviewer or through poor questionnaire design. Practice obtained by interviewing colleagues and volunteer groups of people before starting the survey would at the same time expose any inadequacies of the questionnaire. Agreement between observers can be checked by arranging for each observer to listen to a tape-recorded interview given by another observer and to re-record the answers given by the respondent.

Observer variability can be reduced by using a standardized questionnaire. Users of the Medical Research Council questionnaire on respiratory symptoms are able to obtain special recordings of interviews which illustrate the way in which questions should be asked and the method of dealing with certain difficult answers.

Variation between observers in the reading of radiographic films can be high even among experienced physicians. In the study of the pneumoconioses an effort to standardize radiological diagnosis was made by a committee of the Union Internationale Contre le Cancer (UICC) who introduced an international classification, the UICC/Cincinnati classification, with standard films to illustrate the various categories of abnormality. After slight amendment by UICC the classification was adopted by the International Labour Office

(ILO) as the ILO U/C International Classification of Radiographs of the Pneumoconioses, 1971.

The measurement of blood pressure is affected by several sources of variation apart from the true variation of a person's arterial pressure. Systematic errors may occur both between observers and between measurements taken by the same observer. Additional error can arise through a tendency to terminal digit preference, and it has been found that in recording to the nearest 5 mmHg the digit 5 is often under-recorded at some points on the scale, preference being shown for multiples of 10. Another serious source of error is the tendency for an observer to discriminate for or against certain key pressure values. Rose, Holland and Crowley (1964) describe a modified sphygmomanometer to replace the conventional instrument in epidemiolgical studies and eliminate the more important sources of observer bias.

The electrocardiogram (ECG) is an objective test, valuable in screening for coronary heart disease, but its interpretation is subject to considerable observer variation. To minimize this defect a standardized classfication system, based on precisely defined criteria, was introduced by Blackburn *et al.* (1960). This has become known as the Minnesota Code and is recommended for the initial classification of ECGs. An assessment of its reproducibility and validity was made by Higgins, Kannel and Dawber (1965).

Bias caused by observer variation can arise in tests of lung function unless observers are trained to obtain the maximum response from the subject. It is important to standardize the performance of the tests. Differences could arise in ventilatory capacity if some subjects were tested standing and others sitting. In some tests of lung function it is advisable to introduce a short period of rest before testing if workers have had a long or uphill walk from their department. The testing of smokers should not be carried out until at least one hour has elapsed since their last smoke.

A check can be made on inter-observer agreement in physiological tests by allocating at random a group of people to each of two observers who first test their own group and then, on a second occasion, retest the other group. This method of comparison can be extended to include several observers and also to measure independently the variation between different instruments.

In a countrywide survey observers should work equally in all areas to prevent observer variation from invalidating comparisons between working populations in various geographical areas. Within a workplace observers should interview comparable groups of workers.

Special problems of observer variability can arise in international surveys, since language difficulties usually restrict observers to examining only the workers in their own country. In a study of byssinosis and chronic respiratory symptoms in English and Dutch cotton workers (Lammers, Schilling and Walford, 1964) four observers were used in each country. To detect differences in interview technique between observers in the two countries a small group of English cotton workers were examined by one of the Dutch observers. The interviews were tape-recorded and interpreted by one of the English observers and the results compared. Similar comparisons were made between each of the four observers in their own country. To ensure that the English and Dutch questions on respiratory symptoms and disease were identical in meaning the original questionnaire was translated into Dutch by two physicians taking part in the survey and then back into English by another Dutchman, who was not a physician. Differences from the original version were then corrected.

Variation in biological measurements can occur due to different methods of analysis and different ways of using the same method. Samples should be analysed by the same laboratory.

Non-response

Provision should be made at the planning state for dealing with non-response. A certain number of lapses are usually unavoidable, but to avoid bias in the survey results it is important for non-response rates to be kept low, and a response rate of at least 90 per cent is desirable. It is known that people who are reluctant to be examined often differ from those who co-operate readily (Cochrane, 1951). Workers who are away sick may cause the prevalence of a disease studied in the survey to be underestimated, but in some factory surveys it has been possible to interview absentees in their homes, and even carry out physiological tests, such as simple tests of ventilatory capacity. It is often possible to obtain some information about the non-responders, such as age, length of service and sickness experience, which may give some idea of how important their exclusion from the sample might be (*see* Mitchell, Schilling and Bouhuys, 1976).

A high response rate can only be achieved by making an effort to gain the confidence, support and interest of management and workers. The object of the survey should be explained and information given on the type of examinations, physiological tests, etc. involved, and the amount of time that would have to be spent on them. Reassurance should be given that the subsequent findings would be treated as confidential and would not be used in any way that could cause concern to the workers. A preliminary discussion with employers and trade unions is of great value and is usually essential.

Pilot surveys

It is often an advantage to carry out a miniature or pilot survey of the population before commencing the main survey. Pilot surveys are useful in many ways. They may provide guidance on the adequacy of the sampling frame. If the payroll of workers in a factory is used as a sampling frame the pilot study may reveal that it excludes certain classes of workers, or that some cards are temporarily removed from a card index when required for administrative reasons.

Pilot surveys provide the observers with practice in the use of instruments and questionnaires, and enable estimates to be made of the cost and duration of the main survey. They also provide a check on the adequacy and clarity of the questionnaire and its coding. Some questions may be found to produce limited or unreliable information and can be omitted from the final questionnaire. There may be opportunities for examining the validity of diagnostic tests and criteria.

If only a sample of the population is to be examined in the main study it is important to be able to estimate the size of sample needed to give results of a desired precision. The estimation of sample size requires knowledge of the variation from person to person of the diagnostic indices of main interest, and a pilot survey can be used to obtain an approximate measure of this variability. It can also give some idea of how well the objectives of the main survey are likely to be fulfilled.

Sampling error and sample size

Medical surveys of industrial populations usually aim not only at measuring the extent of a particular ailment or disease among workers in the industry but also at relating the prevalence of disease to various occupational factors, such as type of job, length of exposure to a suspected health hazard, and level of exposure obtained by sampling the working environment. The prevalence of disease and the mean values of physiological and biological tests, or other diagnostic indices, are often compared between occupational groups subjected to various levels of exposure, including a control group. Apart from errors of measurement, the innate variability of human characteristics makes it likely that prevalence rates and averages calculated from sample data will differ to some extent from those that would have been obtained had the whole population of workers been examined. If two samples were selected from the same population it is likely that there would be differences in the results between the two samples, and if one sample was examined again on a later occasion the results would be unlikely to agree exactly with those obtained on the first occasion.

This sampling variability obviously introduces an element of uncertainty into the conclusions that can be drawn from sample data. The sample means and prevalence rates are only estimates of the real means and rates occurring in the population and it is important to know how good these estimates are. It is necessary to know if an observed difference between occupational groups in the prevalence of a disease, or in the mean value of a measurement, is a reflection of a real difference between these occupations in the population or if it has occurred by chance as a result of sampling fluctuations.

It is not possible to elaborate on the statistical methods that enable the problem of sampling variation to be resolved. They still do not allow conclusions to be drawn from sample data with certainty, but they do enable a statement to be made of the limits within which the true population mean or prevalence will lie with a degree of confidence expressed in numerical terms, and enable the calculation of the probability that an observed difference in indices between two or more groups has occurred by chance.

Assuming that the data are free from systematic error, or bias, the extent to which a sample mean or proportion differs from the true population value is known as sampling error. It is not possible to measure sampling error directly since this would require knowledge of the true population values, but its probable magnitude can be estimated from the sample data by calculating what is called the standard error. This is a measure derived from statistical theory by considering the theoretical variation of sample means (or other indices) about the true population mean that could arise if an infinite number of samples, all of the same size, were selected at random from the population. The smaller the variation of the sample means about the population mean, the smaller the standard error and the greater the precision of the sample mean as an estimate of the population mean. In comparisons between groups, where there is a real but small difference between the mean values in the populations from which the samples have been drawn, the smaller the standard error the better the chance one has of detecting this difference from the evidence given by the sample means.

Statistical theory shows that the standard error depends on the variability of the individual observations in the population and on the size of the sample. The more variable the measurement the greater the standard error, the larger the sample size the smaller the standard error. Precision can be improved by using a special sample design that will minimize the standard error by keeping random variation to a minimum, but it is obviously important in planning a survey to consider selecting a sample that is large enough to provide sufficiently precise estimates of population averages or the prevalence of disease, and permit sensitive tests of hypotheses to be made. At the same time, it would be wasteful of time and money to select a sample that is larger than necessary to achieve the precision required. Precision is dependent on the actual size of the sample and not on the proportion of the population included in it.

Methods of estimating the sample size needed to give results of a desired precision are described in various statistical textbooks (Mainland, Herrera and Sutcliffe, 1956; Moser and Kalton, 1971; Armitage, 1971). The formulae require an estimate of the standard error, and assume that a measure of variation in the population is known since this is a component of the standard error. In practice the population variation is usually unknown, but an estimate may be available from a previous study or can be calculated from a pilot study. For the estimate to be reasonably accurate the pilot study should not be too small. The formulae are strictly only appropriate for the simple sampling design, but can be used to provide a preliminary estimate of sample size when more complex sampling methods are used.

Before using the formulae the desired degree of precision must be determined. For example, an investigator may decide that he needs to estimate the prevalence of byssinosis in cotton workers to within ±5 per cent of the true prevalence; or, in a comparative study, he may specify the smallest difference between the mean values of a physiological measurement that he would like to be able to detect from his sample data, if such a difference exists in the population from which the samples were drawn.

In determining sample size, consideration must be given to the possible classifications required in the analysis of the data. Apart from total sample size it is likely to be necessary to estimate the numbers required to give adequate precision in the subgroups.

Diagnostic indices and individual characteristics that can be quantified are often known as variables. In most surveys several variables are to be studied, and the sample size should be estimated separately for the most important variables. If these estimates are similar it may be reasonable to accept the largest number. If there are large differences between the numbers it may be better to use a special type of sampling (called multiphase sampling, and described later) where the variables requiring a large number of observations for the desired precision are observed on all members of the sample, while those requiring a smaller number of observations may be observed only on a randomly chosen subsample of the original sample.

Apart from the question of precision it should be remembered that in an investigation where the prevalence of a disease is likely to be low the sample must be large enough to include an adequate number of persons suffering from the disease.

TYPES OF SAMPLE DESIGN

For statistical methods to be valid a sample should be chosen in such a way that selection is not influenced by human choice. It should be a random sample, where 'random' refers to a method of selection compatible with the laws of probability. A sample consisting of volunteers, or persons expected to co-operate, is obviously unlikely to be representative of the population, but random selection also gives protection against unsuspected sources of bias and is incorporated in all formal sampling designs.

In some surveys of industrial populations formal sampling methods may not be practicable. It may be administratively difficult to select a random sample of workers throughout the industry, and the survey may be restricted to the examination of all workers in one or two factories which have not been chosen at random. The observations will nevertheless show random variation, and it is possible to consider the sample as having been generated from a hypothetical population consisting of an infinite number of similar random observations. Inferences can be made from the sample to the hypothetical population, but cannot be extended rigorously to the true population of workers in the industry.

Only a brief description of sample designs can be given in this chapter. In formal sampling, the method of estimating the sampling error varies according to the complexity of the sample design. The standard error of a mean, or proportion, commonly given in statistical textbooks is only appropriate when simple sampling methods are used. Details of the more complex designs and the statistical problems involved in their use are described by several authors (Moser and Kalton, 1971; Yates, 1960; Cochran, 1977). These designs are appropriate when emphasis is on obtaining accurate estimates of disease prevalence or other population characteristics rather than testing hypotheses.

Simple random sampling

In many surveys simple random sampling is used. A simple random sample is defined as a sample selected in such a way that each member of the population has an equal chance of selection. More specifically, every possible sample of a given size has an equal chance of selection.

Randomness of this kind can only be ensured by using an objective method of selection, and there are tables of random sampling numbers which have been specially devised for this purpose. The tables consist of a collection of digits, 0–9, arranged in random order, and have the property that in a long series each digit, pair of digits, triplet, etc., will appear with equal frequency. Using the tables is equivalent to drawing numbered slips of paper in a lottery. There are several published tables available (Fisher and Yates, 1963; Documenta Geigy, 1970) which include instructions for their use. The general procedure is to number the individuals (or other units), in the defined population from 1 to N, the total number in the population, then to pick an arbitrary starting point in the tables and, using the same number of columns as there are digits in N, go down the columns of digits recording all the numbers in the population as they occur in the table until the number of individuals required for the sample have been selected.

This procedure of random selection should be followed when the payroll, or some other nominal roll, is used as a sampling frame. Roach (1959) found that a sample consisting of the names at the beginning of a list which was in order of works number would have been biased, since the men were to some extent in order of seniority. Similarly, a sample consisting of the names at the beginning or end of a list in alphabetical order would have been biased, since foreign names were more common among the last few letters of the alphabet and there was a tendency for foreign workers to be placed in particular jobs, or placed together because of language difficulties.

Systematic sampling

If the population is large it may be laborious to number all the individuals and select a sample using random numbers. An alternative is to decide on the fraction of the population that is to be included in the sample, called the sampling fraction and expressed as 'one in k', and then choose every kth individual starting with a randomly chosen number between 1 and k inclusive; for example, if a one in 10 sample is being selected, k = 10 and a number between one and 10 is chosen at random to determine the first member of the sample. If the number is five the sample will consist of numbers 5, 15, 25 and so on. These could be selected from an existing list or card index of workers.

This method is not equivalent to simple random sampling unless the list itself is in random order. It does not enable all possible samples of a given size from the population to have an equal chance of selection since the selection of the starting point determines the whole sample.

It is a convenient method of sampling, but care should be taken that the ordering of the list is not related to the subject of the survey in such a way that bias could be introduced by selecting units that were equidistant.

Stratified sampling

In stratified sampling, the population is divided into a number of groups, or strata, determined by factors related to the variables studied in the survey. A random sample is then selected within each stratum. In a survey of respiratory disease in an industrial population one might stratify by the geographical region in which the factories are situated, the type of factory and size of factory, and by occupation, age, sex and perhaps ethnic group of the workers. This would ensure that the sample is representative of the sex, age and occupational structure in the population, and factories in areas with differing levels of atmospheric pollution would be adequately represented.

In addition to increasing the representativeness of a sample, stratification also increases the precision of sample estimates. The total variation in a population consists of variation between strata and variation within strata. In stratified sampling the variation between strata does not enter into the estimate of sampling error since this component of variation in the population is exactly reflected in the sample. Sampling error arises only from the sampling within strata. To achieve the maximum increase in precision the stratification factors should be chosen so that the strata differ as much as possible from each other, but the population within each stratum is as homogeneous as possible.

It is only possible to stratify by a factor for which the population distribution is known, and which can be identified for each sample member. Sometimes it is possible to stratify after selection of the sample if relevant information is not given in the sampling frame; for example, exact occupation may not be given in a list of workers but can be recorded later.

Surveys are usually concerned with a number of variables, and a stratification factor that is related to one variable may not be related to another. Age is an important factor that is related to most indices of disease. Sex is not related to the prevalence of byssinosis but is related to the prevalence of chronic bronchitis and to ventilatory capacity. In this case, factors that are relevant to most variables may have to be chosen, or it must be decided which variables have the highest priority. The use of stratification factors unrelated to the variables will not result in loss of precision, but there will be nothing gained. A pilot study is often valuable in determining the most useful stratification factors. Apart from the gain in representativeness and precision achieved by stratified sampling, the results for the separate strata are often of interest in themselves.

Variable sampling fraction

Stratified sampling can be carried out using the same sampling fraction within each stratum, but in the following situations, it may be preferable to use a variable sampling fraction.

1. If the variation between persons is greater in some strata than in others precision can be increased by using a larger sampling fraction in the more variable strata. This requires prior knowledge of the relative variability in the strata, which is not always available.
2. If cost per sampling unit is expected to be greater in some strata than others a smaller sampling fraction can be taken in the most expensive strata.
3. If the results in the separate strata are of interest in themselves it is important for each stratum to have a sufficient number of observations so that estimates of means or prevalence can be made with reasonable precision. It may therefore be necessary to use a larger sampling fraction in the smaller strata. In a small occupational group it may be better to examine all workers than to leave out two or three, who may feel neglected.

Optimum precision is obtained if the sampling fractions are made directly proportional to the amount of variation in the population strata and inversely proportional to the square root of the cost per unit in the strata. The common measure of variation is known as the standard deviation, which is based on the deviation of the individual values of a series of observations from the mean of the series. The population standard deviation would be calculated by squaring each deviation from the mean, dividing the sum of all the squared deviations by the number of observations and then taking the square root. An *estimate* of the population standard deviation can be calculated from a sample of observations in the same way, but for theoretical reasons the sum of the squared deviations is divided by one less than the total number of sample observations. The size of the standard deviation in each population stratum is usually unknown but an estimate can sometimes be obtained from a previous study or a pilot survey.

When the cost per unit is the same in all strata, the sampling fraction can be made proportional to the measure of variation only. The use of variable sampling fractions does make the statistical calculations more laborious, and where little is known about the population variation it may be better to use a uniform sampling fraction.

Another difficulty in choosing variable sampling fractions is that when several variables are being studied the fraction that is best for one variable is not necessarily best for another. If one variable is more important than the other this could determine the choice.

In stratified sampling the estimates of population means and proportions are weighted averages of the sample estimates given by each stratum. When a uniform sampling fraction is used, the stratum *sample* numbers are used as weights, the mean (or proportion) in each stratum being multiplied by the number of sample observations in that stratum. The results for each stratum are then summed and divided by the total number of observations. When a variable sampling fraction is used the stratum mean, or proportions, are weighted by the *population* number in the stratum.

Multistage sampling

A population can often be regarded as consisting of a hierarchy of sampling units. A large industry spread over the country could be first classified by geographical regions (first stage units), within each region there might be several factories (second stage units), within each factory several occupations (third stage units) and so on, down to the individual workers.

Instead of selecting a sample over the whole industry it may be more convenient, and save time and expense, to select first a random sample of regions, and then either include all factories in each chosen region or select a sample of factories. All the workers in the selected factories may be examined or only a sample taken. Sampling in any stage may be with or without stratification. When a 100 per cent sample is taken at the second stage it is known as cluster sampling. In a survey of respiratory disease in foundry workers in Great Britain (Lloyd Davies, *et al.*, 1971) 2427 foundries were stratified by size into four groups, those employing 1–9, 10–49, 50–249, and 250 or over. A one in 40 random sample of foundries was then selected within each size group, and all men aged 35–64 in the sample foundry asked to co-operate.

In two-stage sampling, when the first stage sampling units vary greatly in size, it is useful to select them with probability proportional to size. Suppose the first stage units consist of five factories and it is desired to select two of them with probability proportional to size. The factories are listed with the number of employees in each factory recorded in one column and the cumulated number of employees calculated in another column as illustrated in *Table 15.2*. Two numbers between 1 and 4647 inclusive are then chosen from a table of random numbers and the two factories that include these numbers in the cumulated figure are selected. (The first factory includes numbers 1–255, the second factory numbers 256–1455 and so on.) The probability of a factory being selected is proportional to its size since

Table 15.2 SELECTION WITH PROBABILITY PROPORTIONAL TO SIZE

Factory	Number of employees	Cumulated number of employees
1	255	255
2	1200	1455
3	512	1967
4	180	2147
5	2500	4647
	4647	

size is proportional to the range of the cumulated figure. If the two chosen numbers were 294 and 3486, factories numbers 2 and 5 would be selected for the sample.

The main advantages of multistage sampling are the saving of time and cost, since the field work can be concentrated in a limited number of regions or factories instead of being scattered all over the country. It also eliminates the need for a sampling frame to cover each individual worker in the industry, since lists of workers would only be required for selection in the final stage.

The great disadvantage of this method is that there is a loss of precision. The sampling error is higher than that found in simple random sampling. This is because there is an increased risk of the sample being unrepresentative. If, in a survey designed to assess the health risk in a particular industry, factories are chosen as first stage units, some will be excluded from the survey. Since working conditions may vary considerably between factories there would be some danger of high exposure groups not being represented.

In stratified sampling the aim is to make the strata differ as much as possible from each other while the units within the strata are as homogeneous as possible. In multistage sampling this policy is reversed. The more heterogeneity there is within the early stage sampling units the smaller the loss in precision.

Multiphase sampling

In multiphase sampling some information is collected from everybody in the sample, while additional information is collected, either at the same time or later, from subsamples of the original sample.

This method of sampling reduces the burden on the investigator and on the people examined, and can result in considerable saving in time and costs. It can be used satisfactorily when there are some variables for which high precision in estimates is less important, or where adequate precision can be obtained with a small number of observations. If there is non-response in the later phase sample because of refusals or absentees, the information collected in the original sample can be used to see if the non-responders differ materially from the rest of the subsample in any basic characteristic that might be relevant to the results of the survey.

Environmental sampling

It is important to determine the relation between the prevalence and severity of an occupational disease and the level of exposure to a toxic substance. This information can be used to determine or revise the recommended permissible level of exposure, such as the threshold limit value (TLV). It is difficult to construct accurate indices of individual exposure since length of the exposure should be considered as well as its level. To calculate an exposure index for an individual requires a detailed occupational history and knowledge of average levels of exposure in the various situations in which he has worked during his employment. Except in prospective surveys this information is rarely available, and environmental measurements taken at the same time as the medical survey may have to suffice. However, computers are now beginning to play an important role in storing data for routine environmental monitoring and, together with the use of personal samplers, should provide a source of more accurate data based on measures of individual exposure rather than group or workplace averages. Rickards (1978) describes a computerized control and data recording system used in routine monitoring of an asbestos textile factory, where every employee working with asbestos is sampled once every three months, using readings from membrane filter samplers. Record cards are printed for each individual from the company payroll and sampling data are added to the card which is returned to the computer for processing.

There are many sources of variation involved in the measurement of occupational exposures to airborne contaminants (*see* Chapters 17 and 18): systematic and random errors in the sampling instrument and in the analytical procedures, fluctuations in concentration between and within work days, in different parts of the work area during a day, and even between workers doing the same job. These sources of variation make the use of statistical methods and random sampling methods just as necessary in estimating exposure levels as in estimating disease prevalence. Statistical methods may also be used to help make decisions about compliance with hygiene standards. The statistical approach enables allowance to be made for random variation and sampling error, but cannot deal with inaccuracies due to systematic error.

To sample the workforce, knowledge of the processes should make it possible to identify workers with similar levels of exposure, and stratified random sampling can be carried out within exposure groups. If sampling cannot be carried out during a whole shift the total period can be divided into intervals and a random sample of intervals selected. If a regular pattern of exposure is known to exist over the shift, the intervals could be stratified by expected exposure level before selection. Balanced sampling plans can be devised to minimize the effects of systematic variation between days or shifts. An exception to random sampling is when the purpose of measuring exposure is to assess compliance with ceiling standards of hygiene. In these circumstances samples are taken during periods of expected maximum concentration.

The number of workers to sample and the number of samples to take depend on the purpose of the sampling, the nature of the processes and the range of concentrations. Theoretically, to obtain a desired level of precision in exposure estimates, the more variable the measurements the greater the number of samples that should be taken; but in practice the additional cost of sampling and analysis may outweigh the statistical considerations. Sampling procedures designed to assess the health risk to employees are likely to differ from those designed to detect changes in the engineering control of the work environment. The National Institute for Occupational Safety and Health (1977) has published a manual which describes statistical sampling strategies to assist in developing efficient monitoring of occupational exposures to airborne concentrations of chemical substances. Environmental monitoring is discussed in more detail in Chapters 17 and 18.

FIELD SURVEYS IN PRACTICE

Studies of dose-response

One of the most important uses of the field survey method of investigation is to provide information that will enable the effect of the working environment on the health of the workers to be assessed in terms of a dose-response relationship that may lead to the establishment of a permissible level of exposure.

In morbidity studies most attempts at measuring dose-response relationships have been based on cross-sectional surveys where health status and current levels of exposure have been measured at the time of the survey, but there has been difficulty in obtaining an accurate measure of an individual's past exposure, even if environmental sampling has been carried out for the duration of his or her employment. Various indices of exposure were constructed by Berry *et al.* (1979) in a study of men in an asbestos textile factory where dose-response relationships were considered between the prevalence or incidence of crepitations, possible or certified asbestosis and the level of dust exposure. This study was not a field survey since the clinical information was mainly extracted from records of routine medical surveillance and the examinations of the Pneumoconiosis Medical Panel, but the methods used in the construction of the exposure indices are just as relevant. For the number of years that dust measurements had been recorded a dust level was calculated for each year and for each job description by taking the average level measured at the static dust sampling locations in the area where the job was carried out. Dust counts were estimated for the years when no dust measurements were taken. The simplest measure of exposure considered was the cumulative dose, where each dust concentration was weighted by the duration of exposure to that level and the components summed. This gives equal weight to early exposure and recent exposure, and ignores the possibility of developing disease after the exposure has ended. Since asbestosis is more dependent on early exposure than recent exposure, and may develop after exposure has ended, another measure (Jahr, 1974) was used which weights the dust concentration at any time by the length of time that has elapsed since exposure occurred. A generalization of this measure, assuming that the weighting factor could be regarded as the time that the dust has been in the lungs if elimination has not taken place, produced a family of exposure measures, with each member of the family defined by the half-life time. A refinement of this measure included a lag period, on the assumption that disease is not observed until some time has elapsed since the start of exposure.

Dose-response curves based on each of the exposure indices were used to estimate the prevalence of possible asbestosis after 30, 40 or 50 years' exposure to 2 f/cm^3, and to estimate the concentration which would limit prevalence to 1 per cent. The variation in the results illustrated the difficulty in drawing conclusions on the validity of present standards from data relating to the dustier conditions which had existed in the past.

In many dose-response studies estimates of past exposure levels are not available, and sampling of the working environment is carried out at or around the time of the medical survey. In a study examining dose-response relationships to help establish permissible levels of exposure to cotton dust Merchant *et al.* (1973) measured byssinosis prevalence, and the change in forced expiratory volume (FEV_1) after six hours exposure, in nearly 1800 men and women employed in mills processing cotton or cotton and synthetic blends. They took 730 dust samples in these mills, using the verticle elutriator cotton dust sampler designed to sample dust with a mass median aerodynamic diameter of 15 μm and less.

Workers were divided into two main groups: those employed in the preparation and yarn processing areas, and those employed in the slashing and weaving areas. In the second group sizing added to the yarn increases the concentration of dust, giving these areas the highest dust levels but with a large proportion of biologically inert dust.

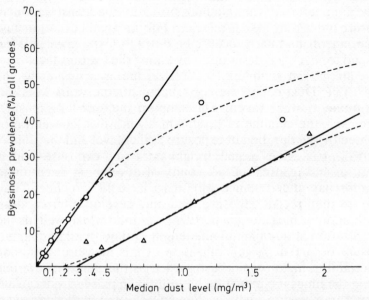

Figure 15.2 Association between byssinosis prevalence and median dust level among cotton preparation and yarn workers (○) and cotton slashing and weaving workers (△) showing linear regressions and fitted probit dose-response curves
Source: Merchant *et al.* (1973)

Workers in the preparation and yarn processing areas were subdivided into four more groups consisting of current smokers, non-smokers, those with grades 1 or 2 byssinosis and those with grade 2 byssinosis only. In each of the six groups data were tabulated according to 10 arbitrary dust levels based on the median dust concentration (mg/m³) in the work area of each individual. The mean ages in the groups were similar.

For dust levels up to 1 mg/m³, which represented the exposure of approximately 95 per cent of the whole population, strong linear associations were found between median dust level and the prevalence of byssinosis, and between dust level (*Figure 15.2*) and percent change in FEV_1. Byssinosis prevalence tended to flatten out beyond the 1 mg/m³ point, showing a curvilinear trend over the complete dust range up to 2 mg/m^3 which was best described by taking the probit of the prevalence and relating it to the $\log_{10}$ of the median dust level. The resultant probit dose-response curves fitted the data well (*Figure 15.2*) and were used to estimate the dust level associated with eight arbitrary levels of byssinosis prevalence as well as to predict byssinosis prevalence for particular dust concentrations. This enabled comparison to be made between any two groups of workers by means of an index of relative toxicity, based on the ratio between the dust levels which would produce a given prevalence of byssinosis in the two groups.

The conclusion drawn from the probit curves was that for preparation and yarn processing workers a reasonably safe level of cotton dust is 0.1 mg/m^3, a level at which nearly 94 per cent of the exposed population were found to have no symptoms of byssinosis. A separate level of 0.75 mg/m^3 was suggested for workers in slashing and weaving areas.

A different approach to the construction of exposure indices was made in a survey designed to measure the prevalence of cataract in steel workers and to investigate the relationship between cataract prevalence and heat exposure (Wallace *et al.*, 1971).

A full ophthalmic examination was carried out on 845 men aged 40–59 years, exposed to varying degrees of heat radiation, and a control group of 93 men with minimal or no heat exposure. All but two cases of cataract were of a mild form (type 1), with anterior and posterior subcapsular opacities lying behind the iris. The analysis of the data was therefore confined to men with bilateral normal lenses or bilateral type 1 cataract.

Among the heat-exposed workers a prevalence of bilateral type 1 cataract of 63 per cent was found in those employed in the steel plant and 52.9 per cent in men working on coke ovens, blast furnaces and hot mill combined. Comparison with a prevalence of 47.7 per cent in the control group showed these differences to be statistically significant ($P<0.05$). The probability of the differences having occurred by chance due to sampling fluctuations is less than 0.05.

In order to relate the prevalence of type 1 cataract to levels of heat exposure two methods of estimating exposure were used. One was a subjective method based on the research team's assessment of the heat exposure associated with a particular job. Four grades of exposure were established and allotted the factors 0, 1, 2 and 3. The years spent in the job were then multiplied by the appropriate heat factor to give a 'years exposed' score. For the second method, measurements were made by a globe thermometer at strategic points related to the various jobs. Heat factors varying from 1 to 7 were used to cover the range of radiant temperatures from 50°C to 350°C in 50° increments. These factors were also multiplied by the number of years spent in a particular job.

For both methods a statistically significant ($P<0.01$) linear relationship was found between the prevalence of cataract and 'years exposed'. The two regression equations were remarkably similar, both showing an increase from a prevalence of 50 per cent at zero 'exposure years' to over 70 per cent at 80 'exposure years'. To eliminate partially the effect of age, the relation between the prevalence of cataract and exposure years was considered within four five-year age groups. Positive trends were observed in all age strata for both methods of exposure assessment, and a composite test of the strata trends showed statistical significance at the 0.01 level for the subjective method and 0.001 for the black body radiation method.

From the findings in this study it was concluded that there was a need for further research, including the development of better indices of personal exposure to infra-red rays.

'Normal' values of lung function

Field surveys of large communities can provide valuable information on the characteristics of healthy men and women in the general population, from which 'normal' values for various physiological and biological functions may be estimated to provide control values for persons exposed to a known or suspected health risk.

The prediction of 'normal' values of the various indices of lung function usually involves the use of simple linear regression equations to describe the relation of lung function to age and height. These are not entirely satisfactory since the assumption of linearity is unlikely to hold for young adults and the elderly; also, the equations have often been derived from data on specially selected groups of men and women who may not be representative of the general population. Although it is known that ventilatory function varies between ethnic groups most published prediction equations have been based on men and women of European descent.

In a recent cross-sectional community study of black and white populations in the United States (Schoenberg, Beck and Bouhuys, 1978) maximum expiratory flow volume curves (MEFV) were recorded in a population of over 3000 lifetime non-smokers, free of respiratory symptoms, aged seven years and over. For each of six lung function measurements (FEV_1, FVC, FEV_1/FVC, PEF, MEF 50 per cent, and MEF 25 per cent) the relation between lung function and age, height and weight was analysed by multiple regression. In addition to the simple terms of age, height and weight the regression equation included their quadratic and logarithmic functions, as well as interaction terms and obesity indices. For each race, separate equations were calculated for girls, boys, women (age 15+) and men (age 18+). Objective statistical criteria were then used to select the optimal equations for each group.

The resultant prediction equations were more complex than those in general use, but compared to the simple linear equations they gave more precise predicted values and narrower confidence limits for both adults and children. In adults, with an age range from 15 years to 80 years, the simple equations gave predicted values that were too high at the ends of the age range and too low in the middle. The optimal prediction equations are likely to be more sensitive in detecting early adverse effects on ventilatory capacity from exposure to airborne contaminations, and should lead to more accurate discrimination between normal and diseased individuals or groups.

The annual decline in ventilatory capacity predicted by linear equations derived from cross-sectional studies has been compared with that found in longitudinal studies. Cotes (1976) describes the results in a study of 79 clinically healthy employees at a medium engineering factory who were seen in 1963 and again in 1973. The mean annual decline in ventilatory capacity over the 10-year period was measured by the forced expiratory volume in one second (FEV_1), the forced vital capacity and the FEV percentage of vital capacity. The results were similar to those predicted by the linear equations obtained from cross-sectional studies, which are commonly used to calculate normal values.

REFERENCES

Armitage, P. (1971) *Statistical Methods in Medical Research.* Oxford and Edinburgh: Blackwell

Bennet, A.E. and Ritchie, K. (1975) *Questionnaires in Medicine.* London: Oxford University Press

Berry, G., Gilson, J.C., Holmes, S., Lewinsohn, H.C. and Roach, S.A. (1979) 'Asbestosis: a study of dose-response relationships in an asbestos textile factory.' *British Journal of Industrial Medicine,* **36,** 98–112

Billewicz, W.Z. (1964) 'Matched samples in medical investigations.' *British Journal of Preventive and Social Medicine,* **18,** 167–173

Blackburn, H., Keys, A., Simonson, E., Rautaharju, P., and Punsar, S. (1960) 'The electrocardiogram in population studies; a classification system.' *Circulation,* **21,** 1160–1175

Bortner, R.W. (1969) 'A short rating scale as a potential measure of pattern A behaviour.' *Journal of Chronic Disease,* **22,** 87–91

Bouhuys, A., Barbero, A., Schilling, R.S.F. and Van de Woestigne, K.P. (1969) 'Chronic respiratory diseases in hemp workers.' *American Journal of Medicine,* **46,** 526–537

Cochran, W.G. (1977) *Sampling Techniques,* 3rd edn. New York: Wiley

Cochrane, A.L. (1951) *The Application of Scientific Methods to Industrial and Service Medicine.* London: HMSO

Cochrane, A.L., Davies, I., Chapman, P.J. and Rae, S. (1956) 'The prevalence of coalworkers' pneumoconiosis; its measurement and significance.' *British Journal of Industrial Medicine,* **13,** 231–250

Cochrane, A.L. and Holland, W.W. (1971) 'Validation of screening procedures.' *British Medical Bulletin,* **27** (1), 3–8

Cotes, J.E. (1976) 'Serial data over 10–22 years for detailed lung function of working men.' *Scandinavian Journal of Respiratory Diseases,* **57,** 316–317

Cotes, J.E. (1979) *Lung Function: Assessment and Application in Medicine,* 4th edn. Oxford: Blackwell Scientific Publications

Crooke, Morgan, D., Pasqual, R.S.H. and Ashford, J.R. (1964) 'Seasonal variations in the measurement of ventilatory capacity and in the answers of working coal miners to a respiratory symptoms questionary.' *British Journal of Preventive and Social Medicine,* **18,** 88–97

Documenta Geigy (1970) *Scientific Tables,* 7th edn. Basle: Geigy

Eisinger, J., Blumberg, W.E., Fischbein, A., Lilis, R. and Selikoff, I.J. (1978) 'Zinc protoporphyrin in blood as a biological indicator of chronic lead intoxication.' *Journal of Environmental Pathology and Toxicology,* **1,** 897–910

Fisher, R.A. and Yates, F. (1963) *Statistical Tables for Biological, Agricultural and Medical Research,* 6th edn. Edinburgh: Oliver and Boyd

Heller, R.F. (1979) 'Type A behaviour and coronary heart disease.' *British Medical Journal,* ii, 368

Higgins, I.T.T., Cochrane, A.L., Gilson, J.C. and Wood, C.H. (1959) 'Population studies of chronic respiratory disease.' *British Journal of Industrial Medicine,* **16,** 255–264

Higgins, I.T.T., Kannel, W.B. and Dawber, T.R. (1965) 'The electrocardiogram in epidemiological studies: reproducibility, validity and international comparison.' *British Journal of Preventive and Social Medicine,* **19,** 53–68

International Labour Office (1972) *ILO U/C International Classification of Radiographs of the Pneumoconioses, 1971.* Occupational Safety and Health Series No. 22 (revised). Geneva: ILO

Jahr, J. (1974) 'Dose-response basis for setting a quartz threshold limit value. A new simple formula for calculating the "lifetime dose" of quartz.' *Archives of Environmental Health,* **29,** 338–340

Lammers, B., Schilling, R.S.F. and Walford, Joan (1964) 'A study of byssinosis, chronic respiratory symptoms, and ventilatory capacity in English and Dutch cotton workers with special reference to atmospheric pollution.' *British Journal of Industrial Medicine,* **21,** 124–134

Lloyd Davies, T.A., Euinton, L.E., Ritchie, G.L., Trott, D.G., Watt, A., West, G.J.S. and Whitelaw, R. (1971) *Respiratory Disease in Foundrymen: Report of a Survey.* London: HMSO

Mainland, D., Herrera, L. and Sutcliffe, M.I. (1956) *Statistical Tables for Use with Binomial Samples – Contingency Tests, Confidence Limits and Sample Size Estimates.* New York: University College of Medicine

Medical Research Council (1976) *Questionnaire on Respiratory Symptoms and Instructions for its Use.* London: MRC

Mekky, Siza, Schilling, R.S.F. and Walford, Joan (1969) 'Varicose veins in women cotton workers. An epidemiological study in England and Egypt.' *British Medical Journal,* ii, 591–595

Merchant, J.A., Lumsden, J.C., Kilburn, K.H., O'Fallon, W.M., Ujda, J.R., Germino, V.H. and Hamilton, J.D. (1973) 'Dose response studies in cotton textile workers.' *Journal of Occupational Medicine,* **15,** (3), 222–230

Mitchell, C.A., Schilling, R.S.F. and Bouhuys, A. (1976) 'Community studies of lung disease in Connecticut: organization and methods.' *American Journal of Epidemiology,* **103,** 212–225

Moser, C.A. and Kalton, G. (1971) *Survey Methods in Social Investigation,* 2nd edn. London: Heinemann

National Institute for Occupational Safety and Health (1977) *Occupational Exposure Sampling Strategy Manual.* Eds. Nelson A. Leidel, K.A. Busch and J. R. Lynch, Cincinnati, Ohio: US Public Health Service, Dept. of Health Education and Welfare

Rickards, A.L. (1978) 'The routine monitoring of airborne asbestos in an occupational environment.' *Annals of Occupational Hygiene* **21,** (3), 315–322

Roach, S.A. (1959) 'Measuring dust exposure with the thermal precipitator in collieries and foundries.' *British Journal of Industrial Medicine.* **16,** 104–122

Rose, G.A. (1962) 'The diagnosis of ischaemic heart pain and intermittent claudication in field surveys.' *Bulletin of the World Health Organization,* **27,** 645–658

Rose, G.A., Holland, W.W. and Crowley, E.A. (1964) 'A sphygmomanometer for epidemiologists.' *Lancet,* i, 296–300

Rose, G.A., McCartney, P. and Reid, D.D. (1976) 'Self-administration of a questionnaire on chest pain and intermittent claudication.' *British Journal of Preventive and Social Medicine.* **31** (1), 42–48

Schoenberg, J.B., Beck, G.J. and Bouhuys, A. (1978) 'Growth and decay of pulmonary function in healthy blacks and whites.' *Respiration Physiology,* **33,** 367–393

Thorner, R.M. and Remein, Q.R. (1961) 'Principles and procedures in the evaluation of screening for disease.' US Public Health Service, Department of Health Education and Welfare, *Public Health Monograph* No. 67

Wallace, J., Sweetnam, P.M., Warner, C.G., Graham, P.A. and Cochrane, A.L. (1971) 'An epidemiological study of lens opacities among steel workers.' *British Journal of Industrial Medicine.* **28,** 265–271

Yates, F. (1980) *Sampling Methods for Censuses and Surveys.* 4th edn. London: Griffin

16

Sickness Absence — Its Measurement and Control

P.J. Taylor and S.J. Pocock

Physicians have sometimes been employed in industry as a direct result of management concern at high rates of sickness absence among employees. Such appointments are usually based on the following argument: this troublesome (and costly) waste of human resources is authorized by the medical profession. Thus, in accordance with the old adage 'set a thief to catch a thief', the company has its own doctor, who will be able to counteract this wastage by identifying malingerers and retiring on medical grounds those with bad absence records. Few occupational physicians would accept these as their primary objectives and thus conflict can, and sometimes does, arise between manager and doctor. Absence control and disciplinary action associated with it is first and foremost a responsibility of line management; the role of an occupational health service and indeed that of the personnel department in this context should be supportive and advisory. This chapter sets out some guidelines for the recording and measurement of sickness absence, considers the wide range of factors which can influence it — including the problems of medical certification — and indicates how an occupational health service may help.

Although the term 'sickness absence' is widely used and we would not suggest that it be dropped, it is nevertheless essential that all who are concerned with it appreciate that it is more accurately defined as *absence from work attributed to incapacity*. This emphasizes the fact that the phenomenon is first and foremost one of absence, and that 'attributed to incapacity' is the subordinate clause. Sickness absence must never be thought of as a reliable indicator of true morbidity, and in certain circumstances the links between the two may be tenuous. This fundamental point is illustrated by the employee who insists on coming to work despite a serious disease just as much as by the unenthusiastic worker who stays at home on sick leave when he develops the first sniff of a winter cold. One matter is clear: sickness absence is of considerable importance in industry because it is both disruptive and costly. Levels in the order of 8 – 12 per cent of normal working time are now common in manufacturing industry and the cost to the taxpayer also is substantial, not only from payment of social security benefits but also from the loss of taxation on earnings and the reduction in the gross national product. Rates of sickness absence have risen markedly in every industrialized country over the past twenty years and those in Britain are far from being the highest (Taylor, 1972).

THE RECORDING OF ABSENCE

Although most organizations make some attempt to record absence among their employees, the details of what is collected and how this is done often differs widely from one firm to the next. From basic principles, however, there can only be two ways of approaching this problem, and these differ greatly in reliability. First there is the recording of attendance; this assumes that an employee is absent unless attendance is shown (from clock cards, and so on) and it is the method commonly used where wages are paid for hours worked, generally applied to manual or 'blue collar' workers. The other method assumes attendance and only *absence* is recorded, as and when it is reported; this has been the system usually adopted for staff or 'white collar' workers and is inherently less reliable. Indeed the reliability of absence records usually falls as status rises, and few episodes may be recorded for senior executives unless they or their colleagues are unusually scrupulous about such matters.

Most organizations that keep such records also make some attempt to classify absence in various types, and some have developed elaborate systems for doing this. An American management study has listed over 70 different categories that have been used. In addition to sickness (either certificated or uncertificated) there may be absence with prior permission, justified, with pay, with adequate excuse, for personal reasons, family ill-health or bereavement, and so on. However, the decision to allocate an absence to one category or another is often subjective and arbitrary, and in larger organizations, where these decisions are taken by many different individuals, comparisons between different groups will be meaningless. Fortunately, for sickness absence the situation is less complicated since it is usual to note whether or not a medical certificate has been provided to explain the absence. Nowadays, many organizations do not require a medical certificate for short spells of sickness absence (usually up to three days). Nevertheless, it is customary to classify such uncertified absence as due to incapacity and employees are usually required to provide their own informal diagnosis. In some places this has been formalized by the use of a 'self certificate' which the employee is required to sign (Taylor, 1969).

Information required

The occupational physician who wishes to study the problem of sickness absence in his or her population is well advised to begin by finding out what information is already being collected. In general, this will depend upon the use to which it is being put. Thus, an accounts department will want to know how much an employee should be paid in basic rate, overtime, bonuses, and perhaps sick pay. Line management or personnel departments may need information for estimating manning requirements, for individual assessments, or for disciplinary reasons. The occupational physician may not be concerned with absence due to jury service or a local sporting fixture, but in order to monitor the health of the workforce it is necessary to know about all spells attributed to sickness or injury and, wherever possible, the nature of the condition that caused it. Even so, it is advisable for the physician to be aware of absence for non-medical reasons since it forms part of 'the withdrawal from work phenomenon' which underlies a substantial amount of sickness absence.

Occupational medicine is concerned with the health of working groups as well as that of individuals. Sickness absence records must be so designed as to provide epidemiological as well as personal information (*see* Chapter 13). As in any scientific study, the value of results can never be better than the quality and reliability of the information initially recorded. An important function of the occupational physician is to advise on what should be recorded and how it may be analysed, and it is here that an adequate understanding of epidemiology and the principles of statistics is of inestimable value. Thus it is essential that information be obtained for both the population at risk and each spell of absence attributed to ill health or injury. Of these, it is the first that provides the greatest difficulty in practical terms, but without it the most comprehensive records of sickness are only of value on an individual basis.

The population at risk

While most occupational health services have records about employees who have attended for examination, consultation, or treatment, few have records on all employees. The basic data on who is employed and where they work can usually be obtained from the payroll, and if the organization has a computerized payroll the problem is relatively easy to solve. In general it is well worth making use of classifications by department or occupation already used for accounting purposes rather than trying to devise one's own.

The minimum information required about the whole population at risk to allow subsequent analysis of sickness absence is the numbers of employees by sex and employment status ('blue' and 'white collar') in the larger departments or divisions, each subgroup being divided into five or ten yearly age groups. Although for analysis it is usual to classify by age (for example, 20 – 24, 25 – 29, 30 – 34 years), there is great advantage in obtaining the year of birth from the payroll so that one can, if necessary, combine absences from more than one calendar year. There is usually no need to know the day or month of birth since it is reasonable to assume an even distribution and allocate the population into groups achieving the age in question at some time during the year. Where labour turnover is relatively low and the total population at risk is reasonably static, it is enough to use the mid-year population as the denominator for all annual analyses of sickness absence. Problems may arise, however, during periods of rapid growth or redundancy when the number employed changes by more than 10 per cent during the year. In these circumstances it may be advisable to measure man-months at risk before calculating the total man-years. High rates of labour turnover (over 25 per cent in one year) may also affect absence rates since it is known that those about to leave tend to have more absence, while new employees may have below average rates—particularly if there is a qualifying period of employment before sick pay becomes available. In these circumstances one should consider dividing the population into those who were at risk throughout the year and those employed for less. This adds to the complexity of calculation but improves the validity of results and the conclusions that may be drawn from them.

Distinction between the sexes is most important since women are usually found to have an absence rate approaching twice that for men. The explanation however is not necessarily that women have more illness, for as Isambert-Jamati (1962) has demonstrated, the difference can also be explained by the lower occupational status of most jobs available to women in industry. In organizations employing a large proportion of women, it is worth while to distinguish between those who are married and those who are not. The nation-wide survey of the then Ministry of Pensions and National Insurance (1965) showed that while married women's sickness rates were much higher than those for men, 'other women' had a similar *severity rate* (*see* p.347) to men. Since the introduction of equality of opportunity and equal pay legislation, sickness absence experience of men and of women in the same jobs and on the same rates of pay have been compared (Taylor, 1979). This has shown that in several occupations women have consistently higher rates than men.

The spell of absence

The information obtained in the category of spell of absence provides not only the personal sickness absence record of each individual for whom the occupational physician has a responsibility, but also the numerator for subsequent calculation of sickness absence rates.

The maintenance of personal sickness absence records is essential for the proper practice of occupational health since they are necessary to enable the medical staff to identify individuals in need of health care or counselling and perhaps special treatment or change of occupation. The only practicable means of doing this is to have a card for each employee (whether or not they have yet been sick) on which the basic details of each episode of absence are entered. For the large organization with ready access to computers the information may be stored on magnetic tape, but this can only replace the personal sickness card if the information can be readily retrieved. A magnificent print-out of a patient's sickness record is useless if it takes days to obtain from an overworked (and reluctant) computer department, when the patient is in the health department today.

On this personal record card (an example is illustrated in *Figure 16.1*), the following personal details should be recorded: name, company identity or National Health Service number, date of birth, sex and marital status, occupation, department, hours of work, and date of engagement. Any changes in the personal details, such as job transfers, should be included on the record cards. For each spell of sickness absence there should be recorded the date of onset and duration, the final diagnosis with an indication whether caused by work or not, and whether the cause was certified by a doctor. Other details about the individual or the spell of absence may be recorded at the discretion of the physician. These may include facts about smoking, immunization status, known chronic disablements, family responsibilities, place of residence, or the identity of the general practitioner. For subsequent analysis this additional information may also be required for the whole population at risk or, when this is very large, for a carefully drawn sample. After all, an observation that three quarters of all spells of bronchitis occurred in smokers immunized against influenza with two or more dependants, may mean little unless one knows how many without bronchitis had these characteristics.

An alternative to the 'list of spells' type of personal record card in *Figure 16.1* is to have a series of annual calendar records for each

NAME SMITH		Christian Names JOHN HENRY		D of B 30-7-33		Company Number 682731	NHS Number OCEK 3174.1
Sex M	Marital Status M	Address 3 ACACIA ROAD			General Practitioner DR JONES		Registered Disabled NO

OCCUPATIONS/WORKING HOURS

Date	New Job etc.
15.3.76	Storeman Days
24.10.77	Machinist. Alt. Day/Night Shift
15.10.79	Clerk Safety Days

SICKNESS ABSENCES

Date off	Date return	Days lost	Certif. Uncertif.	Diagnosis	ICD	Occupational Non-occup
3.1.77	10.1.77	5	C	Influenza	487.1	NO
15.6.77	6.7.77	15	C	Sprained lumbar spine	847.2	O
13.2.78	14.2.78	1	UC	Cold	460	NO
18.9.78	19.9.78	1	UC	Diarrhoea	009.3	NO
15.5.79	18.5.79	3	C	Pharyngitis	462	NO
21.8.79	25.9.79	25	C	Lumbar Disc Lesion	722.1	O
7.1.80	21.1.80	10	C	Influenza	487.1	NO

Figure 16.1 An example of a personal sickness record card

employee in which every working day has a box which remains empty if he attends for work but which on days of absence has recorded a letter code for type of absence (e.g., C = certified sickness, U = uncertified sickness absence, A = absence for other reasons). Holiday (H) and lateness episodes (L) can also be recorded. Behrend (1978) shows an example of such a calendar record. It can give a more immediate visual impression of the amount and pattern of individual absence in a year. However, this system involves more paper (one sheet per employee per year), may be difficult to follow for seven-day shift workers and cannot give the degree of detail on medical diagnosis and the long-term perspective of sickness absence experience that the 'list of spells' records provide. Although calendar records may be of value in a personnel department, cards such as *Figure 16.1* are to be preferred in an occupational health department.

Two aspects of the information about the spell of absence require more detailed discussion. First, it is often alleged that the inaccuracy of the diagnosis as stated on the medical certificate invalidates any serious study of the causes of sickness absence. This problem has been studied by government agencies and by individual physicians in industry. While it must be accepted that a few diseases are certainly under-reported on certificates handed to the patient, malignant tumours providing the best example, the majority are sufficiently reliable to allow study in broad diagnostic groups such as upper-respiratory infections, urinary conditions, cardiovascular disease, and so on (London Transport Executive, 1956). The main problem arises from the substantial proportion of vague and ill-defined conditions which now form about 10 per cent of all cases of certified absence in reports from many countries. The practice of seeing all employees who have been off sick for a minimum period of a month or so (as discussed in Chapter 9) will enable the physician or occupational health nurse to clarify a number of indecipherable or vague diagnoses, and an effective liaison with general practitioners will allow the physician to discuss the few problems that remain. It is reasonable to conclude that the information on diagnosis—provided it is handled with circum-spection and generally grouped under main headings—can provide useful information for monitoring the health of working groups. For recording and analysis, however, it is necessary to code the diagnosis, and this should be done to the rubric of the International Classification of Disease (WHO, 1977). The practice of simplifying diagnostic classifications to a dozen or so main diagnostic groups or bodily systems prohibits the study of specific diagnoses and, therefore, is not recommended. Although for most factory populations broad groups suffice for routine annual analyses, the need may arise to combine several years' figures to look at single conditions such as asthma, back pain and so on. Only if diagnoses have been coded to the three digit rubric could this type of *ad hoc* study be done.

The other matter that should be discussed is whether the duration of absence from work should be recorded in terms of calendar days or working days (or shifts) lost. The medical literature on sickness absence has almost invariably used the calendar day system. A number of reasons have been put forward in support of this, the most telling being the point that a man may be unfit for work for a period of, say, two weeks, whether or not he was supposed to work for more or less than 10 days during that period. It does however overstate the *absence,* and with the preponderance of shorter spells which are now found in virtually every organization, this overstatement can be substantial. Thus a man who takes a day or two off work at the end of a week for a minor ailment but returns to work on the following Monday will be recorded as having had perhaps three or four calendar days of sickness, even though he may have been fit over the weekend.

The protagonists of the system for recording working days (or shifts) lost are found among line managers, accountants, and personnel officers. They argue that it is absence from work that really matters and it is incontrovertible that it is this that costs the organization money. Also, since it is well known that sickness absence is a very different phenomenon from true morbidity, it is only realistic to record the known amount of absence rather than the unknown, and sometimes questionable, period of 'incapacity'. Ill health and incapacity during a weekend off may be unfortunate for the individual, and the physician or nurse may wish to know about it, but in practical terms there is no way of ensuring that they are informed about such incapacity when no work is lost. Thus, where a new system of sickness absence recording is being set up and where the information about *working days lost* is readily available from the organization's personnel or accounts records, this is more useful than calendar days. Many occupational physicians and nurses will disagree with this view, but the reasons discussed above justify it, particularly since the objective is to measure sickness absence and not morbidity.

ANALYSIS OF SICKNESS ABSENCE

The most simple and usual method adopted by personnel departments for expressing the amount of absence in an organization is the 'percentage time lost'. For any single day this involves the number of absentees (usually including all forms of absence) expressed as a percentage of the total number of employees expected to work that day and can give quick information which is of most value for estimating manning requirements. Such daily percentages can then be averaged over weekly, monthly or annual periods to give an overall measure of the time or severity of absence in a workforce. This measure is a *point prevalence rate*. It has the limitation that it gives no indication of the frequency of absence. For example, an annual lost time rate of 10 per cent in a group of 10 employees could be due to one employee being off sick for the whole year or at the other extreme could be due to a single day absence by one of the men on each working day throughout the year. The two situations are utterly different from a medical viewpoint; they also present very different problems to the line manager who could cope with the former situation a great deal more easily than the latter. A measure of frequency would show up this difference, the first situation having an annual average of 0.1 spells per employee, the second some 25 spells per employee (assuming 250 working days in a year).

Rates commonly used for sickness absence analysis

There are two rates which are fundamental in any absence study:

Mean days per person *A severity rate* $= \dfrac{\text{Total days of absence in period}}{\text{Average population at risk in period}}$

Mean spells per person *A frequency rate* $= \dfrac{\text{Total no. of new spells of absence in period}}{\text{Average population at risk in period}}$

The former is a measure of severity of absence and is equal to the percentage time lost multiplied by the number of working days in the period, while the latter is a measure of absence frequency. The time period used is often one year, although shorter or longer periods can be used. One could calculate these rates separately for certified and uncertified sickness, but it is often simpler and more informative to combine the two. Evidently in any investigation of sickness absence one should exclude absence for other reasons; although, if company policy or recording systems make it difficult to make this distinction, the resultant overall absence rates should still be informative, since absence from other reasons is usually very low.

Two other rates may also be useful. The mean number of one-day spells per person can be of especial value in assessing morale in an organization. Also the average length of spell (= total days/total number of spells) is sometimes quoted.

Comparisons between rates

If analyses of sickness absence rates have shown substantial differences between two or more groups of employees, and departmental comparisons within one workplace will almost certainly show this, it then becomes necessary to assess their importance and try to find explanations for them.

Although one may be looking for an occupational factor to explain a high rate in one group of employees, there are several other factors that must first be considered. It is not possible in this chapter to describe all the factors which are known to influence sickness absence rates, but *Table 16.1* shows that these can be numerous and varied. It is not suggested that every one of these factors must be allowed for

Table 16.1 SOME FACTORS KNOWN TO INFLUENCE SICKNESS ABSENCE

National/regional	*Organization/department*	*Personal factors*
Geography	Type and size	Sex, age and status
Race	Management attitude	Occupation and length of service
Season	Supervisory quality	Working hours and wage rates
Health Service	Personnel policy	Job satisfaction
Insurance benefits	Sick pay	Journey to work
Pension age	Working conditions	Medical conditions
Epidemics	Medical service	Family responsibility
State of economy	Labour turnover	Personality

whenever a comparison is required between absence rates. They may well be irrelevant if the two groups are from the same workplace. Nevertheless, the influence of age, sex, and occupational status is so strong that these at least must always be considered. Increasing age is generally associated with a decrease in the frequency of short spells of absence but the average length of spell tends to rise sharply after middle age and there is usually an overall rise in days of absence per person at risk. The relationships with sex and marital status have already been mentioned; those employees of higher occupational status invariably have a great deal less absence than others in unskilled manual grades. It is also worth noting that of all the factors listed in *Table 16.1* only two, 'epidemics' and 'medical conditions' are strictly concerned with morbidity. Most of the other factors are either demographic or are susceptible to management action or personnel policy.

Absence and the individual

The main problem associated with the interpretation of these sickness absence rates is that they cannot express the immense variability between individuals in the amount and pattern of absence. For example, *Figure 16.2* shows the distributions of days, spells and one-day spells of absence in one Scottish factory over a six-year period 1969–74 for the 610 men continuously employed during that period (Behrend and Pocock, 1976). Incidentally, all types of absence are included here, although sickness absence predominates. It illustrates the typical skewed distribution of absence behaviour found in any organization with the great majority having fairly moderate amounts of absence while a few employees have large numbers of days and/or spells of absence. The skewness is usually most marked in days' absence: here there are two employees with no absence in six years, 86 per cent under 200 days, and two men with over 600 days' absence. In the above example, the mean days per man in 1969–74 is 109 (i.e. 18 days per annum) whereas only a third of the men actually had so much absence. The *median* days per man, defined as that amount exceeded or equalled by 50 per cent of the population, has a value of 79 days. The median may often be preferred to the mean for summarizing days, spells and one-day spells since it is less affected by the few employees with large amounts of absence and gives a better idea of the 'average employee's' absence. Of course, the immense variability in absence implies that any concept of 'average' has its limitations.

The full display of absence experience as in *Figure 16.2* every time one undertakes an absence study would be very tedious, especially if as is usual, one also compares different subsets of employees (e.g. departments, age groups). Instead, one can focus attention on a few particular levels of absence, say, the amounts exceeded or equalled by 90, 75, 50, 25 and 10 per cent of the population, with results as shown in *Table 16.2*. These reference levels can then be used for further statistical analysis (*see* Behrend and Pocock, 1976). For instance, one might wish to see whether particular departments or perhaps some other personal factors are associated with an unduly large proportion of high absentees.

Table 16.2 CLASSIFICATION OF INDIVIDUAL ABSENCE RECORDS FOR THE SIX-YEAR PERIOD 1969–74

Workforce distribution*	Absences		
	Days lost	Total spells	One-day spells
Lower decile	16	5	2
Lower quartile	40	9	5
Median	79	18	10
Upper quartile	151	28	17
Upper decile	228	40	24

*Lower decile, lower quartile, etc., denote the amount of absence exceeded or equalled by 90, 75, 50, 25 and 10 per cent of employees respectively.

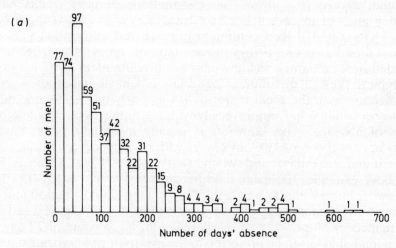

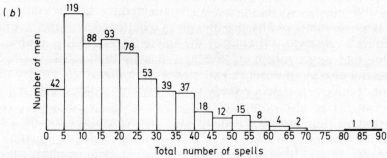

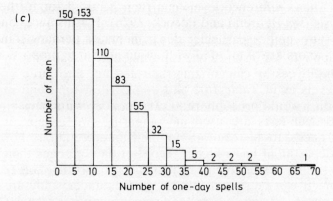

Figure 16.2 Distribution of days' absence, total spells and one-day spells, 1969–74. (a) Distribution of days' absence in intervals 0–19, 20–39, etc. (b) Distribution of total number of spells in intervals 0–4, 5–9, etc. (c) Distribution of one-day spells in intervals 0–4, 5–9, etc.

Another use for the upper deciles in particular is that individual employees who appear to have a sickness absence problem can be identified and further investigation, perhaps including an interview with the occupational physician, might reveal the underlying cause.

The example used here is somewhat exceptional in being based on six years' absence experience. It will often be worthwhile to carry out similar analysis based on one year's absence and it is usually beneficial to both occupational health and personnel departments if such analyses and their interpretation can be placed on a regular annual basis.

The use of statistical methods

The highly skewed distributions of sickness absence in a population mean that absence rates calculated from small numbers of employees can be seriously distorted by the odd employee with exceptionally high absence. For instance, one employee off sick for a whole year as a result of a heart attack will cause the mean days per person in a group of 25 to be increased by 10 days. As a rule of thumb, it is usually not very informative to calculate absence rates for populations smaller than 50 (or preferably 100) man-years at risk. Even with larger samples, however, the usual tests of statistical significance for comparison of means (i.e. *t*-tests) cannot be rigorously applied to mean rates of frequency or severity, and especially not to the latter. Instead, it is more appropriate to base statistical comparisons on the percentages of employees exceeding certain levels of absence as described in the last section, using chi-squared tests of significance. Taylor, Pocock and Sergean (1972) applied such techniques in a comparison of shift and day workers. More complex statistical methods can be applied, for instance one can compare days' absence in two groups of employees by categorizing their absence into three or more different levels as in *Table 16.2* and doing a test for trend in proportions, but this would probably require help from a statistician.

Whatever statistical tests are used in comparing two groups of employees, one needs to be cautious in their interpretation. For instance, if one department has a higher record of days absence than another this may simply be due to having a greater proportion of older employees. Where it can be shown that the groups under study are comparable with regard to age and sex (and sometimes other relevant factors such as occupational status) then direct comparisons can be made. In many comparisons, however, it may first be necessary to use standardization techniques which are discussed in Chapter 13 and more fully by Bradford Hill (1977). The indirect method of standardization is the most appropriate. These methods were originally developed for mortality studies and are usually still described in those terms but the adaptation to absence rates is straightforward.

However, one should beware of getting too involved in complicated statistics since errors and confusions are liable to result. Senior executives in industry are seldom interested in lengthy explanations, so that the occupational physician may prefer to report some simple statistical information with an interpretive phrase such as 'these differences, after allowing for age and sex are (or are not) of statistical significance'. Scientific reports should, however, contain enough information to make clear what techniques were used and what precise conclusions are to be drawn.

Lastly, when planning absence analysis one should not underestimate the amount of data processing involved. A few organizations have elaborate computer-based individual absence record systems whereby every spell of absence is computerized as it happens. For most organizations it may be better to transfer to computer at the end of each year each employee's number of days, spells and one-day spells of absence, plus any relevant personal information (e.g. department, occupation, age, sex, marital status, length of service, etc.) and hence base an annual analysis on these individual summaries. More detailed analyses of absence by day of week, time of year or over longer periods will require more specialist knowledge and careful planning. It is perhaps better not to do them as a routine.

Cross-sectional and longitudinal studies

Much of the above discussion concerning the comparison of absence rates is based on the assumption that a cross-sectional study is to be undertaken, that is, a fixed population of employees is to be observed for a fixed period of time, usually a year, with the objective of discovering those personal or environmental factors which are associated with the various measures of sickness absence. One common error in such an investigation is to conclude that such factors are actually causing differences in absence behaviour. For example, the discovery by Taylor, Pocock and Sergean (1972) that shift workers have less absence than day workers after allowing for age and occupation, does not imply that shift work causes less absence. It could be that those employees who are liable to be frequently sick avoid having to do shift work, or management may select only the most reliable employees to do shift work. Thus, the structure of an industrial population is affected by selection criteria on the part of both management and the employees themselves, and this can lead to spurious relationships.

Cross-sectional studies are essentially concerned with a static situation so that, in order to investigate the effect on sickness absence of changes in the employees' environment, a longitudinal approach is more appropriate (*see* Pocock, Sergean and Taylor, 1972, for an example regarding change in shift pattern.) This will normally entail the calculation of sickness absence rates over several consecutive equal time periods. Subsequent analysis may be complicated by the existence of fluctuations in the incidence of infectious disease, labour turnover, seasonal variations or inconsistency of recording. However, if one is interested in the effect of a particular modification to the environment (for instance, new working methods or a new pay scheme) it may be possible to compare absence rates in a control group of employees not affected by the modification.

Medical certification of incapacity

It is one of the anomalies of industrialized societies that doctors are required to certify fitness or unfitness for work when their training as medical students rarely, if ever, includes this topic. Doctors are taught to listen to (and accept) what their patients tell them and to search for and recognize symptoms and signs of disease. Fitness or unfitness for work is a totally different problem. Even if doctors had the equipment to measure physiological parameters such as maximum oxygen uptake and so on, this information would not necessarily be of relevance when deciding whether a patient was or was not fit to do one particular job. Also the wording of certificates requires a 'fit' or 'unfit' opinion and does not allow for the large area of partial fitness.

Despite this, there is an obligation for doctors in all countries of the world to certify incapacity for work. In the United Kingdom it forms part of the terms and conditions of service with the National Health Service, but in some countries, such as the Netherlands, this responsibility is allocated to a special cadre of 'insurance doctors'. The arguments for and against medical certification of incapacity were considered at length by the Fisher Committee (1973) and this most interesting report includes evidence by the British Medical Association admitting that in the majority of cases such certificates are given more or less on demand. Nevertheless, the Committee considered that, despite its drawbacks, the existing system was probably the best available, and wryly observed that most patients seemed to believe that their doctors could tell whether or not they were speaking the truth.

Unfortunately the medical profession has long been a most convenient scapegoat in the eyes of managers to explain the rising trend of sickness absence. This view brings with it a number of advantages since it absolves the manager himself from any blame.It has also led to the (quite erroneous) belief that no employee can be disciplined or dismissed because of undue sickness absence and it has generated an apathetic acceptance of high levels of absenteeism in many organizations.

The facts of the matter are clear. In all but a small minority of illness or injuries when incapacity for work is usually obvious to all, it is the patient who decides whether or not to consult his doctor and whether or not he feels able to go to work. Similar considerations apply at the convalescent stage after a serious illness or injury when it is, in effect, the patient who decides whether or not he feels able to return to work. Some general practitioners do indeed make serious efforts not to award certificates unnecessarily, but in the United Kingdom this can sometimes result in the patient (and his family) transferring to the list of a more compliant practitioner. In general, however, most doctors are inclined to support the view of their own patients and all are aware how difficult it is to prove a negative — in other words to disbelieve a patient who complains of pain in the back or head and so on. They inevitably accept the patient's own description of his or her job, and here a close liaison between occupational physician and general practitioner can be of great help to both.

Most occupational physicians would agree that it is virtually never worth challenging a *current* certificate of incapacity, but an informal discussion with the general practitioner can often influence whether or not he will issue a further certificate. Managers too must realize that even under recent employment protection legislation it is possible to retire or dismiss an employee on the grounds that his or her absence has been unacceptable and incompatible with continued employment. The essential point is that the employee must be warned and given the opportunity to improve before the final sanction is applied.

THE CONTROL OF SICKNESS ABSENCE

The primary responsibility for the control of all forms of absence from work must lie with line management. On occasion, however, some managers attempt to pass some of their less pleasant tasks to staff functions such as personnel, and, in the case of sickness absence, the occupational physician may be asked to undertake a disciplinary role. This must be resisted since not only does it conflict with the accepted code of medical ethics but also it would rapidly erode employees' confidence in the physician's work and integrity. Despite this, there is a great deal that the physician can do to assist both management and employees in the reduction of absence due to incapacity.

The activities of the occupational physician straddle the fields of both community and personal medicine and thus under each of the traditional medical headings of diagnosis, treatment and prevention, both the working group and the individual employee may be considered.

Diagnosis

By diagnosis is meant an understanding of the real cause of the absence both in individuals and in groups. In broad terms there are four underlying categories: occupational, medical, social and behavioural. Absences covered by a medical certificate citing most of the usual clinical diagnoses may have their ultimate cause in any one of these four categories. Since sickness absence is not synonymous with morbidity, one must decide not only what clinical condition may exist but also why it occurred, and why, when the condition is not totally incapacitating, the individual chose to stay at home. Although overt occupational injury or disease may be relatively uncommon as a cause of sickness absence, study of group records may reveal previously unrecognized occupational stresses, perhaps rarely toxicological, but not infrequently due to the physical or psycho-social stresses generated in the working group. A crop of minor illnesses or even injuries arising in one department may really be due to friction with supervision or a change in production schedules. There is one more strictly medical problem that can also make itself apparent by frequent episodes of absence often attributed to various ailments. This is alcoholism, or problem drinking. The condition is now believed to affect at least 3 per cent of the working population even though it is rarely recognized. The occupational physician or nurse must be constantly aware of the possibility of alcohol as an underlying cause of frequent episodes of absence.

While the strictly medical or physical causes of incapacity should present no great difficulty in diagnosis, it is important to appreciate that an individual's threshold for deciding that he is unfit for work can be altered by his attitude to work and those of his colleagues. Similarly, social or family problems such as sleepless nights with a new baby, illness (particularly psychiatric illness) of his wife, or other pressures at home can also significantly lower a man's threshold. Finally, in every organization there will be a few whose personality and attitude to life and its problems are so immature that consistent and regular attendance at work is almost impossible. Such a person will also seize on any minor malaise as a state of 'ill health' sufficient to justify absence, and it is not difficult for such people to convince their family doctor that they need a medical certificate for a few days off due to their headaches or other minor complaints. Accurate diagnosis can only be made with the help of adequate sickness absence records of individuals and by the frequent monitoring of groups.

Treatment and prevention

The primary role of an occupational health service is prevention. Although therapy in the traditional medical sense may form only a minor part of its activities, treatment is used here with a rather different meaning.

Taken in logical order, the main aspects of this can be considered as:

1. Prevent inception of spells of absence.
2. Reduce the duration of those that do occur.
3. Ensure that those who return can be effective.
4. Assist in the resettlement or retirement of those whose condition makes them unfit to return to their previous occupation.

Each of these four aspects may be considered both in relation to the individual and to the working group. An effective occupational health service must be concerned with each one.

Prevention of inception of absence

From an adequate analysis of the records of their own organizations occupational physicians and nurses must identify the problem areas and decide an order of priority to make the most effective use of available resources. A detailed discussion of each point would be out of place in this chapter but some indication can be given of the methods that may be considered. At the individual level there is little doubt that an efficient initial treatment service staffed by experienced nurses and first-aiders can appreciably reduce a tendency for those with minor conditions to take time off to visit their doctor. Since many now have to make an appointment for such a visit and may feel that they have to elaborate their symptoms to justify troubling their general practitioner, it can be seen that a few days away from work can easily ensue. At the group level, the case for preventive measures, such as influenza inoculation, should be carefully considered if the appropriate antigen is available in time. Mass immunization can be very expensive, but with protection from appropriate vaccines now running at about 70 per cent, there may be a case for inoculating key staff and those at special risk. Health education may also affect the attack rate of some illnesses and it can certainly assist employees in taking a more informed and realistic attitude towards minor as well as major ailments.

Reduction in the duration of absence

Reduction in the duration of absence can be achieved in various ways. Effective liaison between the occupational physician and local general practitioners and consultants can ensure that the patient's absence is not unduly long, and where medical surveillance or continuation treatment can be arranged this can lead to earlier return to work.

Effective return to work

The first day or so back at work is often particularly tiring and the prospect of this, with rush hour travel, tends to delay matters. An arrangement between the doctors for shortened hours or modified duties can reduce this problem. The role of the occupational health nurse in sick visiting is controversial; it is sometimes resented, but many organizations have found that, properly introduced, this can create a feeling among those off sick that their organization is concerned for their health and wants them back as soon as possible. After the more serious conditions that cause absences lasting for a couple of months or more, a mild reactive depression is not infrequent due to the social isolation that has occurred. Sympathetic encouragement and proper understanding of this complication can also shorten what might otherwise become a very long absence indeed due to an irrational fear of returning to work.

Assistance after return to work

The problems of effective rehabilitation and resettlement are most important, and the recent extension of productivity deals involving job flexibility has made the problem more difficult to solve. These have already been considered in Chapter 3 (Work and Health), Chapter 8 (Treatment and First Aid Services) and Chapter 9 (Preliminary, Periodic and Other Routine Examinations).

In conclusion, we would emphasize that to provide effective support to management on the perennial problems of sickness absence the occupational physician and nurse must have access to adequate absence records of both individuals and the group. Only then can they hope to identify those who can benefit from their help. Without such records they must rely on clinical impressions or, even worse, the uninformed prejudices of others.

REFERENCES

Behrend, H. (1978) *How to Monitor Absence from Work*. London: Institute of Personnel Management

Behrend, H. and Pocock, S.J. (1976) 'Absence and the individual: a six year study in one organization.' *International Labour Review*, **114**, 311–327

Fisher Committee (1973) *Report of the Committee on Abuse of Social Security Benefits*. Cmnd 5228. London: HM Stationery Office

Hill, A. Bradford (1977) Chapter 17 in *A Short Textbook of Medical Statistics*. London: Hodder and Stoughton

Isambert-Jamati, V. (1962) 'Absenteeism among women workers in industry.' *International Labour Review*, **85**, 248–261

London Transport Executive (1956) *Health in Industry. A contribution to the study of sickness absence*. London: Butterworths

Pensions and National Insurance, Ministry of (1965) *Report on an Enquiry into the Incidence of Incapacity for Work: Part II*. London: HM Stationery Office

Pocock, S.J., Sergean, R. and Taylor, P.J. (1972) Absence of continuous three-shift workers: a comparison of traditional and rapidly rotating systems.' *Occupational Psychology*, **46**, 7–13

Taylor, P.J. (1969) Self certification for brief spells of sickness absence. *British Medical Journal*, **1**, 144–147

Taylor, P.J. (1972) 'International comparisons of sickness absence.' *Proceedings of the Royal Society of Medicine*, **65**, 577–580

Taylor, P.J. (1979) 'Aspects of sickness absence.' Chapter 21 in A. Ward Gardner ed. *Current Approaches to Occupational Medicine*. Bristol: John Wright

Taylor, P.J., Pocock, S.J. and Sergean, R. (1972) 'Absenteeism of shift and day workers.' *British Journal of Industrial Medicine*, **29**, 208–213

WHO: World Health Organization (1977) *Manual of the International Statistical Classification of Diseases*. Ninth rev., Vol. 1, Geneva

17

Airborne Contaminants

M.K. Molyneux

FACTORS RELATING TO ALL CONTAMINANTS

Many occupational diseases and related problems arise from contact with airborne contaminants. As the effects can be irreversible, preventive action is essential. Air sampling has become an integral part of control and various authorities prescribe hygiene standards or control limits. In most cases the lung is the principal route of entry and from a knowledge of the aerodynamic characteristics of contaminants in air and their behaviour in the lung it has been possible to design suitable methods of sampling and evaluation.

Knowledge of the hygiene standards, their numerical values, documentation, and application is essential, and examples of the authorities concerned are given in *Table 17.1*. The values are expresed as concentrations of contaminants in air, as parts per million (ppm), and milligrams per metre cubed (mg m^{-3}) and fibres per cubic centimetre (fpcc). They are aimed at preventing both harmful irreversible effects and nuisance to persons who are exposed daily as part of their work. The American Conference of Governmental

Table 17.1 EXAMPLES OF HYGIENE STANDARD AUTHORITIES

Health and Safety Executive (United Kingdom)	H & SE
National Institute of Occupational Safety and Health (USA)	NIOSH
Occupational Safety and Health Administration (USA)	OSHA
American National Standards Institute	ANSI
British Occupational Hygiene Society	BOHS
American Conference of Governmental Industrial Hygienists	ACGIH

Industrial Hygienists (ACGIH, 1979) publish annually the Threshold Limit Values which have widespread use in America and Western Europe. Such authorities have influenced the practical approach to measurement and evaluation.

Reference is made to Time Weighted Average (TWA) concentrations for 8 hours, to Short Term Exposure Limits (STEL) for 15 minutes, and to Ceiling Values (C) for instantaneous exposure, and attention is drawn to substances which present an additional risk by skin absorption (S). The contaminants are also classified as mineral dusts, nuisance particulates, simple asphyxiants and carcinogens. There are biological and practical reasons for varying the sample 'averaging' time. For example, the British Occupational Hygiene Society (1968) specified a TWA based on 13 weeks for chrysotile asbestos, as a means of controlling early signs of asbestosis, while a recent review by the Advisory Committee on Asbestos (1979) has led to the recommendation for an averaging time of four hours, with an option to take shorter time samples. Similarly, for more rapidly acting substances such as carbon monoxide and cyanide there are reasons for adjusting the sample averaging time to fit in with the short biological half life which may be minutes or seconds (Roach, 1977).

There is a need to choose an instrumental approach to fit in with the specified averaging time and this can be regarded as a basic feature of air sampling. Where this cannot be achieved the data is treated mathematically as described in a later section.

Contaminants appear in particulate form which embraces dusts, fumes, smokes and mists, and in the form of gases which include gases and vapours. These occupy a very wide size spectrum ranging from less than 0.001 to 1000 micrometres (μm) (*Figure 17.1*). Since biological effect is related to site of deposition there is a need to investigate the

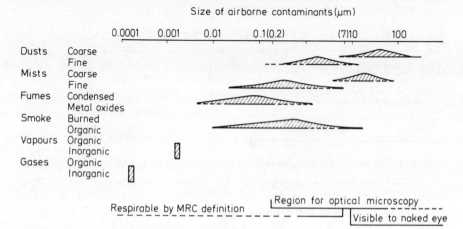

Figure 17.1 Size characteristics of airborne contaminants

appropriate 'particle' size, hence the choice of instruments which can differentiate between total airborne material and the 'coarse' and 'finer' constituents (International Labour Organization, 1967).

Physical characteristics

Dusts, fumes, smokes and mists occupy that part of the size spectrum which is under the influence of gravity which causes sedimentation. The particles have a'terminal velocity' which can be determined by using Stokes law where:

$$V = \frac{2gr(S_1 - S)}{P}$$

where V is terminal velocity, r is radius, S_1 is particle density, S is air density, g is acceleration due to gravity and P is viscosity of air.

This can be applied to a range of particles in size from 1000 μm (very coarse dust) down to 0.1 μm (freshly condensed fume). The associated terminal (or falling) velocities are given in *Table 17.2*. Particles of less

Table 17.2 FALLING VELOCITY OF SPHERES OF UNIT DENSITY

Diameter μm	Velocity cm sec^{-1}
100.00	25.0
50.0	7.2
10.0	0.3
5.0	0.078
1.0	0.0035
0.5	0.0010
0.1	0.000085

than 0.1 μm have diameters which correspond to the mean free path of gas molecules and are subject to Brownian movement. Particles of about 0.25 μm occupy an area where the terminal velocity is minimal and where Brownian movement is also minimal. As a result they tend to remain airborne with little movement relative to the surrounding air. So above 0.25 μm particles settle out at a rate determined by their

terminal velocity, while below 0.1 μm they are subject to the laws of diffusion.

Particles of more than 0.1 μm exhibit other effects which are relevant: they can be electrically charged, they can move along a thermal gradient, they are subject to impingement due to their inertia, and they have optical properties. These features have been used in instruments to collect, measure and visualize particulates in air. Because of settlement, particulate clouds change in concentration and size range with time and with distance from the source. This can be influenced by flocculation which is observed in freshly condensed fume (initially about 0.1 μm in diameter) causing particles to coalesce as they age, due to collisions while in Brownian movement.

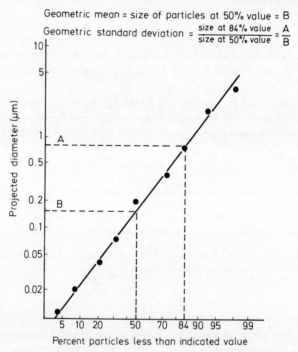

Figure 17.2 Graphical representation of particle size distribution using log probability method

Because of their physical size the particulates can be characterized by optical or electron microscopy and they can be described in terms of particle number, size, weight and surface area. So that the characteristics of different particulates can be compared, the data may be presented as a log probability graph as shown in *Figure 17.2*. Since gases and vapours exhibit none of these physical properties the methods of collection are based solely on absorption or chemical reaction.

Aerodynamics of the respiratory system

There is a reasonable understanding of the factors which influence the entry, deposition and retention of particles in the respiratory tract. For practical purposes the tract has been considered in three sections: the nasopharynx (NP), trachea and bronchial tree (T-B), the respiratory bronchioles, alveolar ducts, atria, alveoli and alveolar sacs (i.e. the pulmonary region; P). These divisions were used by the International Commission on Radiological Protection (ICRP, 1966) to formulate models for deposition in, and clearance from, the tract and diagrammatic representation of the deposition curves is given in *Figure 17.3*. The values

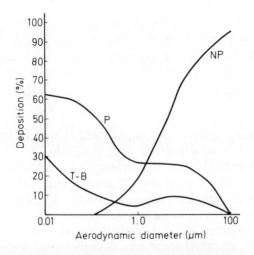

Figure 17.3 Deposition in the respiratory tract. NP = nasopharynx; T–B = tracheobronchial tree; P = pulmonary region: respiratory bronchioles, alveolar ducts, atria, alveoli, and alveolar sacs
Source: ICRP (1966)

are affected by such factors as rate of respiration and hygroscopic properties, but show that the largest size to enter the tract is about 100 μm, that the NP region is penetrated by particles of less than about 10 μm, and that P is penetrated only by particles of less than about 7 μm.

The factors which are operating have already been noted: terminal velocity, inertial effects, Brownian movement and diffusion, so that inhaled particles leave the air stream by impingement on the walls of the NP and T-B regions, and by elutriation in all parts, including the pulmonary region. The particles with the greatest terminal velocity are deposited early in their passage through the tract, and the smaller particles survive the journey to the pulmonary region. A high proportion of the particles of 0.25–0.1 μm are exhaled. A critical feature of this process is that only particles with a terminal velocity of less than about 0.2 cm/second can enter the pulmonary region (P), and the terminal velocity has become an accepted method of defining the so called *respirable* fraction.

Particle size

For practical reasons the term 'particle size' or 'particle diameter' is used to define respirability, whether the method of separation is by terminal velocity or by optical size. This requires brief explanation. Where the diameter is measured in terms of terminal velocity the quoted size refers to the diameter of a unit density sphere which has a terminal velocity equal to that of the particle being measured. This is referred to as the Stokes Diameter.

When the particle size is measured in terms of its optical image, as seen microscopically, the results relate to such parameters as diameter or fibre length. Examples of equivalent indices for asbestos are given by the British Occupational Hygiene Society (1968).

Sampling

If air sampling is required to compare with a control limit, the method adopted should be compatible with the data on which the limit is based. It is essential to have knowledge of the origin, the harmful or related effects, the units of measurement, the size range, the active constituent(s) if present in a mixture, the appropriate sample averaging period (e.g. 8 hours, 4 hours, 15 minutes, instantaneous), the relevant notations (e.g. skin absorption) and the sample position (breathing zone or fixed location). Samples may need to be counted (as for asbestos), weighed (as for welding fume) or analysed (as for lead, silica).

In general control limits are based on relatively simple methods and equipment which can be obtained or reproduced. Hence the conventional powered mechanical pump and filter system can be adapted for collecting particulates which can then be counted, weighed or analysed. Similarly the gas/liquid absorber, or active adsorber, or sampling bag can be used to collect gases and vapours for subsequent analysis. In a later section these basic methods are reviewed, together with more recent developments which use methods of ionizing radiation, laser, light scattering, infrared and ultraviolet absorption and gas chromatography for the direct measurement of contaminants.

Sampling errors

Air sampling involves a number of operations which introduce error and variation into the data. These need to be quantified and taken into account for the purpose of interpretation. Estimates of errors for sampling and analytical methods are quoted by the National Institute of Occupational Safety and Health, USA (NIOSH, 1975, 1977) for a number of common occupational hygiene measurements, using the coefficient of variation (CV) as follows:

	CV
Detector tubes	0.14
Flow meter on personal sampler	0.05
Charcoal tube sampling, analysis	0.10
Asbestos sampling, counting	0.24–0.38
Respirable dust sampling, weighing	0.09
All dust sampling, weighing	0.05

$$\text{where } CV = \frac{\text{standard deviation}}{\text{sample average}}$$

Errors of these magnitudes and greater can be assumed to exist in regular practice and every situation will generate errors due to individual bias, method, instrumental design and use, and errors in analysis.

Sample variation

A single sample of dust, fume, mist, gas or vapour for a given time is one of a population of samples which will be distributed in some mathematically definable way. This variation depends on the stability of the workplace environment which can be changed within wide limits by relatively trivial and uncontrolled factors such as weather conditions, and also by mechanical ventilation, productivity and production methods. NIOSH quote values for this variation, in terms of Geometric Standard Deviation (GSD), for particulate contaminants as 1.6–1.69 GSD, and for gases and vapours as 1.5–1.59 GSD, referring to interday, intraday and interoperator measurements. A proportion of the data has GSDs in excess of 2, which represents relatively high variability. These values can be derived from plots of cumulative distribution as shown in *Figure 17.2* for any measured parameter (e.g. diameter, length, mass concentration) if it is log normally distributed (NIOSH, 1975).

Accounting for error and variation

Account can be taken of error, based on CV, and variation, based on GSD, in the design of the sampling regimen. Samples may be taken over full shift, partial shift or instantaneously, or they may be randomized. Given values for CV or GSD, and the number and duration of the samples, curves can be constructed which can be used to test compliance with a selected standard. The methods do not have wide use outside North America but can be regarded as a move towards more objective investigation of workplace exposure.

There is no universally accepted criterion on which to base a decision about the number of samples required. Either the error and variation have to be measured, or, if a reasonable analogy can be found, they can be estimated and then used as indicated above. Operational errors can arise from faults in basic handling and techniques; that is, in calibration, sampling rate, damage to sample, position of sample, duration of sampling, counting, conditioning (for weighing) and analysis.

Errors from on-site activities can be reduced by proper training of personnel in the practice of occupational hygiene and by choosing well designed proven equipment. Errors arising from laboratory practice can be detected and corrected by participation in quality control schemes which allow performance to be compared with other laboratories; for example, the interlaboratory comparisons of asbestos fibre counts reported by Beckett *et al.* (1976).

Sampling strategy

The ultimate value of air sampling depends on the way in which the instrumentation is used in practice. Given that error and variation are known and allowed for, and that the correct instrument has been selected, it then becomes necessary to take account of position, duration, and number of samples to identify problems which may be peculiar to particular surveys.

When comparisons are to be made with control limits such as TLVs, samples would normally be taken in the breathing zone by personal sampler (described below) and possibly in fixed positions at a height of 1.5 m (nominal head height).

The term breathing zone can be interpreted in various ways but an ideal position would be in the breathable air stream between nose and mouth and close to the face (e.g. 1–2 cm). In practice many personal samples are taken on one or other lapel. However, there are situations where a substantial concentration gradient can be produced across the front of the worker, as might exist in welding operations or other point sources of emission. This has led to the consideration of a head harness with sampling orifice close to the cheek (British Standards Institution (BSI), 1977).

The duration of the sampling period can be influenced by three factors: duration of the operation being investigated; the sampling time of the instrument; the averaging period specified in the control limit, e.g. 8 hours, 4 hours, 15 minutes, or instantaneous. The primary aim should be to sample continuously for the duration of the specified period of the control limit if comparative data is to be obtained. If this cannot be achieved the aim should be to sample for consecutive time periods during the specified period and then to estimate the time weighted average (TWA) by the formula:

$$\text{TWA Conc.} = \frac{C_1\, T_1 + C_2\, T_2 + C_3\, T_3 ----}{T}$$

where $C_1\, C_2$ are concentrations for $T_1\, T_2$ etc., $T_1\, T_2$ are individual sample times and T is specified time (e.g. 8 hours, 4 hours, etc.).

Different sampling regimens are indicated in *Figure 17.4.*

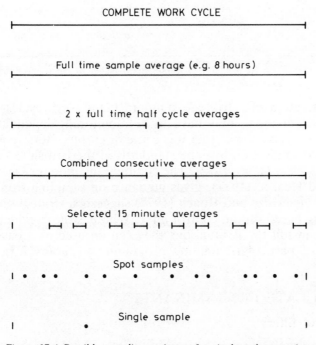

Figure 17.4 Possible sampling regimens for single-cycle operations

If the term 'work cycle' is used to denote a representative series of events, the number of cycles which need to be sampled may be self-evident or an arbitrary decision based on observation and experience may have to be taken. For comparison with control limits a small number of work cycles, possibly one, may yield the essential information. An example is the testing for chromic acid mist arising from plating tanks in accordance with the Chromium Plating Regulations (1931). For evaluation of a novel work situation it may be necessary to sample cycles, say 5–10, in order to obtain a meaningful picture of the average and range of exposures. This should be adequate to differentiate between stable and unstable working conditions.

Table 17.3 MINIMUM FREQUENCY OF MONITORING PERSONAL EXPOSURE (AFTER ROACH, 1977)

Man-shifts covered by personal sampling (per 10 employees)	$Ratio \dfrac{personal\ exposure}{TLV}$
1/month	1–2
1/quarter	0.5–1 or 2–4
1/annum	0.1–0.5 or 4–20
None	<0.1 or >20

Note. Monitoring is less frequent for exposures much higher or much lower than the TLV.

For derivation of a hygiene standard many work cycles (10^2–10^3) and many individuals may have to be monitored, depending on the type of the response, whether it is acute or chronic. Reference can be made to documents on hygiene standards of the British Occupational Hygiene Society (1968, 1972). The National Institute for Occupational Safety and Health, (1977), gives guidance on sampling from populations of a given size and Roach (1977) suggests a strategy for routine monitoring based on the control limit. Whereas these guidelines may not apply in full to all situations they can be used as a basis for the planning of particular sampling programmes (*Table 17.3*).

PARTICULATE CONTAMINANTS

Pumps and filters

There is a requirement in all active sampling methods to aspirate air through either a filter or a cell at a known rate for periods up to 8 hours or more. For fixed position samplers this has been achieved by simple adaptation of mains powered electric pumps which can sample over a wide range from a few millilitres per minute to one cubic metre per minute. Typically the air movers are in the form of diaphragm pumps, rotary pumps or high speed axial fans and the flow rate is controlled by adjustment of stroke (for diaphragm), by air bleed or by critical orifice (as in the Hexhlet sampler).

For personal breathing zone sampling the need is for small, powerful long-running devices which are controllable, and this has led to the development of miniature sampling pumps which are now quite commonly used. Air flow can be selected or controlled to operate effectively over the range 10–5000 ml min^{-1}. An example is shown in *Figure 17.5*. A review by Wood (1977) illustrates the important aspects of

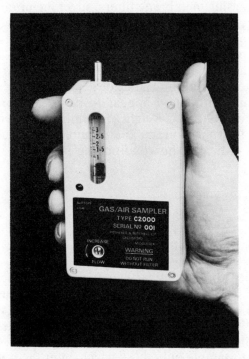

Figure 17.5 Personal sampling pump. (Reproduced by courtesy of Rotheroe and Mitchell Ltd.)

pump selection and shows that where a filter is used, the resistance is related to pore size (for membrane type) and diameter, and, at a rate of 2 ℓm^{-1} the resistance can range from 10 to 500 mm water gauge (wg).

In most applications it is essential to begin the sampling period at a specified flow rate and to maintain that flow rate throughout so that the total volume of air sampled is known. Air flow can be calibrated by reference methods, such as the bubble flow meter and gas meter, and by calibrated air flow meters. Some samplers have built in air meters which can be used to check flow while in use. They are not usually adequate for calibration purposes. Some have revolution counters which can be used to compute total volume of air displaced. With the

Apart from resistance to air flow, filter media have other characteristics which need to be considered. These are: particle retention, electrostatic effects, and water retention. The preference is for filters which retain 95 per cent of particles having a physical size of about 0.2 μm, and in practice acceptable results may be obtained with cellulose, glass fibre or membrane filters. Due to high resistance a practical lower limit to pore size on membrane filters is usually about 0.6 μm.

For weight differences of less than 1 mg, as required for instance for respirable mass using membrane filters, silver membranes are relatively more stable, but loss of deposit is possible (as from all types of membrane) due to low surface adherence. Mark (1974) found weight differences of 0.1–1 mg for a humidity range of 35–70 per cent depending on the type of filter. Cellulose fibre filters are most prone to changes in water content and are not generally suitable for accurate weight determination. There appear to be no authoritative publications on the reliability of weight determinations within the range 0.01–0.1 mg as it applies to regular air sampling practice and each user needs to view critically the conditioning system.

Sampling of particles of 0.2–100 μm

In general terms two types of particulate need to be measured, i.e. the type which can exert a harmful effect at any site in the three areas of the respiratory tract, N-P, T-B and P, and the type which exerts its effect only in the pulmonary region. For the first there is a need to sample all particles which can enter the N-P region, for the second there is a need to sample only those which enter the pulmonary region (P). For descriptive purposes the term *inhalable* can be used to refer to the former (i.e. N-P + T-B + P) and *respirable* can be used to refer to the latter (P only). The term *total* particulate also requires consideration since it is commonly applied to samples collected by open filtering devices.

Respirable particulate

Respirable criteria based on equivalent diameter are the easiest to identify. Reference standards are specified by the Medical Research Council (1959) and the American Conference of Governmental Industrial Hygienists (1979). They are compared in *Figure 17.6*. The MRC selection curve is specified so that 'measurement of dust in pneumoconiosis studies should relate to the respirable fraction of the

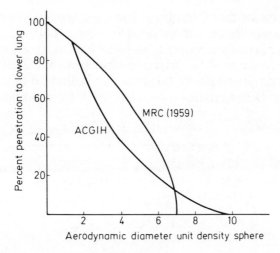

Figure 17.6 Aerodynamic criteria of respirable particulate

dust cloud, this fraction being defined by a sampling efficiency curve which depends on the falling velocity of the particles which pass through the following points: effectively 100 per cent efficiency at 1 μm and below, 50 per cent at 5 μm, and zero efficiency for particles of 7 μm and upwards; all the sizes refer to equivalent diameters. The Hexhlet (Wright, 1954) was the first instrument to be designed to meet these criteria, using horizontal elutriation to achieve selection. The design was based on the formula:

$$A = \frac{Q}{V}$$

where A is surface area, Q is volumetric flow and V is terminal velocity at cut off size.

The surface area (A) of the prefilter and the volumetric flow were matched to give 100 per cent removal of 7 μm upwards before collection. The ACGIH criteria differ in that the curve passes through 3.5 μm at 50 per cent and is zero at 10 μm but the difference between the two criteria for typical fibrogenic dusts is considered to be of little practical importance.

Inhalable particulate

Much less attention has been given to the definition of inhalable particulate and there are no widely accepted criteria which can be quoted which differentiate between the terms inhalable and total. However, the

ICRP data indicate that the entry efficiency of the N-P region is likely to be zero at above 100 μm, and Ogden and Birkett (1978) show that the efficiency for particles at 30 μm is 30–50 per cent at 20–40ℓ min^{-1} and this has led to the concept of the ORB inhalable dust sampler. The efficiency curve of this device might be taken as one definition of inhalable dust (*Figure 17.7*). Other criteria have been used, for example, to separate

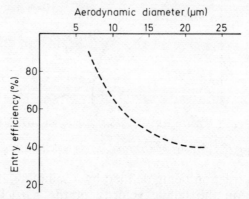

Figure 17.7 Airborne characteristics of inhalable dust
Source: Ogden and Birkett (1978)

airborne cotton dust into: fly (>2 mm physical length); medium fraction (2mm physical size to 7 μm equivalent diameter) and respirable (by the MRC criteria) for research purposes (Molyneux and Tombleson, 1970). Thus by using physical size, equivalent diameter or other means total airborne particulate can be fractionated within a range of specified limits.

Presentday practice

When samples are collected for comparison with control limits it is essential that the methods should achieve a high degree of reproducibility and specificity. The methods referred to in the following section are expected to meet these requirements. However, experience shows that the understanding and evaluation of hazards can be greatly assisted by the use of low cost relatively imprecise and non-specific but informative procedures.

Air sampling instruments which have a current practical application can be considered under the headings of:

1.   Those which yield useful semiquantitative information.
2.   Those which yield quantitative data for comparison with recommended limits.
3.   Those which yield quantitative data for research purposes.

A classification of methods and examples of instruments is given in *Figure 17.8*.

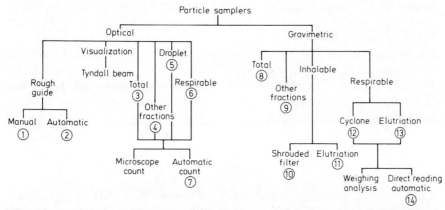

Figure 17.8 Classification and examples of particle samplers. 1 = konimeter; 2 = optical counter; 3, 4 and 6 = open faced filter or shrouded; 5 = cascade impactor; 7 = Royco etc. precision counters; 8 = open glass fibre; 9 = dust fractionator, e.g. Anderson; 10 = United Kingdom Atomic Energy Authority head; 11 = ORB (Ogden and Baskett, 1978); 12 = British Cast Iron Research Association cyclone etc.; 13 = Hexhlet; 14 = Piezoelectric on β absorption

Semi-quantitative methods

Instruments which fall into this category include the konimeter, the optical dust indicator and the precision hand pump. The konimeter is a hand-held instrument consisting of a hand-primed plunger, a jet orifice, an impaction plate, a microscope and an eyepiece graticule with size markings. Several samples can be collected on one plate and these can be viewed after each sample. The advantages are portability and rapid evaluation; the main disadvantages are small sample volume and error due to particle fragmentation which is a feature of impactor type instruments. The optical dust indicator draws air through an elutriated optical system and the scattered light is measured by optical means. The readout is proportional to the scattered light, which in turn is related to the concentration of airborne dust. The system can be battery operated and has the advantages of portability and direct continuous readout. The disadvantage is that the readout is not directly related to absolute mass concentration for all materials, due to their different optical properties.

The precision hand pump combined with a membrane filter provides a ready method of taking spot samples of 100 ml volume, or multiples over a short period, for microscopic evaluation. Because the collection is by filtration the particles are not liable to fragmentation and in this respect the method is more valid than the konimeter.

Where there is less need for hand-held instruments but there is still a need for portability, the advantages of more sophisticated particle counters can be used. With these there is the ability to fractionate dust by size in micrometres (μm) so that a number of size ranges can be scanned. The basic detector and counting units can be combined with computing and recording facilities, covering a particle size range from 0.5 μm upwards.

Instruments such as Royco automatic dust samplers have been used to monitor asbestos processes (Rickards, 1978). All these instruments have the ability to sample close to the breathing zone either by holding the instrument (for the konimeter and the hand pump) or by fitting a probe tube.

Methods for comparison with hygiene standards

The most commonly used unit of measurement for particulate concentration is mg.m^{-3} i.e. mass concentration. The most notable exception is asbestos which is measured in units of fibres ml^{-1}. Limits for other dusts can be specified in terms of millions of particles per cubic ft (mppcf; ACGIH, 1979). The units of fibres ml^{-1} may be applied to other fibrous materials such as man-made mineral fibre (Health and Safety Executive [HSE], 1978). The midget impinger has been the reference instrument for the ACGIH limits while the pump-membrane filter method applies to the HSE asbestos standard. For mass concentration the limits are based on pump filter methods using open filter, enclosed filter or size selective systems.

Counting methods   The pump-membrane filter system can be outlined as an example of current practice in relation to the limit for asbestos. An appropriate sampling package would consist of a miniature pump, either diaphragm or rotary, in line with a smoother filter head, and membrane filter of pore size 0.8 μm. A bubble flow meter or precision air flow meter is connected to the filter assembly and the flow rate is set to a steady 2 $\pm$ 0.1 ℓ min^{-1}. (Pulsation should not exceed $\pm$ 5 per cent of the mean flow rate.) The filter would be covered between calibration and sampling. The pump would be attached to the belt (or other convenient position) of the operator and the filter holder would be attached close to the orinasal region, usually on the lapel. The time of operation would be noted and the corresponding readings on the integral counter and/or air flow meter would be recorded. Details of the operator, process, type of asbestos, control methods, protection and precautionary work methods would also be noted. On completion the package would be removed from the operator, the filter would be recovered and transported in the holder to the counting laboratory. In the laboratory the filter would be removed with forceps and placed on a clean glass slide and cleared with glycerol triacetate or an equivalent method, and covered with a coverslip. The transparent filter would be viewed, using an eyepiece graticule, at ×400–500, under phase contrast illumination. Rules of overlap, sizing and fields or fibres counted would be followed. A purpose-designed graticule is described by Walton and Beckett (1977). The fibres would be counted according to a standard method, e.g. length, 5 μm and above; diameter, 3 μm and below; aspect ratio, 3:1 and above. The microscope would be calibrated by stage graticule for the given eyepiece graticule. Using the appropriate dimensional units the fibre concentration would be calculated from the formula:

$$ \mathrm{fm}^{-1} = \frac{DN}{4AnV} $$

where D is diameter of filtered area, A is area of graticule, V is volume of air sampled, N is number of fibres and n is number of fields

The aim would normally be to count 200 fibres or the same number of fields whichever is the least, which gives a fibre count with an acceptable error.

There are differences in filter-holder design. The open-faced filter of 25 mm or 37 mm diameter has had widespread use, but protection to the filter can be provided by a shroud, as specified, for example, by the AIHA-ACGIH (1975). Reference can be made to the documentation of the British Occupational Hygiene Society (1968), the Health and Safety Executive (1978) and the Asbestosis Research Council (1971) for operational details. Although these are described to within quite fine limits there is still scope for modifying the approach to fit the circumstances; for example, by reducing sampling rate to allow for extending the sampling period while still maintaining a satisfactory fibre distribution and count.

In all counting methods, fibre or other particles, there is an error due to overlap which increases with the number per unit area, therefore dense deposits are to be avoided. The magnitude of the error for coal dust, as an example, is given in *Figure 17.9* (Hodkinson, 1963)..

Gravimetric methods   Methods used for gravimetric sampling can be divided into purpose-built fixed position samplers, and miniature instruments which are suitable for personal breathing zone sampling. Each division can be split into respirable and other samplers (*Figure 17.8*). The important aspects of each type can be illustrated with reference to:

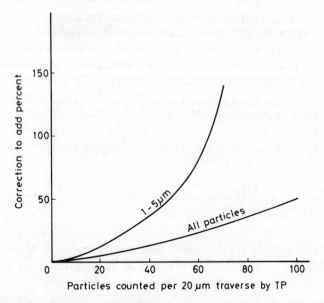

Figure 17.9 Error from overlap in particle counting for coal dust with thermal precipitation
Source: Hodkinson (1963)

1. The basic fixed-position sampler
2. The Hexhlet as a fixed-position respirable sampler
3. The open-face personal sampler
4. The enclosed-face personal sampler
5. The British Cast Iron Research Association (BCIRA) cyclone personal sampler.

Since types 1 and 3 are the same in principle they can be discussed together. Both draw air through an open-faced filter at a known rate, but they differ in the rate of flow, and in the position of sampling. An example of type 1 is given in *Figure 17.10* which is one of a range of instruments covering flow rates of 5–100ℓ min^{-1}. Neither of these instruments has a specified particle selection efficiency, nor do they provide filter protection. Orientation has an important effect on particle capture: facing upwards the filter can collect by sedimentation as well as by air entrainment; facing vertically it can collect by entrainment with reduced efficiency due to vertical elutriation. When the sampler is attached to a moving man there is limited control over the orientation.

It is possible to exercise greater control over particle sampling efficiency by enclosing the filter and locating it downstream of the sampling orifice. This exists in the form of the United Kingdom Atomic Energy Authority sampler (Sherwood and Greenhalgh, 1960) which

Figure 17.10 Static sampler 100 litre min⁻¹ (Reproduced by courtesy of Rotheroe and Mitchell Ltd.)

Figure 17.11 United Kingdom Atomic Energy Authority sampling head (Reproduced by courtesy of Casella London Ltd.)

samples through a 4 mm orifice at a velocity of 265 cm s^{-1} (*Figure 17.11*). This reduces entry of large particles which may be of less toxicological significance and creates more stable sampling conditions. The method has been recommended for the sampling of cadmium (British Occupational Hygiene Society, 1977) and might be expected to have an application to other toxic materials. This represents one possible approach to the sampling of definable inhalable dust, which is also the purpose of the more recent ORB sampler (Ogden and Birkett, 1978).

Where the restricted orifice sampler is not acceptable an alternative is to use a shroud, as referred to above.

The Hexhlet in its modified form (50 ℓ min^{-1} and 4.5 cm diameter filter disc) provides an effective means of collecting respirable particulate by the MRC criteria, for direct weighing and for chemical analysis. The power requirement is relatively high but extended running is feasible and it is possible to retrieve dust from the elutriator in order to estimate the total and the respirable fractions (Higgins and Dewell, 1960). The selection characteristics are affected by attitude and flow rate, hence the need to maintain the instrument within 5 degrees of the horizontal and to monitor the flow rate. The instrument, because of the horizontal plate elutriator, is not well suited to the sampling of fibrous dust, and for any dust it is essential to clean the elutriator between sampling periods.

The development of the miniature cyclone samplers for respirable dust was an important development on a par with that of the Hexhlet since it provides a reasonable means of measuring respirable dust in the breathing zone. SIMPEDS (Harris and Maguire, 1968) (*Figure 17.12*) and the BCIRA sampler (Higgins and Dewell, 1968) are two examples. Both these cyclones sample at a rate of 1.9 ℓ min^{-1} and have a penetration curve similar to that of the Hexhlet. Unlike a horizontal plate elutriator the cyclone is relatively insensitive to orientation and, like the Hexhlet, both the coarse and the respirable fractions can be retrieved for analysis. The respirable sample is collected on a disc and the coarse fraction is deposited in a pot. Methods for the direct evaluation of silica on silver membrane samples are available. The elutriation efficiency changes with flow rate but is practically acceptable over a flow range of 1.7–2.1 ℓ min^{-1}. Previous comments on the need to minimize pulsation and to standardize weighing procedures apply particularly to this mode of sampling because of the errors associated with small weights of sample.

Direct reading instruments The methods quoted above have long-standing recognition and are likely to continue to be used for the foreseeable future, but developments in direct reading mass concentration are worth noting. Two types of instrument can be considered:

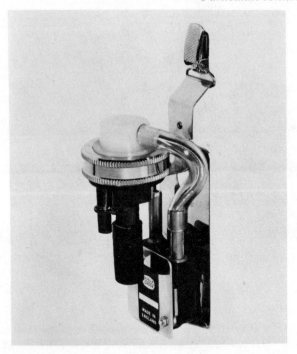

Figure 17.12 SIMPEDS cyclone respirable sampler (Reproduced by courtesy of Casella London Ltd.)

those which use absorption of beta-radiation or change of resistance in a piezoelectric crystal; and a light-scattering laser technique. Examples of the first are the particle mass monitors which collect airborne particles on to a greased plate and the mass collected absorbs beta-radiation from a carbon 14 source. The response is theoretically independent of chemical composition or physical form and the sampling interval can be extended to several hours. The instruments can be used in a particle size selecting mode which conforms with the ACGIH cutoff curve, or the MRC cutoff curve, by combining it with a horizontal elutriator. There is a recording facility for short-term fluctuations, as might be related to the ACGIH STEL and the TWA.

An example of the second type is SIMSLIN, described by Blackford and Harris (1978), which consists of a rotary vane pump which draws laden air through an elutriator, through a light beam and then through a filter. The filter collects the same dust that produces light scattering and can be used for calibration by weight. Light scattering takes place in infrared radiation which is emitted from a laser diode (*Figure 17.13*). SIMSLIN is designed to give instantaneous readings over two ranges 0–199.9 mg m^{-3} and 0–19.99 mg m^{-3} and data may be stored in an internal digital recording system (SIMSTOR), and can be read by pen recorder.

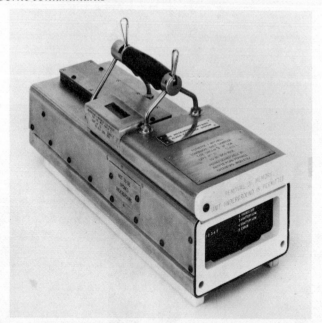

*Figure 17.13 SIMSLIN laser light-scattering automatic dust sampler
(Reproduced by courtesy of Rotheroe and Mitchell Ltd.)*

The penetration curve of the horizontal elutriator conforms to the MRC criteria. The current status of these newer devices in occupational hygiene practice is not altogether clear. They are unable to distinguish between dusts of different chemical and physical composition but the response is such that, given these data, they may be calibrated with enough accuracy to make them quantitatively acceptable in a variety of work situations.

In some respects the more compact beta-radiation and piezoelectric devices also qualify for use as direct reading semiquantitative measurement, as discussed above.

Research methods   Early instruments such as the thermal precipitator and the Hexhlet initially fulfilled a research function; then, as control limits emerged, they assumed a routine sampling role. At the present time there are instruments which have a research or special application, and which may, in time, become tools for everyday practice. Their primary purpose is to characterize airborne particulate in terms of physical size, equivalent diameter and concentration. Examples are the Cascade Impactor as introduced by May (1945); and the more recent Anderson fractionating samplers.

The Cascade Impactor deposits particles at relatively low velocity on to glass slides for microscopic sizing and counting. The instrument provides four stages of separation over a range of 0.3–20 μm equivalent diameter. Similar separation can be achieved with the Anderson samplers, and these are adaptable to breathing zone sampling. This ability to select particle sizes is of value in the initial evaluation of novel hazards when the methods of control might be fitted to the size range. It is also of value in relation to the site of deposition in the respiratory tract. Further reference to the significance of particle size and site of deposition in the lung is made in the section on ionizing radiation in relation to dose from internal radiation (*see* Chapter 18).

GASES AND VAPOURS

Since gases and vapours exist in the molecular state (i.e. size range 0.0001–0.001 μm) and will easily penetrate filtration systems which are used to collect particulates, they are collected by absorption in, or adsorption on to, suitable media. As gases tend to be more reactive than dusts, there is also more scope for direct on-site measurements. Because they exist in the molecular state, they spread by diffusion through the environment relatively quickly (i.e. there is no sedimentation), and for highly toxic agents the presence of high concentrations for quite short periods of time can lead to acute poisoning and/or fire and explosion, hence the need for rapid feedback of data.

There needs to be a flexible approach to air sampling. This requirement is met by numerous methods and instruments. As for particulates, there are semiquantitative and basic laboratory methods for comparison with hygiene standards and there is a choice of direct-reading instruments. There is scope for comparison with daily (8 hours) and short-term (15 minutes) standards, for testing confined spaces for risk of intoxication, fire and explosion, for continuous air monitoring, for alarm systems and for leak detection.

Different sampling requirements

Because of overlap in specifications and in apparent areas of application, there can be difficulty in identifying the best instrumentation for a given task. Some instruments will serve a variety of purposes but others are purpose-built for one function only. There are differences in sampling position, specificity, response time, limits of detection and portability. For instance, the evaluation of time weighted average exposure of a worker requires the use of a miniature sampler in the breathing zone; the method of evaluation needs to be specific and quantitative to levels above and below the hygiene standard (such as the TLV and STEL). Measurements in confined spaces may have to be taken for two distinct purposes, i.e. for toxic or fire/explosion risk, but the instruments are not usually interchangeable. One type is needed to respond to specific toxic agents at TLV concentration, the other to respond to all combustible constituents at concentrations above 1 per cent Lower Explosive Limit (LEL), which would be many times greater than the respective hygiene standard. The requirements for fixed location sampling can be met by quite bulky equipment fitted with probe lines; this gives quantitative but not necessarily specific results. An example is the flame ionization meter which responds to a wide range of hydrocarbons.

The requirements for leak detectors are the least demanding of all since they need only respond to relatively high concentrations of gases and vapours of known composition. A 'yes/no' answer is all that is required. Some overlap in application is possible where, for instance, miniature and portable samplers can function as static samplers (but not vice versa), and the results of a fixed recording monitor can be used to confirm the impressions gained from personal monitoring. Short-term chemical indicator tubes also have wider application than testing for entry to confined spaces, but their scope is limited by inherent features of the device. A classification of methods and examples of instruments are given in *Figure 17.14*.

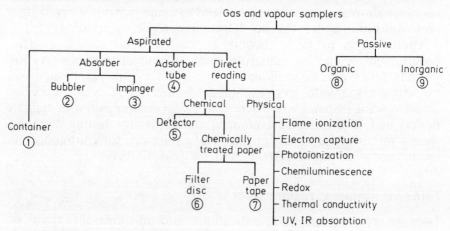

Figure 17.14 Classification and examples of gas and vapour samplers. 1 = vacuum orifice; 2 = bead/sinter/Arnold; 3 = midget impinger; 4 = charcoal silica gel etc.; 5 = Short/long-term tubes; 6 = activated paper; 7 = activated paper tape + optical reader; 8 = charcoal etc.; 9 = active absorber

General practical considerations

There are certain practical problems which are common to many sampling systems where pretreatment of the sampled air is necessary, where adaptation of samplers to the personal monitoring mode is required, and where data has to be presented in an understandable form.

Pretreatment

Almost without exception, there is a need to remove particulate matter from the sample prior to detection or analysis. This can be achieved by commonly used filtering media in the form of a disc or plug, except when specially inert materials are required because the gases or vapours being sampled are highly reactive. In welding processes, for instance, it would be necessary to remove the particulate matter (fume) before measuring the oxides of nitrogen and ozone. Where gases other than the measured contaminant are present, it may be necessary to eliminate interference before analysis, as for example, in a sulphur dioxide sampler which is fitted with a phthalate prefilter. There is also a need in some situations to convert a contaminant to a detectable chemical state, as for NO_2 which needs to be converted to NO before estimation of total oxides of nitrogen (NO_x) by chemical or physical means.

Inert materials

Some gases and vapours react vigorously with, or may be readily absorbed by, the filters and sample lines through which they are drawn. The obvious (and undesirable) result is undersampling (i.e. the measured concentration is less than the real concentration). Hydrogen fluoride will, for instance, attack glass sampling equipment, and ozone will react with commonly used plastics such as PVC and polyethylene.

There is no universally acceptable material, and one which is suitable for one application may not be suitable for another seemingly related task; for example, ozone can be sampled quantitatively by using polytetraflorethylene (PTFE) lines and accessories, but the same material may not be acceptable for chromatographic applications because of permeability effects. Workable answers can, however, be found for most situations and the moral, generally, is to check the compatibility of the equipment with inorganic and organic contaminants before sampling.

Adaptation to personal monitoring

Instruments which aspirate air through a collector or detector can usually be fitted with probe lines (the maximum length is usually specified by the manufacturer of the sampler) which can be attached to workers whose movement is restricted to a small area. In this way personal breathing zone monitoring is feasible for up to eight hours, given small diameter, flexible and inert tubes. In the absence of miniature equipment the method is feasible and has been applied to the investigation of welding processes (Molyneux *et al.*, 1979).

Presentation of data

Few direct reading instruments display data directly comparable with the TWA or STEL, usually because the time scales do not match. If a continuous record is available, the time-weighted averages (TWAs) may be estimated by planimeter or by statistical method, but neither is well suited to repeated assessment. The problem becomes particularly marked when there are wide fluctuations for short periods of time, as for NO/NO_2 concentration in welding processes. The aim should be for continuous automatic integration for the relevant averaging times (i.e. 8 hours and 15 minutes).

Presentday practice

Semiquantitative methods

The instruments and methods included in this section yield semiquantitative data in one or other mode of use either because the sample is taken for a very small fraction of the total exposure time or because the measurement is subject to substantial error. The short sampling time is not suited to the evaluation of eight-hour exposures but may be used to monitor (i.e. regularly check) known hazards. Examples are as follows:

Electronic direct reading, e.g. mercury vapour meter (Figure 17.15), photoionization meter These are relatively complex devices which can be used for quantitative measurement (*see below*) but when used for short periods can be considered to yield semiquantitative data.

Sample collectors, e.g. evacuated flask, gas bag These need to be used in conjunction with suitable laboratory analytical procedures. The error is largely dependent on the analytical method and in general is likely to be less than for the chemical tube and stain methods. However, the sampling is short.

Chemical direct reading, e.g. chemical stain techniques, chemical detector tubes Chemically treated filter papers through which contaminated air is drawn form the basis of some manufactured samplers and can be used for a variety of purposes. A wide range of detector tubes which sample over 1–3 minutes are available, and the error, repeatability and calibration can vary between types of tube and between batches. Their performance is of particular importance when

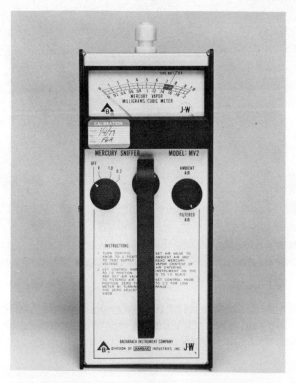

Figure 17.15 Direct reading ultraviolet mercury vapour meter (Reproduced by courtesy of Shaw City Ltd.)

they are used for checking the environment of confined spaces prior to entry. The reliability was studied by the Technology Committee of British Occupational Hygiene Society (1973) as a result of which tubes were classified as follows:

A – conforming in all respects to the proposed specifications.
B – small divergencies from specifications but can be used quantitatively with confidence.
C – suitable for approximate assessment of concentration or semi-quantitative use only.
D – unreliable; not recommended.

The short-term chemical detector tubes are used in conjunction with a precision hand pump, as shown in *Figure 17.16*.

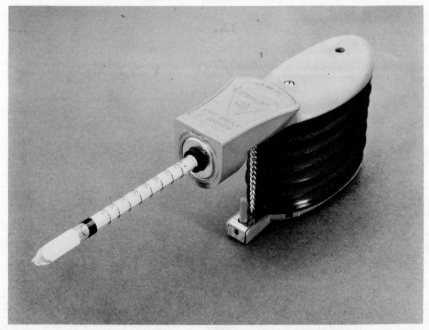

Figure 17.16 Chemical indicator tube and precision hand pump (Reproduced by courtesy of Draeger Safety)

It is very much in the interests of the user to assess the performance of tubes before use and to check calibrations with standard gas mixtures before serious measurements are undertaken.

Methods for comparison with hygiene standards

The aim is to sample for the correct time interval, on the man, in the breathing zone, and also in selected fixed locations. The methods need to be relatively specific and free from interference, and they need to respond quantitatively to concentrations around the TLV/STEL. A measuring range of, say, 0.2–5 × the standard would be acceptable although a wider range is desirable and can often be achieved.

Chemical methods The methods on which much occupational hygiene practice is based can be summarized as follows:

1.   Single or multiple bubblers or impinger.
2.   Activated charcoal, silica gel or other adsorbant.
3.   Passive or diffusion sampler.
4.   Gas bag, plus the more recent evacuated critical orifice sampler.

They can be combined with laboratory colorimetric, infrared, ultra-violet, or chromatographic methods and can be used to test compliance for many organic and inorganic gases and vapours (Elkins, 1959). Important practical points relating to the four systems are noted below.

Bubblers can take the form of conventional laboratory liquid bubblers through which the sample is drawn at a critical rate to achieve quantitative absorption with modifications including sintered glass and glass beads. The principle is adopted in some of the Methods for the Detection of Toxic Substances in Air described by the Health and Safety Executive; for example, for ozone in the presence of nitrous fumes, Booklet No. 18.

The components are available in most laboratories and the measurements can sometimes be completed on site. This has been facilitated by purpose-made field kits complete with quantitative colour comparison discs (e.g. the Lovibond Comparator discs of Tintometer Ltd). Gases such as chlorine, hydrogen cyanide, hydrogen sulphide, and organic vapours such as aniline, isocyanate and toluene are included. Purpose built personal sampler/bubbler kits provide one method of self-contained portable on-site sampling (*Figure 17.17*).

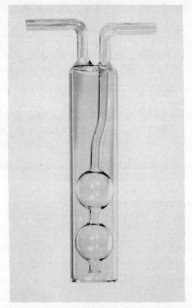

Figure 17.17 Miniature bubbler (Reproduced by courtesy of Casella London Ltd.)

There are practical difficulties with the use of these basic field techniques. Well documented and practised methods can be difficult to reproduce and quality control can be erratic.

Silica gel, activated carbon (charcoal) and porous polymers (such as Tenax) are applicable to the sampling of a wide range of organic gases and vapours, with charcoal having the widest use. The contaminants are adsorbed during sampling and can be chemically or thermally desorbed for analysis. These are simpler to use on site than wet methods, and are well suited to personal sampling in combination with miniature constant flow pumps. The technique is acceptable for measuring personal time-weighted average exposures. Purpose built adsorber/thermal desorber/gas chromatographic systems for on-site evaluation are available. A sampling kit is shown in *Figure 17.18*.

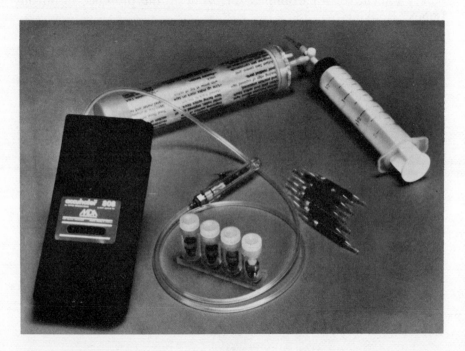

Figure 17.18 Gas/vapour sampling kit with pump, flow control orifices, charcoal tubes and calibrator (Reproduced by courtesy of MDA Scientific (UK) Ltd.)

Diffusion, or passive samplers, are an important development. The operation depends on the diffusion of molecular species due to a concentration gradient. Two sampling methods can be adopted, one is the transfer of gas or vapour across a diffusional barrier in the form of a permeable membrane and on to an adsorbent medium (as in the Porton Diffusion Sampler of Bailey and Hollingdale-Smith, 1977), the other is by transfer across an air space and then on to the adsorbent medium. The adsorbent may be activated charcoal (as for organic gases) or other active adsorber (such as for oxides of nitrogen; Palmes and Gunnison, 1973).

In steady state conditions the passive sampler performs according to Fick's law of diffusion.

i.e. $J = \dfrac{-D\ (P_2 - P_1)}{L}$

where J is the diffusional flux, D is the coefficient of diffusion, P_2 and P_1 are the vapour partial pressure at the outer and inner membrane surfaces (i.e. for the Porton-type sampler) and L is the membrane thickness.

The samplers do not require any electrical or mechanical accessories and can be worn in the same way as a radiation dosimeter. Experience with such devices has been favourable (Molyneux *et al.*, 1977) and they are now available commercially for a variety of contaminants.

Direct presentation of a sample for analysis without desorption or extraction is attractive and the gas bag is used as a standard method for transferring samples from site to analyser. In its usual form it consists of a laminated collapsible bag with tube attached. The inner surface is inert to the sampled gas and the sample is either drawn in by placing the bag in an evacuated container or by passing the sample in through a diaphragm pump. The bags are commonly of the size 5–25 litres suited to direct infrared analysis, but smaller bags can be used for personal sampling and the rate of sampling can be adjusted to fit long- or short-term periods. Like the charcoal adsorber the method has wide application to organic contaminants and can also be applied to inorganic gases.

Direct reading instruments

The basic methods are in many ways adequate for comparisons with standards, through personal and fixed location sampling. Nevertheless, direct reading portable devices offer advantages for economic and technical reasons, and the move towards automatic sampling is inevitable. There are a number of instrumental techniques available for this purpose; important examples are described below.

Ultraviolet methods—absorption and photoionization Ultraviolet (UV) light has an application in two forms, one using absorption and, more recently, the other using photoionization, the ability to induce an electrical charge on a molecule. The phenomenon of UV absorption has been used for many years for the measurement of inorganic mercury vapour (*Figure 17.15*). Mercury vapour strongly absorbs UV light of 253.7 nm wavelength and the absorption is proportional to the concentration. This principle is used in portable instruments, which either aspirate air through a UV/detector cell, or which allow free diffusion through the detection system. Other gases and vapours have the same property of absorption but usually with much lower sensitivity. Jacobs (1967) quotes the concentrations in air to give an equivalent instrumental response as follows:

Mercury 1 ppm
Tetraethyllead 1300 ppm
Perchlorethylene 5000 ppm
Trichlorethylene 100 000 ppm

The technique is also used to determine ozone concentrations in ambient air.

The energy of UV light can be used to ionize the molecules of some inorganic gases (NO, NH_3) and many organic gases and vapours. This principle is used, for instance in photoionization analysers in which a positive potential drives the ions formed by photoionization to a collection electrode where the resultant current is measured.

Infrared analysers   Many gases and vapours, inorganic and organic, have characteristic infrared spectra produced by the absorption of infrared radiation. These can be used to detect their presence and also to determine their concentration in air. The spectrum for styrene in air, for instance, shows strong absorbance of infrared light at 4.2 μm. This principle is adopted in portable infrared analysers such as the MIRAN (*Figure 17.19*).

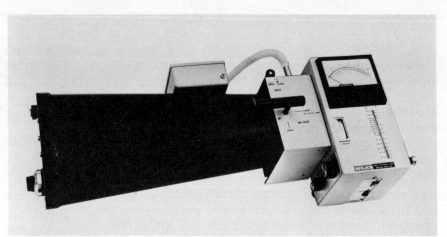

Figure 17.19 Portable infrared gas and vapour analyser (Reproduced by courtesy of Foxboro Analytical Ltd.)

Where a single contaminant is present identification and measurement is achieved with ease, but where a number of absorbing contaminants are present separation is not possible. In such cases the total concentration relating to all absorbing contaminants can be determined, and supporting methods such as mass spectrometry and gas chromatography are required.

Luminescence spectroscopy Chemical reaction can cause excitation of molecules and the resulting radiation emission (light of various wavelengths) is called chemiluminescence. This luminescence can be detected spectroscopically and the spectrum (intensity at different wavelengths) can be used to identify the reacting molecule. This type of luminescence is produced when ozone (O_3) and ethylene oxide react, and when O_3 and nitric oxide (NO) react. This is the principle of the automatic chemiluminescence analysers for O_3, NO and mixed oxides of nitrogen (NO_x). The technique lends itself to automatic continuous sampling, with short response time and high specificity. Where measurements of NO_2 are required, the instruments incorporate a converter (such as dichromate) for the reduction of NO_2 to NO. The technique is well suited to the investigation of the transient peak exposures due to the short response time.

Electrode systems A number of gases can participate in redox (oxidation reduction) reactions and the resultant change in electrical properties (e.g. conductivity) can be used to evaluate the concentration of the contaminant concerned. For most practical purposes, this type of reaction is restricted to inorganic contaminants, (e.g. O_3, CO, NO_2), which in solution, or in membrane diffusion cells, actively produce or remove ions (Bergman, 1975).

Instruments use electrode systems to monitor the change in conductivity from the reaction of SO_2 (sulphur dioxide) and H_2O_2 (hydrogen peroxide) to produce H_2SO_4 (sulphuric acid) in aqueous solution in a bubbler. For a given sampling rate the concentration of SO_2 in the sampled air is related to change in conductivity and time. Since the method is non-specific, other interfering gases such as ammonia must be removed, before measurement. Electrode systems can be compact and 'dry'. The Safety in Mines Research Establishment (SMRE) metallized membrane electrode, for instance, works as a polarographic cell which responds to the diffusion of oxygen or carbon monoxide and other gases across a PTFE membrane.

Gas chromatography Gas chromatographs are available as laboratory and portable units (*Figure 17.20*). When used in conjunction with charcoal tubes or by direct injection the method achieves the separation of components in a gas or vapour mixture. The eluted components may be detected and measured by a number of techniques such as the flame-ionization detector (FID), thermal conductivity or electron capture. Sensitivity depends on the type of detector used but the technique is capable of measuring concentrations down to 1 ppb.

Figure 17.20 Portable gas chromatograph with chart recorder (Reproduced by courtesy of D.A. Pitman Ltd.)

Sulphuric hexafluoride (SF_6) can be quoted as an example of a specific application using an electron capture detector, where by injecting SF_6 into the ambient air the air change rate can be calculated by measuring the decay rate. This provides a useful tool for the evaluation of general ventilation.

Chemical indicator methods   These are relatively simple devices, some of them in a high state of development, however, which use the reactivity of a gas or vapour to produce a quantitative colour change. The principle has much to commend it and in reality it is applied in two forms: the long-term gas detector tube and the chemical paper tape sampler. (Short-term detector tubes are discussed under semiquantitative methods.) The long-term chemical detector tube for eight-hour sampling is a comparatively recent development and enables samples to be taken in relation to the eight-hour standard. Because long-term tubes are intended to function continuously, the hand pump is unsuitable and a miniature air mover is needed to sample at a specified rate of approximately 10–20 ml min^{-1}.

A concentration/time trace is useful for evaluation and for control purposes and this can be achieved by extending the chemical stain principle to the paper tape monitor. In this device, the chemically treated paper tape is drawn at a constant rate over the sampling orifice and the contaminant reacts with the chemical to produce a stain. The intensity of the stain can be measured by reflectometer and the result can be displayed as a measure of concentration. This stain reflectance system is used commercially in personal samplers and in larger instruments. The response time is slow but adequate for eight-hour evaluation and may be adapted to short-term exposure measurements.

CONTROL OF AIRBORNE CONTAMINANTS

The behaviour of particulates and gases is fairly predictable from a knowledge of the particle size, but even so, once airborne, re-entrainment and removal is difficult and costly. The compromise, namely the best practicable means, reflects this difficulty. This focuses attention on the need to prevent release into the atmosphere, and if it occurs, to adopt proven guidelines to minimize exposure. The principles of prevention are discussed in Chapter 25 but brief reference to them is useful to introduce ventilation as a part of the overall control scheme.

General principles

First attention centres on the use of materials which are least likely to become airborne and which are of low toxicity. Materials which are non-volatile and which tend to agglomerate are preferable to those which are brittle and easily divisible (e.g. quartz and crocidolite asbestos); materials which have low toxicity and low irritancy are preferable to those which have a specific harmful effect; good warning properties are preferable to little or no warning properties (e.g. carbon monoxide). For similar reasons low energy processes which do not promote release are preferable to those which use high temperature or vigorous agitation (e.g. welding and mechanical crushing).

When these undesirable features exist the attention moves to secondary control measures which are aimed at preventing entry of contaminants into the workroom. These are segregation, enclosure and exhaust ventilation; these work most efficiently when in combination rather than in isolation. Hence the isolation of large-scale high potential nuclear hazards; the enclosure of small-scale potent sources (e.g. carcinogens), and the use of exhaust ventilation to remove the material from the enclosure (as in the fume cupboard), or to capture the contaminant from an open source (as for open arc welding). Capture and disposal require the application of engineering principles with some finesse if they are to be effective, and they need to be considered in relation to the requirements of make-up air and general ventilation.

The tertiary control methods rely on the worker and his ability to protect himself and his neighbours by personal protection (*see* Chapter 21) and low risk methods of work. This applies particularly where the release is directly under the control of the worker (e.g. for sampling or bench manufacturing methods) and where there is a requirement to enter confined spaces where a source of release may be present. This has led to the development of effective procedures such as work permits, airborne measurement, high grade personal protection and close supervision. Training and education are essential to the tertiary methods of control (*see* Chapter 25).

Control by ventilation

Ventilation systems can be tailored to the needs of individual situations given knowledge, primarily, of the toxicity, the aerodynamic characteristics; the force with which the release occurs, and the operational requirements.

The demands are least for low toxicity materials which can be identified, for example, with reference to the TLV, that is low toxicity (equal to or more than 500 ppm), moderate toxicity (equal to or more than 100 ppm) and high toxicity (less than 100 ppm). This serves to separate the use of general ventilation (suitable only for materials of low or moderate toxicity) from the use of exhaust ventilation (applicable for all levels of toxicity). The former is intended to dilute and remove contaminants which have already entered the workroom air, whereas the latter is intended to capture the release before dispersal occurs (McDermott, 1977).

Types of ventilation system

The types of ventilation are classified in *Figure 17.21,* showing a division between natural and mechanical air movement, and between mechanical dilution ventilation and local ventilation. Many materials can be handled in the open air without harmful effect (showing that mechanical ventilation is not essential for all work situations where toxic materials are present) because of effective dispersal by natural air movement. Similarly, well designed roof and wall louvres and opening windows can effectively ventilate releases within factories and workshops. The system becomes deficient when the rate of removal falls short of the rate of release, when there is a need to distribute air more effectively, and when the release is close to the breathing zone. When this occurs mechanical ventilation is needed to supplement natural dilution and to remove contaminants at source. A mechanical

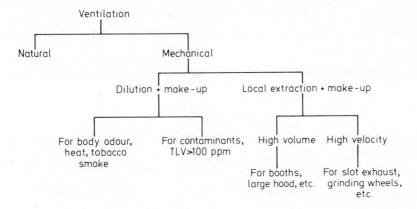

Figure 17.21 Classification of ventilation

system of ventilation introduces other important factors, including make-up air, the conditioning of make-up air, the cleaning of contaminated air, the recirculation of workroom air; all have a direct bearing on the design and cost of the system, and may be weighed against the relative effectiveness of alternative primary, secondary or tertiary control measures.

Dilution ventilation

The effectiveness of dilution ventilation is predictable only within broad limits, hence the restriction to materials of lower toxicity (*see above*). Given knowledge of the rate of release, the effectiveness of the fresh air distribution and the TLV, the quantity of air required can be estimated (ACGIH, 1976) from the formula:

$$Q = \frac{SG \times q \times K \times 100\,000}{M \times TLV}$$

where Q is airflow in $ft^3\ min^{-1}$, SG is specific gravity of the liquid from which the release occurs, q is the rate of use of liquid from which the release occurs (e.g. pints/hour, litres/hour), K is the factor between 3 and 10 (which is low for good distribution and high for poor distribution), M is the molecular weight of liquid and TLV is the ACGIH TLV or equivalent.

The outline design features of a dilution ventilation system are shown in *Figure 17.22*. Even though the airflow may be high, the movement in the workroom air is generally low. It is subject to external climatic factors and movement in the workroom, possibly to the detriment of the designed effect.

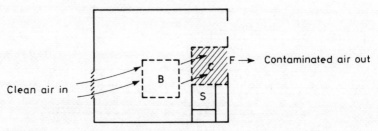

Figure 17.22 General dilution ventilation. B = breathing zone; S = source of contamination; C = contaminated air; F = fan

Local exhaust ventilation

Local exhaust ventilation (*Figure 17.23*) has great potential as a secondary means of control due to the wide range of hood designs, in the form of open pipe or slot, canopy, booth, or more refined purpose-built enclosures. These are summarized in *Table 17.4*. There is a distinction drawn (*Figure 17.21*) between high volume/low velocity open hood, and the low volume/high velocity closely fitted shroud. The latter has been applied to a variety of tools such as grinders, sanders and shot blasters where the need is for a high catchment velocity (*see below*) but not necessarily a high volume.

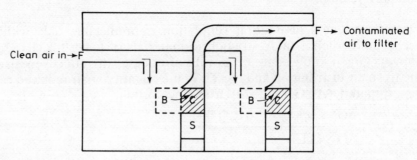

Figure 17.23 Local exhaust ventilation. B = breathing zone; S = source of contamination; C = contaminated air; F = fan in and out

System design   In many workplaces great dependence is placed on local exhaust ventilation to exercise continuous and effective control over irritants, systemic poisons and possible carcinogens. The control may be needed in an isolated hood but may equally well be required at many points in a plant or workroom, and at each point simultaneous, effective control is required. The factors involved in the relatively complex and meticulous design process are summarized below.

Table 17.4 TYPES OF LOCAL EXHAUST HOODS

Type	Form	Application
Plain openings	Open duct with/without flange	Welding fume/gas
Canopy	Suspended over source	Open tanks with liquid
Booth	Surrounding source, e.g. fume cupboard	Confined operations, e.g. laboratory
Side-draught	Extracting to one side or to the rear	Plating tanks; foundry knockout
Lip	Extracting across surface as pull or push–pull	Plating, cleaning tanks
Down-draught	From perforated bench	Fettling bench
Low volume High velocity	Closely fitted around source	Grinding wheels

Pressures in the system are related by the formula:

Total pressure = Static pressure + Velocity pressure
    TP   SP   VP

and VP varies directly with air flow in the system and SP decreases with increased air flow and vice versa. All three pressures are usually expressed in inches of water.

Air flow in the system is related to VP by the formula:

Velocity = 4005 $\sqrt{\text{VP}}$

so that given VP the velocity can be calculated. Velocity is usually expressed in ft min^{-1}.

Pressure losses in the system can be determined for the hood(s), the ductwork, the filter and the stack from standard reference sources (e.g. ACGIH *Industrial Ventilation Manual,* 1976).

Fan static pressure is the amount of static pressure the fan must develop to overcome resistance and provide the required air flow.

Dust velocity is selected from the transport velocity required to transport the contaminant in the system. Minimum values range from 2000 (e.g. for fine light particles) to 5000 ft min^{-1} (e.g. for coarse heavy particles).

Capture velocity ranges from 50 to 2000 ft min^{-1} depending on the generation and type of contaminant (*Table 17.5*). This is a critical factor

Table 17.5 VELOCITY OF AIR REQUIRED TO CAPTURE CONTAMINATION CLOSE TO SOURCE OF EMISSION

Nature of release	Example	Capture velocity ft min^{-1}
Passive movement into still air	Evaporation of cold solvent	50–100
Low velocity/momentum Low air movement	Spraying Welding	100–200
Generation into rapid air movement	Conveyor loading, crushing	200–500
Released at high velocity into very rapid air movement	Grinding Abrasive blasting	500–2000

in all local exhaust systems, particularly open pipes where the velocity at one diameter from the pipe is only 10 per cent of the face velocity. The accumulated losses due to the hood, ductwork, elbows, junctions and expansions can be minimized by following published data on the optimum hood entry shape, on the radius of elbows (i.e. $r = 2.5$ diameter) etc. Fans can be selected to meet particular requirements of horsepower, noise and blade design, as can filters.

Recirculation   The relatively high cost of providing fully conditioned fresh make-up air can be reduced by recirculation; this is acceptable for some contaminants if the recirculated air is cleaned and does not create a health hazard. Recirculation would not, as a general rule, be acceptable for carcinogens or substances with ceiling TLV values. Because of the risk of reintroduction of contaminants into the workplace, warning devices are required with back-up systems for alternative clean air supply.

Measurement and evaluation

The need for regular appraisal of ventilation systems is recognized in factory legislation in the UK, for asbestos (Asbestos Regulations, 1969) and can be applied to all materials where ventilation is a principal means of control. Measurement of pressure and airflow can be made to check conformity with the design criteria and to diagnose faults caused by leaks, accumulation of material, blocked filters or by the fan itself.

The rotating vane anemometer, an example of which is shown in *Figure 17.24*, is one method of measuring linear airflow across the open face of booths or ducts; others use a thermistor bead or

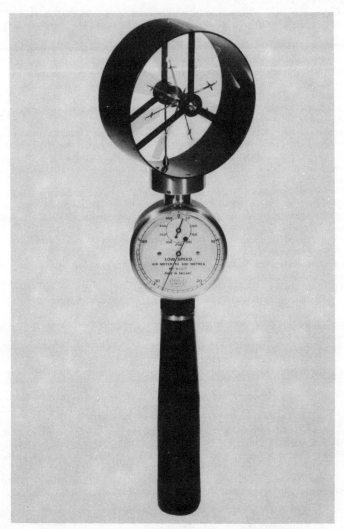

Figure 17.24 Rotating vane anemometer (Reproduced by courtesy of Casella London Ltd.)

deflecting vane to cover a wide range of velocities. The pitot tube can be used to measure total, static and velocity pressures and the smoke tube provides an effective means of visualizing airflow; this is particularly useful for determining the distance of the null point position from a local exhaust pipe or slot.

REFERENCES

Advisory Committee on Asbestos (Health and Safety Commission) (1979) *Final Reports*, Vols I and II. London: HM Stationery Office

American Conference of Governmental Industrial Hygienists (ACGIH) (1976) *Industrial Ventilation*, 14 edn. Committee on Industrial Ventilation, P.O. Box 453, Lansing Michigan 48902, USA

ACGIH (1979) *Threshold Limit Values for Chemical Substances and Physical Agents in the Workroom Environment with Intended Changes for 1979*. P.O. Box 1937, Cincinnati, Ohio 45201

AIHA and ACGIH Aerosol Hazards Evaluation Committee (1975) 'Recommended procedures for sampling and counting asbestos fibres.' *Annals of the Industrial Hygiene Association, J.*, **36**, 83–90

Asbestos Regulations (1969) S.I. No. 690. London: HM Stationery Office

Asbestosis Research Council (1971) Technical note 1. *The Measurement of Airborne Asbestos Dust by the Membrane Filter Methods*. Revised September 1971

Bailey, A. and Hollingdale-Smith, P.A. (1977) 'A personal diffusion sampler for evaluating time weighted exposure to organic gases and vapours.' *Annals of Occupational Hygiene*, **20**, 345–356

Beckett, S.T., Hey, R.K., Hirst, R. Hund, R.D., Jarvis, J.L. and Rickards, A.L. (1976) 'A comparison of airborne asbestos fibre counting with and without an eyepiece graticule.' *Annals of Occupational Hygiene*, **19**, 69–76

Bergman, I. (1975) 'Electrochemical carbon monoxide sensors based on the metallised membrane electrode.' *Annals of Occupational Hygiene*, **18**, 53–62

Blackford, D.B. and Harris, G.W. (1978) 'Field experience with SIMSLIN 11.' *Annals of Occupational Hygiene*, **21**, 301–314

British Occupational Hygiene Society (1968) 'Hygiene standards for chrysotile asbestos dust.' *Annals of Occupational Hygiene*, **11**, 47–69

British Occupational Hygiene Society (1972) 'Hygiene standards for cotton dust.' *Annals of Occupational Hygiene*, **15**, 165–192

British Occupational Hygiene Society: Technology Committee (1973) 'Chemical indicator tubes for measurement of the concentration of toxic substances in air.' *Annals of Occupational Hygiene*, **16**, 51–52

British Occupational Hygiene Society (1977) 'Hygiene standard for cadmium.' *Annals of Occupational Hygiene*, **20**, 215–228

British Standards Institution (1977) Draft Standard DD54. *Sampling and Analysis of Fumes from Welding and Allied Processes*. London: BSI

Chromium Plating Regulations (1931) Amended by Statutory Instrument No. 9, 1973. London: HM Stationery Office

Elkins, H.B. (1959) *The Chemistry of Industrial Toxicology*, 2nd edn. New York: Wiley

Harris, G.W. and Maguire, B.A. (1968) 'A gravimetric dust sampling instrument (SIMPEDS).' *Annals of Occupational Hygiene*, **11**, 195–202

Health and Safety Executive Booklets 1–26. *Methods for the Detection of Toxic Substances in Air*. London: HM Stationery Office

Health and Safety Executive (1978) *Man-made Mineral Fibres*. Report of a Working Party to the Advisory Committee on Toxic Substances. London: HMSO

Higgins, R.I. and Dewell, P. (1960) 'The measurement of airborne dust concentration in iron-foundries using the Hexhlet dust sampler.' *British Cast Iron Research Association Journal*, **8**, 425–436

Higgins, R.I. and Dewell, P. (1968) *A Gravimetric Size-selecting Personal Dust Sampler*. British Cast Iron Research Association Report No. 908

Hodkinson, J.R. (1963) 'Some observations on particle overlap error in dust measurement.' *Annals of Occupational Hygiene*, **6**, 131–142

International Commission on Radiological Protection ICRP (1966) 'Deposition and retention models for internal dosimetry of the human respiratory tract.' *Health Physics*, **12**, 173–207

International Labour Organization (1967) *Dust Sampling in Mines*. Occupational Safety and Health Service, No. 9

Jacobs, M.B. (1967) *The Analytical Chemistry of Industrial Inorganic Poisons.* New York: Interscience Publishers

McDermott, H.J. (1977) *Handbook of Ventilation for Contamination Control.* New York: Ann Arbor, Science

Mark, D. (1974) 'Problems associated with the use of membrane filters for dust sampling when compositional analysis is required.' *Annals of Occupational Hygiene,* **17,** 35–48

May, K. (1945) '*The cascade impactor.*' *Journal of Scientific Instrumention,* **22,** 187–195

Medical Research Council (1959) *Proceedings of the Pneumoconiosis Conference, Johannesburg,* ed. A.J. Orenstein. Boston, Mass. USA: Little, Brown

Molyneux, M.K.B. and Tombleson, J.B.L. (1970) 'An epidemiological study of respiratory symptoms in Lancashire mills 1963–1966.' *British Journal of Industrial Medicine,* **27,** 225–234

Molyneux, M.K.B., Evans, M., Sharp, T., Bailey, A. and Hollingdale-Smith, P.A. (1977) 'The practical application of the Porton Diffusion Sampler.' *Annals of Occupational Hygiene,* **20,** 357–363

Molyneux, M.K.B., Evans, M., Ingle, J., Sharp, G.T.H. and Swain, J. (1979) 'An occupational hygiene study of a controlled welding task using a general purpose rutile electrode. *Annals of Occupational Hygiene,* **22,** 1–18

National Institute for Occupational Safety and Health NIOSH (1975) *Exposure Measurement, Action Level and Occupational Environment Variability.* Washington: US Dept. of Health, Education and Welfare

NIOSH (1977) *Occupational Exposure Sampling Strategy Manual.* Washington: US Dept. of Health, Education and Welfare

Ogden, T.L. and Birkett, J.L. (1978) 'An inhalable dust sampler, for measuring the hazard from total airborne particulate.' *Annals of Occupational Hygiene,* **21,** 41–50

Palmes, E.D. and Gunnison, A.F. (1973) *Annals of Industrial Hygiene Association,* **34,** 78–81

Rickards, A.L. (1978) 'The routine monitoring of airborne asbestos in an occupational environment.' *Annals of Occupational Hygiene,* **21,** 315–322

Roach, S.A. (1977) 'A most rational basis for air sampling programmes.' *Annals of Occupational Hygiene,* **20,** 64–84

Sherwood, R.J. and Greenhalgh, D.M.S. (1960) 'A Personal air sampler.' *Annals of Occupational Hygiene,* **2,** 127–132

Walton, W.H. and Beckett, S.T. (1977) 'A microscope eyepiece graticule for the evaluation of fibrous dusts.' *Annals of Occupational Hygiene,* **20,** 19–24

Wood, J.D. (1977) 'A review of personal sampling pumps.' *Annals of Occupational Hygiene,* **20,** 3–18

Wright, B.M. (1954) 'A size selecting sampler for airborne dust.' *British Journal of Industrial Medicine,* **11,** 284–289

18

The Physical Environment

M.K. Molyneux

This chapter is concerned with the effects on man of the electromagnetic, acoustic, vibration and barometric phenomena occurring in work processes. Several different target organs are involved. They include the ear, organs of balance, eye, skin, fingers and skeletal tissue. In some cases the whole body may be affected. In many occupational environments factors other than those described may be involved. For example, recent interest in the possible risks arising from the use of visual display terminals has focused attention on the interrelationships between noise, illumination, posture, visual acuity, duration of task, degree of concentration and thermal environment. This illustrates the need to avoid taking a narrow view when investigating what may at first appear to be a specific problem. The various components of the physical environment included in this chapter may be classified in broad terms into electromagnetic and other effects (*Figure 18.1*).

The electromagnetic spectrum covers a wide range of wavelengths from 10^{-12} to 10^{10} cm. The short wavelengths at 10^{-12} cm are X-rays and gamma rays which have high energy levels of 10^1 to 10^6 electron volts (eV). They have sufficient energy to cause ionization of living tissue, hence the term ionizing radiation. This property is also shared by alpha and beta emissions, and indirectly from neutron radiation (high energy particles, *see Figure 18.1*). The wavelengths at 10^{10} cm are radiofrequency radiations of low energy, less than 10^{-5} eV, and are too low to cause ionization ($1\text{eV} = 1.602 \times 10^{-19}$ Joule). In total the spectrum consists of the following:

X and gamma radiation	ionizing
Ultraviolet radiation (UV)	
Visible light	
Infrared radiation (IR)	non-ionizing
Radiofrequency radiation (RF)	

Microwave radiation is the short wavelength region of the RF spectrum. Visible light occupies only a small portion of the total

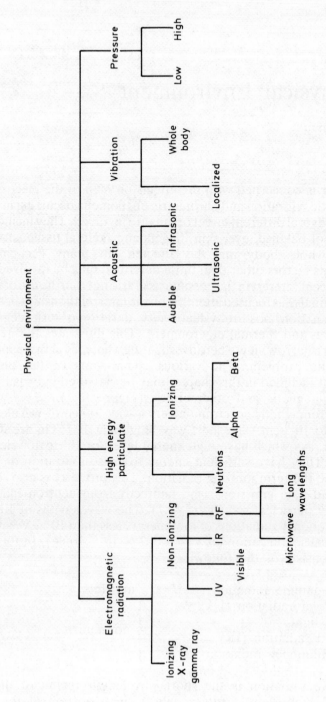

Figure 18.1 Components of the physical environment

spectrum from 400 to 750 nanometres, between the UV and IR, and provides illumination for vision. This form of radiation makes a perceptible contribution to man's biological activity by providing light for seeing. However, when the power of visible light sources are high enough, they can, like the other forms of radiation, be physically damaging.

The physical characteristics of the electromagnetic spectrum are summarized in *Figure 18.2*. The components of the spectrum can be

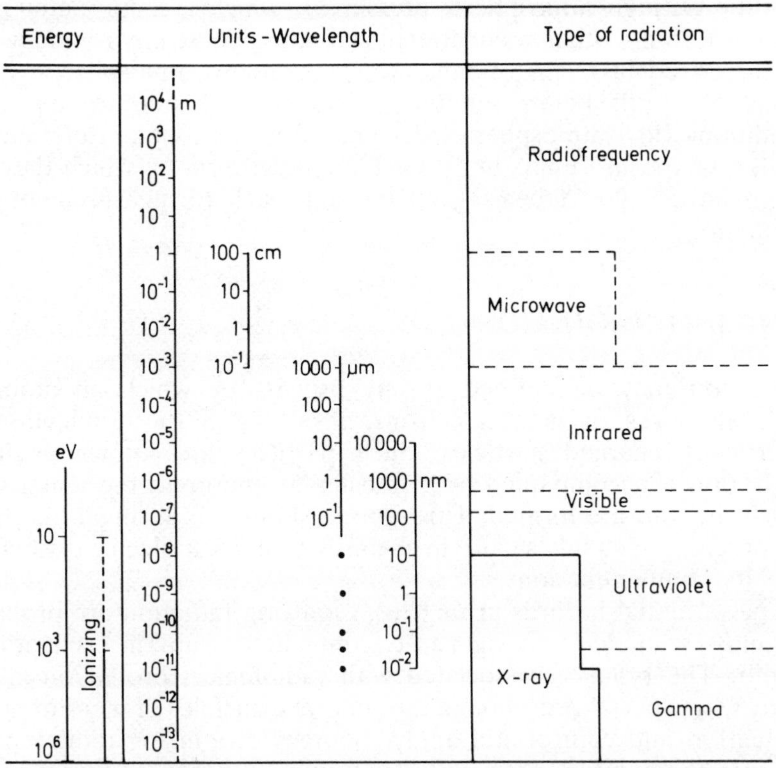

Figure 18.2 Physical characteristics of the electromagnetic spectrum

described as a waveform in terms of frequency (hertz; Hz), wavelength (e.g. centimetres and energy level (electron volts; eV). Wavelength has been chosen as the primary means of comparison based on the metric scale and, for convenience, refers to metres (m), centimetres (cm; 10^{-2} metres), micrometres (microns) (μm; 10^{-6} metres) and nanometres (nm; 10^{-9} metres) depending on the wavelength being discussed. The remaining components of the physical environment which are referred to, are the acoustic spectrum, hand-arm vibration and extremes of atmospheric pressure.

The acoustic spectrum is expressed as decibels (dB), a measure of sound level, and is classified in terms of low frequency (infrasonic), mid frequency (audible) and high frequency (ultrasonic) sound over the range of zero to more than 20 000 Hz. In accordance with standard practice the characteristics are compared in terms of frequency rather than wavelength. Similar principles apply to hand-arm vibration which is characterized by frequency (Hz) and acceleration (metres per second per second: $m \, s^{-2}$).

The section on extremes of pressure focuses attention on hyperbaric conditions (high atmospheric pressures), which are associated with underwater and other occupational activities. These environments are characterized in terms of pressure, as standard atmospheres, or in terms of depth below sea level. The changes due to hypobaric conditions (low atmospheric pressure) are mentioned. Reference is made to exposure limits in as much as the criteria on which they are based have a direct bearing on the approach to measurement and evaluation.

IONIZING RADIATION

Ionization may be defined as any process by which an atom or molecule loses or gains electrons, resulting in the production of electrically charged particles. Such particles are known as ions. Ionization is accompanied by a transfer of energy to the material in which the ions are formed. This process should be distinguished from the process which takes place in chemistry when a molecule dissociates into its component ions.

The potential hazards arising from ionizing radiation are probably the most regulated and vigorously controlled of any in occupational health. The practices associated with radiological protection are in many ways specialized, though the general principles of measurement, evaluation and control still apply. Sources of ionizing radiation are found in the nuclear energy industries, in many other industries and in the natural environment. Examples are given in *Table 18.1*.

Sources of exposure

In broad terms there are two types of source, one is the radionuclide, the other is in the form of high energy electrical devices. The radionuclide contains unstable nuclei which decay and emit radiation in the form of alpha and beta particles, X-rays and gamma rays. The other type of source produces X-rays in an evacuated tube by bombarding a metal target with electrons at a voltage of 16 kilovolts (kV) or more. The photon energies of the radiations are high and can be absorbed by entering the atomic structure of matter, by scattering or by annihilation. Ionization occurs and secondary radiations of X-rays and gamma rays can also be produced.

Table 18.1 SOURCES OF RADIATION

Occurrence	Source
X-ray optics	X-ray emission X-ray diffraction X-ray crystallography

Measurement and detection	Gauges for thickness
	    density
	    level
	Static eliminators
	Lightning conductors
	Smoke detectors
	Analysers

Parasitic	Electron beam welding Electron microscopy Radar transmitters Visual display units Thermionic valves
Luminizing	Radionuclides, e.g. tritium
Medical and dental	X-radiography Bone mineralization analysis
Industrial and research	X and gamma-radiography Neutron generation Linear accelerators Tracer (isotopes)

Table 18.2 SUMMARY OF CHARACTERISTICS OF IONIZING RADIATIONS

Physical form	Radiation		Maximum energy	Penetration	Ionization
Helium nucleus	α		8 MeV	Low	Intense
Electron	β		>4 MeV	Moderate	Low
Electromagnetic*	γ X-ray		10^7 eV	High	High
Neutral particles	Neutron	slow fast	<1 eV >0.1 MeV	High	Nil Produces secondary radiation

*obey inverse square law

Characteristics of these radiations are compared in *Table 18.2* and the transformations which occur in a typical decay chain, for radon (Rn), are show in *Table 18.3*. In this series the photon energy ranges

Table 18.3 ATOMIC TRANSFORMATIONS OF RADON

Isotope	Emission	Energy	Half life
^{222} Rn (Radon)			3.8 days
↓	α	5.5 MeV	
^{218} Po (Polonium)			3.1 days
↓	α	6.0 MeV	
^{214}Pb (Lead)			26.8 min
↓	β	0.7 MeV	
^{214} Bi (Bismuth)			19.7 min
↓	β	3.2 MeV	
^{241} Po (Polonium)			1.6×10^{-4} sec
↓	α	7.7 MeV	
^{210} Pb (Lead)			>20 years

from 0.7 to 77 MeV(10^6eV) and the half life of the isotopes range from 10^{-4} seconds to 20 years. Alpha particles are strongly ionizing even though they have little penetrating power. The potential hazard of radionuclides can be classified into high, medium and low, as shown in *Table 18.4* (abstracted from Ionizing Radiations (Unsealed Radioactive Substances) Regulations, 1968).

Table 18.4 CLASSIFICATION OF RADIONUCLIDES

Class I		High toxicity	Lead	210
			Uranium	230
			Plutonium	240
			Americium	243
			Cerium	246
Class II	Upper subgroup	Medium toxicity	Sodium	22
			Cobalt	56
			Strontium	90
			Iodine	124
			Lead	212
Class III		Medium toxicity	Beryllium	7
			Sodium	24
			Phosphorus	32
			Cobalt	57
			Strontium	85
Class IV		Low toxicity	Tritium	3
			Germanium	71
			Caesium	135
			Thorium	232
			Uranium	238

Harmful effects

Harmful effects may be divided into somatic and hereditary. Somatic effects are those which arise in the exposed individual and hereditary in descendants.

Stochastic and non-stochastic

Radiation effects may be classified as stochastic and non-stochastic. Stochastic means random and stochastic effects are those for which the probability of an effect occurring, rather than its severity, is regarded as a function of dose, without threshold. Once the random event has occurred the subsequent course of events is unrelated to the original dose. The larger the dose, the more likelihood there is of a random event occurring. Examples of stochastic effects are malignant diseases in individuals and mutations in reproductive cells, affecting their progeny. Human cells vary in their radiosensitivity. Those which are undergoing division are more sensitive than those which are not. Blood cells, with the lymphocytes heading the list, are the most radiosensitive. They are followed by the epithelial cells, of which the basal cells of the testes and intestinal crypts are the most sensitive. These are followed in order of diminishing radiosensitivity by the endothelium and connective tissue; tubular cells of the kidneys, bone, nerve, brain and muscle. Children, and in particular the fetus, are regarded as being more sensitive to the effects of radiation than adults.

Non-stochastic effects are those for which the severity varies with the dose and for which there is a threshold. The doses required are many times higher than for stochastic effects. Examples of the former are erythema of the skin, cataract formation in the lens, cell depletion in the bone marrow and gonadal cell damage leading to the impairment of fertility.

The acute radiation syndrome arises following whole body exposures to high levels of radiation, the effects of which depend on the level of dose and may be limited to anorexia, nausea and vomiting or, at the other extreme, death. The effects of high local exposure to radiation depend on the tissue irradiated and, in the case of the extremities, may lead to considerable local damage such as the eventual loss of the digits, but without detriment to general health.

Delayed effects

The term 'delayed effects' can be used in the context of a stochastic effect which is always associated with a latent period, or it may be used in the context of a non-stochastic effect where there is a delay in the damage becoming apparent. This may be exemplified by radiation cataracts where a series of doses, which individually used would not give rise to a clinically detectable lesion, over a period of time result in such a lesion. A second example is that of the basal cells of the testes. The effect on these cells will be immediate, but the effect on fertility may be delayed for several weeks because this is dependent on the already existing spermatozoa which are less radiosensitive.

Internal contaminations with radionuclides

Radionuclides may be inhaled, ingested or absorbed through the skin and so the risk from these nuclides is related to:

1. Their physical characteristics.
2. Mode of entry.
3. The way in which they are handled by the body once they have gained entry.

For example, the risk of carcinoma of bronchus among uranium miners is related to their exposure to radon gas, which they inhale. Strontium is chemically similar to calcium. Strontium-90 is concentrated in bone with a resulting increase in the risk of osteogenic sarcoma. Iodine-131 is concentrated in the thyroid gland with an increase in the risk of thyroid cancer. A preventive measure in the event of the release of iodine-131, e.g. from a nuclear power station, is the issuing of iodine tablets to the population at risk in order to block the uptake of the radionuclide 131 by the thyroid gland.

The dose limits

The dose limits recommended by the International Commission on Radiological Protection (ICRP, 26; 1966a) are based on a whole body dose equivalent per year for occupation exposure of 5 rem. The dose equivalent for members of the public is 0.5 rem. The units associated with dose and other parameters are the curie, the roentgen, the rad and the rem.

Curie (Ci) – This is the unit of source strength or activity of a radionuclide. One curie (Ci) = 3.7×10^{10} disintegrations per second. It is used to measure airborne contamination as μCi m^{-3}; surface contamination as μCi cm^{-2}; specific activity of a radionuclide as Ci g^{-1} or a decimal multiple of it.

Roentgen (R) This is the unit of exposure incident on the measured surface. One R = 2.58×10^{-4} coulomb per kilogram (C/kg) of air.

Rad This comprises the unit of absorbed dose. One rad = 100 ergs per gram.

Rem This is the unit of dose equivalent and is based on the rad. It takes into account the relative biological effectiveness of the absorbed dose. The rem is based on the rad $\times$ the quality factor (Q) and any other modifying factors (N) prescribed by ICRP.

The equivalent SI units are given in *Table 18.5.*

Table 18.5 SI UNITS

Quantity	New named unit and symbol	In other SI units	Old special unit and symbol	Conversion factor
Absorbed dose	Gray (Gy)	J kg^{-1}	Rad (rad)	1 Gy = 100 rad
Dose equivalent	Sievert (Sv)	J kg^{-1}	Rem (rem)	1 Sv = 100 rem
Activity	Becquerel (Bq)	s^{-1}	Curie (Ci)	1 Bq = 2.7×10^{-11} Ci

Measurement and evaluation

The framework for monitoring radiation hazards is based on the standards of the ICRP (1966a) described above, ICRP (1968), and those by government agencies, for example, the Radioactive Substances Act 1960 and Ionizing Radiations Regulations in the UK (*Sealed Sources*) (1969) and (*Unsealed Sources*) (1968). Where direct comparison with the dose equivalent limits cannot be made, 'derived limits' may be set for a workplace relating to contamination of air, surfaces, and materials. These derived limits are based on a defined model of the situation and reflect the basic limits set by ICRP.

The principles of monitoring can be considered under the headings of external and internal dose.

External dose

Exposure to radiation which is incident to the surface of the exposed person can be measured by personal dosimeters and by supplementary techniques, as described below.

Personal dosimeters in the form of a film badge or thermolumines-cent dosimeter can normally be worn on the chest for uniform exposure or in target areas for more localized radiation. The thermoluminescent dosimeter is readily adapted as a finger dosimeter to monitor external dose when radioactive sources are manipulated and the film badge may be made to discriminate between gamma, X-, and beta rays. The thermoluminescent dosimeter is largely indepen-dent of type and photon energy. The film badge can be kept as an original record of exposure, but neither of these dosimeters can be read directly and give no warning of over-exposure.

Supplementary methods can provide more immediate information in the form of the pocket dosimeter, which can be read directly and measures the current discharge from a preset level, and the Geiger-Muller Survey Meter, which can be used to read dose rate for a specific energy range. Others are the ionization chamber, the scintillation detector and the proportional counter.

Exposure to neutrons is relatively difficult to measure but techni-ques based on photosensitive film and other materials followed by microscopic evaluation are available.

The direct reading meters described above can be fitted with both visual and audible warning of high exposure. The advantage lies in rapid overall response at the cost of reduced sensitivity.

The film and thermoluminescent dosimeters have formed the basis of national monitoring schemes such as that of the National Radiolo-gical Protection Board in the United Kingdom. The reproducibility and error can be controlled to within acceptable limits and the data can be collected and retrieved by computer.

All the methods described need rigorous standards of calibration and maintenance.

Internal dose

The ICRP recommend maximum permissible concentrations for airborne radionuclides which can be applied to situations where gaseous or volatile materials such as tritium or iodine are handled, or where airborne particulates are generated from the processing of natural or enriched uranium or plutonium. In such circumstances there is the possibility of contamination of workroom air, of working surfaces, of work clothing and body surfaces, and there is a risk of internal exposure by inhalation and ingestion. Methods of assessment may be required for surface contamination, lung deposition and biological monitoring, and may become part of an integrated programme. Surface contamination can be determined by wipe tests over a specified area, or by direct measurement on representative parts of work surfaces. Examples of standards which apply are given in *Table 18.6*.

Table 18.6 LIMITS FOR SURFACE CONTAMINATION BY RADIONUCLIDES

Surface	Maximum permissible level		
	μ Ci cm^{-2}		
	alpha emitters		Other emitters
	Class I	Class II-IV	
Surfaces, objects in active areas	10^{-4}	10^{-3}	10^{-3}
Body surfaces	10^{-5}	10^{-5}	10^{-4}
All other surfaces	10^{-5}	10^{-4}	10^{-4}

The key to the assessment of lung deposition lies in the model formulated by the International Commission on Radiological Protection (ICRP, 1966b), which states the relationship between the deposition of particulates in the lung and the aerodynamic diameter.

The diameter is expressed as activity median aerodynamic diameter (AMAD) and its relationship with pecentage deposition in the respiratory tract is given in *Table 18.7*. Given the AMAD in micrometres

Table 18.7 LUNG DEPOSITION FOR DIFFERENT ACTIVITY MEDIAN AERODYNAMIC DIAMETERS (AMAD) (AFTER LANGMEAD, 1971)

AMAD micrometre	Deposition (%)		
	N-P*	P	T-B
0.1	13	31	8
1	30	25	8
2	50	18	8
5	75	12	8
10	90	8	8

* *See Table 17.3* for definition

the percentage deposition can be estimated and the corresponding dose in the three lung compartments can be determined. Examples of methods for measuring dose from inhalation include: the United Kingdom Atomic Energy Authority Personal Sampler (Sherwood and

Greenhalgh, 1960); the Personal Centripeter (Langmead and O'Connor, 1969). The UKAEA personal sampler provides little means of particle selection; the Cascade Centripeter is designed to measure AMAD in a static four-stage cascade system; the Personal Centripeter is derived from the Cascade Centripeter as a personal sampler and gives an approximate value of AMAD from a two-stage collector. In practice the UKAEA Personal Sampler provides information on total activity and the lung deposition is estimated on the basis of an assumed AMAD of 1 μm or using information from the other samplers on total activity to give an estimate of AMAD. Langmead (1971) gives AMAD values of plutonium and uranium process materials ranging from 2.5 to 6.0 μm and discusses the practical application of an integrated monitoring system.

Principles of control

The principles of control have been formulated and practised internationally for radionuclides, ionizing and other sources. Legal requirements in the United Kingdom are included in The Radioactive Substances Act (1960), Nuclear Installations Act (1965), Regulations (1968, 1969) and Codes of Practice (Department of the Environment, 1975a and b). Control procedures can be summarized as follows:

1. Notification of relevant activities to the responsible authority.
2. Advance approval for specified activities where preparation is essential.
3. Specification of dose limits and emissions for exposed persons; emergency limits for accidents and incidents.
4. Classification of workplaces depending on risk and exposed persons.
5. Introduction of control measures and monitoring.
6. Introduction of competent practitioners for the specialized tasks of monitoring, evaluation, control.
7. Critical assessment of new plant and projects.
8. Centralization of calibration, dosimeters, records.
9. Medical surveillance of exposed persons.

Exposure can be viewed critically in three stages, first to ensure that the use is justifiable, second that the exposure is minimal (i.e. the balance between application and risk is optimized), and finally that the acceptable dose is not exceeded. Design is of critical importance in many applications, whether it is an installation, process or device, whether it is sealed or unsealed, and it should be considered in relation to operational requirements. Toxicity from activity or other characteristics and the half-life need to be considered; both should be kept to the lowest level. Materials which are dusty, chemically active or volatile should be avoided and where they exist the appropriate measures for segregation, enclosure, ventilation and filtration should be provided. Time of exposure should be kept to a minimum and handling distance should be increased by the use of remote handling techniques. The precautions to be taken for radiography by gamma or X-rays can be used to illustrate some of the engineering principles of control as follows:

1. Enclosures are constructed of lead, brick or concrete to prevent the penetration of radiation, and the required thickness can be calculated. Labyrinths are constructed to provide access but still contain the radiation.
2. Door interlocks are connected to the primary circuit of X-ray machines or gamma source enclosures so that entry can take place only when radiation is at an acceptable level.
3. Warning signals are connected into the energizing system.
4. Areas are marked and distances fixed on the basis of time of exposure and dose rate at the periphery.
5. Lead screens and cones are used to reduce scattered radiation.
6. Gamma sources are stored in containers which can be manipulated by remote means when the source is required for use.

Safe methods of work can be introduced into laboratories where sources need to be manipulated and where volatile or dusty materials are present. Low activity sources may be handled safely using basic laboratory personal protection. Fume cupboards fitted with filters and having sufficient air movement to prevent contamination of the workroom can be used for nuclides of low activity.

Sealed glove boxes are recommended for high activity alpha and beta emitters, manipulated through rubber gloves which are sealed into the box. The inside of the box should be under reduced pressure and radioactive samples introduced via an airlock. Waste material can be segregated into bins or liquid containers for disposal via a controlled route.

Work surfaces and floors should be covered with impervious materials to assist decontamination and work on open benches should be restricted to low activity sealed sources.

ULTRAVIOLET RADIATION

Ultraviolet radiation lies between visible light and ionizing radiation in the electromagnetic spectrum, and can be classified in terms of wavelength as follows in *Table 18.8*.

Table 18.8 CLASSIFICATION OF UV RADIATION

UV region	Wavelength (nm)
Vacuum	100–200
Far	100–280
Middle	280–320
Near	315–400
Actinic	200–320

Sources of exposure

Sources can be described in terms of low emission (e.g. sunlight, low pressure mercury lamps, sun lamps and black light lamps), and high emission (e.g. high pressure mercury vapour lamps, high pressure xenon arcs, carbon arcs, plasma torches and welding arcs). Sources are used in hospitals and laboratories as a bactericide; in industry and advertising, for synthesis and analysis. The main spectral characteristics of UV sources can be summarized as low pressure mercury lamps (253 nm), high pressure mercury lamps (200–230 nm), fluorescent lamps (>320 nm) and black light lamps (366 nm). High temperature open arcs from welding and plasma torches produce an intense broad band emission over the UV spectrum.

Harmful effects

The biological effects can be described broadly with reference to wavelengths as shown in *Table 18.9*. The target organs are eyes and skin and the penetration of and effect of radiation of particular wavelengths

Table 18.9 BIOLOGICALLY SIGNIFICANT WAVELENGTHS

UV region	Wavelength (nm)	Effect
UV–A	320–400	UV fluorescence
UV–B	280–320	Erythema
UV–C	>280	Germicidal

are summarized in *Figure 18.3*. Exposure of the skin causes erythema, which may be associated with the secondary effects of oedema and blistering, and pigmentation. The degree of effect is related to acclimatization. Exposure of the eye causes photokeratitis, such as 'arc eye' which is due to transient exposure to high intensity UV from the

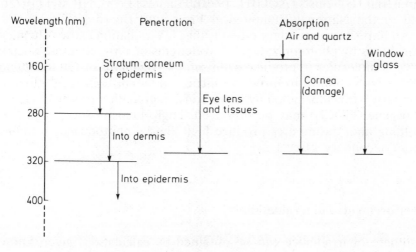

Figure 18.3 Summary of the characteristics of ultraviolet radiation

electric arc. Fluorescence of the ocular fluids may also occur but the effect is temporary and without pathological effects. There is no clear evidence that UV exposure is a primary cause of skin cancer but it may be an important aetiological factor in some susceptible individuals. It is widely recognized that certain coal tar derivatives, in combination with UV radiation, cause severe irritation and blistering.

Exposure limits

These are recommended by various authorities, including the American Medical Association (Clayton and Clayton, 1978; ACGIH, 1979). Important parameters are as follows:

1. The wavelength of the radiation over the range of 200–340 nm.
2. The incidence irradiance expressed in $mWcm^{-2}$ or Jcm^{-2}.
3. The duration of exposure.
4. The biological effectiveness of the irradiance, which depends on the position in the A, B and C spectral ranges. The spectral effectiveness is maximal at 270 nm and this value is used as a reference for the estimation of effective irradiance.

Recent recommendations for exposure of more than 16 minutes allow a total incident radiation on the unprotected eye or skin of $1 \, mW \, cm^{-2}$. For exposure times of less than 16 minutes the total irradiance should not exceed $1 \, Jcm^{-2}$. Further information can be obtained by referring to the list of TLVs provided by the American Conference of Governmental Industrial Hygienists (ACGIH, 1979), Hughes (1978), NIOSH (1972b) and to the National Radiological Protection Board (1977).

An important secondary effect is that UV radiation causes chemical transitions by photoionization. Wavelengths of 170–220 nm dissociate oxygen molecules to produce ozone and wavelengths of 130 to 190 nm break N–N bonds to produce nitric oxide and nitrogen dioxide. Similarly with chlorinated hydrocarbons, degradation products such as phosgene ($COCl_2$) may be produced. High intensity sources such as welding and plasma arcs produce high energy photons which induce these secondary effects.

Measurement and evaluation

Estimate of irradiance can be obtained by calculation, given knowledge of the luminance and wavelength of the source, and distance from the source. Instruments for direct measurement are listed in *Table 18.10*. Measurements need to be related to the relevant

Table 18.10 METHODS OF MEASURING NON-IONIZING RADIATION

Type	Method	Typical application
Thermal	Thermopile	Infrared, ultraviolet and others
	Bolometer	Broad band spectra
	Thermistor	Microwave and others
	Thermocouple	Microwave and infrared
Photon	Photoelectric cell	Visible, ultraviolet and infrared
	Photodiode	Ultraviolet and others
	Photomultiplier	Ultraviolet and others
	Semiconductor	Infrared and others

wavelengths which, for the ACGIH TLVs, are 320–400 nm (near UV) and 200–315 nm (Actinic region) and calibration at the appropriate wavelengths is required. Instrument wavelength selectivity can be increased by use of a monochromator, and the result is expressed as irradiance incident upon the unprotected skin or eye. Airborne contaminants can interfere with the measurements as a result of absorption at the relevant wavelengths.

Principles of control

A considerable portion of the UV spectrum is not visible and control should be based on knowledge of power and wavelength rather than visual assessment.

Access to sources where the limits may be exceeded can be controlled by warning signs and interconnected lights. Warning lights can be used to show when the source is in use. Incident radiation can be reduced by increasing the distance between the operator and the source and the dose can be minimized by reducing exposure time.

Ultraviolet radiation is easily screened and contained, and provision can be made for ventilation for cooling by baffles. Irradiated areas can be screened and direct viewing can be prevented. Non-reflective surfaces should be used inside enclosures and on incident surfaces. Values of reflectance for various materials at different wavelengths are quoted by NIOSH (1972b).

Exposure of skin and eyes can be prevented by clothing and eye protection. Thin open-weave fabrics have relatively high transmissivity and vice versa. *Table 18.11* gives examples; flannelette and poplin are

Table 18.11 TRANSMISSIVITY OF MATERIALS

Material	Transmissivity(%)
Muslin white	50.0
Nylon	27.0
Linen white coarse	12.0
Flannelette	0.3
Poplin	0.0

most satisfactory. Emissions below 160 nm are absorbed completely by air, existing only in a vacuum; emissions of 160–200 nm are poorly transmitted through air or quartz; wavelengths of 200–320 nm are absorbed by domestic glass. Face shields can be used to give full face protection and can be combined with spectacles or goggles if needed. The absorption characteristics of the materials should be appropriate for the radiation. British Standards Institution (BS) (1959) and BS (1960) refer to standard welding visors and filters for protection against the intense broad band UV radiation from welding operations.

INTENSE VISIBLE LIGHT

Sources of exposure and harmful effects

The visible wavelengths between 380 and 750 nm have photon energies of 1.65 - 3.1 eV which are of low hazard and because the emission is visible the reflex responses are normally sufficient to protect the eye. However intense exposure can arise from flash bulbs, spot lights, welding arcs, and viewing of the sun, when the retina may suffer from thermal injury or by photochemical injury, both of which are recognized in the drafting of recommended limits.

Exposure limits and evaluation

Taking the ACGIH (1979) TLV as an example for retinal injury reference is made to the following:

1. Spectral radiance of the source.
2. Burn hazard function which is related to wavelength over the range 400–1400 nm.
3. Viewing duration for constant emission or pulsed emission, limited to the range of 10^{-6}–10 seconds.
4. Band width in nm.
5. Angular subtense of the source in radians.

Similar parameters are used to calculate the allowable exposure time to prevent photochemical injury. It is suggested that luminances of less than one candela (the unit of luminous intensity, equal to 1 lumen per steradian) per square centimetre ($cd\ cm^{-2}$) are not likely to exceed recommended limits, assuming a pupil diameter of 2–3 mm. The principles of measurement and control are as outlined for UV and IR radiation.

ILLUMINATION

This section relates to visible light from artificial sources (i.e. luminaires) or from sunlight. The source of light itself may be the centre of attention, but most likely the light is required as an aid to the visual appreciation of other objects. What is seen depends on the interaction of many environmental factors which include the wavelength, intensity and direction of the light, the colour, texture, shape, size and distance of the object, and the surroundings in which the object is being viewed. All these factors require consideration when assessing visual efficiency (Hopkinson, 1969).

Characteristics of light sources

The characteristics of visible light from artificial sources used for interior lighting are compared with that of daylight over the range of 300 to 800 nm in *Figure 18.4*. All artificial sources differ from daylight

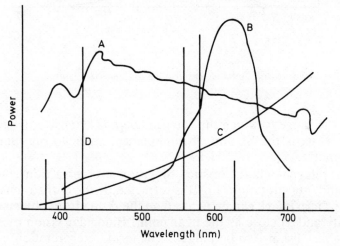

Figure 18.4 Relative power distribution of different light sources. A = sunlight; B = white fluorescent; C = tungsten filament; D = mercury fluorescent

in spectral distribution, extremes being present in the incandescent (tungsten) lamp, which is rich in the red spectrum, and in the high pressure mercury fluorescent lamp, which has a ragged distribution with sharp peaks in the blue, green and yellow regions. These differences in spectral distribution have a bearing on the choice of luminaire and on the measurement of light.

Human response

The spectral response of the eye is shown in *Figure 18.5*. The response is bimodal due to the use of two visual mechanisms, one for low light levels using rod vision, the other for higher levels using cone vision.

The total visual spectrum is from 400 nm (violet) to 750 nm (red) and the greatest sensitivity is within the 500–600 nm band. Intensity of illumination incident on the eye or on a surface is expressed in units of

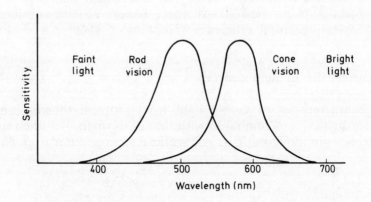

Figure 18.5 Spectral sensitivity of rod and cone vision

lux and the range to which the eye can adapt is extensive, from 10^{-7} to 10^5 lux. (Lux is the SI unit of illuminance, equal to one lumen per square metre.)

Other relevant characteristics of vision are refraction, binocular vision, and interpretation. Light is refracted by the cornea-lens system so that a sharp image can be focused on the retina from objects located at infinity and as close as the near point. Binocular vision by the two eyes enables distance to be judged. The term 'interpretation' refers to the processing of the information on the visual image in the brain, which can be misled to produce optical illusions. These responses are influenced by light intensity, and by such factors as shape, colour, shadowing, texture and flicker.

Levels of lighting

Visual acuity and light intensity are related and levels of lighting can be selected for specific industrial purposes within the range of 20 lux to 1000 lux. Above 1000 lux there is little to be gained in terms of visual acuity, while below 20 lux there may be insufficient light for efficient vision. Within this range of illuminance the needs of individuals have to be considered when choosing levels for specified tasks, but guidance is given in the Illuminating Engineering Society Code (IES, 1977) and basic criteria are quoted in *Table 18.12*. Unless otherwise stated such values apply to a working plane 84 cm (2ft 9in) above floor level. The penalty for insufficient light is poor visual efficiency whereas the penalty for too much light is, among other things, glare and excessive cost.

Table 18.12 BASIC LEVELS OF ILLUMINANCE FOR TYPES OF TASK

Task or work area	Illuminance (lux)
Area with no continuous work	150
Casual work	200
Rough work	300
Routine work	500
Demanding work	750
Fine work	1000
Very fine work	1500
Minute work	3000
Modifying factors	*Adjustment to illuminance*
Low reflectance or contrast	Increase
Errors need to be avoided	Increase
Short task	Decrease
Area windowless	Decrease for casual and rough work

Environmental factors

Guidance on acceptable limits for the other environmental factors which affect vision are described by Lynes (1968) and are outlined below.

Daylight

Unless there is a specific reason for excluding natural lighting from a room a minimum daylight factor of 2 per cent can be applied. This is derived from the formula:

$$\text{Daylight factor } (\%) = FU \frac{Ag}{Af} \times 100$$

where Ag is actual glazed area, Af is floor area, F is window factor and U is utilization factor.

Glare

Direct or reflected sunlight or other light from intense sources can cause discomfort glare. The Illuminating Engineering Society formulates the glare limits as a Glare Index defined as $G = 10 \log_{10} g$, where g is the glare constant. The range of glare index and some recommended values are given in *Table 18.13*.

Table 18.13 GLARE INDEX SCALE AND LIMITS

Glare index		Application
Imperceptible	0–10	Inspection small instruments
Perceptible	10–16	Classroom
Acceptable	16–20	Inspection, general offices
Uncomfortable	22–28	Most industrial tasks
Intolerable	28	

Relative illuminance

To achieve an acceptable balance between the illuminance of different parts of generally lit rooms the following ratios (*Table 18.14*) are recommended by the IES (1977):

Table 18.14 RELATIVE ILLUMINANCE

Ceiling to task	0.3–0.9
Low ceiling to luminaire	0.02 minimum
High ceiling to luminaire	0.04 minimum
Wall to horizontal working plane	0.5–0.8
Wall to task	1 maximum

Reflectance

Internal surfaces of rooms reflect light to varying degrees depending on colour and texture. Values of reflectance are quoted for surface finishes over a range of 0–1, white finishes having values between 0.4 and 0.8, with bricks, timber and concrete between 0.35 and 0.1. Recommended reflectance values for room surfaces are 0.6–0.8 for ceilings, 0.3–0.8 for walls and 0.2–0.3 for floors.

Flicker and stroboscopic effects

The cyclic variation in alternating current of fluorescent and discharge lamps can cause flicker and stroboscopic effects. Visual discomfort may be caused by flicker which occurs near the electrodes at the ends of the tubes and increases with age. Whether or not flicker is apparent the frequency of alternating current can cause a stroboscopic effect on moving parts of machines, lathes and drills, creating the illusion that they are stationary or moving slowly and increasing risk of accident entanglement. Flicker can be reduced by covering the ends of electrodes or the ends of fluorescent tubes and by replacing ageing tubes. The stroboscopic effect can be reduced by regulating the speed of rotation, by using filament lamps or by changing the timing of the discharge by electrical means.

Measurement of illuminance

The illuminance of the working surfaces and other surfaces of a room can be expected to vary with time due to changes in the output of the luminaires and changes in the decor and surface finishes, which in turn alter the reflectance. The aim is to maintain illuminance values which meet agreed design criteria or recommended values such as those quoted in *Table 18.12*. In practice the illuminance over a working area is unlikely to be uniform due to spatial arrangements of the luminaires, windows and the workplaces, and this can be taken into account by adopting a sampling regime based on the area of the room and on the degree of precision required. As an example, using the IES recommended method, the number of measurements required for an area of 20×20 m would be 16 for a precision of up to 10 per cent. Each measurement would be taken at the centre of each square of a 4×4 matrix. The number of measurements would be doubled for a precision of up to 5 per cent. Two types of measurement can be made, planar or scalar. Planar illuminance applies to the luminous flux density in lux, on a flat surface; scalar illuminance applies to luminous flux density in lux measured over the surface of a very small sphere located at a given point. Most design criteria and recommended limits are given in units referring to planar illuminance and the units can be converted to scalar illuminance or vice versa. The photoelectric cell is used in both instruments and an example of a planar cosine corrected photometer is shown in *Figure 18.6*. The cosine correction takes account of light from oblique angles and may be

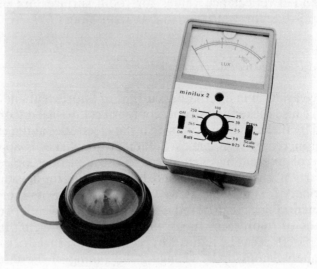

Figure 18.6 Light meter with cosine corrected cell (Reproduced by courtesy of Salford Electrical Instruments Ltd.)

achieved by placing opal acrylic plastic over the cell or by covering the cell with a transparent hemisphere (as shown). The cell can also be colour corrected to allow direct measurement of illuminance of different spectral distributions; otherwise a correction factor needs to be applied. Shadowing during measurement should be avoided and meters should be calibrated annually.

Principles of control

Reference has already been made to numerical values which can be used to exercise some control over illuminance, natural lighting, glare and reflectance. Other practical considerations are described below:

1. Lamps chosen for interior lighting should have the appropriate spectral characteristics particularly where colour judgement is required. Lamps with different colour characteristics should not normally be used together.

2. Luminaires should be fitted with shades or diffusers.
3. Flicker and stroboscopic effect should be reduced, as explained above.
4. Provision should be made for the maintenance of a satisfactory thermal balance particularly where filament luminaires are used.
5. Requirements for safety against explosion and fire should be met.
6. Directional lighting should be used to best advantage for modelling and texture, and for the reduction of unwanted reflections.
7. Contrast should be used to focus attention on the primary task, not detract from it.

INFRARED RADIATION

Infrared (IR) radiation occupies the region in the electromagnetic spectrum between microwaves and visible light, over the range 0.75 to 10^3 μm. The spectrum (*see Figure 18.2*)can be further divided into near IR (0.75–3 μm), mid IR (3–30 μm) and far IR (30–10^3 μm). The energy and wavelength characteristics emitted by a hot body depend on temperature, and the proportion of wavelengths below 1.5 μm increase markedly as the temperature rises through 1000°C (5 per cent) to 2000°C (40 per cent).

Source of exposure

Industrial sources occur in energy intensive industries such as glass and metal industries, in welding and flame cutting. Infrared is present in sunlight and in the emission of filament, fluorescent and other high temperature electrical components.

Harmful effects

Two different types of effect can be identified. One is heat stress caused by the transfer of radiant heat load from a high temperature radiant source. The second effect is aimed more specifically at the eyes and skin as target organs. The generation of heat is the cause of both effects; the energy level of the radiation is too low to cause ionization of tissue.

The biological aspects of radiation to the eyes and skin are summarized as follows:

1. At more than 1.5 μm wavelengths the body is opaque.
2. At 1.3–1.5 μm transmission through the ocular tissue is poor, i.e. there is high absorption.
3. At 1.1 μm a significant proportion of the energy is transmitted by the cornea and penetrates the eye.
4. At 0.75–1.3 μm the skin is transparent to the radiation.

Occupational exposures over 10–15 years to white-hot surfaces with intensities of 0.1 to 0.4 Wcm^{-2} have been associated with glass workers' cataract.

Skin responds to IR radiation by vasodilation and, after repeated exposure, by pigmentation. Skin temperatures of around 45°C cause burns. Retinal burns have been associated with exposure to 20 Wcm^{-2} for 0.1 sec.

Exposure limits

Recommendations to avoid cataractogenesis by infrared radiation differentiate between sources which have a broad spectrum and those which have a low visual stimulus. For the broad band spectrum of 0.7 μm and above a harmful effect is not expected below an incident irradiance of 10 $mWcm^{-2}$. For near-infrared sources the spectral radiance limit can be estimated by taking into account the spectral radiance, the band width and the angular subtense of the source (ACGIH, 1979).

Measurement and evaluation

Instruments applicable to the measurement of IR radiation are given in *Table 18.10*. There are two broad classes, one is based on thermal detection which responds to all wavelengths and has a slow response (e.g. the radiation thermopile and the bolometer), and the other is based on photometric detection of the IR radiation, such as the photodiode, which can be made to respond selectively to parts of the spectrum. Measurements should be taken to represent irradiance in the position of the unprotected eye.

Principles of control

Intense infrared sources are usually perceived and exposure is relatively easy to control. Where the risk is of excessive whole body heat stress reflective screens can be erected between the source and exposed persons. Alternatively havens can be constructed which effectively segregate personnel from the source. Access to exposure zones can be gained by using reflective thermal suits, which in extreme conditions can be ventilated with cooled air. This aspect is covered in Chapter 19.

The risk of cataractogenesis can be effectively controlled using the principles which are outlined for UV exposure. Briefly, this requires the source to be recognized, contained if possible, and where viewing is necessary, to use eye protection with suitable absorption characteristics.

Pigmentation of the skin can be prevented by the use of opaque protective clothing.

LASER

Laser is derived from the term 'Light Amplification by the Stimulated Emission of Radiation'. It is a means of producing sources of great intensity and phase coherence. The wavelength can range from the infrared to the ultraviolet, the power from 10^{-6} W to more than 10 kw; the radiation may be constant width (CW), or pulsed, from 10^{-1} to 10^{-13} sec. Lasers are classified according to power input by authorities such as the American National Standards Institute (ANSI, 1976) along the following lines:

Sources of exposure

Class 1   No risk category.
Class 2   Low hazard potential because of an expected aversion response.
Class 3   May be hazardous by intrabeam viewing of the direct beam.
Class 4   May be hazardous by direct or diffusely reflected radiation.

These classes are related to wavelength and emitted accessible radiant energy in Watts for a time of 0.25 second, which is taken as the elapsed time for the exposed person to avoid exposure by blinking or moving the head. Lasers and devices containing them are now used widely in industry and surgery, and for communication and land survey.

Harmful effects

The target organ of most concern is the eye, though any part of the exposed body surface can be subjected to the thermal or photochemical effects. The photon energies of the ultraviolet, visible and infrared radiations which may be produced are too low to cause ionization in tissue but they cause retinal and corneal damage. The intense visible light may cause a variety of effects, ranging from glare and mild bleaching of the fovea to retinal burns, depending on power.

Exposure limits

Limits are published by a number of authorities, including the American National Standards Institute (Permissible Exposure Limits: PELS) (ANSI, 1976) and ACGIH (Threshold Limit Values: TLVs) (ACGIH, 1979). The ACGIH criteria take account of factors such as wavelength, limiting aperture, intrabeam viewing, repetitively pulsed sources and exposure time. The development of Laser Codes of Practice has been considered by Harlen (1978).

An example of the format, units and order of magnitude of a quoted standard for direct ocular viewing is as follows:

Wavelength (nm)	Duration (s)	Exposure level
Visible 400–700	$10^{-9} - 1.8 \times 10^{-5}$	5×10^{-7} Jcm^{-2}

For full information reference should be made to the relevant standards which are intended to protect against both corneal and retinal effects.

Measurement and evaluation

Detailed information similar to that in the classification may be sufficient to evaluate the hazard but measurements can be made with the instruments indicated in *Table 18.10,* following the principles mentioned in previous sections. The task is exacting because of the wide ranges of wavelength and energy and because of the possibility of both CW and pulsed emission. The beam densities of the two types of laser are expressed differently, irradiance in Wcm^{-2} for CW and radiant exposure, in Jcm^{-2} for the pulsed emission.

Principles of control

Control measures can be applied in the form of shielding or containment of the laser, screening of the operator, reduction of reflection and personal protection. Light-tight enclosures provide adequate containment and where this is not possible opaque non-reflective screens and hoods can be fitted. Absorbent beam stops can be used and entry to the area can be restricted depending on the order of risk. Eyeshields or goggles giving adequate attenuation can be worn. The filters should provide the required attenuation in the relevant UV visible and the IR spectral regions.

MICROWAVE RADIATION

These radiations have relatively long wavelengths within the range of one millimetre to one metre and form part of the overall radiofrequency band which extends from 1 mm to 10^4 metres. The microwave range can be divided into three sections: millimetre wavelengths (0.1–1 cm); microwave radio, radar (1–10 cm); and radar (10–100 cm).

Sources of exposure

Sources of emission are radar and communication, cathode ray tubes, microwave cookers, induction furnaces and electrotherapy devices. The power range extends from megawatts for high power radar to less than 100 watts for diathermy. Nevertheless the photon energy is low and no ionization is likely to occur. Emissions may be pulsed (as for radar) on continuous wave (CW) as in domestic and many other devices.

Harmful effects

The degree of penetration of body tissues is dependent on wavelength. For example, a wavelength of 10–20 cm is absorbed and penetrates deeply, one of 3–10 cm penetrates from 1 mm to 1 cm, while one of less than 3 cm is absorbed by outer skin.

The body is transparent to wavelengths of more than 500 cm which lie in the remainder of the radiofrequency radiation (*see Figure 18.2*). Wavelengths of 10–20 cm present the main risks of deep tissue heating since exposure may take place without the warning sensation of heat. Cataracts have been produced in experimental animals by wavelengths in the region of 10 cm (Carpenter, 1970) and both cataractogenesis and thermal damage to the eye have been caused at a level of 100 mW cm^{-2} experimentally.

Exposure limits

Thresholds of exposure are aimed at a specific absorption rate of energy which approximates to the thermal gain at resting metabolic rate. At this level the thermoregulating mechanisms can dissipate the excess heat without stress. A power density of 100 mW cm^{-2} incident on whole body is thought to be near the threshold and exposure limits are based on a level of 10 mW cm^{-2} with variations depending on time (Lindsay, 1975). In general agreement with this the Medical Research Council (1971) recommends a limit of 10 mW cm^{-2} power density over the wavelengths 1–10^3 cm approximately, and a limit of 1 mW cm^{-2} during any 0.1 hour period for discontinuous or intermittent exposure. The ACGIH (1979) differentiates between continuous wave sources, and repetitively pulsed sources and gives recommended values which take into account factors such as power density levels and time of exposure.

Measurement and evaluation

The introduction of any object into a microwave field causes distortion of the field pattern and shadows, consequently the incident energy on absorbing tissues is difficult to assess. In principle, measurements can be carried out with the instruments listed in *Table 18.10*, such as the bolometer. As for other electromagnetic radiation the bolometer can be calibrated to measure the power absorbed on a temperature sensitive resistance and can be used for peak power emission.

Principles of control

Areas where exposure is likely to exceed the recommended limits should be clearly marked and entry should be restricted. Exposure time should be regulated to minimize exposure. The area of the field may be reduced by reorientation of antennae or other devices which cannot be contained. In addition interlocks and warning signals can be used to prevent accidental exposure. Emissions can be contained by shielding and enclosure. Mesh screens and concrete barriers can be used to attenuate the emission at certain frequencies. The emission from small devices such as microwave ovens can be effectively contained by internal reflection, seals and adequate attenuation. A secondary hazard from X-rays may arise from devices which have voltage in the region of 16 kHz. If so, practices appropriate to ionizing radiation become necessary.

NOISE

The term 'noise' is used to describe unwanted sound which is generated as a byproduct of manmade or natural activities. A simple classification based on frequency of the sound is infrasonic (0–20Hz), audible (20–16 000) and ultrasonic (more than 16 000).

Sources of exposure

Common sources of occupational noise and their places in the overall spectrum are given in *Table 18.15*. All three types of noise are acoustic

Table 18.15 SOUND PRESSURE, SOUND PRESSURE LEVEL AND EXAMPLES OF OCCUPATIONAL NOISE LEVELS

Sound pressure (Pa)	Sound pressure level (dB) (re 20 U Pascals)	Occupational examples
(1)	(2)	(3)
20.0	120	Chipping hammer
	110	Engine test cell
2.0	100	Large circular saw
	90	Power discing
0.2	80	Small electric blower
	70	Typewriter
0.02	60	Normal speech
	50	Small electric compressor
0.002	40	Quiet office
	30	–
0.0002	20	–
	10	–
0.00002	0	Threshold of hearing

in the sense that they are produced by a vibrating source which transmits the vibration to the surrounding air. The resulting changes in air pressure are very small, ranging from less than 10^{-5} to 20 pascals (Pa), and for practical purposes can be described as a sinusoidal wave form. These pressure changes are received by the hearing mechanism and may be sensed as vibrations by other parts of the body as the pressure rises above 2 Pa. All three types of noise are described acoustically in basic and derived parameters.

Basic parameters

Frequency – as hertz (Hz), which denotes cycles per second.
Sound power – as watts, which denotes total sound energy radiated by a source per unit time.
Sound pressure – as micropascals (μPa) which denotes intensity expressed as the route mean square (rms) of the amplitude.

Derived parameters

Sound pressure level – as decibels (dB) which denotes level in defined frequency bands, which is related to the sound pressure.
Sound level – as decibels (dB) which denotes level for linear and weighted networks (*see* Audible noise below).

The term 'sound pressure level' is reserved for measurements in defined frequency bands; the sound level is reserved for linear or weighted values. The relationship between the sound pressure and sound pressure level is given in *Table 18.15,* from which it can be seen that the decibel scale is a logarithmic representation of the sound pressure scale such that:

$$\text{Sound pressure level (dB)} = 20 \log_{10} \frac{Pl}{Po}$$

where Pl = measured sound pressure in μPa
Po = reference sound pressure of 20 μPa (i.e. 0.0002 Pa)

In order that all readings of sound level can be compared they are related to this baseline sound pressure of 20 μPa which approximates to the threshold of hearing at 1000 Hz.

Calculations

The units of sound pressure can be treated arithmetically, but the derived parameters based on decibels cannot. A simple rule can be given for the addition of equal sound levels, i.e. the sum of two equal levels given an increase of 3 dB; for example, 75 dB + 75 dB = 78 dB. When additions or subtractions are to be made for different sound levels, Pl from the above formula is found for each level and the new dB level is calculated from the total of the combined sound pressures.

Audible noise

Noise within the range of 31.5 to 8000 Hz has received most attention because of the effects on the ear. These can take the form of temporary deafness, permanent deafness, tinnitus and mechanical damage to the ear drum, depending on frequency of the sound level and the time over which the total acoustic energy is received. At extreme sound levels the whole body is subjected to intense mechanical stress.

Noise within the same frequency band causes interference with speech at frequencies of 300 to 3000 Hz, and it causes nuisance both in occupational and non-occupational environments.

Limits to reduce occupational hearing loss

Examples of damage risk criteria are: the UK Code of Practice for reducing the exposure of employed persons to noise (Health and Safety Executive, 1972); the Threshold Limit Values of the American Conference of Governmental Industrial Hygienists (ACGIH, 1979); the Hygiene Standards of the British Occupational Hygiene Society (1971, 1976); the Criteria of the National Institute of Occupational Safety and Health (1972a). There are differences in points of detail and it is necessary to refer to the most recent national and international standards for definitive values. All of the standards take account of frequency (Hz), sound level (dB), and duration of exposure, and are aimed at limiting the dose of noise. The parameters involved in the UK Code of Practice (Health and Safety Executive, 1972) are outlined below:

Frequency

Frequency is 'A' weighted over the range of 31.5 to 8000 Hz. This is one of a number of internationally agreed weightings which filters the actual sound pressure level in specified octave bands by an agreed amount. The responses of a sound level meter using the linear, A and C weightings are given in *Figure 18.7*. 'A' weighted readings resemble

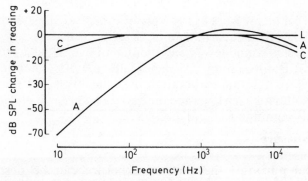

Figure 18.7 The effect of the linear(L), A and C weighting filters of a sound level meter on actual sound pressure level

the response of the ear to the incident noise over the frequencies specified. The Linear and C responses are similar to each other and are used to denote overall (or total) sound level.

Dose

Dose is expressed as dB Level Equivalent (dB(A) L_{eq}) which equates with the steady dB(A) level which would produce the same 'A' weighted sound energy over a stated period of time (eight hours in this case) as a specified time varying sound. It is defined mathematically as follows:

$$L_{eq} = 10 \log \frac{1}{t_2 - t_1} \int_{T_1}^{T_2} \frac{P^2(t)}{P^2 o} \, dt \quad dB$$

The allowable noise exposure and the exposure time are interdependent and follow the 3 dB rule (i.e. a halving of the exposure time may be balanced by a 3 dB increase in dB(A) sound level). Hence, the relationship between sound level and time for a dose limit of 90 dB(A) L_{eq} is as shown in *Table 18.15*. Some other authorities such as ACGIH use a 5 dB rule with a corresponding difference in the sound level–time relationship.

Table 18.15 APPLICATION OF THE 3 dB RULE

Sound level (dB(A))	Times for 90 dBA equivalent exposure	
	Hours	Minutes
90	8	0
93	4	0
96	2	0
99	1	0
102		30
105		15

Types of exposure

For practical purposes the time of exposure can be described as continuous or intermittent, which refers to full-time exposure or part-time exposure. The type of noise can be described as steady or impulsive, an example of the latter being slow repetitive hammering or gun shots. In the Code of Practice noise may be considered steady if the fluctuation on the SLOW setting (see below) of the meter is not more than 8 dB(A).

Meter response

For steady state noise sound level is measured on the SLOW setting 500 milliseconds (ms^{-1}) which in precision grade instruments conforms to such standards as IOS (1979). The FAST setting $(200\,ms^{-1})$ is usually also provided and has an application for other criteria.

Annoyance and speech interference

Noise levels up to and exceeding 90 dB(A) can cause annoyance and interference with speech and telephone communications. There are criteria which apply to annoyance in the community (IOS, 1971; BSI, 1967) and also to speech (IOS, 1973) and telephone interference (Baranek, 1960). Nuisance criteria are denoted by such indices as Effective Perceived Noise Level (EPNL) which is used to classify aircraft noise.

Measurement and evaluation of audible noise

The areas and ranges of measurement where attention is likely to be focused are:
1. For nuisance and speech interference 30–80 dB(A);
2. For occupational deafness 85–120 dB(A);
3. For whole body effects 130 dB upwards.

A wide range of equipment is available to measure and evaluate the parameters of audible noise, to investigate sources of emission and to specify control measures. The differences lie in precision, portability, range of application, and ease of use. The ways in which such equipment might be used in occupational hygiene practice are summarized in *Table 18.16*.

Methodology and instrumentation for audible noise

dB(A) measurements

The precision sound level meter and the noise dose meter together provide an effective means of identifying personnel who are at risk and

Table 18.16 SURVEY AND LABORATORY METHODS OF NOISE MEASUREMENT

Instrument	Application
Sound level meter*	dB(L), dB(C) and dB(A) instantaneous Fast or slow†
Sound level meter* and octave band analyser	As above with octave analysis 31.5–16 kHz
Impulse noise meter	Peak levels as instantaneous or average
Noise average meter	Average noise level for time specified
Noise dose meter	Noise dose relative to predetermined L_{eq} dB(A)
Tape recorder	Recording of noise prior to analysis
Third octave analyser	Detailed analysis from meter or tape
Statistical distribution analyser	Divides noise into level classes
Real time analyser	Gives instantaneous changes in spectra

*Precision to IOS (1969) or Industrial grade to IOS (1979)
†Refers to response time, e.g. fast 200 ms^{-1}, slow 500 ms^{-1}

for quantifying noise dose. For personal measurements the dB(A) values need to be representative of the noise which impinges on the ear and for location measurements the values need to be representative of exposure in the work area. For survey purposes the difference between dB(A) and dB(L) readings can be used to give an indication of the sound level in the frequencies below 1000 Hz. In situations where there are marked and regular gradients in sound level the meter can be used to construct noise contours of, say, 3 dB steps, which provide markers to noise hazard areas. The noise dose meter can be attached to the person or used in a fixed position. It is important that characteristics of the noise dose meter and the chosen criteria are the same. Two types of instruments are shown in *Figures 18.8(a) and (b)*.

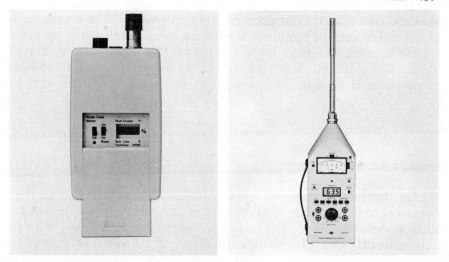

Figure 18.8 Personal noise dose meter (a), and precision integrating sound level meter (b) (Reproduced by courtesy of Bruel and Kjaer Laboratories Ltd.)

Frequency analysis

Neither of the above methods give specific information on frequency analysis, which is required if there is a need to select hearing protection or if there is a need to investigate engineering control methods. The simplest means of obtaining the frequency analysis is by a portable precision sound level meter with octave band analyser, which will measure the levels at the nominal mid frequencies of 31.5, 63, 125, 250, 500, 1000, 2000, 4000 and 8000 Hz. This information can be used to estimate the noise level at the ear when protectors are worn.

Similarly, the frequency analysis can be used to estimate the noise reduction which might be achieved by enclosure of a noise source. All the foregoing comments can be applied to continuous noise and noise which does not fluctuate more than 8 dB on the 'slow' setting of the sound level meter.

Recording and distribution analysis

The precision tape recorder can be used to record noise on site which can then be analysed in the laboratory to give third octave band analysis as dB, dB(A), dB(A)(L$_{eq}$) and statistical distribution. Data can also be obtained on site with a statistical distribution analyser. The technique is useful in the case of randomly fluctuating noise levels when the data cannot be measured from the visual output of a precision sound level meter. This parameter has a specific application in community response to nuisance noise and can be used to evaluate impulse noise.

Impulse noise

Where these fluctuations are exceeded, and for impulse noise, it is necessary to select instruments which have a short response time and a suitable dynamic range. The techniques suggested by BOHS (1976) to determine $dB(A)(L_{eq})$ are outlined in *Figure 18.9,* in order of decreasing

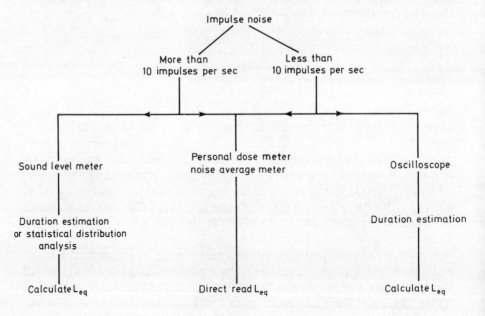

Figure 18.9 Methods of determining impulse noise L_{eq}

ease of use, are: noise dose meter; sound level meter/statistical distribution analysis; oscilloscope. An impulse sound level meter of sufficient sensitivity and dynamic range may be used in conjunction with the ACGIH Threshold Limit Values (see above). The oscilloscope is suitable for both applications but is difficult to use in the field.

Microphones

All the instrumental methods described depend on a transducer which responds to the sound pressure changes and which produces corresponding electrical signals. The aim is to produce accurate and repeatable measurements and to meet the requirements relating to frequency response, dynamic range and directional response, and the availability of suitable microphones may be a major factor in choice of the total instrumental package. The condenser microphone is the most widely used. The frequency response should be linear over the frequency range (in this case 20 Hz to 16 000 Hz), and the microphone should be selected depending on the field conditions. In the variable industrial environment it may be either diffuse field (from all directions) or free field (incident to the microphone). Wind velocity, either artificial or due to weather, produces a response in the microphone which can create significant errors. A wind-shield can reduce this effect at low windspeeds but where there is a noticeable effect satisfactory measurements may not be obtainable. There is less distortion of 'A' weighted readings due to the reduced effect on frequencies below 1000 Hz.

Field technique

The following comments can be made in connection with the measurement of sound levels and with the manipulation of data.

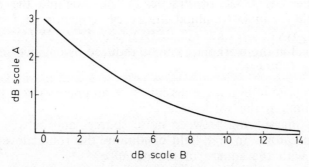

Figure 18.10 Summation of two different sound levels; A = amount to lighter of two noise levels to give total noise level. Scale B = difference between noise levels

1. The characteristics of the sound field in the occupational environment can be divided into the near field, the free field and the semireverberent field. Only in the free field does the inverse square law apply and sound level can be related to distance from the source, i.e. the level decreases 6 dB per doubling of distance due to spherical spreading from the source. In the near and

semireverberent fields other factors interfere, for example reflecting surfaces which make prediction less certain although the conditions may be stable and reproducible.

2. Decibels can be combined by using the formula relating sound pressure level to the ratio of P_1/P_0 (shown above) and the result can be drawn graphically as in *Figure 18.10*. This is a rapid way of making additions by first finding the difference between the two levels (Scale B) then by adding the corresponding correction (Scale A) to the greater of the two levels. Similar data exists for predicting the effect of increasing the number of noise sources; for estimating the attenuation of sound of different frequencies with increasing distance in free field conditions from the source; for predicting the attenuation of screens; for the prediction of sound in semi-reverberent surroundings (Hassall and Zaren, 1979).

Principles of control

Systems of control based on hearing conservation programmes have developed following the publication of the codes of practice and limits for exposure. The programmes relate to the identification of noise hazard areas, the reduction of noise levels and hearing protection. The dose of noise can be reduced by reducing exposure time; reducing noise level; providing personal ear protection. The main considerations for hearing conservation are outlined in *Figure 18.11* and are described below.

Using the UK Code of Practice as an example the allowable exposure time can be doubled for every reduction of 3 dB(A) in exposure between 90 and 110 dB(A) (*see Table 18.15*).

Noise level in the workplace can be reduced by engineering control methods as follows:

1. Keeping the total noise emission from all noise sources below the dose limit in the work area.
2. Regulating spacing between noise sources and between sources and operators. In free-field conditions the sound level roughly varies with the square of the distance.
3. Enclosing the noise sources by a sound reducing (i.e. attenuating) structure which prevents airborne transmission. The theoretical attenuation depends on mass, which is high for materials such as 23 cm brick (51 dB), less for other materials such as 64 mm plywood (17 dB) in the range 100–3150 Hz.
4. Reducing structure-borne transmission by isolation of the source using resilient mountings.
5. Damping of vibrating metal structures or by replacement with materials such as wood.

6. Reducing reflected noise by use of absorbent materials on surfaces such as roof, walls and floors.
7. Placing attenuating screens between the operators and the source.
8. Correcting imbalance and vibration by preventive maintenance.

It is important that the above control procedures should take account of thermal balance and ventilation requirements of operations and machinery.

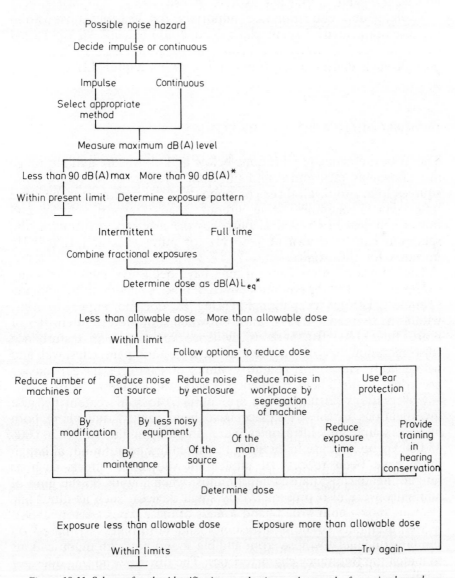

Figure 18.11 Scheme for the identification, evaluation and control of a noise hazard

Ear defenders can be used to reduce noise dose by a substantial amount depending on the type selected and the efficiency of fit (BSI, 1974). The main types are: ear plugs consisting of fine glass wool, mouldable foam, or moulded plastic; earmuffs consisting of ear cups with a soft seal, fitted with a sprung or adjustable headband (*see* Chapter 21). The expected attenuation can be judged from the average sound reductions less one standard deviation as measured by a standardized method such as BS 5108 (1974). Comfort, maintenance and cleaning are important aspects of use.

Training and education are important for hearing conservation because of the many ways in which the operator can influence personal exposure. Audiometry has become widely accepted as an important part of such conservation programmes (*see* Chapter 12).

INFRASONICS AND ULTRASONICS

The airborne acoustic vibrations below and above the audible range are known as infrasonics and ultrasonics respectively. Many noise sources span two or all three components of the spectrum at the same time, typical examples being aircraft, road transport vehicles and power sources. For practical purposes the arbitrary divisions in the spectrum can be drawn at 1–20 Hz for infrasonics and 16 000 Hz upwards for ultrasonics.

The effects of infrasonics on man have been investigated to some extent but the findings do not allow a definitive statement to be made (Tempest, 1976). Occupational hearing loss does not appear to occur within the defined 1–20 Hz range but when subjected to intensities of more than 140 dB there is evidence of vestibular disturbance (disorientation, loss of balance and nausea), aural pain, chest wall and whole body vibration. Tentative limits of exposure have been set according to time and frequency (Gierke and Nixon, 1976).

Observations on ultrasonics are confused to some extent because investigations in the workplace have been carried out where both audible sound and ultrasonics were present. In the range 20 000–45 000 Hz no increase in auditory threshold has been found, although there have been reports of unpleasant subjective effects such as fatigue, headaches, nausea and tinnitus, when sounds on the limit of audibility were also present. Many small devices, such as ultrasonic cleaning tanks, emit broad band sounds of these types (Acton, 1968).

Measurements of both infrasonics and ultrasonics are based on similar principles to those from audible sound but with modifications to match the frequency characteristics. The dB(A) weighting does not apply.

VIBRATION

The body can be subjected to non-acoustic vibrations which are transmitted by contact with moving surfaces, either to the whole body or to a localized area. Both forms of contact are of occupational importance and the latter through the hand-arm system is becoming more common as a topic of investigation in the work place.

Sources of exposure

Whole body vibration is connected with movement in ships, surface transport and air transport, where the components of the driving or power source are transmitted through the structures to the standing, lying or sitting person, and where sea movement or surface roughness (as in road surface) is also a cause. Transmission to the hand arises from tools such as pneumatic hammers, and chisels, which have marked impact characteristics, and rotary discs, swaging machines, grindstones and chain saws. The latter have received detailed investigations as a source of exposure in forestry workers.

The overall frequency range which is of interest is from 0 to 1000 Hz and movement can be described in respect of:

1. Dimension, in three perpendicular planes X, Y and Z (vertical);
2. Rotation in the same planes as roll, pitch or yaw;
3. Acceleration in cm s^{-2};
4. Velocity in cm s^{-1};
5. Displacement in cm.

The extent to which the vibration is transmitted to the whole body or to any localized area depends on such factors as body size, posture and tension, and the effect can depend on frequency, amplitude, duration of exposure, direction of vibration, and clothing as a boundary layer (Griffin, 1974).

Harmful effects

Whole body vibration can cause motion sickness plus a variety of effects which include blurred vision and loss of acuity, loss of efficiency, and discomfort. Disturbing frequencies for eyes, and vision, for chest and other body regions occur in the 2–27 Hz band and, because of natural resonating frequencies, the body can amplify input vibration of around 6 Hz (Gierke and Nixon, 1976).

Hand-arm vibration has more marked pathological effects which cause increased vibrotactile threshold and Reynaud's phenomenon which, in users of hand-operated tools, appears as vibration induced white finger (VWF). This may lead ultimately to permanent disability and can develop from both impact (such as in chipping hammers) and random vibration sources (such as pedestal grinding) (Taylor, 1974).

Exposure limits

Recommended limits for whole body exposure have been formulated by the International Organization for Standardization (IOS, 1978) for sinusoidal (single frequency) and random vibration in the Z axis and X and Y axes combined. These limits apply to three conditions: health and safety, fatigue and comfort, and apply to a frequency range of 1–80 Hz for durations of 1 minute to 24 hours. These standards take account of the fact that the maximum response occurs between 4 and 8 Hz for the Z axis (vertical) and between 1–2.5 Hz for the combined X and Y axes. The lower thresholds of acceleration for fatigue effects are $0.14\,\mathrm{m\,s^{-2}}$ for Z axis and $0.1\,\mathrm{m\,s^{-2}}$ for the XY axes. The standards for health and comfort are 6 dB higher and 10 dB lower than the fatigue values, respectively.

No clear limits for motion sickness can be quoted but vertical oscillations of the order of 0.25 Hz appear to be most effective in producing sickness (Reason, 1976).

Recommended limits for hand-arm vibration are given by the British Standards Institution in the draft for development DD43 (1975). This expresses limits of acceleration over a range of $1\text{--}100\ \mathrm{m\,s^{-2}}$ in 9 octave bands from 44 to 1000 Hz, for two exposure times, 150 min and 400 min. This standard reflects the higher risk associated with hand-arm vibration in the region of 4 to 16 Hz and applies to each of the three perpendicular axes X, Y and Z separately. The lower thresholds of acceleration are 1 cm $\mathrm{s^{-2}}$ for 400 minute exposure and 10 cm $\mathrm{s^{-2}}$ for 150 minute exposure.

Measurement and evaluation

Acceleration and movement can be measured in all the sources by the same techniques, using a sound level meter and accelerometer, or more complex systems based on recording and analysis (the same in principle to acoustic measurements). For impact vibration the equipment needs to be chosen so that the very high peak acceleration levels can be measured. Piezoelectric accelerometers can be used for both applications and may be attached by screw, adhesive, magnet or wax.

Principles of control

Engineering methods can be used to control exposure for whole body and localized transmission. For example, chairs and supporting structures can be suspended by using pneumatic and hydropneumatic springs and damped with viscous dampers. Hand tools may be designed to reduce vibration from the source (e.g. the shaking and ignition forces of the internal combustion engine on a chain saw); to reduce transmission of vibration from the engine to the frame, and from the frame to the handle. Precautions can be taken to control exposure by avoiding long periods of continuous use and by keeping the hands warm and dry by wearing suitably designed gloves. This latter precaution is necessary because the symptoms of VWF are precipitated by cold. For this reason tools such as pneumatic hammers and chisels should be designed so that the exhausted cold air stream is directed away from the hands.

ABNORMAL PRESSURE

Under normal conditions at sea level the body is subjected to an atmospheric pressure which varies within narrow limits around one standard atmosphere of 760 mm of mercury. The partial pressures of the three principle gases are around: nitrogen 596 mm, oxygen 160 mm, carbon dioxide 0.04 mm. Changes occur with height above sea level, and with depth below sea level as shown in *Table 18.17* and also in some manmade situations when the pressure is elevated deliberately.

Table 18.17 DEVIATION FROM NORMAL ATMOSPHERIC PRESSURE

	Altitude depth (ft)	*Pressure: absolute (mmHg)*	*PO$_2$(mmHg)*	*PN$_2$(mmHg)*
Altitude	60 000	54.1	11.5	43
	30 000	226	48	179
	10 000	523	110	409
	Sea level	760	160	596
Depth	33	1520	320	1192
	99	3040	640	2384
	297	7600	1600	59.60

PO$_2$ = partial pressure of oxygen

PN$_2$ = partial pressure of nitrogen

Abnormal situations

Decreased pressures occur, for example, in aviation and in mountaineering when there is a reduction in atmospheric pressure. Quite small reductions in pressure are associated with changes in altitude; the partial pressure of oxygen (PO_2) decreases by 50 mmHg through a height of 10 000 ft, while a large increase in PN_2 occurs in the first 33 ft below sea level, i.e. 596 mmHg (Macmillan, 1968; Miles and Mackay, 1976).

Both changes are of physiological significance. Increased pressures are found in underwater operations which are of great importance in undersea drilling and pipeline operations. They are also commonly found in other industrial and amateur diving operations which are frequently undertaken. Increased pressures are found in tunnelling operations where the pressure is deliberately elevated to reduce the inflow of water from surrounding strata or structures. The abnormal pressures found in these operations vary from 0 to 4 atmospheres for dry tunnelling and from 0 to 30 atmospheres for many diving operations. Experimental dives are undertaken at equivalent depths of 1500–2000 ft, i.e. around 60 atmospheres.

Harmful effects

The primary physiological effects of reduced atmospheric pressure at altitude is hypoxia due to the reduced partial pressure of oxygen. The homeostatic mechanisms can compensate for lack of oxygen up to 10 000 ft (PO_2 110 mm) and acclimatization occurs on prolonged exposure as the oxygen transport system adapts to the reduced oxygen tension. These effects occur in rather remote or specialized circumstances and are not a usual feature of the occupational scene.

By comparison, the effects of hyperbaric exposure (high pressure), have had a marked impact due to acute and chronic disability which may occur. These can be summarized as:

1. *Compression effects* in the joints, on the central nervous system, in the ear, changes in fluid distribution, and increased physical stress.
2. *Toxicity* in the form of nitrogen narcosis and oxygen toxicity due to the high partial pressures of the two gases.
3. *Decompression sickness* caused by the presence of bubbles of nitrogen in body fluids which appear when the tissue fluids are decompressed and nitrogen leaves the solution as a gas. This is recognized by a variety of clinical conditions which reflect pulmonary, circulatory, vascular and neurological effects.

Measurement and evaluation

The circumstances in which these effects occur do not allow for measurement and evaluation in the traditional sense. The known relationship between altitude, depth and pressure allows for acceptable prediction of total and partial pressure. Where required, pressure can be measured by gauge relative to ambient pressure.

Principles of control

Taking wet or dry hyperbaric conditions as being of most concern, precautions can be taken to prevent acute and chronic effects by controlling the rate of decompression and by modifying the respirable gas mixtures.

It is known that the solubility of nitrogen in body tissues is governed by Henry's law, such that

$$\frac{Vg}{Vl} = aP$$

where Vg is volume of gas dissolved at STP (dry), Vl is volume of liquid, P is partial pressure of N_2 and a is solubility coefficient.

Consequently as man dives deeper more nitrogen will dissolve in the body fluid, and during the ascent the dissolved gas is released. By controlling the rate of decompression or the ascent the volume of gas released can be controlled and the presence of gas bubbles may be kept within tolerable limits.

Since nitrogen is present only as a diluent it can be omitted by using pure oxygen or replaced by helium or other inert gas of lower toxicity. Compressed air is the most commonly used gas but there are applications for oxygen, and helium–oxygen mixtures. There are secondary problems because of the acoustic properties of helium which causes speech distortion, and heat loss due to the high thermal conductivity, however it is used for dives of more than 200 ft.

The control of decompression sickness depends largely on the use of tables which state the following parameters:

1. Depth of dive.
2. Duration at a specified depth.
3. Total time for decompression.
4. Stay times at different depths during ascent.

Authorities such as the British Admiralty (BR2806) and the US Navy (Navyships 250–538) have such diving tables, which are adopted internationally and form the basis of commercial and amateur regulations (*British Sub-aqua Club Diving Manual*). There are specific decompression tables for each type of gas mixture because of the need to compensate for the different partial pressures. All of the tables are based on the principle that the period of decompression increases with stay time, up to saturation, at a given pressure.

The principles of control can be illustrated with reference to shallow dives of short duration, dives of extended duration, and dives of prolonged duration.

Shallow dives of short duration

It is probable that gas bubbles arise after decompression from all hyperbaric exposures but that they are acceptable up to a certain threshold. Consequently, for shallow dives the stay time can be regulated so that there is no need for controlled ascent, and in practice a stay time of several minutes can be tolerated, for example, at 60 ft. This applies to many amateur and industrial operations in which the risk is generally low and where compressed air is used for breathing purposes.

Dives of extended duration

For extended stay times it is necessary to regulate the ascent rate by keeping to specified values typically expressed in feet or meters per hour and to specified hold times. Similar guidelines are applied to the controlled decompression of hyperbaric tunnel personnel. These criteria have led to a high degree of control but do not completely eliminate decompression sickness.

Dives of prolonged duration

The technique of saturation diving has both commercial and research applications making use of the fact that once the body fluids are saturated with inert gas they can remain so for long periods. Equilibrium is reached with the partial pressure of the diluent gas after about 24 hours and the technique is used for both shallow and deep dives. Once saturation is reached activity can be centred on a habitat on the sea bed, or the diver can be transferred from sea bed to a ship decompression chamber via a pressurized transfer chamber.

There have been important developments in technology, work practice and legislation, which focus attention on the inherent risks of hyperbaric conditions. Legislation in the UK relates to diving operations at civil engineering works (1960), offshore installations (1974), submarine pipelines (1976) and diving from merchant vessels (1976). Emphasis is placed on the use of decompression tables, on physical fitness and a log of diving activities. Attention is drawn to the physical stress of work at increased pressure and of the excessive heat loss, particularly when oxygen-helium mixtures are used (Hanson, 1978) which can be controlled by the provision of external body heating and respiratory gas heating in deep diving work or in submersible chambers. Because of the need for access to decompression facilities, chambers exist in the form of fixed surface installations and mobile surface chambers for the transfer of patients in need of extensive treatment.

REFERENCES

Acton, W.I. (1968) 'A criterion for the prediction of auditory and subjective effects due to airborne noise from ultrasonic sources.' *Annals of Occupational Hygiene*, **11**.2, 227–234

ACGIH: American Conference of Governmental Industrial Hygienists (1979) *Threshold Limit Values for Chemical Substances and Physical Agents in the Working Environment*. ACGIH, PO Box 1937, Cincinnati, Ohio 45201

ANSI: American National Standards Institute (1976) *American National Standard for the Safe Use of Lasers*. 2136. New York

Baranek, L.L. (1960) *Noise Reduction*. New York and London: McGraw-Hill

British Admiralty *Diving Manual* (BR2806). London: HM Stationery Office

BOHS: British Occupational Hygiene Society (1976) 'Hygiene standard for impulse noise.' *Annals of Occupational Hygiene*, **19**, 179–192

British Occupational Hygiene Society (1971) 'Hygiene standard for wide band noise.' *Annals of Occupational Hygiene*, **14**, 57–64

British Standards Institution (1959) *Filters for use during Welding and Similar Industrial Operations* (BS 679). London: BSI

BSI (1960) *Equipment for Eye, Face and Neck Protection against Radiation arising during Welding and similar Operations* (BS 1542). London: BSI

BSI (1967 amended 1975) *Method of Rating Industrial Noise Affecting Mixed Residential and Industrial Areas* (BS 4142). London: BSI

BSI (1974) *Methods of Measurement of the Attentuation of Hearing Protectors at Threshold* (BS 5108). London: BSI

BSI (1975) *Guide to the Evaluation of Exposure of the Human Hand Arm System to Vibration* (BSI DD43). London: BSI

British Sub-Aqua Club. *British Sub-Aqua Diving Manual*. British Sub-Aqua Club, 70 Brompton Road, London SW3 1WA

Carpenter, R.L. (1970) 'Biological effects and health implications of microwave radiation.' *Symposium Proceedings*, ed. S.F. Cleary, pp 76–81. Publication BRH/DRE 70–2. Washington DC: Bureau of Radiological Health US Dept of Health Education and Welfare

Clayton, G.D. and Clayton, F.E. (1978) *Patty's Industrial Hygiene Toxicology*, 3rd edn. New York: Wiley

Department of the Environment (1975a) *Code of Practice for the Carriage of Radioactive Materials by Road*. London: HM Stationery Office

Department of the Environment (1975b) *Code of Practice for the Storage of Radioactive Materials in Transit*. London: HM Stationery Office

Diving Operations - Statutory Instruments. London: HM Stationery Office
    Diving Operations Special Regulations (1960)
    The Offshore Pipelines (Diving Operations) Regulations (1974)
    The Submarine Pipelines (Diving Operations) Regulations (1976)
    Merchant Shipping (Diving Operations) Regulations (1976)

Gierke, H.E. and Nixon, C.W. (1976) 'Effects of intense infrasound on man.' *Infrasound and Low Frequency Vibration*. ed. W. Tempest, pp. 115–147. London: Academic Press

Griffin, M.J. (1974) 'Some problems associated with the formulation of human response to vibration.' *The Vibration Syndrome*, ed. W. Taylor, pp. 12–23. London: Academic Press

Hanson, R. de G. (1978) 'Working in cold environments - lessons to be learned from diving.' *Annals of Occupational Hygiene*, **21**, 193–198

Harlen, F. (1978) 'The development of laser codes of practice.' *Annals of Occupational Hygiene*, **21**, 199–212

Hassall, J.R. and Zaren, K. (1979). In Bruel and Kjaer, *Acoustic Noise Measurements*, Copenhagen, 1979.

H & SE: Health and Safety Executive (1972) *Code of Practice for Reducing Exposure of Employed Persons to Noise*. London: HM Stationery Office

Hopkinson, R.C. (1969) *Lighting and Seeing*. London: Heinemann

Hughes, D. (1978) *Hazards of Occupational Exposure to Ultraviolet Radiation*. Occupational Hygiene Monograph No. 1, University of Leeds Industrial Services Limited

IES: Illuminating Engineering Society (1977) *The IES Code. Interior Lighting*. IES, York House, 199 Westminster Bridge Rd., London SE1 7UN

ICRP: International Commission on Radiological Protection (1966a) *Recommendations of the International Commission on Radiological Protection*. ICRP 26

ICRP (1966b) 'Deposition and retention models for internal dosimetry of the human respiratory tract.' *Health Physics*, **12**, 173–207

ICRP (1968) *General Principles of Monitoring Radiation Protection of Workers*. ICRP 12

IOS: International Organization for Standardization (1961) *Recommendation for Sound Level Meters*

IOS (1971) *Assessment of Noise with Respect to Community Noise* (R1996)

IOS (1973) *Measurement of Acoustical Noise and Evaluation of Its Effects on Man* (R2204)

IOS (1978) *Guide for the Evaluation of Human Exposure to Whole Body Vibration* (IOS 2631–1978E)

IOS (1979) *Recommendation for Precision Sound Level Meters*

Ionizing Radiations - Statutory Instruments and Codes of Practice. London: HM Stationery Office

Ionizing Radiations (Sealed Sources) Regulations (1969)

Ionizing Radiation (Unsealed Radioactive Substances) Regulations (1968)

Langmead, W.A. (1971) 'Air sampling as part of an integrate programme of monitoring of the worker and his environment', In *Inhaled Particles III*, vol 2, ed. W.H. Walton, pp. 983–985. Surrey, England: Unwin

Langmead, W. A. and O'Connor, D.T. (1969) 'The personal centripeter - a particle size selective personal air sampler.' *Annals of Occupational Hygiene*, **12**, 185–196

Lindsay, I.R. (1975) 'A review of microwave radiation hazards and safety standards.' *Annals of Occupational Hygiene*, **17**, 315–320

Lynes, J.A. (1968) *Principles of Natural Lighting*. Philadelphia: International Ideas

Macmillan, A.J.F. (1968) 'Physiology of changes in pressure.' *Annals of Occupational Hygiene*, **11**, 321–328

Medical Research Council (1971) *Exposure to Microwave and Radiofrequency Radiations. Medical Research Council Recommendations* (MRC 70/1314) London: MRC

Miles, S. and Mackay, D.E. (1976) *Underwater Medicine*. London: Adlard Coles

NIOSH: National Institute for Occupational Safety and Health (1972a) *Criteria for Occupational Exposure to Noise*. Washington DC: US Department of Health, Education and Welfare

NIOSH (1972b) *Criteria for Occupational Exposure to Ultraviolet Radiation*. Washington DC: US Department of Health, Education and Welfare

National Radiological Protection Board (1977) *Protection against Ultraviolet Radiation in the Workplace*. Harwell: NRPB

Nuclear Installations Act 1965 London: HM Stationery Office

Radioactive Substances Act 1960 London: HM Stationery Office

Reason, J.T. (1976) *Motion Sickness and Associated Phenomena. Infrasound and Low Frequency Vibration*, ed. W. Tempest. London: Academic Press

Sherwood, R.J. and Greenhalgh, D.M.S. (1960) 'A personal air sampler.' *Annals of Occupational Hygiene*, **2**, 127–132

Taylor, W. (1974) *The Vibration Syndrome*. London: Academic Press

Tempest, W. (1976) *Infrasound and Low Frequency Vibration*. London: Academic Press

US Navy. *US Navy Diving Manual* (Navy Ships 250–538). Washington DC: US Navy Department

19

The Thermal Environment

G.W.Crockford

Man, like all mammals and birds, produces heat as a result of metabolic activity. The metabolic heat is then lost to the environment in a controlled manner to maintain the body temperature at about 36.8°C or, if under thermal stress, some value normally within the range of 36.5–39°C. Exposure to heat may lead to elevated body temperatures and such exposure is found naturally in farming and other outdoor activities, during the summer in temperate zones, and in the tropics. In industry, whenever processes add heat to the environment there will be hot working conditions, for example in foundries and steel works and in paper manufacture. As the body has to lose heat all the time, anything which interferes with the loss of heat will result in a heat exposure problem. Such interference may be caused by unsuitable clothing, or in a local build-up of heat in the workers' vicinity, as can happen in enclosed or inadequately ventilated spaces.

Cold exposure can occur in outdoor situations as a result of low temperatures, wind and rain, and also in industry, particularly food storage cold rooms. Wind and water play an important part in creating a cold exposure situation.

Conventionally, an environment is referred to as hot or cold according to the reading on a thermometer. However, the heat content of air consists of two components, sensible and latent heat. The sensible heat component is determined by the temperature of the air and its specific heat, that is the quantity of heat required to raise the temperature of 1kg of air by 1°C, that is, 1.005 kilojoules per kilogram (kJ/kg.) If the temperature of the air is above that of the skin, sensible heat cannot flow from the body into the air. Under these conditions the latent heat content of the air is very important. The latent heat, or insensible heat content of air as it is sometimes called, is proportional to the amount of water vapour in the air, normally expressed as weight of water per unit weight of air, or as a partial pressure, e.g. millibars or mmHg. Each gram of water vapour represents 2.4 kJ of latent heat of evaporation, which is taken up when the water molecules move from the liquid phase (sweat) to the vapour phase. The water molecules can only evaporate and remove heat from the body if the concentration of water vapour in the air is less than the concentration of water vapour in saturated air at skin temperatures. To summarize, air has a definable heat capacity which consists of a sensible non-evaporative component, and a latent heat component which is determined by the water vapour content.

The thermal environment has a pronounced effect on personal comfort. Marked deviations from comfort affect productivity, increase error and accidents, and may eventually affect health. Heat also interacts with other environmental contaminants in such a way as to make their effects more serious or, as is often the case, other factors undermine thermal tolerance. Alcohol is a good example; other drugs, carbon monoxide and minor infections have similar effects on a person's thermal tolerance.

HEAT EXCHANGE

The study of man's thermal environment is to a large extent the study of the thermal exchanges that are taking place between the body and the environment by conduction, radiation, convection and evaporation. The control of thermal stress is the control of these thermal exchanges.

Conduction

Heat will flow by a process called conduction from an area of high temperature in a solid, liquid or gas, to one at a lower temperature. The speed with which a given quantity of heat is transferred by conduction is dependent on the temperature gradient down which the heat is flowing and the thermal conductivity of the material. Some materials, such as metals, are good conductors; others, such as air, very poor conductors. Thermal conductivity varies with temperature, that of air increasing by about 0.28 per cent of its value at 0°C for every degree rise. This can be of some importance at high temperatures as the insulation provided by clothing is dependent on the air trapped in the clothing fabrics.

Radiation

All materials absorb and emit radiant energy, the intensity of the emission being proportional to the fourth power of the absolute temperature (Stefan's law) and the emissivity of the radiating surface.

The emissivity is given as a decimal faction of the emissivity of a 'black body' and is numerically equal to the absorptivity. Some generalizations can be made about the emissivity of surfaces: it is low for highly polished metal surfaces and generally increases as their temperature rises and as the surface becomes contaminated with dirt; most substances other than metals have a high emissivity which also depends on the structure of the surface, e.g. polished or matt, clean or dirty.

The spectral energy distribution is determined by the temperature of the emitting body; an increase in temperature not only causes more energy to be emitted but a greater proportion of it to be emitted at the shorter wavelengths (Wien's law). For low temperature sources such as the human body the predominant wavelengths are about 6–24 μm.

The exchange of radiant heat between the body and the environment can be expressed as an equation derived from basic physical principles and is:

$$R = EK \; Ar \; (Tw^4 - Ts^4)$$

where E is the emissivity of the man; K is Stefan-Boltzman constant; A is the area of the man (1.8 m^2 for the standard man); r is the ratio of effective radiation surface to the Du Bois area (0.78 for a static standing man, 0.85 for a moving man), and Tw and Ts are the wall and skin temperatures in degrees absolute. An emissivity factor is not used for the wall temperature which, due to the way in which it is normally measured, should more correctly be referred to as the mean radiant temperature (MRT) of the surroundings.

Convection

Convective heat exchange is brought about by air moving over the body and removing or transferring heat to the skin. Natural convection occurs when the air movement is induced by the body heat heating up the air next to the skin, and the now warm and buoyant air rising and forming a layer of moving air over the body. This is the normal situation in such environments as offices. Forced convection occurs when a wind sufficiently fast to overcome the natural convection currents blows over the body. This is normally the case when a person is walking or where a deliberate attempt is being made to circulate air. Forced convective heat exchange is directly proportional to the temperature difference between the skin and air and for most of the air velocities met in industry, proportional to the square root of the wind speed. As the whole surface area of the body is not subjected to the same wind velocity and some areas will receive air already heated or cooled, the actual heat transfer is normally calculated using an empirically derived convective heat transfer coefficient which takes these factors into account.

The equation for convective heat exchange between a man and his environment is:

$$Hc = 8.3 \; V^{0.5} \; (Ts - Ta) \; W/m^2$$

where V is the wind speed in m/s, and Ts and Ta the skin and air temperatures in °C respectively W/m^2 is heat flow in watts per square metre. The convective coefficient 8.3 is for a nude man standing facing the air stream.

Evaporative heat loss

For insensible heat loss to occur sweat has to evaporate, and this only takes place if there is a water vapour pressure gradient between the sweat on the skin and the ambient air. The greater this gradient the more rapidly the sweat evaporates. Under normal environmental conditions the evaporation of sweat always removes heat from the body, but this can only occur if the wet bulb temperature is below skin temperature, sweat evaporation causing the skin temperature to drop towards the wet bulb temperature. The factors influencing evaporative heat loss are the vapour pressure gradient between the skin and ambient air and the wind speed, the wind removing the boundary layer next to the skin which quickly saturates with water vapour and impedes further evaporation.

A common term used to describe the water content of air is its relative humidity, the quantity of water vapour the air contains expressed as a percentage of what it would contain if saturated at the same temperature. As it is the water vapour pressure gradient which determines the evaporation of sweat, the term is of limited value unless the air temperature is also stated.

The maximum evaporative heat loss with the skin fully wetted is described by the following equation:

$$E_{max} = 13.7 - V^{0.5} \; (P \; skin - P \; air) \; W/m^2$$

where V is in m/s, P skin and P air are water vapour pressures in millibars of skin and air, W/m^2 is heat flow in watts per square metre. For submaximal sweating a latent heat of evaporation of 2.45×10^6 J/kg (580 cal/g) is generally used, although work by Snellen, Mitchell and Wyndham (1970) indicates that 620 cal/g would be more correct. It helps to get the feel of this equation if it is remembered that an evaporative loss of 1 g/minute is equivalent to about 41 watts.

These equations show that the environmental parameters governing the thermal exchanges of the body are the air temperature, wind speed, radiant temperature and water vapour pressure. All these parameters have to be determined when assessing a thermal environment. A more detailed description of heat exchange is given by Kerslake (1972).

Metabolism

In practical situations, the determination of the metabolic rate or activity level presents considerable difficulty unless there is specific data on the tasks that are being done. The metabolic cost of a wide range of tasks is given by Durnin and Passmore (1967). The work study engineers have developed a sophisticated approach to the problem, in which the different body movements are given specific energy costs. It is then a question of identifying the movements and counting them (Lange Anderson and others, 1971). The TLV for thermal stress (American Conference of Governmental Industrial Hygienists [ACGIH], 1979) contains some guidance on the energy cost of tasks but in practice the energy costing normally means determining the type of work. If the operative is sedentary it is about 100 kcal/hour (116W), doing light work 100 – 200 kcal/hour (116 – 232 W), moderate work 200 – 350 kcal/hour (232 – 407 W) or heavy work 350 – 500 kcal/hour (407 – 580 W). It is important to be able to grade work into these categories as the comfort conditions and heat stress limits are determined by the metabolic heat load. The harder a man works the more heat has to be lost and so lower environmental temperatures or higher wind speeds are required if the heat loss is to take place without undue stress or discomfort.

Heat storage

The body is like a tank of water with a definable capacity for storing heat. Although the normal body temperature is about 37°C, temperature deviations above and below can be tolerated. The heat storage is equal to the change in temperature times the specific heat of the body, which is normally considered to be 3.49 kJ/kg°C. The change in body heat is not quite so easy to calculate as the core (i.e. the central part of the body) and the shell (i.e. the skin and extremities) undergo different rates of temperature change when the body is exposed to hot or cold conditions. However, the equation for calculating body heat storage is:

$$S = 3.49 \, W \, (0.67 \, Tr + 0.33 \, Ts) \text{ kJ}$$

where S is storage, W is body weight in kg, Tr is deep body temperature (°C) and Ts is the mean skin temperature (°C).

The practical significance of heat storage is that it enables people to enter extreme environments for short periods of time without coming to any harm. It also prevents people from fully experiencing the stress of a thermal environment unless they stay in it for at least 30 minutes, by which time the protective influence of the heat sink has been used up and they are 'seeing' the environment as it will affect them in the longer term. Most managers have a reluctance to stay in a hot working area for the 30 – 40 minutes required to appreciate what it feels like and so can underestimate the stressfulness of the conditions.

Clothing and heat exchange

Clothing is obviously used to influence the body's heat exchange with the environment, usually to reduce heat losses to an acceptable value. The clothing places a conductive resistance between the skin and the environment but once the heat has reached the surface of the clothing it is again lost by convection and radiation. Clothing also impedes the loss of water vapour and hence evaporative heat loss. The thermal resistance of most clothing assemblies is proportional to thickness with a value of about $0.25°C\,m^2/W$ per centimetre thickness (conductivity $0.04\,W/m°C$) (i.e. with a $0.25°C$ temperature gradient across the clothing one watt of heat will flow across each square metre). The 'clo', which is the mean insulation of a clothing assembly, is a thermal resistance of $0.155°C\,m^2/W$ and will keep a resting man comfortable in an indoor environment of $21°C$. When calculating the heat exchange between a man and the environment, as is done in the Heat Stress Index, the influence of clothing has to be allowed for in the equations governing convective, radiative and evaporative heat transfer. Clothing commonly worn in industry, such as a coverall and underclothes, may reduce heat exchanges by these mechanisms to about 60 per cent of their calculated value.

Summary

There are six factors influencing heat exchange between the body and the environment

1. Air temperature
2. Water content of the air
3. Mean radiant temperature
4. Wind velocity
5. Metabolic rate
6. Clothing

When assessing the influence of an environment on people, all six factors have to be considered. The heat exchanges can be summarized by the heat balance equation for a man which is:

$$M = E \pm C \pm K \pm R \pm S$$

where M is metabolic heat, E is evaporative heat loss, C is convective heat exchange, K is conductive heat exchange, R is radiation and S is storage.

MEASURING THE THERMAL ENVIRONMENT

Instruments

The most widely used temperature measuring devices are alcohol or mercury in glass thermometers, electronic devices using resistance wire, thermistor beads or thermocouples, and bimetallic strip thermometers

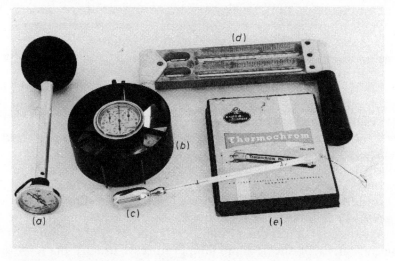

Figure 19.1 Some of the basic instruments used in environmental measurement: (a) wet globe thermometer; (b) vane anemometer; (c) silvered Kata thermometer; (d) sling hygrometer; (e) thermal crayons which indicate the temperature of surfaces by changing colour when drawn on it

(*Figure 19.1*). For use in industry it is important that the scale on the temperature measuring device can be easily read under poor illumination and in difficult positions. An accuracy of ± 0.1°C is of little value if observer error is ± 1°C or more. The calibration of the instruments is also important. The mercury in glass thermometers are seldom calibrated by the user. The bimetallic strip thermometers deteriorate with age and very often show a hysteresis loop when calibrated. Regular calibration of most forms of temperature measuring instruments is required.

A temperature measuring device measures its own temperature, and the assumption is made that it is at the same temperature as the object or material being measured. This is not always the case; for example, if a thermometer is being used to measure air temperature, radiant heat in the environment will also affect it. It is important therefore to protect the thermometer from radiation and to minimize the effect radiation has on it. This can be done by screening the thermometer from radiation with clean aluminium foil and by drawing air over it so that if there is a heating effect from radiation the thermometer is cooled to the ambient condition. Thermometers are frequently placed on walls. The temperature indicated will be a combination of air and wall temperature, which can be misleading. The temperature of the air is normally referred to as the dry bulb temperature (t°db).

The wet bulb temperature is obtained using a temperature sensing head covered with a muslin sock wetted with distilled water and protected from radiant heat. If air is made to flow over the wet bulb at about 4 m/second, thus facilitating the evaporation of water, the reading is referred to as the aspirated or psychrometric wet bulb (t°wb). If the wet bulb is not protected from radiation and is exposed to natural air movements, the reading is referred to as the natural wet bulb (t°nwb).

The difference in temperature between the dry bulb and wet bulb temperatures indicates the amount of water present in the air; the bigger the difference the drier the air. The actual water content of the air (the absolute humidity in kg of water per kg of air), the relative humidity (what the air contains compared with what it would contain if saturated), and the dew point (the temperature to which the air has to be reduced to become saturated), can all be obtained from the psychrometric chart if the dry and wet bulb temperatures are known (*Figure 19.2*). These charts sometimes show the heat content of the air which can also be determined from the dry and wet bulb temperatures.

Measurement of the radiant temperature is usually made with a black globe thermometer, a six-inch diameter hollow copper sphere painted matt black, into which a thermometer is inserted with the bulb at the centre of the globe. The thermometer of choice is a bimetallic strip type, as these are robust, or a mercury-in-glass thermometer mounted in a metal guard. The instrument has a slow response, taking about 20 – 30 minutes to reach equilibrium. It is easy to use but because it is subject to convective cooling for which a correction has to be made, the air movement and dry bulb temperature must be

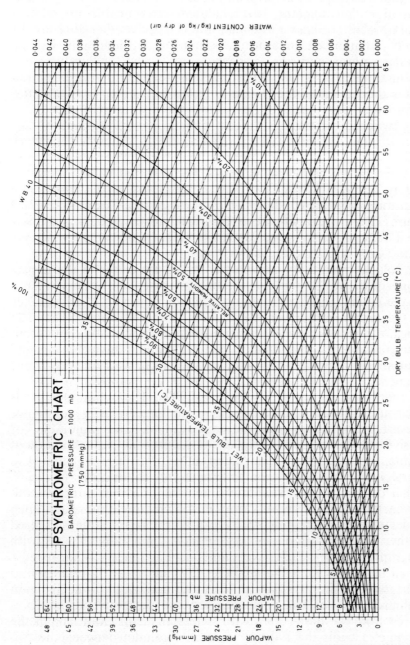

Figure 19.2 Chart for calculating relative humidity, absolute humidity, etc. (see text) from wet and dry bulb temperatures. (Reproduced from Ellis, Smith and Walters, 1972, by courtesy of the Editor, British Journal of Industrial Medicine)

measured at the same time and in the same place in order to derive the mean radiant temperature with the help of nomograms or equation.

Radiation thermometers are available which enable the radiant temperature of surfaces to be measured. They are sensitive to the infrared radiation emitted by the surface and compare it with an internal standard.

An instrument which is widely used for assessing thermal environments is the wet globe thermometer, often called the 'Botsball' after Botsford who developed it. The idea is that a two-inch black globe covered with wet muslin 'sees' the environment in much the same way as a man in that they are both black to infrared radiation, both wet and hence sensitive to the water content of the air, and both influenced by the dry bulb temperature and wind velocity. The wet globe temperature is therefore a thermal index; it does not measure an environmental temperature or provide data which can be used in determining control measures. There are practical problems with the wet globe; it is not easy to see if it is completely wetted and high radiant heat loads or wind may lead to parts of it drying out.

Air velocity measuring instruments make use either of the impact pressure of the wind (e.g. rotating or deflecting vane anemometers), or of the cooling effect of wind on a heated object (e.g. a resistance wire, thermistor or thermometer bulb; Kata thermometer). The main practical difficulties arise when measuring multidirectional or turbulent air movements which tend to occur in outdoor environments and also in maintaining the calibration of the mechanical and electronic instruments. For these reasons the almost omnidirectional silvered Kata thermometer is still widely used. The Kata thermometer is suitable for an occasional survey, but not for detailed or extensive studies. Direct reading equipment is available and preferable for extensive survey work. The problem of maintaining the calibration of instruments is not too serious, as many technical colleges and university engineering departments have calibration facilities. Another drawback of direct reading equipment is that it responds to every fluctuation in air velocity, thus making it difficult to read. A pen recorder or some averaging device can therefore be useful. Silvered Kata thermometers can be obtained in three temperature ranges, with nomographs for the rapid conversion of cooling times to wind speeds.

When using the Kata thermometer the bulb must be cleaned and then heated in a thermos flask of hot water until an unbroken column of coloured alcohol has risen into the top reservoir. The bulb is removed from the flask, dried with a clean tissue or cloth, and as the thermometer cools the fall of the alcohol column between the two temperature marks is timed, using a stop watch. The wind speed is then determined from the Kata factor engraved on the stem, the dry bulb temperature and the cooling time, using the appropriate nomogram. Smoke tubes, although not quantifying air movement, enable a visual picture of velocity gradients and direction of movement to be built up very quickly and are valuable for survey work.

Thermal environment survey

Instruments are only one aspect of measuring the thermal environment. When setting out to measure an environment several questions have to be answered, not least of which is what are the measurements being made for; for example, assessing environmental control equipment or determining the environmental conditions in which people are working. If the latter, it is as well to remember that wind velocity and radiant temperature can vary considerably between points only a matter of inches apart. An instrument set up two feet from where a man works can give quite a different answer when set up in the space normally occupied by the man. Similarly, conditions can change quickly with time. The use to be made of the measurements also needs to be known; for example, if the wet bulb globe temperature (WBGT) index is being used for evaluating the environment, the environmental measurements are made in one way (t°nwb and unscreened t°db) and if the heat stress index is to be used, the measurements are made in another way (t°wb, t°db).

In short, the questions why, how, where, when and for how long have to be answered when planning a survey.

Activity level

In assessing the thermal environment, the metabolic heat load is of considerable importance. Also it is the factor which is under the most direct control of the exposee, who can reduce thermal stress quite simply by reducing the work rate. Since this means a decrease in production, the method of choice is to control environmental factors. In order to quantify the total heat load, the metabolic heat component must be known, but at the same time it is almost impossible to measure it without a considerable investment in time and money. It is therefore customary to etimate the metabolic heat component from tables of values (*Table 19.1*). A more recent method is to use work study techniques which involve breaking the work tasks down into small components, assigning an energy cost to each and then adding them all up to get the total metabolic cost. Work does vary throughout the day, with periods of high and low activity. However, most people doing heavy work will average out at about 5 kcal/minute (350 W) for an eight-hour day. The International Labour Organization standard for work load is an upper limit of 33 per cent of a person's

Table 19.1. METABOLIC RATES FOR DIFFERENT ACTIVITIES (FANGER, 1970)

Activity	Metabolic rate (kcal/h)	Metabolic rate (kcal/m²/h)
Sitting	80	44.5
Standing	100	55.5
Walking 4 km/h	220	122
Standing: light hand work	140–180	77–100
Standing:heavy hand work	180–220	100–122
Standing: light arm work	270	150
Standing: heavy arm work (e.g. sawing)	360–580	200–322
Work with whole body:		
light	270	150
moderate	360	200
heavy	480	266

maximum work capacity of about 17 kcal/minute (1190 W), and is somewhat similar in value. People with less than the average work capacity can only maintain a given work output by using a greater percentage of their capacity; in other words their relative work load is higher, which in turn means that their relative heat load will be higher than that of the stronger members of the group.

Clothing

Clothing has a marked effect on the heat exchange between the body and the environment, which of course is the main reason for wearing it. If the weather changes so that we feel too hot or cold we adjust the clothing accordingly. The metabolic heat output also has a pronounced effect on clothing requirements (*Figure 19.3*) and again it is possible to make suitable adjustments. In this figure the steepness of the slope at 40 watts shows that clothing requirements are very sensitive to environmental temperature at low activity levels. At high activity levels, 200 watts, the slope is shallow, showing that modest changes in clothing thickness can cope with large temperature changes. The reverse of this is that the insulation value of workwear and protective clothing will have minimal effects on comfort if the wearer is sedentary, but will have marked effects when the wearer is active.

In industry there are many working situations where clothing has other functions than heat loss control and must be worn however hot or cold it is and however hard the man is working. Under these conditions clothing will reduce the heat losses from the body by radiation, convection and evaporation, and possibly increase the metabolic cost of the task being done, as the wearer has to move the clothing as well as himself. Normal underwear and a coverall can reduce heat exchanges by about 40 per cent, but the magnitude of the effect depends on the clothing, its thickness and permeability to air and water vapour. The clothing will also reduce heat gain from the environment by a similar amount.

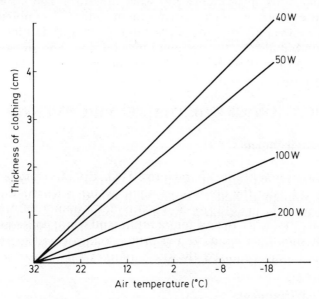

Figure 19.3 The relationship between clothing thickness required for thermal comfort and environmental temperature at four energy expenditure levels

Sensible heat makes its way across clothing fabrics by conduction through the air spaces and along the fibres. Radiation and convection normally play little part in heat transfer across fabrics, except at the surface, in fabrics of low density, or when there are steep thermal gradients across the fabric. Insensible heat transfer is by the diffusion of water molecules through the fabric. Very often this process is too slow to cope with all the sweat produced, so the excess goes into the clothing and evaporates within it or at the surface. Sweat evaporated in the clothing is not very efficient at removing heat from the body, but in hot environments by evaporating in the clothing it reduces the amount of heat getting to the body and so is not entirely wasted. The evaporation of sweat in the clothing can be aided by air movement when the temperature of the fabric tends towards the wet bulb temperature. This increases the temperature gradient across the clothing assembly which aids heat loss. This is a beneficial effect in hot situations, but is potentially lethal in cold environments. The permeability of clothing fabrics to water vapour is an important physical property, but even the most permeable of fabrics cannot handle the volumes of vapour produced by a sweating man. There are only two ways by which the sweat can leave the body; one is to soak into the clothing; the other is to evaporate into the air, which is passing through the clothing micro-environment as a result of a bellows action produced by body movements and wind pressure. The air exchange between the clothing micro-environment and the ambient air is determined not only by the permeability of the fabrics used in construction but also by the design of the clothing. The air exchange is referred to as the clothing ventilation index.

CLASSIFICATION OF THERMAL ENVIRONMENTS

Cold dry environments

Cold dry environments are characterized by a dry bulb temperature of less than $-5°C$ and the absence of liquid water; such conditions are found for example in cold stores. Factors determining the transfer of heat from the body are the air temperature and wind; consequently air movements should be minimized or protection provided against wind, in addition to the required thickness of thermal insulation.

Cold wet environments

Cold wet environments are normally above a dry bulb of $0°C$ but the presence of liquid water from rain or spray on skin or clothing surfaces makes these surface temperatures tend towards that of the wet bulb, which can be a number of degrees lower than the dry bulb; the dry bulb will therefore underestimate the severity of the cold stress in these environments. Penetration of clothing by liquid water will tend to destroy the insulation by establishing evaporative heat transfer within it, and consequently this factor has to be taken into account. The weight of water on and absorbed in clothing can also be sufficient to collapse it on to the skin, so reducing its thickness and hence its insulation value. Cold wet conditions are found in open air occupations such as farming and fishing.

Hot humid environments

Hot humid environments comprise those in which a high dry bulb temperature is coupled with high water vapour pressure, the water vapour pressure being sufficient to have an influence on the man's ability to lose heat by evaporating sweat. Air movement past the body is required to facilitate evaporation under these conditions and reduce the thermal stress. Good ventilation and air exchange are required in any enclosed or partially enclosed space as the occupants are adding water to the air all the time and so exacerbating the condition. The limiting factor in these conditions is often not the production of sufficient sweat for evaporation but the ability to evaporate it from the body surface. Any form of clothing is going to impede evaporation, and the natural tendency in these environments is to discard it.

Hot dry environments

In hot dry environments the ability to produce sufficient sweat may limit a man's heat dissipating capacity, the evaporation being quite efficient due to the large vapour pressure gradient between the skin and air. The dry bulb temperature may rise to values in excess of 40°C and be accompanied by a high radiant heat load. In the industrial context hot dry environments can be classified according to whether they consist of the following:

1. A high dry bulb temperature with little radiant heat.
2. (a) Low dry bulb with a high omnidirectional radiant heat component.
   or
   (b) Low dry bulb with a high unidirectional radiant heat component.
3. (a) Both high dry bulb and omnidirectional radiant heat.
   or
   (b) Both high dry bulb and unidirectional radiant heat.

The comfort range

The comfort range is a zone of environmental conditions in which it is possible to work over a range of metabolic rates without undue thermal strain occurring. The temperature depends on the clothing, work and the metabolic rate but the zone is characterized by the subject not having to wear excessive or special protective clothing. Temperatures within the range 16 – 24°C appear to be acceptable, with the heavier work loads at the low end and sedentary tasks for the warm end of this range. Very hard work or the use of protective clothing or equipment will lower this temperature zone, a point frequently forgotten, and it will be elevated by high air movement.

This simple classification of thermal environments is useful when considering control measures for a particular situation as it indicates the more important or critical component in the heat balance equation. For example, it is difficult to use any protective clothing in hot humid environments because it reduces even further the already limited evaporative capacity of the man. Protection against radiant heat, however, is feasible if the air is dry and at a low temperature, so facilitating convective and evaporative heat loss.

THERMAL INDICES

The thermal environment is normally referred to in subjective terms, such as hot or cold. An alternative is a detailed description in terms of the dry bulb, wet bulb and radiant temperatures, air velocity and water vapour content. Although exact, this is cumbersome and would lack meaning to all but the experienced in this field. In addition, the important factor is the effect of the environment on the thermal economy of the person exposed, and it can be seen from the heat balance equation that air temperature, radiant temperature, wind velocity and water vapour pressure can all be varied in such a way that their net effect remains the same. For example, within limits, changes in air temperature can be compensated for by changes in wind speed, so that the convective exchange remains the same. Similarly, changes in the radiant component can compensate for changes in convective exchange. To overcome these drawbacks, research workers have put a considerable amount of effort into devising methods of describing the environment in terms of a scale or index which represents the effect of the thermal environment on people. Several of these indices enable a given environment to be described by a single figure.

In their simplest form an index consists of the dominant factor such as the dry bulb temperature which is used by most people in temperate zones. However for industrial purposes more sophisticated indices are necessary, such as the effective temperature (ET) and corrected effective temperature (CET) (Yaglou, 1926; Bedford, 1946), the predicted four-hour sweat rate (P4SR) (McArdle *et al.*, 1947) the heat stress index (HSI) (McKarns and Brief, 1966). The ET scale is based on subjective assessments of the environment; observers were asked to say which of a range of climates differing in air temperature, humidity and air movement combinations felt the same as a reference one of practically still saturated air. The P4SR is based on the physiological response of sweat rate and the HSI on a heat balance or engineering approach to heat exchange between the body and environment. Heat stress indices, of which there are over 30, show an evolution in that the more primitive ones are based on subjective assessments, then come the physiologically based ones followed by those developed from a physical or engineering approach. The latest, Fangers' comfort equation, is based on the physics of heat exchange and physiological criteria for comfort. The fact that there appears to be an evolutionary series does not mean that the latest indices are the best at predicting the effects of a thermal environment on man. One reason for this, as well as for the large number of indices, is that an index is normally developed to deal with a rather limited range of environmental conditions, and when using them it is most important that the index being used was designed for the environments under study.

Indices for cold environments have not attracted the same attention as those for hot environments and the wind chill index (*Table 19.2*) appears to be the most commonly used, next to the dry bulb temperature, for assessing cold environments. The main use of the wind chill chart is to determine if there is a danger of frost bite.

Calculation of the indices

Effective temperature

The normal scale of effective temperature for clothed people and the basic scale for those stripped to the waist are show in *Figure 19.4* and *Figure 19.5*. The much more marked effect of wind speed on the effective temperature for the stripped state should be noted. As the effective temperature (ET) is expressed as a saturated still air environment equivalent, the same ETs on the basic and normal scales do not mean the same in terms of comfort. Clothing is obviously going to make the lower ETs more comfortable and the higher ones less comfortable. The procedure for working out the ET is to draw a straight line between the wet and dry bulbs (line AB) and then read off the ET at the point the appropriate wind velocity line cuts across it. The corrected effective temperature (CET) is determined in the same way by substituting the globe temperature for the dry bulb, or by the following procedure.

1. Determine the absolute humidity of the air from the dry bulb and wet bulb temperatures.
2. Determine a pseudo-wet bulb temperature from the absolute humidity and a dry bulb temperature the same as the globe thermometer reading.
3. Determine the effective temperature represented by the pseudo-wet bulb reading and the globe thermometer reading.

These methods can obviously give different values for the CET of an environment but the first method which was suggested by Bedford (1946) is the one generally used in the UK.

Table 19.2. COOLING POWER OF WIND ON EXPOSED FLESH EXPRESSED AS AN EQUIVALENT TEMPERATURE (UNDER CALM CONDITIONS)

Estimated wind speed		Actual thermometer reading (°C)											
mph	ms⁻¹	10.0	4.4	−1.1	−6.7	−12.2	−17.8	−23.3	−28.9	−34.4	−40	−55.4	−51.1
calm	calm	10.0	4.4	−1.1	−6.7	−12.2	−17.8	−23.3	−28.9	−34.4	−40	−55.4	−51.1
5	2.2	8.9	2.8	−2.8	−8.9	−14.4	−20.6	−26.1	−32.2	−37.8	−43.9	−49.4	−55.6
10	4.5	4.4	−2.2	−8.9	−15.6	−22.8	−31.1	−36.1	−43.3	−50	−56.7	−63.9	−70.6
15	6.7	2.2	−5.6	−12.8	−20.6	−27.8	−35.6	−42.8	−50.0	−57.8	−65.0	−72.8	−80.0
20	8.9	0	−7.8	−15.6	−23.3	−31.7	−39.4	−47.2	−55.0	−63.3	−71.1	−78.9	−86.7
25	11.2	−1.1	−8.9	−17.8	−26.1	−33.9	−42.2	−50.6	−58.9	−66.7	−75.5	−83.3	−91.7
30	13.4	−2.2	−10.5	−18.9	−27.8	−36.1	−44.4	−52.8	−61.7	−70.0	−78.3	−86.1	−95.6
35	15.6	−2.8	−11.7	−20.0	−29.4	−37.2	−46.1	−55.0	−63.3	−72.2	−83.3	−89.4	−98.3
40	17.9	−3.3	−12.2	−21.1	−29.4	−38.3	−47.2	−56.1	−65	−73.3	−82.2	−91.1	−100

Wind speeds greater than 40 mph (17.9 m/s) have little additional effect).

Little Danger (for properly clothed person) Maximum danger of false sense of security

Increasing danger Danger from freezing of exposed flesh.

Great danger

Trenchfoot and immersion foot may occur at any point on this chart

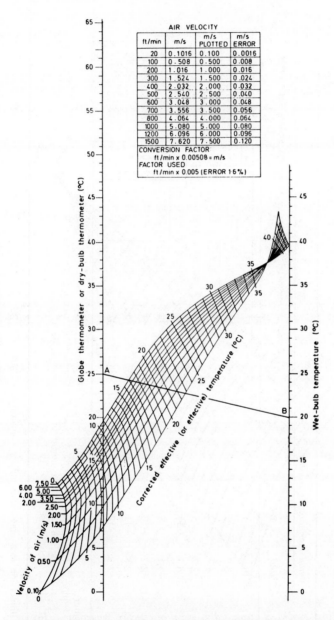

Figure 19.4 Effective temperature chart normal scale (indoor clothing). (Reproduced from Ellis, Smith and Walters, 1972, by courtesy of the Editor of British Journal of Industrial Medicine)

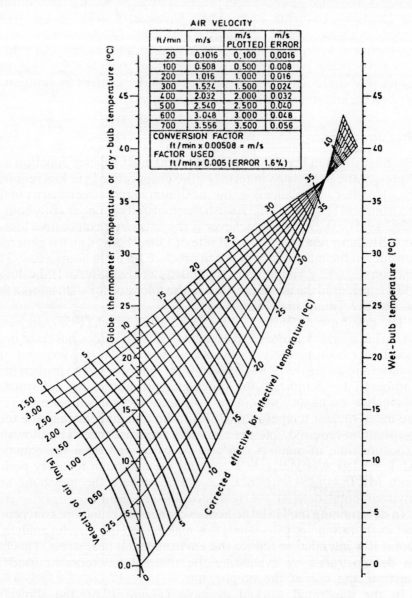

AIR VELOCITY

ft/min	m/s	m/s PLOTTED	m/s ERROR
20	0.1016	0.100	0.0016
100	0.508	0.500	0.008
200	1.016	1.000	0.016
300	1.524	1.500	0.024
400	2.032	2.000	0.032
500	2.540	2.500	0.040
600	3.048	3.000	0.048
700	3.556	3.500	0.056

CONVERSION FACTOR
ft/min x 0.00508 = m/s
FACTOR USED
ft/min x 0.005 (ERROR 1.6%)

Figure 19.5 Basic scale (stripped to waist) of effective temperature. (Reproduced from Ellis, Smith and Walters, 1972, by courtesy of the Editor of British Journal of Industrial Medicine)

If it is necessary to change the environment because it is too hot or too cold, the most effective and practical way can be determined using the charts by changing each of the parameters, wind speed, globe temperature, dry bulb and wet bulb temperature in turn or in combination. Air movement is normally the easiest and least expensive parameter to modify in practice, radiation by shielding comes next and then air temperature and water content by ventilation or air conditioning.

Heat stress index

The heat stress index (HSI) is based on the heat balance equation and expresses the heat stress in terms of the evaporative heat loss required (E_{req}) for thermal balance over the maximum evaporative capacity of the environment (E_{max}) or 2400 British thermal units per hour (Btu/hour = 0.252 kcal or 0.293 W), whichever is the least. An evaporative loss of 2400 Btu/hour represents a sweat rate of 1 litre/hour, which is generally considered the maximum a man can produce for eight hours a day. The fraction (E_{req}/E_{max}) is multiplied by 100 to give the HSI. An HSI value of 100 is considered the maximum that can be tolerated for eight hours a day by healthy young male adults.

The development of the HSI by Belding and Hatch (1955) was based on earlier work by Haines and Hatch (1952). The index has since been modified, most recently by McKarns and Brief in 1966 and the nomograms in *Figure 19.6* are taken from their work. The broken lines numbered 1 – 8 indicate the order and procedures for determining convective exchange, maximum evaporation, wall temperature, i.e. the mean radiant temperature (MRT), the radiant heat gain, the total evaporation required, and the allowable exposure time. The allowable exposure time in minutes can also be calculated from the equation AET = $(250 \times 60)/(E_{req} - E_{max})$ and the minimum recovery period from MRT = $(250 \times 60)/ E_{max} - E_{req})$ where the metabolic and environmental data at the rest location are used.

In determining the HSI, the heat exchanges by radiation, convection and evaporation are calculated; this makes the HSI a useful analytical tool if it is intended to reduce the environmental heat stress. This can be demonstrated by evaluating the thermal environment used to illustrate the use of the nomogram.

In the illustrated worked example (*Figure 19.6*), the allowable exposure time is six minutes with a minimum recovery time of six minutes, a situation which should be improved. By setting the components of the HSI out as shown in *Table 19.3* along with the parameters it is proposed to alter, it is possible to investigate the effect on the HSI of changes in any one of them. (The wind speed is an obvious choice in view of its low value.)

474

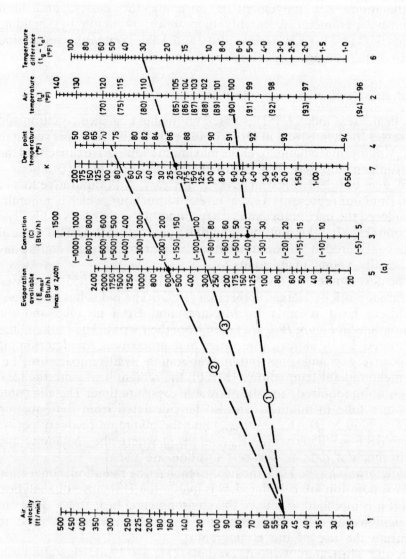

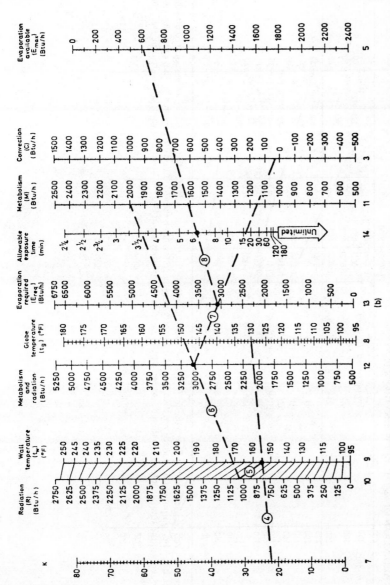

Figure 19.6(a) and (b) Nomograms for calculating the convective and radiant components of the heat balance equation, the mean radiant temperature t_w, E_{max}, E_{req} and the allowable exposure time. The procedure is indicated by the worked example. Note that the convection column has a positive and negative side according to whether the air temperature is above or below 95°F. (Reproduced from McKarns and Brief, 1966, by coutesy of the Editor of Heating, Piping and Air Conditioning)

Table 19.3 EVALUATION OF A THERMAL ENVIRONMENT

Factors determining the thermal load		Wind speed (ft/min)		Shielding only	Shielding V = 200 ft/min	Work load reduced only	All three changes combined
		50	200				
$T_a = 100°F$	C =	+40	+90	+40	+90	+40	+90
Dewpoint 73°F	E_{max} =	620	1375	620	1375	620	1375
$t_w = 155°F$	R =	+1050	+1050	+700	+700	+1050	+700
	Metabolism =	+2000	+2000	+2000	+2000	+1500	+1500
	E required =	3090	3140	2740	2790	2590	2290
	AET min =	6	8.5	7	10.5	8	16
	MRT min =	6					
	HSI =	490	228	440	203	416	166

A thermal environment can be evaluated by listing the factors which are determining the thermal load including the evaporation required and the maximum evaporation available. The component heat loads are then listed for the original environment and for modified environments. The significance of the individual components in the total heat load can then be assessed by inspection. Heat loads are in Btu per hour; allowable exposure time AET; minimum recovery time MRT.

The radiant load may be changed by suitably placed shielding; the work load by introducing some mechanization or altered methods of work. Because the dry-bulb temperature of 38°C (100°F) is close to that of the skin, 35°C (95°F), it would be expected that large increases in wind velocity could be introduced without convective heat gain (C) adding too greatly to the heat load, in fact between 50 and 500 ft/minute, C increases by only 110 Btu/hour while at the same time E_{max} is increased to 2400 Btu/ hour. If the air temperature had been 49°C (120°F) this change in air velocity would have produced an increase of 600 Btu/hour in convective heat gain, substantially reducing the benefit to be derived from the increase in E_{max}.

Considering the allowable exposure time (AET), it can be seen that reducing the components of the heat load by the values suggested in this table will only extend the AET by a few minutes unless all are combined, in which case an AET of 16 minutes can be obtained and the E_{req} falls below the 2400 Btu/hour level. Once the evaporation required has been reduced to 2400 Btu/hour or less it only remains to achieve an E_{max} of this value to enable unlimited exposure to be permitted and this, as already shown, can be done by increasing the wind speed to 500 ft/minute.

Table 19.4. AN EVALUATION OF AN ENVIRONMENT TO DETERMINE IF PROTECTIVE CLOTHING CAN BE USED

	Given example see Figure 19.6	*80% reflection*	*50% reflection*	*50% reflection*	*50% reflection and V = 200 ft/min*
C(Btu/h) =	40	40	40	40	90
E_{max}(Btu/h) =	620	–	–	620	1375
Radiation (R; Btu/h) =	1050	250	525	525	525
Metabolism (M; Btu/h) =	2000	2000	2000	2000	2000
E_{req}(Btu/h) =	3090	2290	2565	1945	2625
AET (min) =	6	6.5	6	7.5	12

From an inspection of the figures it would appear that the maintenance and improvement of evaporative heat loss is essential for its successful application. A reduction in the incident radiant load of about 30 per cent due to the protection of normal clothing is included in the original radiant heat load, hence the true radiant load is about 1500 Btu/h.

In the situation illustrated by this example there is a substantial radiant heat component and this raises the question of heat reflecting clothing. *Table 19.4* sets out the situation for protective clothing which reflects either 80 per cent of the radiant heat, as might be the case with a new garment, or 50 per cent with an old garment, but which prevents evaporative heat loss; it also sets out the data for a garment which reflects 50 per cent of the radiant heat load without reducing the evaporative heat loss and is also able to take advantage of any increase in air movement for facilitating evaporation. The allowable exposure times (AET) show that protective clothing is of no real value unless it does not interfere with evaporation, and then an improvement in the AET is only obtained if the air movement is increased at the same time to 200 ft/minute.

If the radiation is unidirectional it may be quite feasible to reduce the thermal load substantially by clothing which only covers the exposed parts of the body, or by screening; but even if the radiant heat load on the man is reduced to zero in this environment, the remaining evaporative requirement of 2400 Btu/hour is still in excess of the maximum evaporative loss of 620 Btu/hour available to the man. An increase in wind speed to 500 ft/minute is therefore required if an E_{max} of 2400 Btu/hour is to be achieved and so allow the HSI to be reduced to 100 or less.

In designing the nomogram, the radiant heat load component on the man has been reduced by about 30 per cent to make allowance for the effect of clothing on radiant heat pick up (Hertig and Belding, 1963). The reflection or reduction in radiant heat load used in this example therefore applies to the radiant load acting on the man as given by the nomogram and not the total incident radiant load. The 80 per cent and 50 per cent reflection of this heat load would require surfaces which reflected about 84 per cent and 66 per cent of the total incident radiation.

From this study of the environment the air velocity seems to be the parameter which holds the key to alleviating the thermal stress. However, a work rate of 2000 Btu/hour (500 kcal; 580 W) is well above the level that can be maintained for an eight hour shift, and this should attract the investigators' attention and a reduction to 1500 or even 1250 Btu/hour should be the aim.

The wet bulb globe temperature index

The wet bulb globe temperature index (WBGT) is calculated from the formula: WBGT = 0.2tg (globe) + 0.1t°db (dry bulb) + 0.7 t°nwb (natural wet bulb); it was originally developed by Yaglou and Minard (1957) for measuring outdoor environments with a solar radiation component. The simple weightings of just three measurements have enabled instruments to be produced which give a direct reading in the WBGT index and this simplifies environmental measurements considerably. The scale is therefore becoming widely used. A modified version of the formula is used for measuring indoor environments: 0.7 nwb + 0.3 tg.

The unventilated or natural wet bulb and an unscreened or non-silvered dry bulb thermometer are used when determining the WBGT index. These readings are therefore of no value and in fact must not be used for determining other indices or the water content of the air; particularly where radiant heat is present, the natural wet bulb and dry bulb thermometer can be heated up by many degrees leading to ridiculous HSI and ET values.

Fanger's comfort equation

Fanger's comfort equation (Fanger, 1970) is based on the heat balance equation for man, but also takes account of the influence of clothing on heat transfer and respiratory heat loss. The heat balance is then worked out for a state of balance when certain physiological criteria defining comfort are met; these are skin temperature and sweat rate. Both are defined in relation to metabolic rate. The harder a person works the lower the skin temperature and the higher the sweat rate for comfort. In addition to identifying comfort conditions in terms of all six factors (i.e. dry bulb, wet bulb, wind velocity, radiant temperature, work rate and clothing level), Fanger has developed his approach to define the performance of a given environment in terms of a 'predicted mean comfort vote' and predicted percentage dissatisfied. The comfort equation is a major development which makes it possible to define comfort conditions for a wide range of work rates and clothing levels. The reader is referred to Fanger (1970) for details of a new and useful approach to the thermal environment.

Index for cold environments

The only index relating to cold environments appears to be the one by Siple and Passel (1945), who introduced the wind chill scale as a result of some studies in Antartica. The index is based on the time it takes 250 mℓ of water to freeze; this is influenced by air temperature and wind speed in an environment that is far from constant and with metabolic heat production that can vary from 50 to 500 W/m^2. Thermal stress can therefore be environmental or metabolic in origin.

In temperate zones, the main problem is one of limiting heat loss to that produced by metabolism. The method normally employed is to adjust clothing to the required level (behavioural adaptation), but this can be backed up by restriction of the blood supply to the skin and extremities, referred to as the shell, so keeping the heat in the core. Although the latter is the natural mechanism it does lead to discomfort, loss of manual dexterity and grip strength. For a sedentary person these changes can occur at temperatures as high as 15°C and are very noticeable to people doing fine work demanding a high degree of precision and dexterity. Once a person is cold, it takes a long time for the vasoconstriction to ease unless deliberate action is taken to warm the hands and arms. The effectiveness of increasing the thickness of clothing to achieve comfort will depend on the activity level (*Figure 19.3*). When doing sedentary work in low temperatures it is quite possible to feel comfortable but for the hands to be sufficiently cold to interfere with typing and similar activities.

The cutaneous vasoconstrictor response to cold does not take place over the head; consequently, unless protected, large quantities of heat leave the body through the head. Protection of the head is therefore vitally important in cold working conditions if body heat is to be conserved.

The discomfort associated with excess body heat loss is normally sufficient to ensure that workers take the necessary precautions, but instruction is required if extreme environments are going to be entered. The workers should also be made aware of the behavioural changes associated with hypothermia, so that they can recognize them in their fellow workers, and the need for first-aid measures.

Sweat loss

Environmental heat stress arises when it becomes difficult for the metabolic heat to move from the body to the environment. The response of the body is to pump more blood through the skin so increasing skin temperature and hence the gradient between skin and environment. This is quickly followed by the production of sweat and

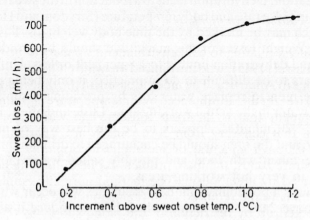

Figure 19.7 The sweat loss is plotted against the body temperature increment above sweat onset. Saturation of the sweating mechanism occurs at just over 1°C above the starting temperature which is normally about 37°C

recourse to evaporative heat loss. The stimulus for sweating is an elevation in deep body temperature, a proportional control system appearing to be involved which balances the heat load with the required evaporative heat loss. There is obviously a limit to the amount of sweat that can be produced and the limit is reached when the deep body temperature has risen by about 1.2°C (*Figure 19.7*). The sweat glands are then working maximally and no more sweat can be produced. The World Health Organization has used this aspect of the thermoregulating system to set a physiological limit for work in the heat of 38°C deep body temperature. Deep body temperature is not always easy to measure accurately, and a heart rate of 110 bpm may be used. These standards are for prolonged periods of work in the heat and it is possible to exceed them without harm for short periods. A realistic upper limit for deep body temperature is 39.2°C (Iampietro and Goldman, 1965).

ACCLIMATIZATION

It is possible to minimize the adverse effects of hot environments by allowing the workers time to acclimatize, a process of physiological adaption characterized by an increased sweat output and a lowering of the pulse rate and deep body temperature in response to the thermal stress. Acclimatization to a particular environment develops quickly, being almost completed in ten days, but it is also lost equally quickly, two to three days without heat exposure leading to a marked loss of tolerance. Any layoff due to holidays or illness should therefore be followed by a period during which the man is allowed to reacclimatize, and on moving to an even hotter job a period must be allowed for further acclimatization. Dehydration leads to a reduction in the sweat rate and an elevation in pulse rate and body temperature (Strydom and Holdsworth, 1968) which may be marked by the time body weight loss has reached 3 per cent. Weight losses of this magnitude should be avoided if at all possible and dehydration limited to 1.8 per cent or less. Unfortunately this appears to be difficult to do in practice, it only being possible to encourage men to replace water loss by providing a supply of cool palatable water in the actual working area. Little and often, 100–200 ml every 15 – 20 minutes, appears to be the best way of maintaining hydration, and the men should be encouraged to drink in this way. Salt should be taken with food and possibly salted water (1 g/ℓ) made available in very hot working areas.

It is always a possibility that heat casualties will occur when the wet bulb is above 24°C. The factors which help to induce it are clothing (particularly if it is impermeable), high work loads, rigid routine or discipline which prevent behavioural adaptation such as a voluntary reduction in work load, dehydration and lack of fitness. There is also some evidence that other environmental factors like carbon monoxide, alcohol and drugs may reduce thermal tolerance.

One important social pressure which is very often overlooked, but nevertheless results in heat casualties, is that produced by the working group or team. People like to pull their weight and will often try to continue doing so when they know they should withdraw from the heat. Workers are most vulnerable on return from holidays or sick leave, when they have lost some or all of their acclimatization. The working group should be aware of the hazards associated with a loss of acclimatization. Above all, men should be left in no doubt that they are their own best thermometer which allows for infections, dehydration, etc. The inbuilt thermometer also allows for the greater proportional load that a small man has to deal with compared with a larger man if they are doing the same work in absolute terms.

EVALUATION

With six factors determining the exchange of heat between man and his environment it is not easy to lay down environmental standards. Fanger has done it for comfort but it took him over 200 pages. Some generalization is therefore necessary.

Working limits

The preferred thermal environment for work is generally referred to as 'comfortable', which expresses satisfaction with the thermal environment, but, as clothing and posture influence the body's heat exchanges, and the work rate the quantity of heat that has to be lost, there is a wide range of environmental conditions which can be so described. For the sitting person dressed in light clothing (0.5 clo) an air temperature of about 25° with a relative humidity between 40 and 60 per cent and an air movement of 0.15 m/second will feel comfortable. If heavier clothing is worn (1 clo: e.g. a full suit with vest and shirt), the air temperature will have to be decreased by 2 to 3°C in order to achieve comfort, and to 18°C if the activity level is increased to about 4.5 kcal/minute.

The commonest cause of complaint with environments that the occupants consider should be comfortable is inadequate air movement. The 0.15 m/second air movement used as the standard in the UK appears to be too low, as the air movements generated by body heat is of the same order of magnitude. A more alive and less stuffy environment is produced by air movements in the range above 0.15 – 0.25 m/second. However, if the air speed is increased, the dry bulb temperature has to rise to compensate for the increased convective cooling.

Although it is customary for most sedentary tasks to be done under comfortable conditions, the same does not apply to manual tasks, and the question arises as to what the environmental limits are for physical work undertaken for eight hours a day five days a week. The World Health Organization (1969) recommends that the following environmental limits be applied:

Sedentary work	(2.6 kcal/kg/h)	30°C	CET
Light work	(4.3 kcal/kg/h)	28°C	CET
Heavy work	(6.0 kcal/kg/h)	26.5°C	CET

These work loads correspond to 3, 5 and 7 kcal/minute (200, 350 and 490 watts) for a 70 kg man. Acclimatization to the hot conditions may increase these values by 2°C CET (corrected effective temperature).

Environments with these CET values are uncomfortable, but they are physiologically safe in that men can work for many hours in them without showing any ill effects. However, CET values well in excess of 30°C are quite common in industry and in these environments the working time is determined to a large extent by the body's ability to store heat. The quantity of heat stored is limited by the deep body temperature and by the pulse rate reaching values which are subjectively unacceptable.

The usual limiting value of deep body temperature for day-to-day exposure is about 38°C and exposure times and rest allowances are calculated on this basis. The value of 38°C as the upper limit is indicated by many investigations including Lind (1963), Löfstedt (1966) and Wenzel (1968). The basic physiological reason for the limit is probably that the main avenue of heat loss, sweating, becomes saturated at about this value.

One drawback to the use of thermal indices as predictors of thermal stress or strain is that they work best for the conditions under which they were devised, and there may be considerable errors if they are used for assessing conditions which differ markedly in character from the original ones. Some care is therefore required when using indices, and some discrepancies between observations and calculations are almost certainly due to differences in clothing and variations in work rate and actual heat exposure experienced as the exposee moves about the working area.

In a review of thermal limits for industrial workers Bell and Watts (1971) have suggested that for sedentary workers the lower comfort limit for 80 per cent of workers in winter is 15.5°C CET and the upper limit in summer 21.8°C CET. For skilled work the upper limit beyond

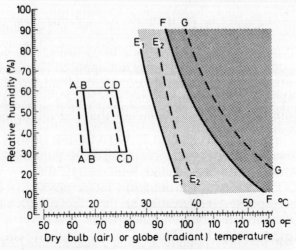

Figure 19.8 Chart showing limits of environmental severity for sedentary workers, winter comfort (A-A to C-C and summer comfort B-B to D-D) for 80 per cent of the population, recommended range of air speeds 0.05 to 0.15 m/s. The working efficiency limit of 16.7°C CET is indicated for a wind speed of 0.5 m/s by the line E_1-E_1 and for 5.1 m/s by E_2-E_2. The line F-F represents the limit at which about 5 per cent of fit young men working at 150–300 kcal/h may suffer adverse physiological effects. For 8 hour shifts at 400 kcal/h the line E_1-E_1 represents the limit of physiological safety and for short exposures at 100 kcal/h the line G-G (Reproduced from Bell and Watts (1971) by courtesy of the Editor of British Journal of Industrial Medicine)

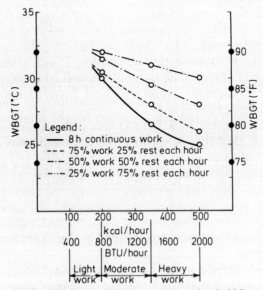

Figure 19.9 Permissible heat exposure: threshold limit values. (Based on Assessment of Heat Stress and Strains, 1971, Industrial Health Foundation Inc., 5231 Centre Avenue, Pittsburgh, Pennsylvania 15232, also in ACGIH (1979) p. 66)

which efficiency would be expected to deteriorate significantly is suggested as 26.7°C CET. Details of the upper limits of the physiologically safe environments are also given. The authors have summarized their work in a chart which is reproduced in *Figure 19.8*.

In recent years the WBGT index has become widely used for assessing heat stress although the Botsball thermometer is considered by some to have advantages. A TLV has been developed, based on the WBGT index and physiological criteria, which is easy to use (*Figure 19.9*). Full details are available in the ACGIH table of TLVs (ACGIH, 1979); There is also a *Criteria Document on Occupational Exposure* published by the US Department of Health Safety and Welfare (1972).

The WBGT index is not a precision index predicting exactly the physiological responses; it overestimates the effect of high humidity and underestimates the effect of high wind speeds. It is nevertheless a very useful working tool which enables 'in works' thermal standards to be developed.

The real advantage of the thermal indices lies not in having a figure to tell workers if they are too hot or not but in determining the optimum method of control and for planning in advance the courses of action to be taken as temperatures rise, as for example during a heat wave. The index itself is a figure which it is believed comes closer to representing the thermal impact of an environment on the workers than the obligatory thermometer. Provided the index is measured in the right place(s) it is reasonable to use it for determining when action such as rest pauses should be taken.

CONTROL

Control consists of either reducing the heat load on the operative, increasing the ease with which heat can be lost, or both. Control is achieved when the heat balance equation for the operative is balanced at a selected and acceptable point.

When considering control measures it is useful to think in terms of the heat source; the operative's work procedures and position in relation to the source; the heat exchanges by conduction, convection, radiation and evaporation; activity level; rest pauses; and the use of clothing.

The source

If the source is spatially definable in the sense of being a furnace or some other heat emitting body it should be examined with a view to reducing heat loss into the working environment. Screening of radiation and exhausting hot air should be considered as well as insulation and aluminium foil coverings. If water vapour is emitted in sufficient quantites to elevate the wet bulb of the working environment, control at source should be considered.

Work procedures, heat exchange and rest pauses

The operative's work procedures should be examined with a view to minimizing thermal exposure by facilitating work at a distance and reducing the time of exposure. The redesign of tools may make it possible to work from behind screens, or at a greater distance. If observations have to be made, closed circuit television may be an effective way of moving the worker away from the heat source. In some situations, heat can be conducted to the body from hot surfaces, particularly when laying on or against a surface. When this happens, not only is heat conducted through clothing to the body but convective and evaporative loss from the involved area is blocked so adding to the stress. Radiation travels in straight lines but is absorbed by surfaces and reradiated. The control of radiation by screens and keeping surfaces cool by providing adequate ventilation is therefore an important control measure. Screens must not reduce air velocity in the working area. Screens do not have to be solid; chains, wire mesh and water sprays all block a percentage of the radiation. Convection can produce a positive or negative heat load depending on whether the air temperature is above or below that of the skin or clothing. Because air movement also plays a critical role in determining evaporative heat loss, when air temperatures are above skin temperature, convective exchange and evaporative loss may have to be considered together. However, provided the dry bulb temperature is not too far above that of the skin, the increase in evaporative loss gained by increasing wind speed is greatly in excess of the convective gain. When air temperatures are below skin or clothing temperature, increases in wind speed are beneficial. In some situations (e.g. inside a furnace or oven) fans can be used to blow cold air into the working area, creating an area of low dry bulb temperature in an otherwise hot environment. The air, if moving vigorously, may penetrate the clothing and facilitate evaporation within the clothing micro-environment. The use of fans should never be overlooked for even mildly stressful environments.

The most widespread source of wild heat in industry is the body itself and the heat and water vapour given off rapidly make any enclosed or poorly ventilated area uncomfortably hot and even potentially dangerous. Heat stress can of course be reduced by lowering the work rate and by introducing rest pauses in cool areas. The TLV for thermal stress provides recommendations on appropriate rest pauses.

Clothing

Clothing is normally used for reducing heat loss, but in industry garments may have a protective function and cannot be removed when the wearer becomes hot. It is therefore most important that such clothing be loose fitting and allow air to circulate under it. There are some situations where such clothing can be dangerous and where ready access of hot air to the clothing micro-environment must be prevented. Clothing is also used to reduce heat flow into the wearer in extreme conditions. Used in this way the clothing will normally only extend the tolerance time but if the dry bulb is low and there is a vigorous air movement past the body (i.e. the heat load is from radiation and/or metabolism) convective and evaporative losses may be sufficiently high for clothing to turn an intolerable environment into a tolerable one.

PROTECTIVE CLOTHING

Protective clothing against heat can be divided into two groups: conditioned and unconditioned. When *conditioned* garments are used the wearer is independent of the environment, whereas *unconditioned* garments are normally only intended to extend the tolerance time of the wearer to the environment.

Unconditioned garments

Some of the principles underlying the design of heat protective clothing are as follows:

1. Where environments permit, protection should be provided to the minimum surface area of the body in order to maintain unimpeded heat exchange, particularly by evaporation.

2. Radiant heat can be reflected back by shiny metallic surfaces such as aluminium.
3. High resistance to heat flow can be obtained by the use of thick layers of insulating material.
4. The thermal capacity of fabrics can be utilized to provide a heat sink between the wearer and the environment.
5. The insulation layers should have a high resistance to compression and good compressional resilience.
6. Permeable materials with good wicking properties will facilitate the absorption and evaporation of sweat.
7. Vapour barriers can control the movement of water and water vapour when such control is required.
8. Materials that do not burn but degenerate into a solid layer, thus retaining some mechanical integrity, will provide a physical barrier between the skin and heat source after they have been destroyed.

The actual choice of design and materials for a heat protective garment is determined by the features of the environment, the value of the thermal load and the type of work the wearer will be doing. If the heat load is mainly radiant and unidirectional, only the exposed areas of the body should be protected – preferably with heat reflecting clothing. If the radiant load is omnidirectional, then total body protection may be required, allowing circulation of air under the garment to facilitate evaporative heat loss. If the air temperature is high then thick insulating clothing is necessary with aluminized surfaces to reflect any radiant component of the heat load.

Conditioned garments

The commonest form is air conditioning, probably followed by electricity for heating and water for heating or cooling. Conditioned garments are used not only for heat but for protection against toxic dusts and gases. Irrespective of its use, as soon as a barrier is placed round a man his thermal balance has to be considered.

A conditioned garment assembly consists of the garment itself, a distribution system, a circulating pump or a source of power to drive the conditioning medium round the garment, a cooling or heating unit and in many cases a connecting hose or wire to an external source of power, air or water.

The absence or presence of a connecting link between the suit and an external source of power is the feature which divides conditioned assemblies into two main groups: self-contained and tethered.

Figure 19.10 A ventilated radiant heat reflecting suit. The use of an aluminized suit may extend the tolerance time of the wearer, but to achieve environmental independence the cooling capacity of the air supply to the suit must balance the environmental and metabolic heat loads. The environmental heat load is also acting on the air hose. (Reproduced by courtesy of the Editor of Occupational Health)

Conditioned heat protective assemblies (*Figure 19.10*), including the air or water hose in tethered assemblies, should be designed to cope with a specified heat load.

CONCLUSION

Methods of controlling the thermal environment are reviewed. It is possible, using the thermal stress indices, to set an environmental quality target and then calculate the best course of action to achieve it by modifying the different components of the heat balance equation. In practice, air movement is usually the most important factor.

REFERENCES

ACGIH: American Conference of Governmental Industrial Hygienists (1979) *Threshold Limit Values for Chemical Substances and Physical Agents in the Workroom Environment.* ACGIH, PO Box 1937, Cincinnati, Ohio, USA

Bedford, T. (1946) *Environmental Warmth and Its Measurement.* Medical Research Council War Memo No. 17. London: HM Stationery Office

Belding, H.S. and Hatch, T.F. (1955) 'Index for evaluating heat stress in terms of resulting physiological strains.' *Heating, Piping and Air Conditioning*, **27**, 129 – 136

Bell, C.R. and Watts, A.J. (1971) 'Thermal limits for industrial workers.' *British Journal of Industrial Medicine*, **28**, 259 – 264

Durnin, J.V.G.A. and Passmore, R. (1967) *Energy, Work and Leisure*. London: Heinemann Educational Books

Ellis, F.P., Smith, F.E. and Walters, J.D. (1972) 'Measurement of environmental warmth in SI units.' *British Journal of Industrial Medicine*, **29**, 361 – 377

Fanger, P.O. (1970) *Thermal Comfort*. Copenhagen: Danish Technical Press; reprinted New York: McGraw-Hill, 1973

Haines, G.F. and Hatch, T.F. (1952) 'Industrial heat exposures – evaluation and control.' *Heating and Ventilating*, **49**, 93 – 104

Hertig, B.A. and Belding, H.S. (1963) 'Evaluation and control of heat hazards.' In *Temperature – Its Measurement and Control in Science and Industry*, Vol. 3, Pt 3, ed. J.D. Hardy. New York: Reinhold

Iampetro, P.F. and Goldman, R.F. (1965) 'Tolerance of man working in hot, humid environments.' *Journal of Applied Physiology*, **20**, 73 – 76

Kerslake, D.McK. (1972) *The Stress of Hot Environments*. Cambridge University Press

Lange Andersen, K., Shephard, R.J., Denolin, H., Varnouskas, E. and Maseroni, R. (1971) *Fundamentals of Exercise Testing*. Geneva: World Health Organization

Leithead, C.S. and Lind, A.R. (1964) *Heat Stress and Heat Disorders*. London: Cassell

Lind, A.R. (1963) 'A physiological criterion for setting thermal environmental limits for everyday work.' *Journal of Applied Physiology*, **18**, 51 – 56

Löfstedt, B.E. (1966) 'Human heat tolerance.' Dissertation. University of Lund

McArdle, B., Dunham, W., Holling, H.E., Ladell, W.S.S., Scott, J.W., Thomson, M.L. and Weiner, J.S. (1947) 'The prediction of the physiological effects of warm and hot environments: the P4SR index.' *Medical Research Council (London) Royal Naval Personnel Report* 47/391

McKarns, J.S. and Brief, R.S. (1966) 'Nomographs giving refined estimate of heat stress index.' *Heating, Piping and Air Conditioning*, **38**, 113–121

Siple, P. and Passel, G.E. (1945) 'Measurement of dry atmospheric cooling in subfreezing temperatures.' *Proceedings of the American Philosophical Society*, **89**, 177–199

Snellen, J.W., Mitchell, D. and Wyndham, C.H. (1970) 'Heat evaporation of sweat.' *Journal of Applied Physiology*, **29**, 40–44

Strydom, N.B. and Holdsworth, L.D. (1968) 'The effects of different levels of water deficit on physiological responses during heat stress.' *Internationale Zeitschrift angewiss Physiologie einschluss Arbeitsphysiologie*, **26**, 95–102

US Department of Health, Safety and Welfare (1972) *Criteria Document on Occupational Exposure to Hot Environments*. Health Services and Mental Health Administration, National Institute of Occupational Safety and Health

Wenzel, H.G. (1968) 'Pulse rate and thermal balance of man during and after work in heat as criteria of heat stress.' *Bulletin of the World Health Organization*, **38**, 657–664

World Health Organization (1969) *Health Factors Involved in Working Under Conditions of Heat Stress*. Technical Report Series No. 412. Geneva: WHO

Yaglou, P. (1926) 'The thermal index of atmospheric conditions and its application to sedentary and to industrial life.' *Journal of Industrial Hygiene*, **8**, 5–19

Yaglou, P. and Minard, D. (1957) 'Control of heat casualties at military training centers.' *American Medical Association Archives of Industrial Health*, **16**, 302–316

20

Ergonomics

E.N. Corlett

The study of people engaged in industrial work has a relatively long history. Work study pioneers sought increased productivity by a mechanistic analysis of movement, using time as a criterion and introducing judged allowances to match their analyses to the real world. During and after the First World War, industrial psychologists, particularly in Great Britain and the United States, examined the effects of working hours and overtime on output, and related working temperature to accidents. In Britain a series of wide-ranging studies were published by His Majesty's Stationery Office, under the aegis of the Industrial Fatigue Research Board and later the Industrial Health Research Board.

There had been considerable scientific activity in areas of work physiology and anthropometry in the years preceding the Second World War, particularly in Britain, Germany, Belgium and Sweden. During that war, as a result of military requirements, research workers dealing with problems of military performance recognized the inherent possibilities of using fundamental principles drawn from a broad basis of the human sciences to attack practical problems. As the editorial to the first issue of *Ergonomics* (1957) said; 'The essential unity between functional anatomy, physiology and experimental psychology became very clear, and the need was emphasized for close and sustained co-operation between workers in all three disciplines, and between these and physical scientists and mathematicians.'

The term 'ergonomics' was coined in the late 1940s to emphasize that this new subject was a unified study of 'the laws of work'. Unknown to those who devised the term, a similar word, 'ergonomia', had been proposed in Poland nearly 100 years earlier. Although the term and its concept were recognized in Russia at the time, this initiative was lost until its rebirth in the United Kingdom in the late 1940s.

The thrust of the applications of these associated sciences in the form of ergonomics is to study and design work situations from the standpoint of the person involved. This position requires that the range of the available population and the full range of the anatomical, physiological and psychological variabilities in the population are taken into account. The objectives of this approach to the study of work are that disease or ill effects are prevented at the design stage and the work is arranged to provide the minimum of obstructions to the doer of it. The work needs to be optimally matched to the requirements and characteristics of the worker to achieve a high quality of performance.

Two simple examples illustrate this fact. In a study of industrial forklift trucks, widely used in industry for the transportation of materials from one area to another, it was observed that the functions of accelerate, brake and reverse were achieved by pedals with layouts that varied according to the make of the truck. Most of them were arranged in a different sequence from the internationally accepted arrangement for pedals in the motor car. The result was frequent errors in driving by those who changed trucks during the working day and an increased risk of incorrect pedal movements if emergency actions were taken; an already hazardous industrial device was made more hazardous.

The second example relates to information-handling by inspectors, in this case of television screens. A complex product is prone to many faults after manufacture, each of which may arise independently of the others. Within the limited time available in a production cycle, the scanning of a product for all possible faults and the maintenance of effective levels of vigilance for the whole of a working period, cannot be done successfully. It was clearly not being done during the inspection of television screens, with considerable loss to the manufacturer. Changes were introduced. The major change was to have sample inspection earlier in the process so that the particular mix of faults could be estimated. This estimate was passed to the inspectors, who then concentrated primarily on these faults. Their information load was, therefore, better matched to human capacities. Feedback knowledge of their success enabled them to assess and control their performance. Quality to the customer improved to such an extent that a loss-making line became profitable.

Performance variables have long been important in psychology and they are also valuable to ergonomists. Time and errors are the classic research indicators of performance change. The complex issue of motivation, however, is a variable influence, the recognition of which, over the last twenty years, has brought increased attention to the work of industrial sociologists and others who examine the influences of work content and organizational relationships on performance and on attitudes to work. A primary objective of ergonomics is the creation of jobs which are possible and appropriate for people. Because it helps the workers concerned, this in itself increases the level of motivation.

If the job has no aspects to which people can become attached, then almost by definition they will become alienated and remove themselves, physically or psychologically, from the workplace. Only now are the long-term effects of such imposed work stresses being recognized (Cooper and Payne, 1978).

THE PHYSIOLOGICAL COSTS OF WORK

Many people believe that industrial work is no longer a physically arduous experience, the increase in automation and power-assisted tools having taken over from human muscle. Most occupational health practitioners are aware that this is an over-optimistic view, applicable to a relatively small sector of industry. Foundries, forges, many machine and assembly shops, most of the building industry, road transport and breweries are only a few of the industries where large numbers of people live by their muscular output.

If physical effort is part of the job it is possible to use indirect calorimetry, or measurements of the heart rates of workers previously assessed for their work capacity, to estimate the work load and to propose suitable rest pauses. Such methods allow the effects of environmental heat to be taken into account as part of the work load. Simplified methods exist for these measurements, one of the better known being due to the physiologist Brouha (1967). He devised a technique of estimating the slope of the heart rate recovery curve for the first three successive minutes after stopping the task, and using indices from these measures to estimate the work load and adequacy of the rest times taken by the worker.

The graphs of *Figures 20.1–20.3* illustrate the relationships utilized by Brouha. The first graph (*Figure 20.1*) shows the total recovery cost (in total heart beats from cessation of work until recovery) incurred for various levels of work load, demonstrating the change in recovery cost with increasing work loads. This graph has a high variability, but the second graph (*Figure 20.2*) demonstrates that when the variability due to subjects is controlled, by plotting the total cardiac cost against the recovery cost for each individual, a high correlation clearly exists. *Figure 20.3* presents the three values used by Brouha to estimate the work load, from which it will be noted that an apparently modest increase in heart rate *during the recovery phase* represents a heavy load on the individual.

More recently the availability of small heart-rate recorders using long-running tape cassettes makes it more possible to gather heart rates continuously over many days. These, when compared with event recorders which give evidence of the activities carried out over the same period, identify those tasks which make major contributions to an excessive work load and enable them to be dealt with.

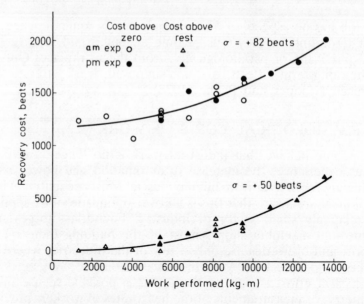

Figure 20.1 Influence of increasing work load on cardiac cost of recovery in a constant environment
Source: Brouha (1967)

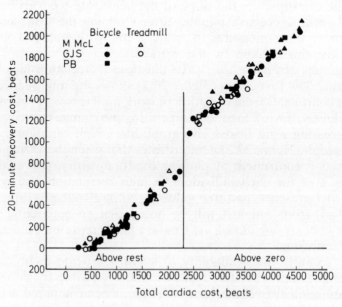

Figure 20.2 Linear relation between total cardiac cost and recovery cardiac cost. Left: above resting rate. Right: above zero
Source: Brouha (1967)

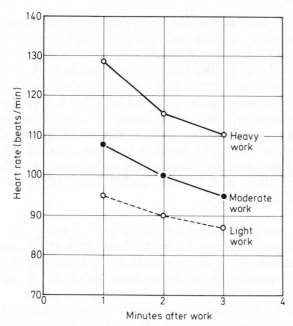

Figure 20.3 Higher level of heart rate recovery curves with increasing work load
Source: Brouha (1967)

Small changes in arrangements of a workplace can make surprising differences to the effort needed. To lift something from a platform about 600 mm high to one 1000 mm high involves half the energy cost that is needed to lift from floor level. It is sometimes possible to aggravate a situation by incomplete attention to materials handling. For instance, mechanizing the sand-feeding to foundry moulding machines may increase the output, but if the moulding boxes are not placed at a suitable height for transfer to and from the machine, the increased number of the heavy iron boxes to be lifted may cancel out the gain from better sand provision, leaving the operator as overloaded as before. Obvious though this point may appear when stated in these simple terms, it is made evident by considering the work of the moulder rather than the problems of handling sand. Solutions to production problems, arrived at by technologists, may often be technically effective but lead to human, and consequently further production, problems which are unforeseen, although not unforeseeable.

A less well recognized, but potentially more damaging area of physiological loading is that arising from working postures. Many studies of sitting comfort and seat design have endeavoured to provide designers with reliable rules and data for workplace seating, and there is no doubt that there have been major advances. The needs of heavy goods vehicles and tractor drivers are much better understood, and modern commercial vehicle seats provide a high degree of safety and comfort for the majority of their users. Industrial and office seats, while improved over earlier years, leave much to be desired, probably because, unlike vehicle seats, they are usually designed independently of the rest of the work space. Coupled with this is a widespread belief in the importance of avoiding standing at work, resulting in workplaces which require a worker to be seated to do his job. The pooling of the blood in the lower limbs, when experienced for most of a working lifetime, must be an important contributor to the leg and foot problems of office and factory workers engaged solely in bench activities.

The necessity to adopt one posture, or a very limited range of postures, is one example of a fault present in many jobs. Although none of the postures may be in the least difficult or harmful when adopted occasionally, their repetition many hundreds of times a day without the opportunity to redistribute the loads to other muscle

Table 20.1 BAD POSTURES VERSUS PROBABLE SITES OF SYMPTOMS

Bad postures	Probable site of pain or other symptoms
Standing (and particularly a pigeon-footed stance)	Feet, lumbar region
Sitting without lumbar support	Lumbar region
Sitting without support for the back	Erector spinae muscles
Sitting without good foot rests of the correct height	Knee, legs and lumbar region
Sitting with elbows rested on a working surface which is too high	Trapezius, rhomboideus; and levator scapulae muscles
Upper arm hanging unsupported out of vertical	Shoulders, upper arms
Arms reaching upwards	Shoulders, upper arms
Head bent back	Cervical region
Trunk bent forward; stooping position	Lumbar region Erector spinae muscles
Lifting heavy weights with back bent forward	Lumbar region Erector spinae muscles
Any cramped position	The muscles involved
Maintenance of any joint in its extreme position	The joint involved

(Reproduced from *Applied Ergonomics*, **Vol. 1, No. 5,** 1970, by courtesy of P. van Wely and the publishers, IPC)

groups can lead to severe bodily distortion and musculoskeletal damage. A press operator, for instance, may make 8000–10 000 components per day. A pilot study by van Wely (1970), urgently requiring more extensive replication, demonstrated how industrial surgery data could reveal postural problems arising from the job which could then be used as strong evidence to support the modification of jobs in the interests of both the health and performance of workers (*Table 20.1*).

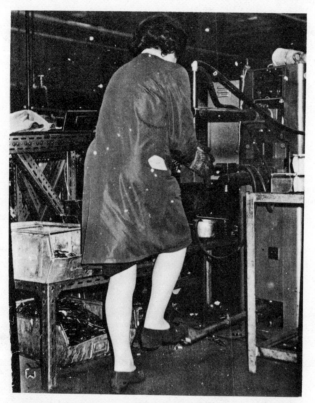

Figure 20.4 Demonstrates the typical posture of a spot welding machine operator. The weight is always carried on the same foot because practice has given efficient co-ordination of pedal and work positioning with right foot operation. The low work point is evident; the top electrode obstructs a view of the work and causes a further lean to the left

In studies of posture, ergonomists are increasingly using reports of perceived pain or discomfort experienced by an operator during the working day. Studies of spot welders (*Figure 20.4*) by Corlett and Bishop (1978) using this technique, and taking measures of the sizes of operators and equipment, as well as the working and non-working periods, illustrated how changes in machines reduced the discomfort levels (*Figure 20.5*). The reductions in discomfort levels were brought about by changes to the equipment. These afforded the operator a range of more balanced and upright postures and reduced loads on back, neck and shoulder muscles caused by working in one particular position.

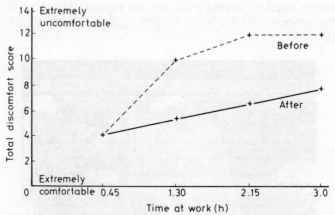

Figure 20.5a Average total discomfort scores before and after machine change in three spot welders
Source: Corlett and Bishop (1978)

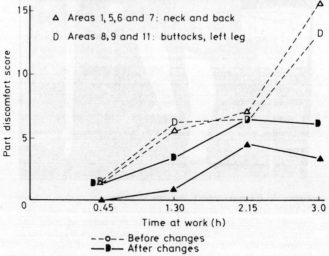

Figure 20.5b Part discomfort scores. △: neck and back. D: buttocks and left leg.
Source: Corlett and Bishop (1978)

The perception of stress by the worker is a useful complement to the measurement of stress by instrumentation. Extensive studies by Borg (1977) and others have demonstrated the regular nature of the perception of physical work load in relation to measures of oxygen consumption, heart rate and working activities. While medical practitioners have used patients' responses as an important input to diagnoses and social scientists have asked about attitudes to work and other matters, it is curious that physical work load studies have only recently taken to incorporating subjects' opinions as part of the measurement of work load and effort.

THE PHYSICAL ENVIRONMENT

As with other aspects of work, the ergonomist looks at the physical environment from the worker's standpoint. Increasingly, comprehensive data from occupational hygienists and others specify levels of exposure to environmental factors which are harmful to man or which show changes in human function. The understanding of the impact of such changes on the performance of the working human being, and the design to eliminate or counteract such effects, are of particular interest to the ergonomist.

During the last decade industrial noise has received an enormous amount of attention and its effects are now widely recognized; industrial deafness is identifiable (*see* Chapter 12) and can be prevented. Its prevention may require the use of personal ear protection, and recent work (*see* Chapter 21) has shown that comfort must be taken into account if the most effective protection is to be achieved. Effective protection, however, does not mean only that the operator is protected from industrial deafness. His purpose in the workplace is to do a job and this usually means communicating with his equipment and other people. Control information from the equipment is very often auditory, from the simple level of using the click from a push button to recognize that it has been successfully operated to more complex interpretations of the state of a process by using the changes in the spectrum of the noise emitted by various parts of it. If noises such as these are masked or attenuated to sub-threshold levels and are not replaced by equivalent and compatible information through other sensory channels, the work will be affected and the operator will be under a greater strain. In such circumstances he or she is likely not to wear the hearing protection. He will also discard it if it limits speech communication or if he fears that emergency signals may be missed. Obviously, talking can be more than just social chit-chat but even where it fulfils no evident job-related function it is still important as a human requirement in the workplace and must be taken into account in the design of working situations.

Special clothing or equipment to protect the wearer from environmental effects has had most consideration by the military services, but recently the needs of civilian occupations have been more widely recognized (*see* Chapter 19). The inclusion of comfort as a parameter in the investigations and design acknowledges that the worker must often wear the clothing for all the activities of a working day and that protection by itself is not enough. By this criterion much of present-day protective clothing is inadequate and people such as building and road workers, delivery roundsmen and many factory workers, people whose need for protection is less dramatic than the pilot or nuclear equipment operator but is no less real, have available to them clothing designs which could be bettered if designs were based on a more comprehensive study of their needs.

The other major environmental influences on a worker are light and heat. Lighting is needed to see and it is 'seeing' that the ergonomist is interested in. Perhaps the more precise statement is that it is perception which is important, for it is what the worker perceives rather than just the stimulation of the retina which affects his behaviour. Lighting levels, contrast and glare now lend themselves to reasonably precise specification for general decisions regarding work-place lighting (*see* Chapter 18). Where controllers must use large display panels or television screens for long periods the problems are more complex. Their work includes the identification of particular parts of the process, of the state of a part when related to other parts, and the recognition of emergency states. Emergencies are often indicated by visual warnings presented in the midst of many indicators showing otherwise satisfactory process situations, and the design of these warnings presents problems of intensity and colour, of symbols and layout for the panel designer. The user of a panel does not necessarily see the process he controls in the same way as the engineer who designed it, just as many who match the needle to a mark in the viewfinder of an automatic camera do not recognize that they are trading-off between shutter speed and lens aperture. The panel presentation devised by the engineer, therefore, may not provide the process operator with the most effective means for controlling the process.

The lighting of a situation such as the one described above is not separate from its design in other respects. To add to the designer's difficulties the increased availability of visual display units (VDUs) and their attachment to minicomputers has led to many office workers' jobs involving long periods of reading VDU screens. This has given rise to complaints of visual discomfort. Subsequent studies have shown that careful control of the symbol size and contrast, 'refresh' rate* of the screen, colour, viewing distance and surrounding lighting is necessary. Work requiring the continuous use of a VDU or other

---

*The 'refresh' rate is the frequency with which the electronic system scans the screen, thereby renewing the image, and the small fluctuations in intensity give rise to effects similar to flicker which increase the fatigue experienced when using these instruments.

screen-based reading device such as a microfiche reader is undesirable and the introduction of periods of other work is a necessary protection against visual discomfort and reduced efficiency.

INFORMATION HANDLING

The ability to make qualitative judgments, to take account of a wide variety of influences and merge them and to transfer their attention and mode of responding to a situation as it changes, make the human contribution to work unique and irreplaceable. Such abilities require the manipulation and interpretation of information from many disparate sources before it can be used to provide decisions. In some cases it has proved possible to define the amount of information handled and to indicate rates of handling above which errors have occurred. Low rates of information use have also been found to lead to errors and this has been related to the level of 'arousal' of the person concerned. As with more physiologically active tasks there is an optimum range for mental activity if attention to and performance of the job are to be optimum.

Much early work in ergonomics concerned control and display relationships, sometimes unkindly referred to as 'knobs and dials ergonomics'. This aspect is by no means unimportant even today, although control and display problems now arise with electronic systems rather than mechanical ones. This early work demonstrated important relationships between control and display responses for low error performance. It also showed the importance of stereotypes for control design, that is the expectations people have as to how equipment will respond in relation to the direction of movement of a control, as well as defining the desirable sizes for controls and displays. For perception of detail there is a minimum visual angle (angle subtended at the eye by the object being looked at) which is needed for given conditions of illumination, and where that object is a dial it would seem that two minutes of arc is the minimum size (*Figure 20.6*). However, if an object is moving, as in the inspection of components on a conveyor, then there is an immediate fall-off in acuity. Where a person is being vibrated, the ability to see detail or read is seriously reduced. The modern procedure of using digital displays rather than a pointer or other analogue, which can be interpreted from its position rather than by the number it points to, can lead to poorer performance when an operator is exposed to vibration.

The increasing responsibility borne by such workers as air traffic control operators, and the time pressures on inspectors and process controllers, have led to increasing interest in decision-making as an information-handling process. The information the controller needs and

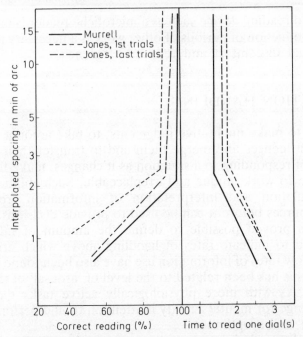

Figure 20.6 *The relationship between reading accuracy, reading time and angles subtended at the eye by an interpolated spacing (Reproduced from Murrell, 1958, Ergonomics, by kind permission of Chapman and Hall Ltd.)*

Figure 20.7 *A touch-sensitive screen in use. The operator selects the part of the computer program he next wishes to employ by touching the point on the screen which displays the relevant number. Thus he eliminates a 'coding' step (i.e. typing in the number), reduces the likelihood of errors and increases the speed of use of the instrument.(© British Crown Copyright, 1979)*

its presentation to him is defined by our understanding of the manner in which people may best retain and manipulate a complex of interacting relationships. For several aircraft approaching one airport a complete listing, even if updated at frequent intervals, is less useful than a presentation related to the geographic space with certain identifying symbols. The control of the situation is improved if the operator can handle the display itself rather than a separate terminal and so touch-sensitive VDU screens have been developed (*Figure 20.7*). Pointing to a list of functions and then to a particular aircraft symbol allows the function (e.g. height) to be immediately displayed for the aircraft. Such control systems are likely to become more usual since they reduce the learning needed for special controls and allow more 'natural' behaviour to be adopted.

Decision problems are not limited to the context of equipment control but are evident, for example, in management. Production engineering and operational research investigations have developed computer procedures for the simulation of production control, and ergonomists have attempted to adapt their presentation so they may be used as aids to decision making. The little success to date appears to be due more to a lack of understanding of the controller's precise needs rather than to any inherent infeasibility in the procedure. Again it seems likely that in management, decisions will soon be aided by ergonomically presented data via flexible VDU systems.

PERSPECTIVES ON WORK

Ergonomists have an approach to design — or redesign — of work situations which might be summed up as 'designing from the man out', rather than the common technological approach of 'design the man out'. In the short term the criteria to be satisfied are an increased facility to do the work, increased safety and, where feasible, increased performance. In the longer term the maintenance of good health is a primary target, together with the maintenance of optimum working stresses from the point of view of the worker.

This last point requires some enlargement. Work contributes to life in both positive and negative ways and is a major influence in shaping the individual. An underloaded or overloaded biological system will adapt so that it is less effective under 'normal' loads. In the case of underloading, for example, consider the changes in the load-bearing capacity of bone structure under weightlessness or the severe mental strain imposed on a previously isolated individual when he returns to everyday living. It is not just in the areas of physical load or excess information-handling that the ergonomist seeks to modify situations to enable the individual to cope. Social scientists are demonstrating links between job content and leisure patterns suggesting that people who are under-stimulated at work, for example, by being employed as machine loaders or members of an assembly line, are likely to have non-involving leisure activities requiring minimal human interaction.

In the design of work, therefore, ergonomists are increasingly concerned with attempts to harness a wider spectrum of the behavioural sciences to their needs. As well as experimental psychology, they look to industrial sociology, psychiatry and other areas which can contribute an understanding of both how and why people work (*see* Chapter 23).

These considerations, of course, are more relevant to ergonomists concerned with industry than they might be, for instance, to ergonomists concerned with the design of consumer products, although social behaviour norms are important in certain areas, such as the study of household accidents. The industrial ergonomist's interest in work design may be described as a process of matching the work to the person doing it, in terms both of making it more possible, that is suitable, and providing the opportunity for the person to have some commitment to it.

The earlier discussion has described many points on which the quality of modern work as a physical and mental activity is questioned. What is in doubt is whether it is work for which a human is suitable, and whether it places undue strain and risk on him. For example, industry is introducing more and more automatic processes and robot devices for handling material. Product variety, however, still demands the involvement of people in manufacture and assembly, and is likely to continue to do so in the foreseeable future. The design of work for the positive maintenance of physical and mental health is not a relatively shortlived activity catering for a declining need. It may become *more* important as service functions, both for equipment and for people, increase in proportion to 'conventional' manufacturing jobs. Because the variety of activities undertaken is increased the difficulties of an occasional task can be overcome. As tasks become more individual they become more critical; a damaged component on assembly is one less good product from thousands but a faulty service does not lead to a change in the proportion defective but to total failure. The loading on the worker concerned may change from a predominantly physical to a mental load, making it more difficult to deal with but none the less essential to achieve the objective of making work more 'possible', i.e. suitable for people.

The question of 'attachment' to the work is more controversial. 'Attachment' is used in the sense of being contrary to 'alienation', a state reported by many social scientists as existing among large numbers of the working population. The alienated worker does a job only as a source of pay and very little else. The aim of attachment is not to design for motivation, since this could be used as a more innocuous term for brain-washing people into doing the job, but to provide the job with those factors which enable people to find interest in the work and to become voluntarily involved. These factors appear to be encompassed by tasks requiring some skill, with a measure of complexity and an ability to check and be responsible for what has been done. In addition, it is important for people to work in an organizational relationship with others, which enables the work to be done under the worker's own authority. This means that he initiates and controls those tasks and activities which are associated with the production of material (or whatever the job output is) and the maintenance of the work system (machinery, organization, etc.).

Such considerations may seem remote from studying the physiological cost of shovelling sand and may well be of little importance to many of the ergonomists engaged on research. Performance, however, is one of the measures used by an ergonomist to examine the effectiveness of his intervention, and where this intervention is in a work system he is bound to take note of such factors since they may be major performance inhibitors or a primary source of strain for the worker.

METHODS OF WORKING

In response to an enquiry as to whether a company uses ergonomics, it is not unusual to be told that they do, and the evidence presented is a set of anthropometric tables, sometimes based on United States servicemen. Data covering the appropriate population is essential to the ergonomist's work, but ergonomics knowledge is not sufficiently far advanced to rely solely on such material. The ergonomist is more concerned with individuals than with averages and tries constantly to propose dimensions to suit the largest possible range, often aiming for 95 per cent of the total. He also endeavours to understand precisely the functions required of the people in the system and those provided by (or proposed for) the equipment. A different distribution of these functions may be the appropriate route to a better system rather than the provision of measures to alleviate some inadequacy in the work such as training, selection or rest pauses.

An understanding of the functions and the tasks needed to carry them out enables recommendations to be made for the design of equipment and environment, but these often require testing before being confirmed. Mock-ups of working relationships can be made simply and tested for reach, movement, access, etc., to see if the expected population range can use the equipment. If data are not available, laboratory-style tests may have to be done to confirm selected relationships.

An example of the need for a study of this kind arose during the investigations of open-fronted presses. *Figure 20.8* shows an extreme but not unusual problem of access to these machines through the opening of the guard. Not only is the width of the guard too small, but the vertical height of the opening is also too small and too low for adequate vision. The other aspects of this thoroughly inadequate workplace are obvious.

Figure 20.8 Press operator. This is a particularly bad example of the inadequacy of the arrangement of machines. Two-handed loading of the bars is necessary, visibility is severely obstructed by the guard. The height of the machine, and the poor guard design, require the operator to work in the manner illustrated

To determine the appropriate dimensions for a guard opening in relation to the working population, laboratory studies on a test rig were conducted. From these studies new guards were built and fitted to presses in industry, to be used by the operators in their normal work. The evaluation of the results took place over several months and was not made on the results from the laboratory trials alone. However, it was found that the trials had predicted correctly the industrial requirements.

Where groups of people are together, as in a control room, the testing phase should include simulations where the control personnel run through their normal activities. Even modest tests of this kind have been shown to avoid design errors which would otherwise have had to be dealt with by more expensive changes to the equipment after its installation.

It is the lack of trials of this sort, where proposed equipment is brought together and worked under an expert eye, that is responsible for the poor results of so much well-meaning industrial expenditure. As pointed out at the beginning of this chapter, industrial seating does little for industrial comfort, primarily because it is not appreciated that sitting is not an activity separate from working. A seat and workplace must therefore be arranged together and if a workplace is already pre-empted, in the sense that a machine exists, then even more care is needed in seat specification.

Ultimately the success of the ergonomist must be measured by the health and satisfaction demonstrated by the working population, objectives much in harmony with those of occupational medicine. Ergonomics is not the sole prerogative of the practising ergonomist, however, but should be part of the understanding and expertise of engineering and industrial designers, production engineers and methods personnel. All these people, in the course of their regular activities, create or limit the working activities of others. Once they have fixed their designs or proposals they may also have fixed the tasks, environments and relationships for other workers for years to come. It is unsatisfactory and inefficient to modify what could have been done correctly in the first place, and the increasing trend, in engineering courses, to teach the relevant areas of ergonomics needed for each discipline is to be welcomed.

POSTSCRIPT

The subject of ergonomics has been presented in this chapter in relation to an advanced technological society. Obviously, it is just as relevant to other societies, its objectives being of universal importance. Work situations need to be designed for *the population which will use them*. In many parts of the world, for example, squatting is a common posture for relaxation and it may be appropriate to reorient the presentation of equipment to the operator to take account of this rather than assume that chairs, as in Western countries, are appropriate. These matters should not go by default, neither should they be decided by guesswork, for there is no reason why people should be forced into needless cultural or social changes. The transfer of, say, European equipment to Africa will not necessarily be satisfactory. The dimensions of equipment, the forms of presentation of data for control or maintenance, the hours of work and organizational structure may have to be modified to suit physical and cultural aspects of the receiving country.

As emphasized earlier, the primary target of the ergonomist is to make work match the needs and requirements of people, and certainly not to see people in the designer's own image. There is only one designer who has done this, and ergonomists should be well content if their work can leave the image as He made it.

REFERENCES

Borg, G. (1977) *Physical Work and Effort.* Wenner-Grens Center International Symposium series, Vol. 28. Oxford: Pergamon Press
Brouha, L. (1967) *Physiology in Industry,* 2nd edn. Oxford: Pergamon Press
Cooper, C.L. and Payne, R. (1978) *Stress at Work.* Chichester: J. Wiley and Sons
Corlett, E.N. and Bishop, R.P. (1978) 'The ergonomics of spot welders.' *Applied Ergonomics,* **9,** 23 - 32
Ergonomics (1957) Editorial, **1,** 1
Van Wely, P. (1970) 'Design and disease.' *Applied Ergonomics,* **1** (5), 262 - 269

FURTHER READING

Edholm, O.G. and Murrell, K.F.H. (1973) *The Ergonomics Research Society, a history.* The British Ergonomics Society
McCormick, E.J. (1976) *Human Factors in Engineering and Design.* New York: McGraw-Hill
Grandjean, E. (1980) *Fitting the Man to the Task.* London: Taylor and Francis Ltd

21

Personal Protection

D. Else

In this chapter the term *personal protection* is used to encompass protective clothing as well as equipment such as eye and hearing protectors and respirators. Personal protection can be used to safeguard people from dirty conditions or to protect the product from contamination. Its most common use is to keep the user healthy and safe, and it is within this context that personal protection is discussed here.

Controlling dangers at source, or making the place of work safe, have long been considered preferable to relying on personal protection. Once the decision has been made to use personal protection rather than control at source, the fact that it is regarded as the last line of defence should not obscure the vital need for the allocation of competent people and adequate resources to ensure its effectiveness. Unfortunately thought is often focused on the personal protection hardware itself and insufficient consideration is given to the systems which must underpin any personal protection scheme. Seldom are the costs of cleaning, maintenance and training evaluated when decisions are being made to opt for personal protection rather than control at source. Personal protection is therefore often seen as an inexpensive solution because no consideration is given to the true ongoing costs of achieving a high usage of well maintained equipment.

The approach taken in this chapter is to explore the key elements that are vital to the success of personal protection schemes generally. A more detailed analysis of hearing, respiratory and eye protection is presented to illustrate the use of equipment and the assessment of schemes. In conclusion, a list of questions is given which provides a basic framework for assessing the adequacy of any personal protection scheme.

KEY ELEMENTS IN PERSONAL PROTECTION SCHEMES

Choice of protection

Personal protection is selected initially to provide protection against the particular danger against which it is being used. To assess its adequacy and to ensure that, in theory at least, it is capable of providing an adequate safeguard, it is necessary to obtain the following information:

Nature of the danger

Qualitative and quantitative information is needed about the particular danger. For example, the chemical structure and the concentration of any airborne contaminants need to be known before respiratory protection can be selected. Similarly, gloves cannot be chosen for work with acid baths unless the type of acid and its concentration are known.

Performance data for personal protection

Data are needed from which to deduce the ability of the personal protection to reduce the particular danger. Often these data will have to be obtained from the manufacturers of the protective clothing or equipment. Usually the data will be the result of tests performed under standardized conditions conforming to the relevant national or international standards for the particular type of protection.

Acceptable level for exposure to the danger

For many dangers the acceptable level is zero, for example, protection of the eyes against projectile impacts or the skin from attack by caustic solutions. The hygiene limit (e.g. Threshold Limit Value), or some fraction of the hygiene limit may be used as the acceptable level for other dangers such as the protection of hearing against noise exposure, or eyes from laser radiations.

Degree of protection

The calculated degree of protection will not be achieved in practice unless:

1. The fit of the personal protection on the user is as good as the fit achieved during the original performance tests (i.e. the national or international standard tests).
2. The equipment is cleaned and maintained.
3. Those involved in the scheme receive training.
4. The scheme is monitored.
5. Its effect on job performance is assessed.
6. Management is committed to the scheme.

User trials

The manufacturer's performance data are usually the results of tests with a relatively small sample of people; often males of only one ethnic group. It is therefore vital to check the fit of the protection when it is initially issued to the user. If manufacturers do not produce the item of personal protection in a range of sizes it will be necessary to obtain a range of similar items from different manufacturers, so that different sizes can be made available from which the best fit for the individual user can be selected. The user should then be trained to fit the personal protection correctly every time he or she makes use of it.

The calculated degree of protection will not be achieved in practice unless the personal protection is worn at all times when the person is at risk. Therefore, the item of personal protection must be compatible with the work which has to be done and it must be compatible with any other personal protection which has to be used at the same time. Comfort, or at least lack of undue discomfort, is therefore another vital factor. Personal protection which theoretically provides a very high degree of protection, but is uncomfortable and therefore removed for part of the time, may provide less protection in practice than items which in theory are slightly less effective but which are more comfortable and therefore worn for a higher proportion of the time.

Wherever possible users should be given a choice from a range of items, all of which must be theoretically capable of providing adequate protection. This increases the likelihood that the form of protection finally chosen, will be compatible with the work and with other personal protection which has to be used. Leaving the final decision to the user also increases the likelihood of acceptance of the equipment because the user has taken part in the decision making.

It is seldom practical to keep more than a small range of items because of the problems associated with stocking spares and replacement items. User trials can be used initially to decide on the most appropriate small range of personal protection from the vast range that is usually commercially available.

Guidance in selection of the range of most acceptable and effective protection can be obtained during the user trials. The items should be tested by a wide range of people engaged in all the work operations, and the items must be checked for compatibility with other personal protection the workers have to employ. The purpose of the trials should not be to select the two or three types of personal protection voted acceptable by the largest number of people, because this could result in three items being chosen which only suit average-sized people. The purpose of the trials should be seen more as need to encompass the first or second choice of as many users as possible.

Maintenance

The calculated degree of protection will not be achieved in practice unless the personal protection is cleaned and maintained. Cleaning can often be left to the users, provided that adequate provisions are made for them to have the necessary cleaning materials available. However, it is seldom wise to rely on users to check or maintain their own personal protection because the decreases in performance are usually insidious and therefore unlikely to be noticed by the user. The use of disposable items can reduce the need for maintenance, but it must be remembered that systems relying on disposable personal protection may still have a maintenance requirement, for example, maintenance of dispensers for disposable earplugs.

Training

Personal protection schemes are unlikely to be successful unless all people involved receive adequate training. Users and those responsible for supervision need to know what the equipment is capable of protecting against, the consequences of non-use, how to fit the equipment, how to clean it, and when to have it replaced. The people responsible for maintaining and issuing the equipment also require adequate training to perform their functions in the scheme. A problem which is commonly encountered is that although training is given when the scheme is initially introduced, no thought is given to refresher training or the training of all new starters or people transferred from other work areas.

Inspection

The use of personal protection should be monitored, for example by visual inspection during random visits. Procedures should exist for investigating the reasons why some individuals do not use the protection. Marking areas where personal protection has to be used can help to remind people entering these areas, but care should be taken to limit the boundary of an area. Small personal protection areas which relate closely to people's subjective impressions of where the hazards exist are likely to be more effective than making complete workshops into personal protection areas.

Effect on job performance

A common reason given by people for not wearing personal protection is that it makes the job more difficult to perform, or that they feel less safe when using it. Such complaints should not be dismissed without thorough investigation. Personal protection is often worn on one or more of the major routes for information input to the body, and in certain circumstances the flow of information can be significantly impeded by the protection: gloves, for example, reduce tactile information; respirators reduce olfactory information; and eye protectors, respirators or helmets can reduce the peripheral visual field. However, people often acclimatize to the wearing of personal protection and either learn to make use of the information which is slightly transformed by the protection or they learn to make use of cue information from another modal input: they begin, for example, to make greater use of their eyes rather than rely on hearing the direction from which sounds originate.

Management commitment

The commitment of management at all levels can be a vital ingredient for a successful scheme. Management not only need to wear the personal protection themselves when they are in the hazardous areas, they also need to underpin the scheme with written systems and procedures for updating it to cope with changes to the process; changes to materials used in the process; reductions in the hygiene limits; change of personnel; the availability of new types of personal protection or the lack of availability of current types of personal protection.

HEARING PROTECTION

There are two main types of hearing protectors:
1. *Earplugs*, which are inserted in the ear canals. These are usually disposable items made from materials such as glass down, plastic coated glass down or polyurethane foam. Re-usable earplugs are

made from semi-rigid plastic or rubber and these can be obtained in either a range of sizes (often three or five sizes), or in one size designed to fit all shapes of ear canals. Re-usable earplugs have the disadvantage that they require frequent washing.

2. *Earmuffs*, which cover the external ears. These consist of two rigid cups held together by a headband. Annular cushions, made from envelopes of plastic filled with polyurethane foam or a liquid, are used to seal between the cups and the sides of the head around the ears.

The reduction in sound level provided by a hearing protector, usually referred to as its attenuation, depends on the frequency of the noise in which it is worn. The frequency dependence of attenuation can be seen in *Figure 21.1* which displays the results of attenuation tests on four

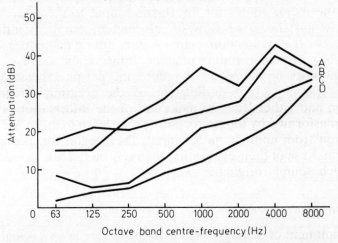

Figure 21.1 Comparisons of the attenuation data for four hearing protectors. A = high attenuation earmuff; B = diposable expanding polyurethane foam earplugs; C = low attenuation earmuff; D = disposable earplugs

types of hearing protectors. Attenuation is usually determined by tests in which the hearing thresholds of subjects are tested with and without hearing protectors being worn. The tests are performed at a series of frequencies throughout the audible frequency range on a small group of subjects: in the USA three determinations are made of attenuation throughout the frequency range for each of 10 subjects (American National Standards Institute, 1974); in Great Britain two determinations are made for each of fifteen subjects (British Standards Institution, 1974b).

Hearing protectors should be chosen to reduce the noise level at the wearer's ears to below the recommended limit for unprotected exposure to noise, e.g. 85 decibels (dB(A)). They cannot be chosen from simple measurements of the A-weighted level of the noise because the reduction in sound level they provide depends on the particular frequency spectrum of the noise. An octave band analysis of the noise is therefore necessary to provide a measure of the frequency spectrum of the noise from which the A-weighted sound level at the user's ears is computed from the attenuation data obtained from the manufacturers (*see* Chapter 18). The calculation procedures are described in detail in the Code of Practice for reducing the exposure of employed persons to noise (Department of Employment, 1972). Unfortunately, hearing protectors are often purchased without the necessary calculations being made and as a result the user may be inadequately protected because, for example, an earmuff might reduce the noise level from a high frequency noise by as much as 35 dB(A) but the same earmuff worn in a low frequency noise in a compressor room might only provide 10 dB(A) reduction in sound level.

Several investigators have shown that the degree of protection provided by earplugs (Padilla, 1976) and earmuffs (Regan, 1977) may be considerably less than has been calculated during the selection of the protectors because in practice the users do not fit the protectors as effectively as do subjects during laboratory attenuation tests. There are two stages at which hearing protectors can be incorrectly fitted: initially when the protectors are selected and fitted to the individual man or woman; and on each occasion that the individual fits the hearing protector prior to entry into the noisy area. If earplugs are not supplied in a universal size, then supervision will be necessary during the initial fitting to ensure that the correct size of earplug is selected; some manufacturers provide a tool for gauging ear canal diameters while others expect the correct size to be found by trial and error from the range that is available. Clearly, the users must be given adequate instruction to enable them to fit the earplugs correctly as a matter of routine.

Earmuff users should be given an opportunity to select from a range of earmuffs, all of which should be capable of reducing the level of noise to below recommended limits. The earmuff which is most suitable for the majority of users does not necesarily provide an adequate and comfortable fit on the minority of users. Some earmuffs are larger than others and some have more adjustment in the headbands. Ivergard and Nicholl (1976) have shown that allowing people to compare earmuffs by trying them on for a couple of minutes enables people to choose those which are most acceptable to them after wearing them for sometime.

The wearing of spectacles has been shown to reduce the protection provided by earmuffs by as much as five decibels (Nixon, 1974). This problem can be solved by careful selection of both spectacles and earmuffs. The spectacles, which will probably need thin sidearms, will have to be fitted by a trained person.

The removal of hearing protectors for very short periods during exposure to noise has been shown to reduce substantially the protection afforded (Else, 1973). The relation between the protection and the percentage of the exposure duration for which the hearing protector is worn is shown in *Figure 21.2*, in which the hearing protector is assumed

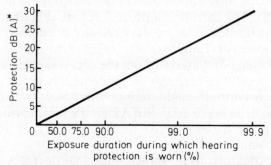

Figure 21.2 The effect of removing a hearing protector for short periods of time. The theoretical case of a hearing protector which provides infinite attenuation. (*Protection is defined as the reduction in daily equivalent-continuous sound level afforded when the hearing protector is worn for less than the total duration of noise exposure.)

to provide infinite attenuation; in this simplified case it is assumed that no noise reaches the ears while the hearing protectors are worn. As can be seen from *Figure 21.2*, which is based on a three decibel trading relation between noise level and exposure duration, no hearing protector could provide more than six decibels of protection if worn for less than 75 per cent of the exposure duration. Similarly, if it were necessary to protect a person exposed continuously for eight hours per day to a sound level of 115 dB(A), no hearing protector could provide adequate protection unless worn for at least 99.9 per cent of the exposure duration; i.e. removing the hearing protectors for one minute would result in the person receiving twice the recommended daily maximum noise dose. (The recommended maximum noise dose is taken to be 85 dB(A) for eight hours per day.) In practice hearing protectors do not provide infinite attenuation and therefore some noise energy is received while the protectors are worn. The relation between the protection provided and the percentage of the exposure duration for which the protectors are worn is shown in *Figure 21.3* for hearing protectors which reduce the instantaneous sound level by 10 dB(A) and 20 dB(A) respectively (Else, 1976).

Clearly if a high attenuation hearing protector is removed for part of the noise exposure then the same degree of protection, or greater, could be achieved with one which provides less instantaneous reduction in sound level, provided that it would be worn for a sufficiently high percentage of the exposure duration.

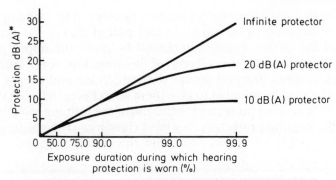

Figure 21.3 the effects of rmoving hearing protectors for short periods of time. Comparisons of the protection afforded by hearing protectors which reduce the instantaneous ssound level by 10 decibels and 20 decibels respectively. (*For definition see Figure 21.2)

How comfortable a hearing protector is and whether it can be worn for the full duration of exposure by people who may also have to wear other forms of protective equipment are vital factors that should be considered during the initial selection. User trials are usually the only way of discovering how acceptable, comfortable or compatible a particular type of hearing protector will be for a particular application. Unfortunately, it is quite common to find that people exposed to noise levels only slightly in excess of recommended noise limits have been issued with high attenuation earmuffs; the high attenuation is often obtained at the expense of comfort and as a result high usage of earmuffs is not achieved.

Earmuffs require maintenance. Earmuff cushions are likely to need replacement frequently because most cushions become less compliant as a result of contact with natural hair oils, perspiration and hair dressings. *Figure 21.4* shows a comparison of a new cushion with one that had been used for three months in a foundry. Many manufacturers now make available spare cushions or replacement parts which include cushions and the inner foam inserts for earmuff cups. The complete earmuff or parts of the earmuff are likely to have to be replaced as a result of wear in the adjustment mechanisms; damage resulting from the earmuff being dropped; distortion resulting from exposure to intense heat; or as a result of normal wear and tear.

High usage of hearing protectors is unlikely to be achieved unless an effective training programme has been designed and incorporated within the hearing protection scheme. All people exposed to high noise levels, or responsible for such areas, must receive training at the start of their employment because hearing damage occurs rapidly during the initial period of noise exposure. All hearing protectors are uncomfortable to some degree, and the training, therefore, has to arouse sufficient concern for the wearer to want to cope with the slight discomfort rather than suffer the consequences of hearing damage. The training should explain the structure of the ear and which part of the ear is damaged by noise, and the person concerned should be given the opportunity to listen to examples of the effects of hearing loss, e.g. from tape recordings or films. If small groups of between ten and twenty people are trained it may be possible to give them all a choice between three or four different hearing protectors at the end of the training session. Their usage of the hearing protectors *they* have chosen can then be monitored during the couple of weeks following the training session.

Figure 21.4 Comparison of new and used earmuff cushions

One of the questions likely to be raised during training sessions is the possible effect of hearing protectors on the safety of the wearer. Potential wearers, or people who have worn hearing protectors for short periods of time, may express concern about speech communication or the perception of warning sounds or monitoring sounds. Hearing protectors can in some circumstances present a further hazard to the user, especially if the user has already suffered significant hearing impairment (Burns, 1973; Else, 1976). Many studies have shown that hearing protectors reduce the user's ability to detect the direction from which sounds originate (Atherley and Noble, 1970; Else, 1976). Earmuffs have a much greater effect than earplugs, though even earmuff users seem to cope by making greater use of visual cues.

The wearing of hearing protectors tends to reduce the wearer's voice level by approximately three decibels (halving the sound energy output) and this has been shown to reduce speech intelligibility when both listener and talker wear either earplugs or earmuffs in noise levels of 85 dB(A) and above (Howell and Martin, 1975).

RESPIRATORY PROTECTION

Respiratory protection equipment is divided into two broad classifications:

1. *Respirators*, which purify the air by drawing the contaminated air through a medium that removes most of the contaminant;
2. *Breathing apparatus*, which provides air or oxygen from an uncontaminated source. Respirators do not protect against oxygen deficient atmospheres and they should not be used in atmospheres immediately dangerous to life.

Table 21.1 APPROXIMATE NOMINAL PROTECTION FACTORS FOR RESPIRATORY PROTECTION EQUIPMENT

Type of respiratory protection equipment	Nominal protection factor*
Respirators	
single use filtering facepiece respirator	5
half-mask (cartridge) respirator	10
full facepiece (canister) respirator	500–1000
powered air-purifying respirator	500
powered visor respirator	10–20
Breathing apparatus	
fresh air hose apparatus	50
compressed airline apparatus	1000–2000
self-contained breathing apparatus	2000

* For definition see p. 521

Respirators

Respirators can be subdivided into five main types (*Table 21.1*):

1. *Filtering facepiece respirators*, sometimes called disposable or single-use respirators, in which the whole facepiece, designed to cover the nose and mouth, is manufactured from filtering material. Maintenance problems are reduced by using these devices because the entire respirator is discarded after use. These respirators look similar to the so-called nuisance dust masks which have been available for many years. Care should be taken not to confuse the two types because the simple nuisance dust masks are designed only to exclude large particles.
2. *Half-mask respirators*. These usually consist of a rubber or plastic facepiece designed to cover the nose and mouth and have replaceable filter cartridges.

3. *Full-facepiece respirators.* These consist of rubber or plastic facepieces designed to cover eyes, nose and mouth and have replaceable canisters to the facepiece connected either directly or via a flexible tube which allows the canister to be worn on a harness.
4. *Powered air-purifying respirator.* This usually consists of a battery-powered fan which draws air through a filter element to remove the contaminants. The purified air is then blown into a half-mask, full-facepiece, or hood or blouse. The great advantage of the powered air-purifying respirator is that it usually supplies air at positive pressure so that any leakage is outward from the facepiece during most of the breathing cycle.
5. *Powered visor respirators.* This type of respirator is a development of the powered air-purifying respirator in which the fan and filters are usually mounted in a helmet. The purified air is blown down behind a protective visor, past the wearer's face.

All the above types of respirator are available with filters suitable for protection against harmful dusts and fibres; most of them are also available with special adsorbing or absorbing filters for protection against vapours or gases. Filters for combating combinations of dusts and vapours can also be obtained.

Breathing apparatus

Breathing apparatus can be subdivided into three main types (*Table 21.1*):

1. *Fresh air hose apparatus,* in which air suitable for respiration is drawn from an adjacent uncontaminated area along a large-diameter air hose by the breathing action of the wearer, or by a manually operated blower or bellows, or by a motor operated blower.
2. *Compressed airline apparatus,* which provides compressed air suitable for respiration through a flexible hose attached to a compressed airline. Filters may be included in the airline to remove undesirable contaminants, such as oxides of nitrogen from the compressor. The apparatus requires a continuous supply of clean compressed air either from a compressor or from cylinders. The use of special compressors for breathing apparatus is preferable to the use of general compressed air which is usually available in factories. The compressed airline can be attached via pressure-reducing valves to half-masks, full-facepieces, hoods, blouses or protective visors.
3. *Self-contained apparatus.* This provides air or oxygen from a cylinder, cylinders, or other container which is carried on a harness attached to the user's back or chest. Most self-contained apparatus is

designed to have a duration of approximately thirty minutes; some small escape sets are available with cylinders which last for only ten minutes, and special recirculating equipment can be obtained with a duration of up to three hours. Self-contained apparatus is used mostly for rescue purposes but it can be used by people going into hazardous areas to complete repairs or inspections.

Nominal protection factors

The nominal protection factor (npf) is used in the United States and Europe for the selection of respiratory protection equipment. The nominal protection factor is defined as the ratio of the concentration of the contaminant present in the ambient atmosphere to the calculated concentration within the facepiece when the respiratory protection is being worn:

$$\text{npf} = \frac{\text{concentration of contaminant in atmosphere}}{\text{concentration of contaminant in facepiece}}$$

An npf of 20 therefore implies that only one-twentieth of the contaminant outside the respirator finds its way into the wearer's breathing zone. Some of that contaminant would be expected to have passed through the filter medium and the remainder would have found its way into the facepiece through leakage pathways around the periphery of the facepiece. The npf factors are derived from tests in which samples are taken from within the facepiece of respiratory protection while it is being worn by subjects in a test chamber in which known concentrations of respirable size-range particles or gases are generated.

Table 21.1 indicates the approximate nominal protection factor that could be obtained from the various forms of respiratory protection. Estimates differ between countries and are subject to revision; reference should therefore be made to current national guidelines such as the recommendations of the National Institute for Occupational Safety and Health in the United States (Pritchard, 1976) or the British Standards Institution (1974a).

The selection of respiratory protection has to be based on a knowledge of the type of contaminant, its concentration in the atmosphere and the hygiene limit for that contaminant. When these three items of information are available a calculation can be made to determine the factor by which the concentration of the contaminant has to be reduced to limit the exposure to below the hygiene limit. A type of respiratory protection, capable of providing a sufficient npf, can then be chosen.

Several investigators have shown that the degree of protection provided by half-mask respirators (Toney and Barnhart, 1976) and full facepiece respirators (Hounam *et al.*, 1964) may be considerably less in practice than has been calculated during the selecting of the respirators. The respirators often do not fit the users in practice because they have facial shapes unlike those of the subjects used in laboratory tests: for example they may be women, men from other ethnic groups, or they may just have a facial shape which the respirator does not fit. It is therefore vital to ensure that when the person is initially fitted with a respirator a test is performed to ensure adequate fit. Some respirators are available in a range of two or three sizes, but it will probably be necessary to stock a range of at least three different manufacturers' respirators in order to be able to fit most individuals. A range of qualitative and quantitative tests for fitting respiratory protection are described by Prichard (1976) and Douglas (1979). The most simple test adopted in Britain is described below.

'To ensure proper protection, *the facepiece fit should be checked by the wearer each time he puts it on*. This may be done in this way: *Negative pressure test.* Close the inlet of the equipment. Inhale gently so that the facepiece collapses slightly, and hold the breath for 10 seconds. If the facepiece remains in its slightly collapsed condition and no inward leakage of air is detected, the tightness of the facepiece is probably satisfactory. If the wearer detects leakage, he should readjust the facepiece, and repeat the test. If leakage is still noted, it can be concluded that this particular facepiece will not protect the wearer. The wearer should not continue to tighten the headband straps until they are uncomfortably tight, simply to achieve a gas-tight face fit (British Standards Institution, 1974a).'

The negative pressure test also serves as an excellent method of checking for leaks caused by factors other than poor facepiece fit. For example, leaks caused by poor sealing of an exhalation valve or an incorrectly fitted cartridge would be identified by the negative pressure test. The negative pressure test cannot, however, be used with loose fitting respiratory protection such as visor respirators, hoods or blouses, although for some of these devices the fit is less important so long as the device maintains the pressure inside slightly in excess of the ambient pressure to ensure that any leakage is outward.

The calculated degree of protection can only be achieved in practice if the respiratory protection is worn during the total period of exposure. If it is slipped down on to the chest even for a small percentage of the time it may no longer fully protect. This can easily happen in practice because wearers of respiratory protection often do not realize that the very small particles, capable of penetrating to the depth of the alveoli in the lungs, will be present in the air long after the large visible particles have settled. Harris *et al.* (1974), after a comprehensive study of the use of the half-mask respirator in four coal mines, concluded that virtually every miner wore his respirator only intermittently. They derived an 'effective protection factor' (epf) to describe the actual factor by which the daily dose of contaminant is reduced when a respirator is worn for less than the total duration of exposure.

The concept of 'effective protection factor' can be extended to compare the effectiveness of various forms of respiratory protection. In *Figure 21.5* the epf has been calculated for respiratory protection equipment of various nominal protection factors assuming that they are worn for less than the total duration of exposure. As can be seen from *Figure 21.5*, if a respirator which is capable of reducing contaminant

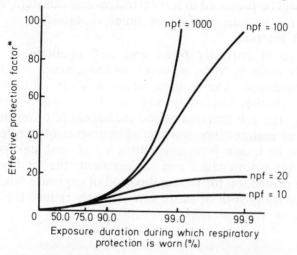

Figure 21.5 The effect of removing respirators for short periods of time – shown for respirators of various nominal protection factors (npf). (*The effective protection factor (epf) is defined as the factor by which exposure is reduced when a respirator is worn for less than the total duration of exposure.)

concentrations by a factor of 1000 is taken off for 10 per cent of the exposure duration, then in practice the exposure will only be reduced by a factor of 10, not the calculated 1000. Clearly a device with a low nominal protection factor which is worn all the time may provide greater protection than a higher nominal protection factor device, if the latter is less comfortable and consequently worn for a much lower percentage of the time.

Respiratory protection cannot provide the calculated degree of protection beyond the first day it is used unless it is adequately maintained. Respiratory protection, unless it is disposable, needs to be cleaned each day and adequately stored when not in use. It needs to be inspected while it is being cleaned, and defective parts replaced whenever necessary. If the users themselves are expected to clean and maintain their respiratory protection they must be trained to perform these tasks and they must be provided with the necessary facilities. However, if more than a couple of people have to issue respiratory protection, employers would be better advised to develop a central system for cleaning and maintenance. Such a central system provides a much greater degree of control; one or two people charged with specific responsibilities for maintenance and cleaning would also be more likely to notice gradual deterioration in equipment which might go unnoticed by the individual user. If respirators with replaceable canisters or cartridges are used it will also be necessary to have a procedure for deciding when these filters should be replaced. Some manufacturers of respirators will be prepared to test cartridges and canisters, which have been used for various periods of time, to establish the optimum replacement interval.

High usage of correctly fitted and well maintained respiratory protection is unlikely to be achieved in the absence of an effective training programme. The training needs will vary with the type of equipment; whether the equipment has to be worn frequently or infrequently; the job function of the individual (e.g. respirator user, supervisor, or maintenance worker). The respiratory protection user may have to be taught how to perform a face-seal leakage test; the procedures for maintenance and replacement; the vital necessity of wearing the protection for the total period of exposure; and that even one or two days growth of beard can seriously reduce the protection afforded by most respiratory protection.

EYE PROTECTION

The selection of eye protection relies on qualitative assessment of the hazards rather than the quantitative assessments required for respirators or hearing protectors. A comprehensive survey of eye hazards is essential to enable the correct type of eye protector to be selected to combat the particular hazard or combination of hazards. Spectacle-type eye protection is designed to protect the wearer against impacts or relatively low energy projectiles (e.g. small particles of metal swarf ejected during the machining of metal on a lathe). If protection is required against high energy projectiles, molten metal, chemical splash, dusts or gases, spectacles would be unlikely to prove adequate. More robust goggles or faceshields would have to be used; and these would have to be of an appropriate type for the hazard (e.g. chemical type goggles with protected ventilation ports: *Figure 21.6*). Similarly, protection against radiations, such as from gas or arc welding or lasers, requires special eye protection (British Standards Institution, 1960).

Reference should be made to the standards against which the various eye protectors are tested to ensure their adequacy (British Standards Institution, 1967; American National Standards Institute, 1968; Standards Association of Australia, 1974).

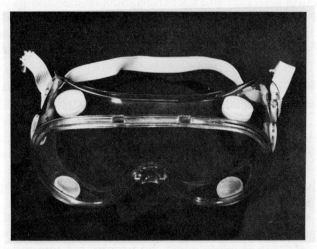

Figure 21.6 An example of a chemical type goggle with protected ventilation ports

Eye protection will be rejected if it does not fit properly. Poorly fitting eye protection can cause localized high pressure points; discomfort due to misalignment of the axes of the wearer's eyes and the eye protector lenses respectively; or unnecessary restriction of the peripheral visual field. There are two methods of providing the range of sizes of eye protector necessary to fit people with different facial shapes and sizes. A device can be chosen which is available in a range of sizes; some manufacturers, for example, produce safety spectacles in a range of frame sizes to accommodate different distances between the eyes; various 'bridge' fittings; and with a range of lengths for the sidearms. Alternatively, the range of sizes can be produced by offering a choice from a range of eye protectors made by different manufacturers. Each manufacturer produces goggles in only one size but some manufacturers' designs tend to be large while other designs tend to be much smaller.

An eye protector which is only worn for 50 per cent of the time the person is at risk cannot reduce the risk by more than a factor of two, regardless of how effective the protectors are in theory. Every effort should therefore be made to provide people with the most comfortable eye protectors. Facilities should be available for cleaning and storing, and procedures should exist for inspection and replacement of defective items. As well as the obvious training requirement for all people exposed to eye hazards, it may be necessary for someone to be trained to fit eye protectors. Some manufacturers provide courses specifically for this purpose.

The potential for increasing other risks should be considered during eye protection selection and user trials. Spectacle sidearms can reduce the effectiveness of hearing protectors and respiratory protection, but attention to these factors during the initial selection can largely overcome many of the problems (Wigglesworth and Cole, 1973).

ASSESSING PERSONAL PROTECTION SCHEMES

Set out below is a list of questions which provide a basic framework for assessing the adequacy of any personal protection scheme.

1. Why is personal protection being used to control the danger rather than making the place of work safe?
2. What efforts are being made to phase out the use of personal protection when methods become available to make the place of work safe, for example when the process is being redesigned, or when new machines are being purchased?
3. Does the personal protection theoretically provide adequate protection. Have calculations been made by competent people?
4. Does the personal protection provide the same degree of protection in practice; for example, does it fit all members of the user population and is it worn all the time by people at risk?
5. What provisions have been made for cleaning and maintenance of the personal protection?
6. What provisions have been made for training all people involved with the personal protection scheme?
7. Are written procedures available to underpin the scheme?
8. Is the marking of personal protection areas adequate?
9. What provisions have been made for updating the personal protection scheme?

REFERENCES

American National Standards Institute (1968) *Standard Practice for Occupational and Educational Eye and Face Protection*. ANSI 287.1–1968, New York

American National Standards Institute (1974) *Method for the Measurement of Real-ear Attenuation of Hearing Protectors and Physical Attenuation of Earmuffs*. ANSI 3.19–1974, New York

Atherley, G.R.C. and Noble, W.G. (1970). 'Effect of ear defenders (earmuffs) on the localization of sound.' *British Journal of Industrial Medicine*, **27**, 260–265

British Standards Institution (1960) *Equipment for Eye Face and Neck Protection Against Radiation Arising During Welding and Similar Operations*. BS 1542:1960, London

British Standards Institution (1967) *Specification for Industrial Eye Protectors.* BS 2092:1967, London

British Standards Institution (1974a) *Recommendations for the Selection, Use and Maintenance of Respiratory Protective Equipment.* BS 4275:1974, London

British Standards Institution (1974b) *Method of Measurement of Attenuation of Hearing Protectors at Threshold.* BS 5108:1974, London

Burns, W. (1973) *Noise and Man,* 2nd edn. London, Murray

Douglas, D.D. (1979) 'Respiratory protective devices.' In F.E. Clayton and G.D. Clayton, eds *Patty's Industrial Hygiene and Toxicology,* 3rd edn, pp. 993 - 1057, New York, Wiley

Else, D. (1973) 'A note on the protection afforded by hearing protectors – implications of the energy principle.' *Annals of Occupational Hygiene,* **16,** 81–83

Else, D. (1976) 'Hearing Protectors', unpublished doctoral thesis, University of Aston, Birmingham, England

Employment, Department of (1972) *Code of Practice for Reducing the Exposure of Employed Persons to Noise.* London, HM Stationery Office

Harris, H.E., Desieghardt, W.C., Burgess, W.A. and Reist, P.C. (1974) 'Respirator usage and effectiveness in bituminous coal mining operations.' *American Industrial Hygiene Association Journal,* **35,** 159–164

Hounam, R.F., Morgan, D.J., O'Connor, D.T. and Sherwood, R.J. (1964) 'The evaluation of protection provided by respirators.' *Annals of Occupational Hygiene,* **7,** 353–363

Howell, K. and Martin, A.M. (1975) 'An investigation of the effects of hearing protectors on vocal communication in noise.' *Journal of Sound and Vibration,* **41,** 181–196

Ivergard, T.B.K. and Nicholl, A.G. McK. (1976) 'User tests of ear defenders.' *American Industrial Hygiene Association Journal,* **37,** 139–142

Nixon, C.W. (1974) *Hearing Protection of Earmuffs Worn over Eye Glasses,* Report AD - 785 386, Wright Patterson Air Force Base, Ohio

Padilla, M. (1976) 'Ear plug performance in industrial field conditions.' *Sound and Vibration,* **16,** 33–36

Pritchard, J.A. (1976) *A Guide to Industrial Respiratory Protection,* United States Department of Health, Education and Welfare (National Institute for Occupational Safety and Health) Publication (NIOSH) 76-189, Washington, US Government Printing Office

Regan, D.E. (1977) 'Real-ear attenuation of personal ear protective devices worn in industry.' *Audiology and Hearing,* **9,** 16–18

Standards Association of Australia (1974) *Code of Practice for Industrial Eye Protection,* AS 1336:1974, Sydney

Toney, C.R. and Barnhart, W.L. (1976) *Performance Evaluation of Respiratory Protective Equipment Used in Paint Spraying Operations.* United States Department of Health, Education and Welfare (National Institute for Occupational Safety and Health) Publication (NIOSH) 76-177, Washington, US Government Printing Office

Wigglesworth, E.C. and Cole, B.L. (1973) *Vision and Its Protection. A symposium on visual efficicency and eye protection at work.* Sydney, Australia Optometrical Publishing Company

FURTHER READING

Aucoin, T.A. (1975) 'A successful respirator program.' *American Industrial Hygiene Association Journal,* **36,** 752–754

British Standards Institution (1959) *Filters for Use During Welding and Similar Operations,* BS 679:1959, London

British Standards Institution (1969) *Respirators for Protection Against Harmful Dusts, Gases and Scheduled Agricultural Chemicals,* BS 2091:1969, London

British Standards Institution (1970) *High Efficiency Dust Respirators.* BS 4555:1970, London

British Standards Institution (1971) *Positive Pressure Powered Dust Hoods and Blouses,* BS 4771:1971, London

British Standards Institution (1974) *Part 1. Closed Circuit Breathing Apparatus,* BS 4667:1974, London

British Standards Institution (1974) *Part 2. Open Circuit Breathing Apparatus,* BS 4667:1974, London

British Standards Institution (1974) *Part 3. Fresh Air Hose and Compressed Airline Breathing Apparatus,* BS 4667:1974, London
British Standards Institution (1974) *Part 4. Escape Breathing Apparatus,* BS 4667:1974, London
Brown, D.T. (1975) *Phase I in the Development of Criteria for Industrial and Firefighter's Head Protective Devices.* US Department of Health, Education and Welfare Publication (NIOSH) 75–125, Washington, US Government Printing Office
Coletta, G.C., Arons, J.J., Ashley, L.E. and Drennan, A.P. (1976) *The Development of Criteria for Firefighter's Gloves:* Vol. I. *Glove Requirements;* Vol. II. *Glove Criteria and Test Methods.* US Department of Health, Education and Welfare Publication (NIOSH) 77–134 A and B, Washington, US Government Printing Office
Coletta, G.C., Schwope, A.D., Arons, J.J., King, J.W. and Sivak, A. (1978) *Development of Performance Criteria for Protective Clothing Used Against Carcinogenic Liquids.* US Department of Health, Education and Welfare, Publication (NIOSH) 79 - 106, Washington, US Government Printing Office
Crockford, G.W. (1967) 'Heat problems and protective clothing in iron and steel works.' In C.N. Davies, P.R. Davis, and F.H. Tyrer, eds, *The Effects of Abnormal Physical Conditions at Work,* 144–156, London, Livingstone
Crockford, G.W. (1971) 'Heat protective clothing.' In *The Encyclopaedia of Occupational Health and Safety.* Geneva, International Labour Office
Home Office (1974) *Breathing Apparatus and Resuscitation,* Book 6 in the *Manual of Firemanship* series, London, HMSO
Industrial Safety Protective Equipment Manufacturers Association (1978) *Reference Book of Protective Equipment,* 5th edn. London, Industrial Safety Protective Equipment Manufacturers Association
Kvoes, P., Fleming, R. and Lampert, B. (1975) *List of Personal Hearing Protectors and Attenuation Data.* US Department of Health, Education and Welfare, Publication (NIOSH), 76–120, Washington, US Government Printing Office
Martin, A.M. and Whitham, E.M. (1976) *The Acoustic Attenuation Characteristics of 26 Hearing Protectors Evaluated Following the British Standard Procedure.* Institute of Sound and Vibration Research Memorandum No. 546, University of Southampton, England
National Institute for Occupational Safety and Health (1978) *NIOSH Certified Equipment List,* United States Department of Health, Education and Welfare, Publication (NIOSH) 79–107, Washington, US Government Printing Office
Smoot, D.M. and Smith, D.L. (1977) *Development of Improved Respirator Cartridge and Canister Test Methods.* US Department of Health, Education and Welfare Publication (NIOSH) 77–209, Washington, US Government Printing Office

SOURCES OF INFORMATION

American National Standards Institute, 1430 Broadway, New York 10018, USA
British Standards Institution, 2 Park Street, London W1A 2B5, England
Canadian Centre for Occupational Health and Safety, 435-150 Main, Hamilton, Ontario L8P 1HB, Canada
Department of Safety and Hygiene, University of Aston in Birmingham, Gosta Green, Birmingham B4 7ET, England
Health and Safety Executive, 25 Chapel St., London NW1 DT, England
Industrial Safety (Protective Equipment) Manufacturers Association, 69 Cannon Street, London EC4N 5AB, England
National Institute for Occupational Safety and Health, 994 Chestnut Ridge Road, Morgantown, West Virginia 26505, USA
Standards Association of Australia, 191 Royal Parade, Parkville, Victoria 3052, Australia
TUC Centenary Institute of Occupational Health, London School of Hygiene and Tropical Medicine, Keppel Street, London WC1E 7HT, England

22

The Contribution of Occupational Health Services to Safety

A. Ward Gardner

Most doctors and nurses in occupational health services consider health and safety at work as part of the same function, but health hazards and injury hazards are commonly regarded as separate issues. *Safety problems* are those which arise when cause and effect are closely related in time, for example, cases of mechanically induced trauma, acid burns and gassings, and in the United Kingdom these are classified as 'accidents'. *Health problems* are those which arise when cause and effect are not so closely related in time, for example, dermatitis from repeated skin contact with a dust or liquid, or pneumoconiosis which may occur after many years of exposure to dust. Thus safety problems are seen as those which produce *immediate* effects and health problems as those which result in damage to health after varying periods of exposure.

Such distinctions are artificial when viewed in terms of human suffering, economic cost, and inefficiency. The occupational physician and nurse should therefore be aware of the implications both of injuries at work and of occupationally induced disease. Although most workforces have much higher mortality and morbidity rates from non-work-related than work-related diseases and injuries, many *non*-work-related deaths and diseases are preventable, for example those related to smoking, road injuries and alcohol. Occupational health services are concerned with the prevention of all kinds of injury and disease (*see* Chapter 7), but it is the work-related diseases and injuries that they are best equipped to prevent. Several conclusions follow:

1. Knowledge of the risks of disease and injury is not enough to ensure that action is taken to prevent them.
2. Health and safety problems, although similar in many respects, are viewed and dealt with differently, both by individuals and by organizations because injuries have immediate effects, whereas in

disease the effects are delayed. For this reason the prevention of occupational disease and ill-health is covered separately in Chapter 25.
3.   As the largest amount of preventable disease and injury is not related to work, the health professions at workplaces deal with a small but important part of the total problem.
4.   Work injuries engender more emotion and more effective preventive action than non-occupational injuries largely because they involve arguments about who is at fault.
5.   Nearly all injuries are somebody's fault, that is preventable, and the principles of safety management are similar at work and elsewhere.

MANAGING SAFETY

Multiple causes of injuries

Theories of multifactorial aetiology of injuries and the epidemiological approach to prevention are now well understood by those experienced and trained in occupational health, but this cannot yet be said of the majority of physicians and nurses, nor of all but a few managers, supervisors and workpeople. An example may clarify the problems involved. A person painting a factory places a ladder against the wall, ascends the ladder and falls off. The incident could be described in a number of ways: the ladder slipped, the person fell, the ground was uneven, the nearby machinery was damaged by the falling ladder or by the falling person. These descriptions of the events are often stated as causes: the ladder slipped, the person fell, and so on. But these are incomplete views because they fail to ask *why* the ladder slipped and *why* the person fell.

The end results of the incident could be a broken leg (injury), damage to nearby plant and machinery (property damage), or no personal injury and no property damage. How we investigate an incident which *by chance* results in a broken leg or in costly property damage may be quite different from what we would do if we 'got away with it' and no injury or no damage resulted. The critical event was similar in each case (*Figure 22.1*): the person fell off the ladder. It follows therefore that investigation and corrective action should be similar regardless of the consequences which occur by chance. Important preventive opportunities can be missed if the no injury, no damage situations are not investigated. The questions which now present are related to why the incident occurred – was the ladder placed against a slippery surface, was the ladder lashed, or did the person behave unsafely by leaning out too far, or did he place the ladder at an unsafe angle? These events can be

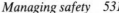

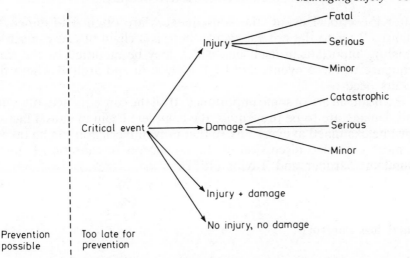

Figure 22.1 The consequences of the critical event

described as unsafe conditions (slippery surface, unlashed ladder) or unsafe acts (leaning out too far, placing the ladder at an unsafe angle). But there are still questions to be asked *why* did the person not spot the slippery surface? *why* did he not lash the ladder? *why* did he lean out? and *why* did he place the ladder at an unsafe angle? These questions deal with education, knowledge and attitudes, and reflect both on the kind of management and supervision of the person (why was he allowed to do these things at all?) and on the concern of the person involved to behave safely (did he know that what he was doing was safe or unsafe, and why?)

The chain of causation is therefore as shown in *Figure 22.2.*

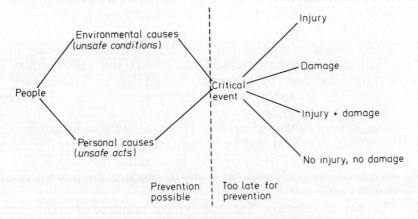

Figure 22.2 Causation, prevention and consequences

After the critical event, the consequences are often determined by chance. Prior to the critical event, there is a chain of causes each of which is important in itself and which may be modified by the time sequence of these events and by the place in and around where the events occurred.

Each and all are of some importance. If all the consequences of injury and damage are to be prevented, the complete chain of causal factors must be examined at every link so that preventive action can be taken. A more extended discussion of these concepts of causation is to be found in Gardner and Taylor (1975).

Total loss control

The concept of trying to prevent injuries, health risks, fires, property damage and spoiled work or products by examining causation is known as *total loss control*.

Against this background of multifactorial causation, there are five principles in the practice of loss control (Petersen, 1978):

1.   An unsafe act, an unsafe condition, an injury, property damage, or a 'near miss' are all indications that *something is wrong with the management system*. Each of these can be identified and recorded. Common factors can be sought. The evidence can then be evaluated so that action can be taken to control the problem(s).
2.   We can predict that *certain sets of circumstances will produce severe injuries or serious damage. These circumstances can be identified and controlled*. Searches should be made to try to identify potential hazards so that preventive action can be taken. It is not enough to react to things which have happend. A positive seeking out of new, known or potential hazards and problems should be made.
3.   Safety should be managed like any other important function in an organization. Management should direct the safety effort by setting *aims* and by *planning, organizing* and *controlling* to achieve the desired goals. The setting of aims should be precise and specific. Pious hopes are not a good target, whereas an aim to have 20 per cent less injuries and damage in the same group/area/ workforce next year may be a tough but realistic target for action. The effectiveness of the group in meeting the targets can be monitored monthly and the plans to produce the desired results can be reviewed at intervals.
4.   The key to effective line (safety) performance is management procedures that fix *accountability*. Managing safety is therefore no

different from good management of any other function. Performance should be measured and both personal and group responsibilities should be fixed so that line accountability is possible. It follows
that the management of safety is a line function and that the safety
expert is only an adviser to the line. These concepts enable people
to see and sustain their own parts in any safety activity and allow
those above them to exercise control by accountability.

5.   The function of safety is to *define and locate the operational errors
     which allow umplanned, unwanted and unforeseen events to occur.*
     This function can be carried out in two ways: by asking why the
     unplanned and unforeseen events occurred; and by asking whether
     known and effective controls are being utilized and if not, why not.

Petersen's principles therefore demonstrate that the management of
safety is similar to the management of other functions in an organization
and is *a part of* normal management not *apart from* it, as is often and
mistakenly supposed.

The same principles apply to prevention of disease – *see* Chapter 25,
which deals in detail with the training of management, supervisors and
workers to enable them to undertake their responsibilities for health
and safety. The skills both of the health and of the safety professionals
will supply technical input to the analysis of problems and to the
possible technical solutions to these problems. However, in all cases the
control of these problems involves line management, from the board to
the shop floor supervisor, and most important, the *active* participation
of all the workpeople if progress is to be made and success achieved.

LEGISLATION

Health and safety

Recent trends in world legislation recognize that health and safety are
inseparable. The United States Occupational Safety and Health Act of
1970 (OSHA) and the United Kingdom Health and Safety at Work Act
1974 (HSWA) are examples of the trend. Another innovation in these
Acts was to extend their application to all places of work, and to include
people and places of work which had been excluded; for example in the
United Kingdom mines and quarries were previously covered by special
Acts, and the self-employed are now covered for the first time. Thus the
scope has been widened and made more uniform by removal of
anomalies and exceptions.

Legislation imposes duties and standards. Standards are often set after a *discussion document* has been issued. The discussion document tries to anticipate what standards are desirable or necessary. This procedure allows employers, employees and their respective experts in particular fields, and anyone who has either a view to express or who has an interest in the matter to argue their case before more definitive steps are taken by issuing a code of practice.

The *code of practice* should embody all the desirable factors and, so far as is possible, should result from what is generally seen as good practice by those experienced in the matter in question, and should take notice of the comments made arising from the discussion document.

Duties can be spelled out by legislation or by a code of practice. Legislation is difficult to amend and takes time. Codes of practice can be swiftly updated and will, in time, acquire considerable status because it would be a rash employer or employee who deviated from a code of practice without cogent reason. The duty of employees to conform to safety practices and to co-operate in safety activities, both by using safety devices and by wearing protective equipment which is issued, is mandatory. Participation by members of the workforce through safety representatives is another recent trend which appears to be beneficial, both by creating greater awareness and by generating commitment.

In Britain, formal statements of safety policy are required under the HSWA and this, together with the representational and consultative provisions has undoubtedly heightened awareness of health and safety both in managements and workforces. Responsibilities have been laid down both for workpeople in terms of co-operation and compliance with safety procedures and for management with regard to the need for ensuring such procedures and compliance with them.

Compensation for injury

In most countries the law of tort exists. It allows an employee to take an action against his employer for damages for a work injury. Many who are concerned with occupational health and safety regard it as counterproductive to accident prevention. The prospect of litigation makes it difficult to frame, maintain and enforce effective health and safety regulations (Robens Committee, 1972). Tort may delay remedial measures for fear that this may be taken as an admission of negligence. Tort may also have adverse effects on relations between employer and employee (Royal Commission on Civil Liability: Pearson Committee, 1978).

While most countries, like the United Kingdom, have a dual system of tort and no fault compensation (e.g. industrial injuries benefit), some, notably Canada and New Zealand, have abolished tort for work injuries. The advocates of tort claim that it encourages stringent surveys by insurance companies and discourages unsafe practices by employers through fear of exposure in court. There is no evidence, however, that its abolition has lowered safety standards. The main objection to its abolition comes from two sources: the trade unions, because it would deprive the injured workman of a source of additional compensation available to other injured persons; and those who, on moral grounds, object to abandoning the common law principle of making up to the injured party, so far as money can serve, the loss and injury which has been sustained. The remedy may lie in the socially desirable objective of a no-fault scheme, with similar levels of payment for both sickness and injury whatever the cause. The case for such a scheme was put by Lord Beveridge in 1942.

'If a workman loses his leg in an accident, his needs are the same whether the accident occurred in a factory or in the street; if he is killed, the needs of his widow and other dependants are the same however the death occurred. Acceptance of this argument and adoption of a flat rate of compensation for disability, however caused, would avoid the anomaly of treating equal needs differently, and the administrative and legal difficulties of defining just what injuries were to be treated as arising out of, and in the course of, employment. Interpretation of these words has been a fruitful cause of disputes in the past. Whatever words are chosen, difficulties and anomalies are bound to arise. A complete solution is to be found only in a completely unified scheme for disability without demarcation by the cause of disability.'

INJURIES AND THE STRATEGY FOR THEIR PREVENTION

In choosing measures for preventing *injuries* a 'mixed strategy' should be employed which takes account of three phases of injury control (Hadden and Baker, 1980):

1. Preventing potentially injurious events (*the pre-event phase*, seen as the fields of safety and accident prevention);
2. Minimizing the chances that injuries will result (*the event phase*, exemplified by wearing eye protection and car seat belts);
3. Reducing the unnecessary consequences of injury (*the post-event phase*, which includes good first aid and skilled medical care).

Contribution of health professionals

The third phase of injury control – that of reducing the unnecessary consequences – is the one most frequently omitted in a safety programme. In many work populations a competent occupational health service, which generally includes good first aid, skilled nursing services and medical back-up, can reduce loss of time in the workforce from sepsis, badly treated injuries and other complications (*see* Chapters 7 and 8).

Another example of the effectiveness of this third phase is that the number of deaths from road traffic injuries in the United Kingdom over the last 25 years has declined appreciably. Although there is more traffic on the roads and the number of incidents is not lessening, fewer people die because of the greater skills in carrying out first aid, in transporting injured people safely to hospital, and in hospital treatment.

Health professionals have an important part to play in the pre-event phase when they carry out pre-employment screening of people who are in potentially dangerous jobs such as train drivers and airline pilots (*see* Chapter 9). Good records both of absences from work and their causes and of positive findings at health screenings can contribute to safety (*see* Chapters 9 to 13).

As alcohol consumption continues to rise in most countries, the predicted number of alcohol-dependent people is expected to escalate exponentially (de Lint, 1974). Alcohol-dependent people are well known for their destructive potential both to themselves and to others, both at work (Grant, 1979) and elsewhere. Many occupational groups are without information about the detection of alcohol-dependent people; in most work situations active cover-up of known heavy or occasional drinkers is common. Cover-up merely delays the onset of remedial measures and encourages the onset of more serious dependency and further acts of destruction. Alcohol-dependent people will always contribute a share of injuries and damage which is much greater in proportion than their number in the workforce. There is an increasing necessity to train managers and supervisors in how to detect and manage people who may have an alcohol-dependence problem. This is an area where health professionals can make a significant educational and diagnostic contribution to injury and damage prevention which will also have useful social by-products for the organization and for family life.

Choice of countermeasures

Doctors and nurses trained in preventive medicine recognize the contribution which epidemiology can make in the analysis of causation and in prevention (*see* Chapters 13 – 15). But this contribution is not seen by the majority of those who deal with injuries and damage. Their concepts of aetiology are generally in the class of 'the ladder fell', mentioned earlier in this chapter. Ross McFarland (1962) showed the value of epidemiology in relation to motor vehicle accidents and points the way towards the wider uses of epidemiology in promoting work safety and evaluating countermeasures so that priority can be given to those which *most effectively* reduce injury. Effective measures may not be the ones which are easiest to accomplish. For example, driver behaviour leading to unsafe acts is a common contributory factor to many road injuries. So psychological screening of drivers could be useful – but a seat belt strapped securely in place before a crash will be more effective in the prevention of most injuries.

SUCCESSFUL SAFETY PROGRAMMES

Management commitment

Without guards, machines are unsafe, but an effective safety programme depends on management commitment and good communications within the organization. Recent work by Cohen (1977) has demonstrated that strong management commitment to safety and frequent close contacts between workpeople, supervisors and managers on safety matters are the most important factors in successful safety programmes.

Physicians and nurses can stimulate management commitment to safety by talking with managers, unions and safety representatives about injuries which have happened, by discussing health and safety policy with management, with health and safety committees, and by stimulating the conscience of the organization with whom they work. Health professionals are in a position to talk to anyone in the organization and to ask questions like 'How many injuries will your group have next year?' and 'What will *you* be doing to ensure that your group will have less injuries in future?' Health professionals have also an important role in educating the workforce in the risks to which they are exposed and how to prevent them, and in ensuring that *everyone* in the organization knows how to carry out life-saving first aid and first aid for other injuries such as chemical burns. These have to be done correctly and immediately if lives are not to be lost needlessly and if serious chemical injuries are to be averted.

Workforce stability

Other factors related to successful programmes are workforce stability and personnel practices that promote stability, such as careful selection procedures, skilful job placement and promotion schemes which recognize the importance of the safety performance of managers. Early career training in health and safety, indoctrination in good safety practices and follow-up training have also been found by Cohen to contribute to safety performance, in addition to the more obvious factor of stringent standards of housekeeping and environmental control. The adaptation of conventional safety practices to the special needs of individuals and individual workplaces are also a significant factor in good safety at work.

Simonds and Shafai-Sahrai (1977) studied factors apparently affecting injury frequency in eleven matched pairs of companies and concluded that those related to lower injury frequency rates were: top management involvement in safety, better injury record keeping systems, use of accident-cost analysis, smaller spans of control at foreman level, recreational programmes for employees, higher average age of employees, higher percentage of married workers, longer average length of employment with safety devices on machinery. This list, which illustrates the multifactorial aetiology of incidents, injuries and damage, agrees well in principle with Cohen's work.

Education, information and discussion

There is a fear in the minds of some employers that the truth about risks and hazards should not be told in blunt and forthright terms in case it leads to demands for more money or causes apprehension among workpeople. Dishonesty in relation to the hazards of work is a recipe for mistrust and ultimately for disaster of one kind or another. Workpeople are for the most part adults who, given the necessary information in ways which they can understand, will be able to use that information to help themselves and others, and to co-operate in the measures which may be required to deal with hazards. Attitudes towards a safety programme can be turned sour by behaviour which fails to recognize the need for full and free communication of information about risks and the possible human error which may arise from a job. Positive encouragement to work safely can be effective only if both the workforce and the management have a full understanding of the problems and the desire to put active prevention into practice. *The safety policy of the organization must be understood by all.* Managers and supervisors must recognize their responsibilities with regard to safe working, and all the members of the workforce should be aware of the likely hazards of every job which he or she is likely to perform and how to prevent any unfortunate outcome.

Although this may appear to be both self-evident and obvious, the steps which convert these concepts into practical training are not always readily accomplished. Time, money and effort are always required and without such steps no health and safety programme can work and no serious and motivated participation in health and safety can be expected either from managers and supervisors or from the workforce. Training in health and safety can lead to improved industrial relations since it is an area of common interest to the management and the shop floor where polarization will be counterproductive both to the well-being of the workforce and to the organization as a whole. Indeed, health and safety are too important to disagree about. It follows therefore that discussions which lead to agreement and understanding are a *sine qua non* of any well directed health and safety effort in any organization.

CONCLUSIONS

In the fields of safety and in the prevention of incidents, injuries and damage, health professionals can stimulate management interest by asking questions about why injuries occur and about what are *tolerable numbers of unplanned events and of injuries in the organization.* By helping management to formulate practicable plans which will reduce or eliminate these unwanted and unforeseen events, good intentions can be made to work in practice. While there is a need for technical experts, it is also necessary to educate the whole workforce into taking positive steps on a day-to-day basis to reduce the number of injuries. If a management group cannot manage safety then it probably cannot manage anything well. The safety record of an organization is, and should be, recognized as an indicator of the capacities and calibre of its management.

A safety programme should be based on a mixed strategy which examines the pre-event phase, the event phase and the post-event phase so that the measures applied reduce both the number and the severity of injuries and damage.

Priorities will vary and as one problem is reduced, so others will emerge, requiring regular reviews of the situation so that changes can be recognized and corrected.

So-called accidents are, with very few exceptions, manageable phenomena, not visitations of ill chance.

REFERENCES

Beveridge, Sir William (1942) *Social Insurance and Allied Services*, Cmnd 6404. London: HM Stationery Office

Cohen, A. (1977) 'Factors in successful occupational safety programs.' *Journal of Safety Research*, **2**, 168–178

de Lint, J. (1974) *Alcoholism: a Medical Profile*, pp. 75–108. London: B. Edsall

Gardner, W. and Taylor, P. (1975) *Health at Work*, pp. 113–126. London: ABP

Grant, M. (1979) *Current Approaches to Occupational Medicine*, pp. 281–192. Bristol: John Wright & Sons

Hadden, W.J. and Baker, S.P. (1980) *Injury Control in Preventive Medicine*, 2nd edn, ed. D. Clarke and B. MacMahon. Boston: Little Brown & Co.

McFarland, R.A. (1962) 'The epidemiology of motor vehicle accidents.' *Journal of the American Medical Association*, **180,** 289–300

Petersen, D. (1978) *Techniques of Safety Management*, 2nd edn, pp. 16–27. New York: McGraw Hill

Robens Report (1972) *Report of the Committee on Safety and Health at Work*, Cmnd 5034. London: HM Stationery Office

Pearson Report (1978) *Report of the Royal Commission on Civil Liability and Compensation for Personal Injury*, Cmnd 7054–1. London: HM Stationery Office

Simonds, R.H. and Shafai-Sahrai, Y. (1977) 'Factors apparently affecting injury frequency in eleven matched pairs of companies.' *Journal of Safety Research*, **9,** 120–127

23

Mental Health of People at Work

Alexis Brook

Concepts about the mental health of people at work are changing. The traditional role of occupational psychiatry is to identify the mentally ill, to refer them for treatment and assist in their rehabilitation, or to arrange for their discharge. Various studies, however, have led to the realization that the way an organization is administered, the degree of work satisfaction experienced by its employees and the quality of its morale may affect the mental health of the individual (Jaques, 1951, 1970; Bridger *et al.*, 1964). Increasingly, therefore, attention is given to studying those factors in an organization which may contribute to mental ill-health and those which promote mental well-being (Tredgold, 1949; Kearns, 1971; McLean, 1974). The broader concept of occupational mental health is related to recently developed ideas about 'the quality of working life' and 'the social responsibility of industry'.

In assessing a person's well-being much attention is given to eliciting information about such aspects of his personal environment as the nature of his family life and the quality of the relationships within the family. Comparatively little, however, is given to understanding the working environment although the average adult spends about half his waking time during the week at the place of work. The way the organization that employs him is managed can have a profound influence on his general well-being and mental health and, indirectly, on that of his family. Many organizations pay considerable heed to general working conditions and provide, for example, congenial canteens, generous holidays and good pension schemes. However important, these are not the factors that actually provide the setting for the achievement of work satisfaction and a state of good morale, which are not only essential elements for the maintenance of mental health, but are also the factors that motivate an individual to work to the best of his ability (Herzberg, Mausner and Snyderman, 1959).

The primary function of most organizations is to produce goods or provide services. Over the past ten years there has been increasing concern to reconcile the main objectives of an organization with the human needs of the people who work in it. This concern is reflected in three ways: how to make work safer, how to make work physically healthier and how to make work more satisfying. When managements give consideration to these problems they usually do so in that order with little, if any, attention being given to mental health. Occupational physicians and nurses, whose main duties are to promote safety and the physical health of the workforce, have, however, recently taken an interest in the mental health of people at work and in problems of work satisfaction.

THE FIELD OF MENTAL HEALTH

The field of mental health of people at work can be regarded as a pyramid, with the tip representing the less common but severe forms of mental illness, and the base the more common but blurred boundary where normal emotional difficulties merge into ill-health. These common disorders may usefully be considered in two categories which are not mutually exclusive and are probably about equal in size. The first contains those unconnected with the work situation, which possibly stem from personal or family difficulties but which reveal themselves at work. The second contains those which arise specifically from the interaction between the individual and his work situation, and which are known collectively as 'stress disorders at work'.

The term 'stress' has been employed for many years to describe the interaction between an individual and his environment. It does not, however, have a precise scientific meaning and is used in different ways. The terminology in this field has been described as a semantic mess (Murrell, 1978). There are two commonly accepted but diametrically opposed meanings to the term 'stress'. One, derived from engineering, is that the stimulus is 'stress' and the resulting effect on the individual is 'strain'. The other, introduced by Selye (1957) is that the stimulus is known as the 'stressor' and the reaction it causes is 'stress'. In practice the term 'stress' is used simultaneously in both senses, and this leads to confusion. 'Stress at work' describes the whole range of difficulties that may occur in the interaction between the individual and his working environment, its only value being to direct attention to general problems in this area. In some cases organizational factors are more important than individual factors.

It is not possible to give an exact figure for the incidence of stress disorders at work. Although statistics of spells of sickness absence and of days of incapacity, indicate 'mental, psychoneurotic and personality disorders' as a group, this leaves out those illnesses categorized under such headings as 'respiratory disorders', 'rheumatism and allied conditions', 'disorders of the digestive system', 'skin diseases', or 'ill-defined conditions' where psychological factors have been significant in their aetiology. Nor does it allow for those disorders which present with behavioural symptoms. Fraser's (1947) detailed survey of a large number of factory workers indicated that over 25 per cent of all sickness absence was due to neurosis. This figure is probably still a fair indication of the incidence of absences due to psychological causes.

PRESENTATION OF PSYCHOLOGICAL DIFFICULTIES

Psychological difficulties may manifest themselves in the individual and in the organization. There is no consistent correlation between the external factor and the type of disorder. Symptoms in the individual can be mental, somatic or behavioural. The common mental symptoms range from feeling worried, apprehensive or dissatisfied (which have nothing to do with being ill), through feelings of strain, tension and irritability to acute anxiety and severe depression. Somatic symptoms or physical illness may be the outcome, these being entirely or partly the expression of the underlying psychological difficulty. In this group are symptoms such as headaches, muscular aches and indigestion, and bodily ailments such as upper respiratory tract infections, skin conditions and some cardiovascular disorders. Behavioural symptoms may take such forms as quarrelsomeness, poor work, accident-proneness, absences from work or frequent job changing. The common manifestations in the individual of psychological difficulties between him and his working environment are poor work, a general diminution in his feeling of well-being and various forms of ill-health.

Psychological difficulties may manifest themselves in the organization as a whole, usually by a rising incidence of the various symptoms described above, by a general deterioration in morale, or by an increasing rate of absenteeism or of people leaving the organization.

The problem may first come to the attention of the department manager or foreman, the shop steward, the personnel manager, the training officer, welfare officer, occupational nurse or physician or the individual's general practitioner. The involvement of the occupational physician usually begins when a worker is referred to the medical department. This may have been done by his line manager, personnel manager or training officer with a specific request for help; he may have been referred following formal assessment procedures; he himself may have asked for help; or the medical department may have observed the need for it. It is helpful for the doctor to bear in mind three aspects of assessment.

1. The individual.
2. The individual and the organization.
3. The individual as the presenting symptom of problems of the organization.

ASSESSING THE INDIVIDUAL

The way an individual reacts to his external environment depends on such factors as his hereditary endowment and his life experiences. If his stability is undermined by neurotic conflicts or personality problems he may be much more prone than others to react to quite minor difficulties. Depending on the sort of person he is he may interpret an external frustration as a minor difficulty to be taken in his stride or as a major threat to his whole personality, reacting, for example, with gross anxiety or a paranoid attitude or by withdrawal from work. One reason why people change jobs frequently is because of unresolved internal conflicts which prevent them from coping with external difficulties.

There are numerous individual factors that vary from person to person. Although difficulties may occur at any age people in general are more liable to produce symptoms of stress during phases in life where personal psychological readjustments have to be made. Those that have particular relevance to the work situation are late adolescence and various points in middle age.

Late adolescence

The late adolescent not only has to make the transition from school to work, but also has to cope with the identity problems and emotional readjustments involved in finally relinquishing dependence on parents and becoming an adult. Symptoms that may alert the doctor to conflicts in this area include a history of frequent job changing, difficulty in settling to a job, absenteeism, problems in training and frequent spells of sickness with an acute exacerbation of symptoms at the time of examinations. This difficulty over examinations often presents management with a dilemma about retaining or dismissing the adolescent, particularly if he is intelligent, has ability and if much effort and money has been invested in training. If medical advice is sought a psychiatric assessment often indicates that the basic problem is connected with the internal psychological conflicts of this particular phase of readjustment which may cause a temporary reduction in the individual's capacity for coping. The value of identifying an individual's problems at this phase is that a small amount of help can often not only enable him to surmount this difficulty but it can also facilitate his maturation and emotional growth.

Middle age

From a psychological point of view middle age is a period of about 25 years, starting shortly before the age of 40 years and extending to soon after 60 years. It spans the middle years between early adult life and late adult life, and one can consider early middle age, middle middle age and late middle age.

Early middle age

Middle age can be considered as beginning at mid-life, the phase of transitions from the first to the second half of life. Awareness of this change gradually crystallizes over several years, from about 37 onwards. By 40, or certainly in the early 40s, most men and women sense that some fundamental change has occurred. The more satisfied an individual is with his achievements in the first half of life the more satisfactorily will he be able to adjust to middle age. Anxiety at this age is often related to feelings about fulfilment and becomes closely linked to an intense concern about career development. Earlier hopes and ambitions may, at this time, have to be reconciled with more realistic possibilities. Difficulties may express themselves in various indirect ways, especially if the individual is largely or entirely unaware of his underlying conflicts. He may develop mental symptoms ranging from mild anxiety to severe depression or various psychosomatic reactions. Expressions of dissatisfaction with his spouse may be used to mask disappointment or other painful insights about himself. Alternatively, a difficult situation in real life may be used as a convenient vehicle into which to project inner conflicts in an attempt to be rid of them. The job itself, for example, may then be felt to be the source of his frustration, or his seniors or even the organization as a whole may be felt as the cause of his disillusionment. Examples of these ways of trying to avoid acknowledgement of conflicts stemming from internal doubts and anxieties can frequently be seen in those who are in a midde-management position where they have to cope with the double burden of internal conflicts and the increased pressure of their actual responsibilities.

Middle middle age

The age of 50 is often felt as an important turning point – the last opportunity to make a career change. Symptoms, particularly of anxiety or depression, in the year or two before, are frequently related to conflicts in this area.

Late middle age

A man who has had a very satisfying middle age will usually be better able to deal with the next major phase of readjustment which is from the end of middle age to the beginning of old age. This phase is frequently permeated by feelings about retirement which often loom large from about the age of 57 or so. Many people of this age have reached their goals in life, are well satisfied with their achievements and looking forward to retirement. Others may be very apprehensive about the adjustments that have to be made. The manager, for example, who has always thrived on having people coming to him with their problems, may dread the time when this will no longer be so and when eventually the situation may be reversed. There may be difficulties over mourning the loss of work. Worries about retirement often mask the deeper fears of old age, dependency, loneliness and, basically, death. Anxieties about these topics contribute to the varied symptoms that develop in late middle age. Many organizations are beginning the feel the obligation to try to help people over this boundary from work to retirement.

THE INDIVIDUAL AND THE ORGANIZATION

Work and mental health

Although work is necessary as a means of livelihood, it also offers the individual the possibility of meeting his interests and fulfilling his personal needs. A real interest in the actual job can in itself be a source of personal satisfaction. Those situations which most people find satisfying have been described by Herzberg, Mausner and Snyderman (1959) as including interesting and challenging work, genuine responsibility, opportunity for achievement by the individual, recognition of that achievement, and scope for individual advancement and growth. A major factor is that the requirements of the job should stretch the person's abilities to the full. He is then receiving constant reassurance of what he can do and it is this that enables him to maintain a realistic adaptation to life. Moreover, if he can identify with the aims of the employing organization, and if it provides a secure framework in which he can work and where his role is clearly defined, this will contribute further to his work satisfaction, particularly if the management is genuinely concerned with his work problems. These are some of the factors that provide the opportunity of developing a sense of achievement and well-being and of experiencing personality growth.

For most people, however, the main or the only gratification from their work is the pay. Particular examples are manual and semi-skilled workers, the majority getting little work satisfaction. Many maintain that the money is their only concern, but their frequently expressed dissatisfactions and complaints, particularly about boredom with the job, tend to belie this. Dissatisfaction and frustration mobilize the individual's greed, envy and jealousy, leading to destructive attitudes and often despair. The mental health of the individual will in turn affect the state of morale of the whole organization.

The interaction between the individual and the organization

Some common situations in the interaction between the individual and his working environment that lead to an impairment in general well-being or mental health are considered below.

Overwork

It is helpful to differentiate between overload and overstretch. Overload refers to a demand for extra work which is within the individual's capacity. Reference has been made to the importance, for mental health, of a person having his abilities stretched to the full. In such a situation overload for a brief period is not stressful but is often felt as a challenge. For a longer period, however, it can be a gradually mounting strain. Overstretch, when the person is stretched beyond his abilities, is different. A common example is over-promotion. If the individual's skills are increasing and the promotion is not excessive and support is adequate, he will usually grow into the job successfully and the temporary strain on him and his colleagues is usually justified. However, if a promotion is excessive or comes at a time when he has reached the peak of his abilities, the effect on him and the result of this on his colleagues will be considerable. He will feel under strain, take work home and sleep badly, his competence at work will decrease and he will become anxious. The situation usually deteriorates until a point is reached when his distress is redefined as illness and he is referred for medical help. The organization has then, in a sense, exported the problem because it is no longer felt as one for the management. Recognizing this type of situation before an individual is promoted, or in the early stages of a mistaken promotion, is not easy, but the price paid for a failure to do so is very high in terms of the distress to the individual and the consequences for the organization.

Under-work

When the individual is understretched he becomes dissatisfied. He will then begin to mistrust his own capacities and gradually become demoralized, and this often expresses itself in poor work or in frequent spells of absence for minor illness. Garland (1966) has given a vivid description of this condition which can lead to a gradual impoverishment of the personality. One of the major tragedies of prolonged unemployment is that the individual can become so demoralized as to become quite unable to work. Overmanning, which leads to under-work, is a fact of working life in many organizations; people work less to enable others to maintain jobs and to avoid redundancy. The increasing application of automation will probably result in people working a much shorter week. Most leisure activities do not meet the same needs that can be satisfied in employment, and very little research, as yet, has been carried out into the other types of activity that will be necessary to fill the gap.

Unsuitable promotion

Almost invariably promotion involves an increasing need for the exercise of managerial skills. Certain personality attributes help a manager to carry out his role. These include the ability to work in situations of conflict, to tolerate disagreements, criticisms, envious attacks and unpopularity. A manager should be able to tolerate uncertainty and to contain anxiety, not only that aroused by specific situations but also that aroused by the anxieties off-loaded on to him by others who are unable to contain them themselves.

A common example of unsuitable promotion is that of a skilled technician who lacks the necessary personality resources for a managerial post. For many years he may have worked well, obtaining satisfaction from using his technical skills and from the appreciation of others for his advice. His promotion to a managerial role not only deprives him of much of the satisfaction obtained in his previous role but also places him in a situation where he is unable to cope. Eventually, some form of breakdown may occur, frequently of a depressive nature. Another example is the promotion of a person with a strong personality to a role which does not contain adequate opportunities for him to exercise it. This leads not only to personal frustration but also to a distressing situation for his subordinates, as he is constantly exercising authority beyond the limits of the role.

The reverse situation, when an individual is appointed to a role which demands more exercise of authority than he has the personality resources to meet, leads to a situation of great difficulty. He is constantly labouring under a feeling of stress and his subordinates usually feel very insecure. As a way of coping he may talk to his subordinates in such a way as to undermine their self-confidence, making himself feel that his troubles are due to his inadequate staff. He will complain about them constantly and come to feel, quite incorrectly, that he can never take a holiday because his department could not manage without him.

The strain of uncertainty

The anxieties aroused by uncertainty are usually particularly hard to tolerate. Some people thrive on anxiety but most have varying limits of tolerance, both for the intensity and for the time over which the anxiety has to be tolerated. Jaques (1961) has described in detail the psychological aspects of the time-span over which uncertainty has to be endured before the outcome of a decision is known. A manager, having considered various alternatives, makes his decision, and then has to contain his uncertainties until it is known whether his decision was the most appropriate. He may not know the outcome for weeks or months, and how he copes depends on many factors. One of the most important is the quality of the support he is given by his own manager. Insufficient support given to middle managers is one of the most potent factors in impairing their capacities, in diminishing their efficiency, in undermining their well-being and, sometimes, in leading to physical or mental ill-health. On the other hand too much support can be resented intensely. Commonly the individual is not only provided with no support by his senior, but his senior — insecure in himself — projects his own uncertainties into that individual.

Withdrawal of elements that lead to work satisfaction

Some people enjoy their work and find that it fulfils them in every way. Most, as already suggested, find work no more than a livelihood. For many there are only a few areas of satisfation and these are then highly valued. A frequently felt area of satisfaction is the opportunity to determine one's way of working. Frankenhaeuser and Gardell (1976) have demonstrated the importance of this type of discretion in leading to work satisfaction. The loss of such a valued area of work satisfaction as a result, say, of a reorganization of work tasks can lead to demoralization and maybe to depression.

Role unclarity and role conflict

Unclarity (or ambiguity) sometimes occurs in the content of the role. More often, however, it is associated with accountability, namely, who is accountable to whom and for what. The work situation can become very stressful if the requirements conflict, particularly if there is too little time in which to complete them. Role conflict is experienced by the individual who is on the boundary of two groups with differing interests and who himself feels equally strongly about the needs of each. A junior manager may find himself identified with the administrative need to implement certain decisions but at the same time identified with the workforce's opposition to them.

Difficulties in relationships with others

More waking time during the week is spent in work relationships than in family relationships. Many interpersonal difficulties may arise in a working group, leading to a degree of stress with which the individual may or may not be able to cope. The most notable are those that a person encounters with his seniors and with his subordinates. He may have a senior who delegates too much work or too much responsibility. More commonly, the senior has problems with delegating sufficiently, the difficulty usually being over delegating responsibility. The senior may regard himself as the personification of responsibility and all subordinates as the embodiment of irresponsibility. In such a situation the subordinate is left frustrated because he is deprived of the full use of his decision-making abilities – one of the essential elements for job satisfaction. On the other hand, the subordinate may have difficulty in accepting the authority of the senior because he himself cannot tolerate the feelings of dependency that can arise in such a relationship. A senior has authority over his subordinates but depends upon them. The subordinate can undermine the confidence of the manager by unreliability so that the manager is never sure if his instructions will be carried out. An already insecure manager will find this situation particularly stressful. A similar type of situation is created when, for example, a middle-aged manager is beginning to break down as a result of over-promotion. A considerate director sometimes tries to help by providing him with an assistant. However, if this assistant is efficient, the strain on the manager can be greater if he has problems of rivalry and fears of his inadequacies being increasingly revealed. On the other hand, the rivalry that a senior may have towards the junior can lead to his unconsciously blocking the latter's advancement. Moreover, the junior may not appreciate the subtle forces at work and may become very frustrated without realizing why.

A particular manager who is a major threat to the mental health of his subordinates is one who cannot tolerate his own inadequacies, and constantly tries to foist them on to his subordinates. He may then talk to them in such a way as to arouse their anxieties and undermine their self-confidence, and then try to use this situation to establish more confidence in himself by implying that it is they who are inadequate and telling them what to do. On the other hand, the manager whose personality is characterized by sincerity can make even potentially unstable subordinates feel secure.

Problems at the shop floor level

The employee has little control over decisions about his life at work. Argyris (1970) has pointed out that at this level abilities more central to self-expression are used minimally, and for many people this results in boredom and frustration. A particular problem is coping with feelings of resentment over having to submit for so much of the time to the decisions of others. This applies particularly to work on the assembly line. The Volvo car factory at Kalmar has gained renown for its production technology, which has eliminated the traditional assembly line in order to give employees greater opportunities to influence their own work.

THE INDIVIDUAL AS THE PRESENTING SYMPTOM OF PROBLEMS IN THE ORGANIZATION

The question to be posed is how much the individual is a patient in his own right and how much he is the presenting symptom of problems in his own unit or in the whole organization. If any one unit constantly shows a high incidence of stress symptoms among the individuals of which it is composed, this is an indication for examining whether the organization itself may not be causing the stress. Equally it might be useful to consider whether a person in any one unit who is showing symptoms is doing so not because he himself is necessarily ill, but because that unit is using him as a vessel to carry the tensions which it has not been able to resolve. This is often shown by the fact that his removal from the unit simply results in the eventual casting of someone else in this role.

What at first may appear to be manifestations of difficult personalities are often reactions to faults in the design of the system or of the way in which it is operated. An example may illustrate this. Two departments of about 30 people each were merged for administrative reasons. One department wanted the merger whereas the other did not. This problem was not adequately tackled and an unsatisfactory structure was devised for the new combined department in which role requirements were unclear and there was confusion as to accountability. Over the next year there was a great deal of dissatisfaction, poor work and a high rate of absenteeism. An enormous amount of time was spent discussing a rota for certain duties (this is usually pathognomonic of other more painful difficulties that cannot be faced). Finally, the manager of the merged department resigned. This brought matters to a head, causing senior management to recognize that it was the organization and its structure that were primarily the cause of the trouble rather than the individuals who worked in it.

The quality of concern of management towards the people they employ is a major factor affecting morale. Revans (1969) in a study of the attitudes of staff towards innovation found that those factories in which workpeople felt that their managements listened to their opinions and suggestions showed general satisfaction with their working conditions. A further important factor is whether the management is felt to be genuinely concerned for the personal well-being of their workers rather than by considerations of motivating them towards greater output.

The quality of management concern is reflected, for example, in the way in which change is introduced, whether as a result of technological discoveries or for administrative reasons. Well-planned changes minimize the number of people who develop symptoms of stress, whereas those that are badly planned have the reverse effect. Brook (1978) has described in some detail the problems associated with the process of change. The effect of being involved in a period of change in an organization depends on several factors: the positive feelings that people have about it, the anxieties that it mobilises in both the individual and the group, and the manner in which it is conducted. This last factor is often the most significant. Experience suggests that a change instigated entirely in the interests of the organization without proper consideration of its effect on the people concerned is a very potent stressful factor.

This underlines the importance of ensuring that as much attention is given to the human consequences of the proposed change as is usually given to its material aspects. This is one of the major factors in ensuring that changes occur with minimal disruption of the well-being, morale and health of the staff and organization. Whether this is achieved depends on the management philosophies of the organization, the personality of the manager who has to administer the change and on the type of change that is planned.

Development of mental ill-health in an organization

There are three stages of development of mental ill-health. If the cause is recognized, the first stage can be the preventive one, the key factor being whether the possible effects on the employees of the way in which they are being managed are considered seriously. No doctor is involved and the responsibility rests at management's door. In the second stage minor symptoms may already have developed in several individuals, or morale may be dropping in the unit as a whole. There is vague general unease but definite symptoms are not yet apparent. In the third stage there are symptoms of mental ill-health in the form of physical, mental or behavioural disorders as outlined above. Problems in the first two stages may be noticed by the manager, the personnel or welfare officer, the occupational health nurse or the doctor. By the third stage a doctor is certain to be involved as clinical symptoms are clearly recognizable. Identification of the problems in the second, or preferably in the first, stage is the responsibility of the management, and increasing attention is now being given to what is called the social responsibility of industry. Commonly, however, managements do not deal with problems until the third stage by which time the responsibility for many of them can be transferred to the medical services.

As already indicated, psychological conflicts at work are largely the result of the interactions between the individual and the organization. Both factors are equally important but some individuals are particularly vulnerable.

Vulnerable individuals

Some vulnerable individuals can be identified:

1. People at phases of readjustment such as middle-age.
2. People who have reached the peak of their abilities, particularly if this is in late middle-age when they are becoming less adaptable.
3. People with personality difficulties, in particular, those with marked obsessional traits for whom change is a threat to their security. Erskine and Brook (1971) have described a group who are conscientious and methodical, who have been in the same job for many years and who have over-invested their emotions in the job which has thus become over-valued; their work situation has become their 'way of life'. The people in this group who were found to be specially vulnerable were men in their mid-fifties who, had there not been a change, would probably have continued in their jobs for another 10 – 12 years, to retire with a sense of achievement and dignity. The need to adapt to a change in methods of working

severely undermined their feelings of security and precipitated a breakdown, which usually took the form of anxiety and depression with a marked tendency to grumbling, partly realistic and partly paranoid.

4. People with family difficulties, the strain of which may lead to chronic tensions at work. Alternatively, work may be used as a compensation for lack of satisfaction at home and then the job or the work situation may become over-valued.

Vulnerable roles

A role is a position in the organization which is occupied by an individual. Some vulnerable roles can be identfied:

1. As already indicated, any role in which there are unclarities over accountability imposes a greatly added burden on its occupant.
2. The role of middle manager is potentially a very vulnerable one. He, in particular, has to reconcile pressures from above to ensure the maximum output in the interests of the organization with demands from below for the personal needs of the people in the work situation to be understood, contained and supported. His position becomes particularly difficult if the organization is undergoing a process of change where there is pressure from above to effect the change in the interests of the organization and a demand from below for the containment of the anxieties stirred up by it. This may be complicated if the manager himself has personal anxieties about the change, feels ambivalent about it and cannot easily accept management policies. If the occupant of such a potentially vulnerable role is also a vulnerable individual his insecurity will increase and so justify his subordinates' feeling that he is inadequate. They may then intensify their criticisms of him which may lead to his eventual breakdown.
3. An unusually susceptible role is created in the vacancy left by a manager who retires after having held his position for a long time. The mixed reaction to his departure usually contains very strong negative feelings which are now liberated. His successor finds himself in a highly vulnerable situation having to bear the brunt of the resentment while coping with the usual anxieties of new responsibilities.

Effect of an individual's psychological difficulties on his group

A group can be seriously affected by the psychological difficulties of the key members. For instance:

1. A manager who has his own problems which he cannot contain.
2. A manager who uses his subordinates for his own needs, such as gratification of power, stimulates feelings of dependency and resentment.
3. A group where two key people at the top, such as the manager and his deputy, are in a disagreement which cannot be verbalized or resolved.
4. Groups which receive inadequate communication from the top; this implies not only lack of factual information but also lack of support to the group's role.

In these situations minor mental, psychosomatic or behavioural symptoms may develop in many members of the group.

GROUP PROCESSES

As soon as someone joins an organization he becomes a member of several groups: for example, the organization as a whole, his department or unit, and his particular work group the members of which are accountable to the same manager; depending on his position he may also belong to a managerial group. Emotional conflicts within the large or small groups or units of which the organization is composed are inevitable, however much the individuals concerned want to work together. When in addition, there are the accumulated effects of the numerous possible 'misfits' between the individual and his organization that have already been described, feelings of frustration or anger may become mobilized and are very difficult to deal with, particularly if the source is not understood. The customary working methods of a group form a social structure which helps to keep its anxieties under control. A vivid and detailed description of the functioning of social systems as a defence against anxiety in an institution has been given by Menzies (1960).

At times group behaviour may predominate over individual reactions. While wanting to deal with their problems realistically, people, without realizing it, often feel trapped in certain modes of group behaviour described by Bion (1961) as 'basic assumption groups'. For instance, they may be in the state of a 'dependency group' and while sounding as if they are making a genuine attempt to solve their problems are really looking for a leader to protect and rescue them. On occasions they may be in the state of a 'fight/flight group' taking action as a means of coping with their conflicts, which they are unable to resolve in any other way. Action, therefore, may take the place of rational thinking instead of being preceded by it.

Individually and collectively the people are struggling with the feelings of distress, anxiety, anger and confusion that have been roused within them and are trying to locate responsibility as a way of experiencing relief. Workers may identify the management as the source of confused and irrational thinking and irresponsible behaviour; management, equally sincerely and equally incorrectly, may come to the same conclusion about the workers. It is the organization that is now sick, each group being paranoid in its attitude toward the other. In these situations it is hard to re-establish meaningful discussion. Occasionally a sufficiently capable leader may then emerge whose aim is to try to re-establish rational thinking and effective communication, and the group may follow him in this course. More often, however, a leader arises whose aim is to meet the irrational needs of the group and to lead it in its fight against wherever the fault is felt to lie. The strength of the wish for such a leader is often so great that one who supports rationality is overthrown. This whole process has been more vividly described in a novel by Golding (1958) than in most textbooks.

ROLE OF THE OCCUPATIONAL PHYSICIAN

The investigation, the arrangement of treatment and the rehabilitation of those designated as sick

The traditional medical model is to look for an illness to treat. However important this is, it frequently reflects only one aspect of the situation. As indicated throughout this chapter, symptoms in the individual are often the outcome of difficulties in the relationship between the employee and the organization. It is important, therefore, for the doctor to resist the pressure that is often put on him to regard his job as simply to look for an illness to treat. An essential part of his assessment is to examine the interaction between the employee and the organization.

Many doctors feel that they should give prime consideration to the best interests of the patient. Others feel that they should be mainly concerned with the interests of the organization that employs them. Usually, however, the doctor finds himself trapped in a conflict of loyalties, an uneasy situation for him in which he feels worried and often guilty, either towards the individual or towards the organization. To reconcile the best interests of the individual and the organization is one of the most difficult tasks confronting occupational physicians.

Most people from time to time develop minor symptoms as a result of psychological conflicts at work, and a few days off work enable them to muster their resources. The clinical state that allows an individual to do this is often one of the minor physical ailments, his lowered state of well-being having led to his developing the illness. The need to be off work is usually connected not only with the actual ailment but with the underlying stress. This type of short sickness absence can be considered as a self-regulatory mechanism which may prevent more serious illness.

Requests for a longer period of sickness absence are more complicated. For example, a patient suffering from a depressive breakdown may have been prescribed a period off work, as well as a specific remedy. After a month or so the patient, although feeling much better, may be reluctant to return to work. It is often not appreciated that his desire to have a further period off work may be as much because of his difficulties at work as because of the more recognized clinical symptoms. It is sometimes difficult in any one case to know whether allowing a person a brief period off work is in this way directly helpful to him in enabling him to muster his resources rapidly, or whether it is quite unhelpful and might even facilitate the development of chronicity. Decisions can only be made by greater understanding of the problems between the individual and his organization that led him to take refuge in absence from work.

If someone is off work for any length of time there are many factors that influence the outcome. A crucial one is the strength of the relationship between the individual and the organization. If they want him back and he wants to return the chances of his recovery from those symptoms resulting from the conflicts at work are high. If the reverse is the case, then those symptoms are much more likely to persist until some alternative outcome is found.

Rehabilitation of those who have been off sick with mental disorder involves not only helping the worker to return, but also helping his unit to accept him back. While it is the patient's state of mind which is important in determining the success or otherwise of rehabilitation, it has frequently been found that the attitudes of his colleagues are equally important. Their attitude will be affected, in turn, by the degree of support they receive. The appropriately trained mental nurse by working both with the patient and with the foreman can exercise a major influence in this field. If a foreman can feel that he himself is fully supported, that his feelings, both rational and irrational, about having back in his unit a man who has had a mental breakdown have been verbalized and clarified, he will be more able to tolerate him and help him. Furthermore, he will also be able to help to provide the climate in the unit which will enable the workmates to support him.

Early identification

Early identification of mental ill-health is in the best interests of the individual because it can lead to early treatment. It also means that the patient is more likely to receive treatment while continuing at work, which is in the interests of the organization as well as of the patient. Early identification is the key to preventive medicine because if the stressful factors in the working environment which result from the way in which the organization is administered can be identified this information can be fed back to management. There are several methods for early identification of problems. One is by training, so that doctors, nurses and personnel managers are well aware of them.

Training and education

Closer collaboration between medical and non-medical staff can lead to earlier identification and treatment of many problems of stress at work. To enable them to make decisions, nursing and non-medical professional staff need training appropriate to their roles; for example occupational health nurses who are trained in the early identification of mental ill-health can often recognize symptoms before they become severe. In a survey carried out by an occupational physician and a psychiatrist who were jointly studying some psychiatric problems in industry (Erskine and Brook, 1971) eight out of 60 patients referred for assessment were found to be suffering from undiagnosed clinical depressive illnesses, some of them being quite severe. The presenting problems had been general unhappiness, irritability, unsatisfactory work and frequent spells off work, the duration being from three months to two years.

The occupational health nurse, as the 'wise woman of industry', often becomes the worker's confidante. Counselling has, for some years, been recommended as part of her role, but she herself is only able to carry this out if she has received appropriate training and has adequate support from the physician.

Collaboration between the occupational health physician and the other professions in the organization

Collaboration develops most effectively when the doctor spends a significant part of his time out of his medical department, learning about the organization, visiting all parts of it and getting to know people in their work settings.

Doctor/manager relationships

Many managers, particularly personnel managers, welcome the help of the doctor in understanding some of the psychological problems they encounter in their work. Others, however, would rather not be informed about such problems. A manager who formulates his problem as being due to subordinates who are suffering from psychological difficulties often only wants to transfer responsibility entirely to the doctor. The doctor, however, on the basis of his knowledge of the organization and its structure and administration, may feel that it is his duty to draw attention to difficulties in the organization that might be contributing to the symptoms. Thus, without realizing it, manager and doctor often have conflicting basic aims and so can quickly lose patience with one another. The problem is more complicated than the question of who copes with an irksome burden: it is linked to the dilemma that confronts anyone in a position of responsibility.

For the manager the dilemma lies in how to overcome both the anxieties of the people over whom he has managerial responsibility and the frustrations, uncertainties and at times sense of helplessness that he may feel in the process. The doctor has to cope both with the distress of the manager and with the worries and uncertainties that the manager may cause in him.

The anxieties of the workpeople, manager and doctor, all become concentrated in the doctor/manager relationship. Problems are certain to arise between doctor and manager if the interpersonal aspects of their work are ignored; this can lead to the rapid breakdown of a potentially valuable working relationship. Responsibility for decisions about administration lies with management and it is only when the doctor/manager relationship is non-conflicting that the doctor's help can be accepted. Even so, doctors and managers may be apprehensive about working together. There may be many areas of misunderstanding and the organization may be worried that a doctor's involvement would seriously disturb its equilibrium. Although personal incompatibility between manager and doctor is sometimes an obstacle to collaboration it usually transpires that such difficulties in fact originate in uncertainties inherent in changing the established relationship between management and doctor from one in which the occupational physician is concerned solely with illness, to one in which the doctor is also an adviser taking a special interest in human relations within the organization. This area has been largely neglected and requires detailed study.

Collaboration between the occupational physician and the psychiatrist

Both the occupational physician and the psychiatrist have a great deal to contribute to human relations within the organization. A combined effort at study may yield techniques that enable doctor and management to work creatively together.

There are many possible patterns of collaboration. One, fully described elsewhere (Erskine and Brook, 1971, 1976), involves the psychiatrist spending half a day a week with the occupational health team with the intention of enhancing its psychological resources. If the occupational physician feels that the individual has symptoms that might be related to psychological conflicts he asks the psychiatrist to interview the patient. The two doctors then meet to examine the problem from their respective viewpoints and in particular to consider together whether any factors in the interaction between the individual and the organization have contributed to his symptomatology. Thus the patient becomes a research tool. In identifying some of the factors in an organization that have led a particular individual to exhibit such severe symptoms as to necessitate medical referral, it becomes possible to recognize the very factors that contribute to lesser, but nonetheless distressing, degrees of impairment of general well-being or mental health. This pattern of collaboration involves the psychiatrist as a true member of the occupational health team, if only on a part-time basis.

Another method of combined study is for about eight or 10 industrial physicians to meet with an appropriately trained psychiatrist to consider the psychological aspects of the problems the physicians encounter in their daily work. Experience of these different groups in other settings has proved their value for training and support (Balint, 1964).

PREVENTION AND IMPROVEMENT

The aim of a preventive approach to mental health is to maintain the well-being of those who might become casualties. The aim of an improvement approach is to go a stage further and to try to augment the quality of working life. The improvement of the quality of working life is one aspect of the social responsibility of industry. With a satisfactory working relationship established between doctor and management the occupational health team can help an organization to give the same consideration to the human consequences of management as they do to its other functions.

REFERENCES

Argyris, C. (1970) *Intervention Theory and Method.* Massachusetts: Addison
Balint, M. (1964) *The Doctor, his Patient and the Illness,* 2nd edn. London: Pitman Medical Publishing Company
Bion, W.R. (1961) *Experiences in Groups.* London: Tavistock Publications
Bridger, J., Miller, E.J. and O'Dwyer, J.J. (1964) 'The doctor and sister in industry.' *Macmillan Journals,* London
Brook, A. (1978) 'Coping with the stress of change at work.' Department of Health and Social Security *Health Trends,* **10** (4) 80–84; and *Management International Review,* **18,** (3) 9–15

Erskine, J.F. and Brook, A. (1971) 'Report on a two-year experiment in co-operation between an occupational health physician and a consultant psychiatrist.' *Transactions of the Society of Occupational Medicine,* **21,** 53–56

Erskine, J.F. and Brook, A. (1976) 'A method of developing the psychiatric resources of an occupational health team.' *Journal of the Society of Occupational Medicine,* **26,** 132–135

Frankenhaeuser, M. and Gardell, B. (1976) 'Underload and overload in working life.' *Journal of Human Stress,* **2** (3), 35–46

Fraser, R. (1947) *Incidence of Neurosis Among Factory Workers;* Industrial Health Research Board Report No. 90. London: HM Stationery Office

Garland, T. (1966) 'Frazer's disease.' *Transactions of the Society of Occupational Medicine,* **16,** 83–84

Golding, W. (1958) *Lord of the Flies.* London: Faber

Herzberg, F., Mausner, B. and Snyderman, B. (1959) *The Motivation to Work.* New York: Wiley

Jaques, E. (1951) *The Changing Culture of a Factory.* London: Tavistock Publications

Jaques, E. (1961) *Equitable Payment.* London: Heinemann Educational Books

Jaques, E. (1970) *Work, Creativity and Social Justice.* London: Heinemann

Kearns, J.L. (1971) *Stress in Industry.* London: Priory Press

McLean, A. (1974) *Occupational Stress.* Springfield, Illinois: Thomas

Menzies, I. (1960) A case-study in the functioning of social systems as a defence against anxiety.' *Human Relations,* **13,** (2), 75–121

Murell, H. (1978) *Work Stress and Mental Strain.* Department of Employment, Work Research Unit, London

Revans, R.W. (1969) 'Managers, men and the art of listening.' In S.H. Foulkes and G.S. Prince eds, *Psychiatry in a Changing Society.* London: Tavistock Publications

Selye, H. (1957) *The Stress of Life.* London: Longmans

Tredgold, R.F. (1949) *Human Relations in Modern Industry.* London: Duckworth

24

Toxicity Testing

E. Boyland

'All substances are poisons; there is none which is not a poison. The right dose differentiates a poison and a remedy' (Paracelsus 1493–1541). Following this principle, it becomes essential to obtain a quantitative index of toxicity in order to estimate the margin of safety for industrial or other chemicals and the therapeutic index for medicinal products.

If occupational hazards are to be avoided or reduced, knowledge of their qualitative and quantitative aspects is essential. It is a function of toxicology to provide this knowledge. Regulatory authorities like to have evidence of safety but it is generally more difficult to demonstrate absence of toxicity than harmful effects. It is often assumed that there are safe levels of exposure to chemicals except for carcinogens, mutagens and teratogens, but there are differences of opinion as to what these levels should be. Russian experts consider that there are thresholds for all types of harmful action, including carcinogenic and mutagenic effects (Sanockij, 1975).

The establishment of permissible levels of exposure as maximum allowable concentrations (MAC) or threshold limit values (TLV), discussed in Chapter 25, must in the first place depend on considerations of chemical and physical properties and toxicity data from animal or other biological experiments. The levels determined from these investigations may be, and frequently are, modified as a result of clinical experience from the surveillance of workers.

Before any biological experiments or tests are carried out on any compound, the chemical and physical properties and chemical structure should be considered. If the material is chemically reactive as an alkylating agent or in other ways then it is likely to be irritating to skin, lungs or eyes, and it could be carcinogenic. Compounds with certain solubility characteristics, but chemically inert, act as anaesthetics and their biological properties are associated with ease of skin and cell penetration. Thus absence of chemical reactivity does not necessarily indicate that a product is harmless. The relatively inert solids, quartz and asbestos, and the relatively inactive liquids, hexane, benzene and carbon tetrachloride, are not harmless. Some types of toxicity can be predicted or at least expected from chemical structures, including sympathomimetic and parasympathomimetic actions (due to inhibition of cholinesterase), carcinogenicity (of many classes of compounds), bacteriostatic activity (of sulphonamides, antifolic acid compounds, phenols and chelating agents); and analgesic power. The relationships between chemical structure and biological action are discussed by Albert (1978); Ljublina and Filov (1975) predict MAC values and other biological activities from chemical and physical properties.

The three main types of toxicity testing are acute, short-term and long-term.

ACUTE TOXICITY

The first essential parameter in toxicity evaluation is the acute toxicity as expressed by mortality following administration by appropriate routes. The value for acute toxicity is usually expressed as LD_{50}, the lethal dose which is estimated to kill half of the treated animals. The slopes of the dose response curves of compounds vary considerably and the relationship between dose and toxicity at low doses is often useful in designing further toxicity tests.

Perhaps more important than the actual figure for the LD_{50} are the time and mode of death. If the death is quick it is useful to know whether it was due to inhibition of respiration (by inactivation of haemoglobin or the cytochrome system), asphyxia due to tetanic convulsions (as with strychnine), or nicotinic actions (with cholinesterase inhibitors), anaesthesia or other effects causing rapid death. All animals used in the toxicity test should be examined *post mortem* to obtain further evidence as to the cause of death. In some cases biochemical investigations into function and blood and urine chemistry may be of value.

Many authorities indicate that rodents used in acute tests should be observed for 14 days. It is recommended that they be maintained and observed for 21 days, as there are some compounds (e.g. busulphan and other radiomimetic substances) which cause death that may occur between 14 and 21 days after treatment.

Oral administration is the obvious route of choice for food additives but it is also useful for products to which workers may be exposed, although they are most likely to be at risk by inhalation or skin absorption. Determination of toxicity by dermal application to animals is difficult because of the small area of shaved skin available and the toxicity may vary with the area of application and the nature of the skin. One advantage of dermal toxicity tests is that any direct skin irritation should be seen; but vesication occurs only with human skin. The toxicity by subcutaneous injection is often measured. Comparison of toxicity by dermal application and by injection gives some indication of absorption through skin and of the degree of hazard from skin exposure.

All compounds, including those that are normally gases or volatile liquids, can often be administered by injection or gavage (gastric intubation) dissolved in a suitable solvent. There are few suitable solvents with low toxicity. If the material is water-soluble this is probably the best medium, but the concentration of aqueous solutions should be as nearly isotonic with physiological saline as possible, and the volume administered to rodents should not be greater than 10 ml/kg body weight. With organic solvents the volume should not exceed 5 ml/kg body weight. Oil-soluble compounds are often given in olive oil, sesame oil or other vegetable oils. Such natural oils are viscous and vary in composition and in toxicity. Pure triglycerides such as tricaprylin are less variable in composition and toxicity, but are so expensive that they cannot be widely used in most laboratories. Dimethyl sulphoxide, N,N-dimethylacetamide and propylene glycol are good solvents for many organic compounds and have low toxicity.

Products which are dusts, aerosols, gases or volatile liquids should also be tested by administration in the vapour phase or as aerosols, either in large chambers or in apparatus by which the nose of the animal is exposed. With larger animals, exposure can be made with masks covering the head. If the whole body of an animal is exposed in a chamber the compound may be deposited on the fur and subsequently ingested or absorbed through the skin, so that absorption is not entirely through the respiratory tract. If the substance is irritant the exposure time should be sufficiently long (say one hour or more), so that the effect of breath holding is unimportant. With anaesthetics it is of value to observe the concentrations and times which cause behavioural changes. The dose administered is best expressed as CT where C is the concentration generally expressed in mg per m^3 and T is the time. In exposure experiments of this kind the concentration of the product under test should be monitored by chemical sampling and analysis. With dusts and aerosols the particle size distribution should also be measured. The temperature and humidity of the chamber or other apparatus should be controlled. With some solvents and other compounds with low vapour pressure and low toxicity the dose which produces any effect may take many hours to administer. Some idea of the toxic dose by inhalation can be obtained from results obtained by injection or involving oral administration. Methods are described by the World Health Organization (WHO, 1967).

Following 'inhalation exposure' particular attention should be paid to the lungs in the *post mortem* examination. Biological experiments with dusts and aerosols are particularly difficult because the site of deposition in the respiratory tract depends on the size of the particles. The noses of most animals are effective filters or screens which remove solid particles and absorb many volatile compounds and so protect the lungs from exposure. It was possible to induce lung cancer in dogs by exposing the lungs directly by tracheotomy tubes (Auerbach, *et al.,* 1967). This may be the best method of determining inhalation toxicity if the effect is due to lung damage.

There is little point in making an exact quantitative determination of the LD_{50} as the value between species can differ by 'one or two orders of magnitude'. It is also dependent on many factors including sex, nutritional state, age, season and time of day affected by biological rhythms, temperature, humidity and type and volume of solvent used. Because of the difference in susceptibility of different species it is advisable to determine the toxicity in more than one animal species. It is not necessary to obtain exact figures for the value; to know whether one compound is twice as toxic as another compound is generally sufficient. The value for the approximate LD_{50} is of use in estimating the magnitude of the doses to be used in other tests.

The value and purpose of acute toxicity tests have been discussed by Morrison, Quinton and Reinert (1968). The main object of an acute toxicity test is not to establish a figure for the LD_{50} with precision, but to learn something about the way in which the drug is acting as a poison (Paget and Barnes, 1964). Determination based on the method of Deichmann and Leblanc (1943) in which six animals are used for estimation of the approximate lethal dose should be sufficient for most purposes.

SHORT-TERM TOXICITY

Having obtained some indication of the acute toxicity of a substance administered by an appropriate route, it is usual to examine the effect of repeated doses. These experiments should indicate the level which is non-toxic or the 'no effect level': this level is often difficult to define but is generally accepted as that which produces no obvious toxic effect in behaviour or function and does not reduce the rate of growth by more than 10 per cent. They should also demonstrate whether the material has cumulative effects, and detect which organs or systems are affected by the substance. The usual accepted period for this test is 90 days and it is frequently called the '90-day test'. This period is so well established and traditional that it would be difficult to change. Ninety days is about one tenth of the average life-span of rats. However, there are few if any effects that are shown in 90 days that would not develop in a shorter time. The Health and Safety Commission (1977) says that although animals are exposed to a compound for 90 days, the test takes about six months to complete and that 'experience gained during 90 days and longer studies suggests that the greater majority of adverse effects are seen in the first 30 days'. In spite of this suggestion some authorities will probably insist on the 90-day test for some time.

Substances that are administered orally are usually mixed with the animals' food but if the material is irritant, carcinogenic or hazardous in other ways the mixing and feeding of the large amounts of diet involved may be dangerous to laboratory workers. In some laboratories special facilities are available to reduce such hazards. As an alternative the material can be given by gavage for five days in each week, but administration in this way can produce effects different from those caused by more continuous dosing that occurs when the material is in the diet. With inhalation studies it is usual for animals to be exposed for 6–7 hours daily for five days in each week.

If a product or substance has high acute toxicity much more can be given by continuous administration in food, drinking water or air than by gavage. Thus with compounds such as cyanide or nicotine that are toxic, but rapidly metabolized or excreted, very much more can be tolerated if they are given continuously than by repeated daily doses administered by injection or force-feeding. It is obvious that long-term effects can be different according to dosage schedule.

On the other hand continuous administration can have disadvantages. If the material has an unpleasant taste or smell, it may reduce the intake of food or water and so reduce growth or cause dehydration. The estimation of the dose is probably less accurate when the material is given in food or drink but the proportion absorbed may be greater. In this subacute toxicity test, groups of 10 or 20 animals of each sex should be treated with four or more dose levels, the highest level being one with which toxic effects would be expected. Rats are frequently used for the test but animals of a second species should also be used if they are likely to reveal some special effect. The dose levels should be chosen so as to indicate the 'no effect level' and should probably be such that there are twofold differences between the levels.

During the test the animals and their food must be weighed at weekly intervals in order to estimate the growth rates and food consumption. The animals should be observed (particularly when being weighed) for any clinical signs of toxicity, and any animals which are sick should be killed so that *post mortem* examination together with observation of clinical signs can indicate the cause of death. Physiological activity such as lung function, kidney function, liver function, electrocardiography, electroencephalography, nerve function, or electromyography should be measured if the products being investigated are likely to affect these activities. Similarly estimations of enzyme activities, e.g., cholinesterase with pesticides and induction or inhibition of microsomal systems that metabolize foreign compounds, can be useful. Blood should be examined for changes in white and red cells and in chemical constituents after 30 days and at the end of exposure. At the end of the exposure all animals must be killed and full *post mortem* examinations made. Some organs (e.g. liver, kidney, heart and spleen) should be weighed, and selected organs fixed and examined by appropriate histological methods.

The World Health Organization (WHO, 1978) considers it 'advisable to select all tissues and organs for fixation in buffered saline'. In range-finding tests, restricted pathology is common practice. For example, the heart, liver, spleen and kidneys and all grossly abnormal tissues are collected for fixation.

LONG-TERM TOXICITY TESTS

Although the estimation of carcinogenic activity is one aim of chronic toxicity tests, other pathological effects should not be overlooked. Delayed changes that are not neoplastic may occur in many tissues including the central nervous system (as with agenized flour and carbon disulphide), in the eye (i.e. cataract with naphthalene or galactose, and retinopathy with chloroquine), the liver, lungs, kidneys and circulatory system. These effects should be seen if the organs listed in *Table 24.1* are examined.

Long-term toxicity tests may reveal a shortening of life which would then require further investigation to find the cause. The chronic toxicity testing is the most expensive and difficult of the battery of procedures required to establish the safety or hazard of a substance or product. Because of the difficulties it is advisable to carry out other toxicity tests and short-term tests for carcinogenicity before the more extensive procedures are started. The data from the short-term or 90-day tests are essential in planning the chronic toxicity tests.

Difficulties of the chronic tests include the possible illness or death of experimental animals from infection or loss of appetite, the chance that the wrong food or compound is administered, the risk that animals which die may be eaten by other animals in the cage and the danger that personnel involved in the work may be exposed to carcinogenic or other toxic hazards.

Carcinogens are slow poisons and it is difficult to know how far the precept of Paracelsus applies to them, because there may be no safe dose or threshold, but it is possible that the right dose of some carcinogenic substances might differentiate between the toxic and useful effects. Many of the products used in the treatment of cancer such as alkylating agents and oestrogens are themselves carcinogenic. Shubik and Sicé (1956) and Boyland (1958) reviewed the methods used

Table 24.1 ORGANS AND TISSUES TO BE EXAMINED IN ROUTINE TOXICITY TESTS

First priority: All animals including controls

Liver (f)	Epididymis
Kidney (f)	Lung
Adrenal glands (f)	Bone marrow
Heart (left ventricle) (f)	Mesenterial lymph node
Spleen (f)	All organs showing gross changes of
Thymus	shape, weight, colour, or structure
Testis	

Second priority: High dose animals and part of controls. Medium and low dose animals only when significant changes observed in high dose group.

Thyroid	Prostate
Parathyroid	Coagulating gland
Pituitary gland	Tonsils
Salivary glands	Brain
Stomach	Spinal cord
Duodenum	Eye
Small intestine	Optic nerve
Large intestine	Peripheral nerve
Pancreas	Urinary bladder
Ovary	Skin
Uterus	Mammary gland
Cervix uteri	Bone-cartilage
Vagina	Skeletal muscle
Seminal vesicles	Gall bladder

Third priority: Organs and tissues not examined in routine experiments unless indicated by clinical observations or motivated by scientific interest.

Heart valves	Oviduct
Purkinje fibres	Vulva
Aorta and other blood vessels	Penis
Thoracic duct	Paraurethral and preputial glands
Tongue	Pineal gland
Teeth	Spinal ganglions and roots
Lips	Sympathetic ganglions and trunk
Gingivae	Nerve fibre endings
Hard palate	Meninges
Nasopharynx	External ear
Larynx	Middle ear
Trachea	Inner ear
Oesophagus	Olfactory organ
Vermiform appendix	Subcutaneous lymph nodes
Rectum	Tendons
Anus	Adipose tissue
Ureter	Intervertebral disc
Urethra	Synovial membrane
	Lacrimal glands

(f)–frozen section stained for fat obligatory
Source: Zbinden (1976)

for the determination of carcinogenic activity. Many of the problems raised in those reviews are still with us. In the International Agency for Research on Cancer (IARC) monographs on the evaluation of the carcinogenic risk of chemicals to man it is frequently stated that 'the available data do not allow an evaluation of the carcinogenicity of this compound to be made, thus showing that it is often difficult to determine carcinogenic activity'.

An important change in the last decade has been the increased awareness of the possible hazards from the handling of carcinogens and other toxic products in laboratories. Many precautionary regulations have been introduced which should protect workers, but they also increase the cost and difficulties of the work. Some of the types of safety precautions are listed in *Table 24.2.*

Table 24.2 SAFETY PRECAUTIONS FOR THE USE OF CARCINOGENIC SUBSTANCES IN LABORATORIES

Permissible exposure limits

Notification of emergencies

Exposure monitoring and measurement

Regulated areas

Protective clothing and equipment

Housekeeping, waste disposal

Hygiene facilities and practices

Medical surveillance

Signs and labels

Record keeping

The type and scale of work carried out in different laboratories vary widely and many institutions have set up regulations for the handling of carcinogens that should reduce the hazard. Many special laboratories have been designed and built to reduce exposure or workers to the carcinogenic substances used.

Much of the work of experimentation and testing in carcinogenicity is concerned with proof that substances are not carcinogenic and therefore safe. The proof of the negative is always difficult and toxicologists have generally found the study of activity more interesting and rewarding than determining what is safe. Thus, because the work seems dull, suitable investigators either by temperament or ability are hard to find. For this and other reasons the numbers of animals used should be the minimum compatible with sensitivity of detection and the maximum amount of information should be obtained from each animal (*see Table 24.3*). In some laboratories 50 different organs are examined (*see Table 24.1*).

Table 24.3 COMPARISON OF RECOMMENDATIONS OF DIFFERENT BODIES FOR CARCINOGENICITY TESTS

	Eurotox	*FAO/WHO*	*British Ministry of Health (1960)*	*Food Protection Committee, USA*
Species and route of administration	Rats – feeding	Rats – feeding	Rats – feeding	Rats – feeding
	Mice – feeding	Mice – feeding	Mice – feeding	Mice – feeding
	Mice – repeated injection		Mice – repeated injection	Dogs – feeding
				Mice – skin application
	Rats – repeated injection		Rats – repeated injection	Mice – single injection
Number of animals of each group which				
(a) should survive	15		12	
(b) used at beginning		20		50
Duration of tests	Rats – 2 years	Rats – 2 years	Rats – 2 years	Rats – 2 years
	Mice – 80 weeks	Mice – 80 weeks	Mice – 80 weeks	Mice – life-span
				Dogs – 4 years

One of the problems is the choice of species. Rats and mice are used widely but it is generally agreed that two species should be used and if possible one of these should not be a rodent. Marmosets have been used for teratology studies but they are not suitable for carcinogenicity tests, because of their life-span of more than five years. Hamsters are sometimes used, and ferrets may become useful when there is more experience of their behaviour. Mice are convenient but have the drawback that tumours occur in untreated mice of many strains. The British Biological Research Association (BIBRA) has published a book (Grasso, Grampton and Moosey, 1977) on the mouse and carcinogenicity testing in which 'data cast considerable doubt on whether the results of this type of testing can be regarded as meaningful to man, whatever the strain of mouse used'. However, it is best to know the properties and limitations of the means used in testing and more is known about the occurrence of tumours in mice than in any other species. This is an advantage. Until more suitable animal species are known, mice must be used.

It has been recommended that species or strains of animals which metabolize foreign compounds in the same way as human subjects should be used for tests. One could ask, which human subjects?, because not all men and women metabolize foreign compounds by the same routes or to the same extent. Metabolism is increased and inhibited by many factors. In most cases it is not known which of several metabolites are the ultimate carcinogens. Until more is known about metabolism of carcinogens and the biochemistry of carcinogenesis, it does not seem possible to choose suitable species on these grounds.

Peto (1974) has suggested that the *in vivo/in vitro* distinction may often be less important than the over-riding need to use as experimental models epithelial cells from species with a life-span and body weight comparable to that of man. It is difficult to follow the reasoning; rats and mice seem to be more appropriate than large monkeys, sheep or goats.

Animals may be treated through pregnancy, as neonates, as young or adults or through some generations. In some cases such as skin application to mice, the response does not vary with age (Peto *et al.*, 1975), but the incidence of mammary cancer in female Sprague–Dawley rats treated with methyl cholanthrene is much less in old than in young rats (Huggins, Grand and Brillantes, 1961).

In some cases there is clear optimum dose of carcinogen; this was marked with dieldrin in the experiments of Thorpe and Walker (1973), and with vinyl chloride (Maltoni, 1975).

Evidence of carcinogenicity is more convincing if there is correlation between the incidence of tumours and the amount of carcinogen applied. The dose response curve for the lethality of potassium cyanide is almost 100 times more steep than for the carcinogenic action of single doses of benzo[*a*]pyrene injected into mice. Because the dose response curve with carcinogens is relatively shallow the intervals between doses should not be too close. Twofold intervals are too close and tenfold too wide, so that intervals of three, four or five are suitable (Weisburger and Weisburger, 1967). Sometimes treatment with a carcinogen inhibits or delays the incidence of cancer; thus the carcinogenic isoniazid decreases the incidence of mammary tumours in mice (Toth and Shubik, 1960).

That part of the dose response curve which is linear can be used to calculate the cancer risk with low doses. The data on injection of polycyclic aromatic hydrocarbons (Bryan and Shimkin, 1942), however, indicate that the response at low levels is not always linear. The effective dose of carcinogens, which vary a millionfold between aflatoxin and saccharin, is only one of the parameters of carcinogenesis. Others are as follows:

1. Latent or induction periods.
2. Maximum tumour incidence.
3. Degree of malignancy.
4. Multiplicity of tumour types.
5. Initiation and promotion.

The British Panel on carcinogenic risks (Dodds *et al.,* 1960) decided that insufficient evidence was at present available for it to express opinion about hazards to man from initiators or promoters. It now seems that cocarcinogens or promoters probably present a hazard and need to be investigated.

Many natural and synthetic products in the environment have partial carcinogenic activity. Cocarcinogens shorten the induction period and their activity may possibily be expressed as the reciprocal of the average latent period when the incidence of cancer is 50 per cent.

$$\text{Promoting activity} = \frac{1}{Lt_{50}} \text{ at the } TD_{50}$$

Where Lt_{50} is the average latent period and TD_{50} is the dose of carcinogen that induces cancer in 50 per cent of animals treated. Alternatively, the activity might be defined as the dose that induces tumours in some standard period such as 100 days for mice or one year for rats.

The induction of cancer is a genotoxic effect and so related to other genotoxic effects such as teratogenicity and mutagenicity. Teratogenicity should be taken as an indication of probable carcinogenic activity. Some substances that are teratogenic are toxic and thus it is difficult to demonstrate that they are carcinogenic.

It is difficult to show that toxic materials such as arsenical compounds, selenium compounds and dioxins are carcinogenic because only small amounts are tolerated. It is possible to obtain a few tumours with saccharin because it is so non-toxic that enormous doses can be given. If substances are as poisonous as mercurials, arsenic or dioxins then the fact of their possible carcinogenic activity is of minor importance. We are agreed that exposure to them should be minimal. The hazards of carrying out carcinogenic experiments with such toxic materials are such that the results which could be obtained would not justify the performance of the experiments.

The transfer of animal data to the human situation is often difficult; an example of this is illustrated with phenobarbitone. Tumours have been induced in mice and rats by feeding diets containing moderate amounts (500 ppm) of phenobarbitone. Clemmesen and Hjalgrin-Jensen (1977) claimed that phenobarbitone was not carcinogenic to inmates of an asylum treated for epilepsy. There was an excess of brain tumours, some of which may have been present before treatment, and of liver tumours, some of which could have been caused by treatment with thorotrast. The authors conclude that: 'To protect man against dangers to which we expose laboratory mice seems hardly worth the effort, and if the evidence, as presented here, on 8078 persons followed for one to two decades should be of no avail to the discussion of the postulated carcinogenicity of phenobarbital, we may as well abandon cancer epidemiology.'

This study shows how difficult it is to demonstrate the safety of a compound by epidemiological studies of man, and how essential animal investigations are. The benefits of phenobarbitone in most cases are such that the slight risk of cancer may be accepted. Recent experiments have indicated that phenobarbitone acts as a promoter of liver tumours in rats and many investigations have shown that saccharin is a promoter of bladder cancer (cf. Boyland, 1979). Earlier workers such as Tannenbaun and Muhlbock have shown the effects of diet and caging on the incidence of cancer. This effect is well illustrated in the data from Roe and Tucker (1973), shown in *Table 24.4* in which groups each of 40 outbred Swiss albino male mice were maintained for 18 months. The effects of diet and housing on tumour incidence are remarkable.   During the last twenty years we have become aware of the hazards and pitfalls of testing for carcinogenicity. In many laboratories the procedures and facilities are elaborate and sophisticated. It is hoped that the necessary results will become available in the future using fewer animals but obtaining the maximum amount of information from each. In the past too little research has been carried out to compare and improve methods of testing. In many cases old methods have been used and give 'data which do not allow evaluation of the compound to be made'.

SHORT-TERM TESTS FOR CARCINOGENICITY

Chemical mutagens were first discovered during the Second World War as a spin-off from research in the chemical warfare agents — mustard gas and nitrogen mustards. The idea that cancer originated through somatic mutations was considered decades before any chemical mutagens were known. Although a relationship between mutagens and carcinogens appeared to be probable, it seems to be difficult to establish because cancers probably only occur in vertebrate species and mutagenic activity was estimated by changes in bacteria, fungi or insects

Table 24.4 CANCER IN MICE – INCIDENCE OF CANCER IN GROUPS OF OUTBRED SWISS ALBINO MALE MICE MAINTAINED FOR 18 MONTHS ON A STANDARD PELLETED DIET.

Groups	Number of mice	Number per cage	Weight of diet/day(g)	Survival to 18 months	Total number of tumours	Liver tumours	Lung tumours	Lympho-reticular neoplasms	Other neoplasms
1	40	1	4	Similar	4	1	1	2	0
2	40	1	5		4	2	0	1	1 testis
3	40	1	ad libitum		32	15	2	11	2 testis 1 kidney 1 thyroid
4	40	5	ad libitum		23	8	6	9	0

(Boyland, 1958). Cancer and many other effects are produced by ionizing radiations and carcinogenic compounds; Dustin (1947) coined the term radiomimetic to denote the effects and the compounds that produce them. There are many radiomimetic effects including mutagenicity, teratogenicity, production of chromosome damage and inhibition of growth (Boyland, 1952).

The determination of carcinogenic activity by long-term animal tests is difficult and expensive but some promising alternative procedures are available. It has been thought for a long time that the carcinogenic process could be somatic mutation (Boveri, 1914), and since the first chemical mutagens were shown to be carcinogenic, evidence of the correlation between mutagenic and carcinogenic activity has accumulated (Boyland, 1952).

Bacterial mutation test

A practical test for mutagenic activity of particular strains of *Salmonella typhimurium* developed by Ames, McCann and Yamasaki (1975) has been widely used. In the bacterial mutation or 'Ames' test specially selected strains of *S. typhimurium,* which are dependent on the presence of histidine for growth, are incubated in a minimal glucose agar medium in Petri dishes with various concentrations of the substance under test and without any addition to the control. The special tester strains grow poorly on this medium so that few colonies are seen, but some mutants become independent of histidine and colonies of the mutants can be seen and counted, and the number of mutant colonies gives a measure of the mutagenic activity. It is usual to add a preparation of microsomes obtained by centrifugation of a rat liver homogenate in order to metabolize those compounds that are not themselves chemically reactive. Some compounds, such as alkylating agents, are directly active and the liver preparation might inactivate compounds that do not require activation. For this reason it is recommended that the mutagenic test be carried out both with and without liver preparations. Ames, McCann and Yamasaki (1975) and Purchase *et al.* (1978) have shown about 90 per cent of compounds known to be carcinogenic in animal tests are mutagenic and only about 10 per cent of compounds which are considered not to be carcinogenic are positive in the test. The test is therefore a useful indicator of possible carcinogenic activity.

The correlation between mutagenicity and carcinogenicity as indicated by this test is good, but regulatory authorities are generally not prepared to accept a negative 'Ames' test as evidence that a substance is not carcinogenic. On the other hand, if a new substance is being developed for some industrial or commercial purpose a positive result with the bacterial mutation test is an indication that it could be carcinogenic. The induction of cancer involves several stages, the first of which is probably initiation in which normal cells are transformed into potential tumour cells. Such cells may be changed into tumour cells by promoters which are sometimes different from initiators, although there are many compounds that are complete carcinogens. It is probable that the stage of initiation is a kind of mutation and the bacterial test measures initiating activity.

There are some compounds that are carcinogenic (particularly in rats) but are not mutagenic and do not react with deoxyribonucleic acid (DNA). Ashby *et al.* (1978) have suggested that such substances, including saccharin, phenobarbitone and DDT, are 'epigenetic'. They are probably promoters (Boyland, 1979); under some conditions promoters may be as important as initiators in carcinogenesis. The bacterial mutation test is a valuable indicator of initiating activity and an analogous short-term test which indicates promoting activity would be useful.

Mammalial cell transformation

Many investigators have shown that viruses and chemical carcinogens can transform normal cells grown *in vitro* into cancer cells (*see* Heidelberger, 1973). Styles (1977) has developed and assessed a method using two lines of human cells and one of baby Syrian hamster kidney cells exposed to chemicals in liquid tissue-culture media with and without rat liver microsome preparations. The exposed cells were allowed to grow in semi-solid agar and the colonies counted. A fivefold increase in colonies over the control culture is regarded as a positive result. The method gave similar results to those obtained with bacterial mutation when tested on the same 120 compounds.

Other short-term tests for carcinogenicity

Other short-term tests for carcinogenic activity which were not so accurate in predicting activity include:

1. The degranulation of rough endoplasmic reticulum from rat liver (Williams and Rabin, 1971).
2. The mouse sebaceous gland suppression test (Bock and Mund, 1958).
3. The reduction of tetrazolium red test (Iversen and Evensen, 1962).
4. The tissue reaction to subcutaneous implants in mice (Longstaff, 1978).

Although these tests have limited application they may be of use in some special situation. Thus the tetrazolium red test is claimed to be of value in identifying possible hazards from North Sea oil products.

Short-term tests for promoting activity

The tests for bacterial mutagenicity and cell transformation probably measure initiating activity and other stages, particularly that of promotion. Mondal, Brankow and Heidelberger (1978) showed that saccharin enhanced cell transformation of mouse embryo cells which had been treated with a non-transforming initiating dose of methylcholanthrene. Saccharin was one thousandfold less active than the most active promoting agent 12-0-tetradecanoylphorbol-13-acetate. This is a promising approach to the measurement of tumour-promoting activity. Other changes associated with promoting activity described by Weinstein (1978) might also be used.

With the increasing costs and difficulties of carrying out conventional animal toxicity tests and the development of short-term tests for carcinogens, it is hoped that such tests will be developed to investigate other types of toxicity. Worden (1974) has discussed the use of tissue culture in toxicity testing. Such methods can be of value in studying mechanisms and in some special problems such as the investigation of solid materials used in dentistry and surgery.

REPRODUCTION AND TERATOLOGY

After the thalidomide tragedy, in which some 10 000 malformed children were born, the Food and Drug Administration issued guidelines for reproductive studies which included three phases: reproduction, teratology, and late pregnancy, lactation and postnatal growth. This means that animals of both sexes must be treated for 60 days before mating and the young examined for any malformations and during lactation. These young animals are then mated and their offspring examined at birth and for 21 days during lactation. The principles of reproductive studies are given in outline by WHO (1967). The embryonic damage may result in gross malformations seen by direct observation, soft tissue anomalies detected by dissection or histology and bone anomalies that can be found by staining the skeletons with alizarin red-S. Parents of deformed animals may destroy or eat the abnormal offspring so that in some cases it is advisable to deliver fetuses by Caesarian section.

The choice of species is difficult. New Zealand white rabbits have been used, but they are herbivores, have small litters, a gestation period varying from 32 to 36 days and are not inbred. Rats have some advantages but are resistant to the teratogenic effects of some compounds. Mice are more sensitive to teratogenic effects, but are more difficult to breed than are rats or rabbits. Damage to the embryo can often be seen in chick embryos in which control of conditions is easy. The disadvantages are that chicks are not mammalian, that the embryos do not metabolize foreign compounds so that lipid soluble compounds accumulate in the yolk sack and that some solvents (e.g. dioxane) can cause malformations.

OTHER INVESTIGATIONS

Some foods, food additives, natural products and industrial chemicals cause allergic reactions when eaten, inhaled or deposited on skin. Animal tests that could predict such reactions are needed but as yet are only being developed.

In some circumstances it may be necessary to carry out other procedures including metabolic studies, investigation of behavioural effects and examination of the possible irritant effects on special organs such as the eye.

The Health and Safety Commission published a Discussion Document as a *Proposed Scheme for the Notification of the Toxic Properties of Substances* in May 1977. This is a valuable short account of the procedures that are needed. It is hoped that the recommendations will be ratified and accepted as a basis for toxicity testing in Britain and the European Community.

In June 1979 the United States Food and Drug Administration regulations for *Good Laboratory Practice* as published in the Federal Register on 22 December 1978 became effective. These regulations have important international implications. They will raise the cost of investigations but should increase the confidence that can be placed on the results.

THE INTERPRETATION OF LABORATORY DATA

The scientific assessment of the results of toxicity tests is often difficult and they are frequently evaluated differently by different experts. The application of the data to legislation controlling the use and exposure to chemicals is even more difficult.

In the United States the detection and control of carcinogens comes under the auspices of several agencies, including the National Cancer Institute (NCI), the National Cancer Advisory Board and Panel, the National Institute for Occupational Safety and Health (NIOSH), the

National Insitute of Environmental Health Sciences (NIEHS), the National Centre of Toxicology Research (NCTR), the Department of Energy, Research and Development Administration (ERDA), the Council on Environmental Quality (CEQ), the Occupational Safety and Health Administration (OSHA), the Environmental Protection Agency (EPA), the Food and Drug Administration (FDA), and the US Department of Agriculture (USDA). These bodies sometimes differ in their recommendations. That there are differences between these organizations, between labour and management, and between different countries and between various groups in public interest movements is not surprising.

From the data available, decisions must be made as to whether a substance can be used or prohibited. If it is allowed to be manufactured and used then the conditions and levels of use must be defined.

The main differences in permitted exposures are between Eastern Europe and the rest of the world. The Russian hygiene standards expressed as MAC values are generally lower than those accepted in other countries. A suggested Russian definition is that 'the MAC of a chemical compound in the environment is the concentration that, by its action on the human body periodically or throughout life, directly or indirectly via ecological systems and also through possible economic losses, will not cause the development of physical or mental diseases (including latent and temporarily compensated conditions) or changes in the state of health that go beyond the limits of adaptive physiological responses, detectable by modern methods of investigation either immediately or in long term, in this or subsequent generations.' The World Health Organization's concept of health is 'a state of complete physical, mental and social well-being and not merely the absence of disease or infirmity'. These definitions seem reasonable but the difference in accepted values in the USSR and USA could be due to the emphasis that Russian experts place on reflex actions, and an American definition of the accepted conditions is that which 'nearly all workers can be repeatedly exposed day after day, without adverse effect', (*see* Chapter 25).

The results of experimental data have undoubtedly helped in the control of hazards, particularly the control of cancer, for example 2-aminofluorene was found to be an effective insecticide but chronic toxicity tests showed it to be a potent carcinogen so that it was never used in commercial agriculture. Similarly, 4-aminodiphenyl, manufactured and used in the rubber industry in the USA, was never manufactured in England because Michael Williams showed it to cause cancer in animals (Walpole, Williams and Roberts, 1952). Soon after the demonstration of the carcinogenicity of this compound in animals, it was found to have caused bladder cancer in American workers and its production was stopped.   An evaluation of the carcinogenic risks of aniline in 1974 stated that 'at the present time, the weight of epidemiological evidence suggests that aniline is not a carcinogen for the human bladder'. Recent results from the National Cancer Institute showed that aniline hydrochloride produced haemangiosarcomas, fibrosarcomas and sarcomas in rats. These findings appear disturbing but indicate that searches should be made to determine whether exposure to aniline induces cancer at sites other than the urinary bladder in man.

With the increase in the number and volume of chemical products in factories and the environment and the growing awareness of possible hazards, there is a great demand for toxicity testings. At present very little of the total effort is in research to improve and simplify methods. Much work is needed in research and development of procedures that would yield more significant results with less expenditure of animals, time and money.

REFERENCES

Albert, A. (1978) *Selective Toxicity*, 6th edn. London: Chapman and Hall

Ames, B.N., McCann, J. and Yamasaki, E. (1975). 'Methods for detecting carcinogens and mutagens with the Salmonella/mammalian microsomes mutagenicity test.' *Mutation Research*, **31**, 347–363

Ashby, J., Styles, J.A., Anderson, D. and Paton, D. (1978). 'Saccharin: a possible example of an epigenetic carcinogen/mutagen. *Food and Cosmetic Toxicology*, **16**, 95–103

Auerbach, O., Hammond, E.C., Kirman, D., Garfinkel, L. and Stout, A.P. (1967). 'Histologic changes in bronchial tubes of cigarette smoking dogs.' *Cancer*, **20**, 2055–2066

Bock, F.G. and Mund, R. (1958). 'A survey of compounds for activity in the suppression of mouse sebaceous glands.' *Cancer Research*, **18**, 887–892

Boveri, T. (1914) *Zur Frage der Enstehung maligner Tumeron*. Jena: Gustav Fisher

Boyland, E. (1952) 'Effects of radiations and radiomimetic drugs.' *Endeavour*, **2**, 87–91

Boyland, E, (1958) 'Biological examination of carcinogenic substances.' *British Medical Bulletin*, **14**, 93–98

Boyland, E. (1979) 'Saccharin: from carcinogen to promoter.' *Nature*, **278**, 123–124

Bryan, W.R. and Shimkin, M.B. (1942). 'Quantitive analysis of dose-reponse data obtained with three carcinogenic hydrocarbons in strain C3H mice.' *Journal of the National Cancer Institute*, **3**, 503–531

Clemmesen, J. and Hjalgrin-Jensen, S. (1977) 'On the absence of carcinogenicity to man of phenobarbital, *Acta Pathological Scandinavica*, Sect. A. Suppl. 261, pp. 38–50

Deichman, W.B. and Leblanc, J.T. (1943). Determination of the approximate lethal dose with about six animals. *Journal of Industrial Hygiene and Toxicology*, **25**, 415–417

Dodds, E.C., Barnes, J.M., Bonser, G.M., Boyland, E., Broom, W.A., Frazer, A.C., Haddow, A., Orr, J.W., Peacock, P.R., Salaman, M.H., Scott, C.M., Goulding, R. and Lake, W.F. (1960) 'Carcinogenic risks in food additives and pesticides.' *Monthly Bulletin of the Ministry of Health*, **19**, 108–112

Dustin, P. (1947) 'Some new aspects of mitotic poisoning.' *Nature*, **159**, 794–797

Grasso, P., Crampton, R.F. and Moosey, J.(1977) *The Mouse and Carcinogenicity Testing*. Carshalton: The British Biological Research Association

Health and Safety Commission (1977) Proposed scheme for the notification of toxic properties of substances: Discussion document. London: HM Stationery Office

Heidelberger, C. (1973) 'Chemical oncogenesis in culture.' *Advanced Cancer Research*, **18**, 317–366

Huggins, C., Grand, L.C. and Brillantes, F.D. (1961) 'Mammary cancer induced by a single feeding of polycylic hydrocarbons and its suppression.' *Nature*, **189**, 204–207

Iverson, O.H. and Evensen, A. (1962) *Experimental Skin Carcinogenesis in Mice*. Oslo: Norwegian University Press

Ljublina, E.I. and Filov, V.A. (1975) 'Chemical structure, physical and chemical properties and biological activity.' In *Methods used in the USSR for Establishing Biologically Safe Levels of Toxic Substances*, pp. 19–44. Geneva: WHO

Longstaff, E. (1978) 'The implant test.' *British Journal of Cancer*, **37**, 954–958

Maltoni, C. (1975) 'The value of predictive experimental assays in occupational and environmental carcinogenesis. An example: Vinylchloride.' *Ambio*, **4**, 18–25

Mondal, C., Brankow, D.W. and Heidelberger, C. (1978). 'Enhancement of carcinogenesis in C3H/IOTI/2 mouse embryo cell cultures by saccharin.' *Science*, **201**, 1141–1142

Morrison, J.K., Quinton, R.A. and Reinert, T.H. (1968) 'The purpose and value of LD_{50} determination.' In *Modern Trends in Toxicology*, **1**, 1–17 London: Butterworths

Paget, G.E. and Barnes, J.M. (1964) 'Toxicity testing.' In *Evaluation of Drug Activities: Pharmacometrics I*, D.R. Laurence and A.L. Bacharah, eds, pp. 135–166. London and New York: Academic Press

Peto, R. (1974) Editorial: 'Guide lines on the analysis of tumour rates and death rates in experimental animals.' *British Journal of Cancer*, **29**, 101–105

Peto, R., Roe, F.J.C., Lee, P.N., Levy, L. and Clack, J. (1975) 'Cancer in ageing mice and men.' *British Journal of Cancer*, **32**, 411–426

Purchase, I.F.H., Longstaff, E., Ashby, J., Styles, J.A., Anderson, D., Lefevre, P.A. and Westwood, F.R. (1978) *British Journal of Cancer*, **37**, 873–959

Roe, F.J.C. and Tucker, M. (1973) 'Recent developments in the design of carcinogenicity tests on laboratory animals.' *Proceedings of the European Society for the Study of Drug Toxicity*, **15**, 171–177

Sanockij, I.V. (1975) 'Investigation of new substances: permissive limits and threshold of harmful action.' In *Methods Used in the USSR for Establishing Biologically Safe Levels of Toxic Substances*. pp. 25–35. Geneva: WHO

Shubik, P. and Sicé, J. (1956) 'Chemical carcinogenesis as a chronic toxicity test.' *Cancer Research*, **16**, 728–742

Styles, J.A. (1977) 'A method for detecting carcinogenic organic chemicals using mammalial cells in culture.' *British Journal of Cancer*, **36**, 558–563

Thorpe, E. and Walker, A.I.T. (1973) 'The toxicity of dieldrin (HEOD) II Comparative long-term oral toxicity studies in mice with dieldrin, DDT, phenobarbitone, βBHC and γBHC.' *Cosmetics Toxicology*, **2**, 415–432

Toth, B. and Shubik, P. (1960) 'Mammary tumour inhibition and lung adenoma induction with isonicotinic acid hydrazide.' *Science*, **152**, 1376–1377

Weinstein, B. (1978) 'Cell culture studies provide new information on tumour promoters.' *Nature*, **270**, 659–660

Walpole, A.L., Williams, M.H.C. and Roberts, D.C. (1952) 'The carcinogenic action of 4-aminodiphenyl and 3.2 dimethyl-4-aminodiphenyl.' *British Journal of Industrial Medicine*, **9**, 255–264

Weisburger, J. and Weisburger, E. (1967) 'Tests for chemical carcinogens.' *Methods in Cancer Research*, **1**, 307–398

WHO: World Health Organization (1967) *Principles for the Testing of Drugs for Teratogenicity*. World Health Organization Technical Research Series, No. 364

WHO: (1978) Environmental health criteria, 6. *Principles and Methods for Evaluating the Toxicity of Chemicals. part I*, Geneva: WHO

Williams, D.J. and Rabin, B.R. (1971) 'Disruption by carcinogens of the hormone-dependent association of membranes with polysomes.' *Nature*, **232**, 102–105

Worden, A.N. (1974) 'Tissue culture' *Modern Trends in Toxicology*, **2**, 216–250

Zbinden, G. (1976) 'Formal toxicology.' *Special topics. Progress in Toxicology*, Vol. 2, p. 10. Berlin, Heidelberg, New York: Springer-Verlag

25

Prevention of Occupational Disease and Ill Health

R.S.F. Schilling

There are four different categories of environmental agents or factors which cause occupational disease or ill health. They are physical, chemical and biological agents and psychosocial factors (*Table 25.1*), and they may act separately or in combination. In the past, control measures were confined almost exclusively to the disabling diseases arising from exposure to toxic chemicals, physical agents and pathogenic micro-organisms. Now prevention is aimed at early detection and control of all kinds of hazards before they cause disability. It is broadly based to encompass work factors that may be of aetiological importance in common disorders such as hypertension, coronary heart disease and bronchitis (*see* Chapter 3). The influence of psychosocial factors on mental health, and their identification and control, present their own particular problems, and are dealt with in Chapter 23.

Fifty years ago the employer had to conform to legal requirements which were couched in general terms without clearly defined environmental standards. Permissible levels which are discussed later in this chapter were adopted by the USA and USSR in the 1930s (Holmberg and Winell, 1977). At first they were regarded as goals or tentative guidelines. In some countries they now have the force of law, and in others they are used as a guide by factory inspectors in deciding whether or not the law is being observed. The world-wide use of permissible levels of exposure has led to more effective control.

Occupational disease and ill health account for a relatively small proportion of all sickness and deaths, but offer more scope for prevention because they are caused by environmental agents or factors which can be controlled. An occupational health service has failed in one of its most important functions if a health hazard in the workplace has not been successfully prevented. This depends, in the first place, on identifying adverse effects and their causes.

Table 25.1 CATEGORIES OF OCCUPATIONAL AGENTS OR FACTORS THAT MAY CAUSE DISEASE AND ILL HEALTH

Category	Examples of types of agents or factors
Physical	Radiations High and low atmospheric pressures High and low temperatures Noise Vibration
Chemical	Drugs Dyes Explosives Fertilizers Fibrogenic mineral dusts Paints Pesticides Plastics Solvents Wood, plant and organic dusts

Biological	Viral diseases:
	  rabies
	  hepatitis, types A and B
	
	Bacterial diseases:
	  anthrax
	  brucellosis
	  leptospirosis
	  tetanus
	  tuberculosis
	
	Fungal diseases:
	  dermatophytoses
	  histoplasmosis
	
	Parasitic diseases:
	  ancylostomiasis
	  schistosomiasis

Psychosocial	Work organization:
	  leadership style*
	  communication*
	  worker participation* and fulfilment*
	  security*
	
	Type of work:
	  repetitive
	  overloaded
	  underloaded
	  shift work

*These factors may help to promote well-being; shortcomings, or lack or them, may cause ill-health

IDENTIFICATION OF RISKS

There are two distinct methods of identification, this being the first step towards control. The more effective way is to predict a risk and prevent it before anyone is exposed. The more common method is to identify a risk by adverse effects on people at work.

New compounds and processes

The hazards involved in using a new compound may be predicted from animal toxicity tests (*see* Chapter 24). In the USSR the production and use of new substances is prohibited by law until they have been subjected to toxicological evaluation. A number of countries like the United Kingdom have, or propose to have, a statutory scheme for manufacturers and importers to notify the government of new substances to be made or used in quantities exceeding a specified limit, e.g. one tonne per annum. Notification includes a profile of chemical, physical and biological properties, and recommendations for safe handling. This kind of information should prevent the use of highly toxic materials and enable appropriate action to be taken from the outset to protect both manufacturers and users.

If the use of a material known or thought likely to be toxic is regarded as essential, protective measures may be introduced at the design stage of an industrial process and backed up by health and environmental monitoring once production has started. Plant design should ensure that exposures can be contained within acceptable permissible limits.

Established processes

In established processes there are three methods of identifying a hazard. These are clinical observation, epidemiological enquiry and environmental monitoring.

Clinical observation

The first indication of an occupational hazard may come from a worker presenting with signs and symptoms of a disease which can be related to a specific occupational exposure. The association will be missed unless a careful occupational history is taken. Enquiries should be made about fellow workers beings similarly affected (*see* Chapter 3).

Epidemiological enquiry

This may be the only way of identifying occupational diseases that occur commonly in the community, such as lung and bladder cancers. Epidemiology also helps to verify the risks identified by clinical observation. Its uses are fully discussed in Chapters 13, 14 and 15.

Environmental monitoring

This may identify a health risk before harmful effects have occurred in exposed workers. It is especially appropriate for irreversible and disabling diseases such as silicosis, asbestosis and the occupational cancers which cause no acute symptoms and may take many years to develop.

INTRODUCING PREVENTIVE MEASURES

The preventive measures to be used depend on the nature of the harmful substance or agent, its degree of toxicity and its routes of absorption into the body. Protection built into the design of the process is preferable to a method which has to rely on adapting existing plant, patching up faulty equipment or continual intervention. Safe maintenance is often as necessary as good design and installation.

The most important preventive measures are as follows:

1. Elimination or diminution of risk by substitution, redesign of process or improved work methods.
2. Total enclosure of a process.
3. Segregation of a process to reduce the number of people exposed.
4. Control by dilution ventilation and by local exhaust ventilation.
5. Suppression of dust by water and wetting agents.
6. Good housekeeping which includes cleanliness of the workplace and disposal of waste.
7. Personal protection including limitation of exposure times, clothing, equipment, barrier creams, immunization and personal hygiene.
8. Training for health and safety.
9. Maintaining control by environmental and health monitoring.

SUBSTITUTION AND REDESIGN

The ideal method of eliminating a risk is to substitute a non-toxic substance for a poisonous one. Classic examples of this type of substitution are the use of phosphorus sesquisulphide instead of white phosphorus, which eventually eradicated phossy jaw among match makers; and the use of freon as a refrigerant instead of methylbromide gas, which is highly toxic and odourless and therefore has no warning properties. Sometimes there is no suitable non-toxic substitute and a less toxic one has to be used. For example a solvent such as benzene can be replaced by toluene, which has a lower toxicity and volatility. Fibrous glass has been used as a substitute for asbestos in many processes, because it seems to be less toxic than asbestos. Although it causes irritation of the skin, eyes and upper respiratory tract in high concentrations, no human cancer risk has been proven. Substitution may not be carried out unless the use of toxic materials is prohibited by law. In Britain, leadless or low solubility glazes in pottery manufacture were not universally introduced until it was made compulsory to do so. In some industries substitution of toxic materials may be impossible or impracticable as in the manufacture of pesticides, drugs or toxic solvents and production of radioactive sources.

Analogous to the substitution of a toxic substance are:

1. The redesign of a process to eliminate physical hazards such as noise or radiation.
2. Improved work methods.

Many machines used in industry were invented at a time when the reduction of impact, or continuous noise, was not considered. The powered weaving loom with its flying shuttle, invented in 1785, is ill-guarded, difficult to clean and produces noise of sufficient intensity to cause permanent hearing loss. In recently developed looms a very high weaving speed is achieved by the almost noiseless 'blowing' of the weft across the warp by a tiny puff of water at high pressure. Completely new standards of guarding machinery parts have been achieved, and static electricity problems from man-made fibres have been eliminated. The most dramatic advance of all is the enormous reduction in noise levels, especially in the design of power engines; for example, the Wankle engine where the reciprocating piston has been replaced by a rotary motion, and the RB211 Rolls Royce aero engine where fan noise from jet turbines has been reduced.

Substitution and redesigning are a matter of technical ingenuity implied in the true origin of the word 'engineer'. A welcome trend is the increased incentive for engineers to eliminate or reduce health hazards at the design stage.

Examples of improved work methods are given in Chapter 17.

TOTAL ENCLOSURE

Total enclosure or isolation of a process can ensure that workers do not come into contact with toxic materials. It involves mechanization, automation or handling materials in confined spaces with gloved inlets. It is a widespread practice in the manufacture or manipulation of toxic chemical substances such as asbestos, pesticide concentrates and radioactive materials (*see Figure 25.1*). Operations involving high

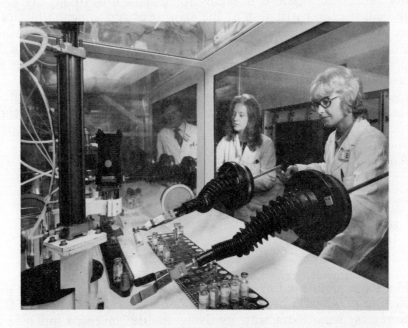

Figure 25.1 Tong-operated dispensing enclosure for radioisotopes. Caps of vials are being crimped and transferred from dispensing boats to storage trays. (Copyright – The Radiobiochemical Centre, supplied by Photographic Library, United Kingdom Atomic Energy Authority, 11 Charles II Street, London, SW1)

exposures to dust or fumes, for instance shot blasting or metal spraying, may be done in totally enclosed cabinets with the worker outside. Similar total protection is used to shield workers from physical agents such as ionizing radiations, intense visible ultraviolet, and infrared or microwave sources.

All equipment, whether enclosed or automated, requires maintenance and repair during which protective measures may have to be removed. In such circumstances safety procedures must be specified, including permits to carry out the work. Other methods of protection such as respirators may have to be used. Maintenance workers must be included in health surveillance programmes.

SEGREGATION

Segregation of a process limits exposure to a small group of workers. By isolating a hazardous job from the rest of the production line, it makes control less costly and more effective. For example arc welding can be confined to a corner of an engineering shop. Engine testing, which is very noisy, can be done in a separate noise proofed room. Any process with a risk of small leaks of toxic contaminants can be isolated and permissible levels maintained by dilution ventilation.

Segregation is less efficient than total enclosure but, combined with other control measures, it can simplify the organization and supervision of prevention.

CONTROL BY VENTILATION

The two main types of ventilation system for control of contaminants are *general dilution*, which is a useful method of keeping contaminants of low toxicity under control, and *local exhaust ventilation*, which is appropriate for the control of a wide range of contaminants released near the breathing zone of the workers (*see* Chapter 17).

SUPPRESSION OF DUST

Dust may be suppressed by water and wetting agents. 'Wetting down' by spraying prevents dispersion of dust in moving or breaking down

Figure 25.2 Directional shearer in USA coal mine, fitted with water spray for dust suppression. (Supplied by Anderson Strathclyde Limited, 47 Broad Street, Glasgow)

rock, or before sweeping dusty floors where vacuum cleaning is not available. Airborne dust in foundries can be controlled to some extent by keeping moulding sand moist. Water may be used to control dust in mining operations (*Figure 25.2*).

Electrostatic precipitation suppresses dust by giving particles a negative charge so that they adhere to positively charged plates. It is used for this purpose in gold mines and to clean chimney effluents for controlling air pollution.

GOOD HOUSEKEEPING AND DISPOSAL OF WASTE

General cleanliness of the workplace, often termed 'good housekeeping', is important for two reasons. It reduces exposure but also encourages tidiness and therefore safer methods of working. Many processes give rise to a variety of contaminants; dust accumulations can produce fire hazards and be continuously displaced into the atmosphere by random currents of air. Premises should have interiors designed to minimize structural projections and ledges. Cleaning methods such as hosing or vacuum cleaning should be used as a routine. Dry brushing or use of air blowing hoses should be eliminated, particularly where there are irritant or harmful dusts.

Toxic waste material may be dangerous to the workforce and to the community. Provisions for their disposal should be made in the planning stage of new plant and incorporated into existing buildings so that workers are protected from exposure to toxic residues, contaminated waste products and containers of no further use. An occupational health service should be prepared to advise management on the disposal of waste, which can give rise to a community health risk.

PERSONAL PROTECTION

Limitation of exposure times

Limitation of time of exposure is used to protect persons exposed to physical agents such as noise and radiation.

Clothing and equipment

The use of protective clothing and equipment is described in more detail in Chapters 19 and 21. It may be essential in the protection of outdoor workers against the elements, or workers in cold storage plants. The use of protective equipment is often required to supplement environmental control measures, particularly where workers have high intermittent exposures; for example, to dust and noise. They may find the necessary respirator or ear protection irksome and co-operation may be difficult to obtain except among those who are highly motivated and disciplined, and who have been trained in safe methods of work.

Protection of the skin is a problem on its own. The worker should be informed about the possible harmful effects of substances being used, and trained to avoid contact with them wherever possible. Where hot materials have to be handled or where there is a danger of friction, suitable gloves of leather or other insulating material should be provided. A variety of polyvinyl chloride and cotton gloves are commonly used. They have the disadvantage of tearing easily, making fine work difficult and often inducing sweating of the palms. This may predispose to the development of skin disease.

Barrier creams

Barrier creams are widely used and popular with workers, management and unions. Usually they are labelled either 'water repellent' or 'oil and solvent repellent'. No cream gives more than limited protection against substances that are harmful to the skin. However, workers often claim that the application of these creams at the beginning of a shift helps the removal of grime at the end of the day. The use of these creams need not be discouraged provided they are known to be non-irritating and non-sensitizing. On the other hand no reliance should be placed on their ability to prevent either irritant or allergic dermatitis.

Immunization

The provision of occupational health services for agricultural workers and the staff of health services is relatively new. Such people are exposed to a wide range of biological hazards from their contact with infected patients, animals and certain materials. Immunization gives some protection against some infections, but it is no alternative to the positive preventive measures outlined in this chapter and others, such as sterilization and use of disinfectants. It cannot guarantee full protection against infection. Diseases for which immunization can give some protection are as follows:

Anthrax

Anthrax may occur in agricultural workers and handlers of animal hair, wool and hides. Active immunization is now widely adopted. This involves three intramuscular injections of vaccine at three-week intervals and a booster dose after six months and then annually.

Rubella

Rubella is an occupational risk of certain groups of women such as school teachers, hospital staff and nursery attendants. An occupational health service should make rubella vaccination available to women of child-bearing age, particularly in the above groups. They need to be screened to determine their immunity, and the non-immune offered vaccination (Faculty of Community Medicine, 1978).

Tetanus

Tetanus may occur following penetration or crush type wounds among farmers, gardeners and those who work with domestic animals and with soil. Active immunization with tetanus toxoid should be a routine procedure for everyone, and essential for those with an occupational risk.

Tuberculosis

Tuberculosis is an occupational risk of health workers in hospitals and sanatoria who care for patients with the disease, and of laboratory workers. A Heaf test should be done as a routine and BCG vaccine offered to those found to be negative. Tests should be repeated every five years to identify those who have reverted to negative.

Viral hepatitis

This is now recognized as an occupational risk to hospital staff and laboratory workers and others caring for the sick. It is caused by two distinct agents referred to as virus A, which is the presumed agent of infectious hepatitis; and virus B, which is associated with serum hepatitis. At present there is no hepatitis B vaccine. No real step has been made towards the development of a conventional vaccine because the virus has not been grown in cell culture. Other preparations for active immunization are being considered (*British Medical Journal,* 1980). Normal human immunoglobin attenuates infectious hepatitis when given during the incubation period. However, there is no reliable evidence that it protects persons after a definite exposure in the laboratory to virus B.

Personal hygiene

A high standard of personal hygiene is essential for persons exposed to toxic substances or handling infected animals or materials. It prevents accidental ingestion from contaminated hands, skin absorption and contamination from door handles and other objects which can be a source of infection or poisoning to others. Washing facilities should be ample and conveniently situated with hot and cold water, soap and clean towels. Baths and showers should be provided for persons doing dirty work or where there is a risk of extensive skin contamination. Where splashes or accidental skin exposures are likely to occur, showers and eye-wash bottles should be readily available in the workroom. Places for smoking, eating and drinking must be outside work areas.

Legal requirements for washing accommodation are often minimal and ignored. It is an important duty of an occupational health service to ensure that these amenities for personal hygiene are provided and maintained where there are health risks from contamination.

TRAINING FOR HEALTH AND SAFETY

The means of control have to be chosen according to the particular hazard. Nevertheless it is preferable to rely on measures such as substitution and total enclosure, which operate continuously, and are independent of the whim of individuals. Sir Thomas Legge (1863 - 1932) expressed this principle in three of his famous axioms:

1. Unless and until the employer has done everything — and everything means a good deal — the workman can do next to nothing to protect himself, although he is naturally willing enough to do his share.
2. If you can bring an influence to bear external to the workman (that is, one over which he can exercise no control), you will be successful; and if you cannot or do not, you will never be wholly successful.
3. All workmen should be told something of the danger of the material with which they come into contact and not be left to find out for themselves, sometimes at the cost of their lives.

Mechanical devices such as locally applied exhaust ventilation and the wearing of protective equipment are not external to the work-man. Mechnical safety devices may be removed, obstructed or switched off because they cause discomfort or interfere with work. For similar reasons, protective equipment may be discarded or not worn correctly. While the installation and maintenance of control measures and the supply of protective equipment are management's responsibility, their proper use depends on the co-operation of workpeople. In modern industry it is increasingly difficult to make work safe without their full participation (Gardner, 1967). Legge's belief that the workman often could not, or would not, take responsibility for his own health and safety, reflected the paternalism which was a prominent feature of social reform at the beginning of this century. In highly industrialized societies today, this philosophy is no longer acceptable and the principle of co-partnership between management and workers, which applies to the prevention of both accident and disease, has been established and may be expressed as follows: 'It is seldom possible to prevent occupational injury and disease successfully without the co-operation of workers. To get their co-operation they must be fully aware of the hazards of their work and how these can be contained. They must also be given some responsibility and incentive to act for their own health and safety.'

In many countries in Europe, health and safety committees are required by law and responsibility is shared by employers and employees. Thus training of management and workers in health and safety is essential. In some countries it is a legal requirement and operates at four levels (Printing and Publishing Industry Training Board, 1977).

All workers

During their induction and job training all workers are instructed in safe and healthy methods of work and their responsibility for the health and safety of themselves and others.

Line management

Middle managers, supervisors and foremen need to have the following:

1. A knowledge of the legal, technical and practical aspects of work activities within their control and the precautions necessary to control identified hazards.
2. A general appreciation of health and safety and legal requirements in workplaces.

3. Skills in communication with senior management and shop floor workers and health and safety committees.
4. The ability to train and control the training of persons in their charge.

Senior management

The training needs of directors and senior managers will be at a more general level except in smaller firms where their responsibilities will be those of line managers. Generally senior managers should have sufficient appreciation of health and safety to be able to assess the relative importance of various issues and determine priorities.

Health and safety advisers and workers' representatives

These are persons who play a special role in a part-time capacity. They act as advisers, or represent workers' interests, to line managers who must always be responsible for health and safety in their own departments. As they investigate potential hazards as well as causes of accident and illness they must have a technical knowledge of health and safety hazards and legislation. They need to have skills in investigating causes of injury and in inspecting and surveying workplaces.

The staff of occupational health and safety services, that is physicians, nurses, occupational hygienists and whole-time safety advisers, should advise management on the content of these training programmes and participate in them at all levels. In addition to this formal training the physician and nurse, on a day-by-day basis, educate informally and individually those who attend for treatment, advice or routine examinations. This type of individual advice may be valuable in the control of hazards which demand a high standard of personal protection such as exposures to dangerous liquids, aerosols and noise levels. The physician and nurse also have to keep all levels of management constantly aware of health and safety problems, and awaken, and keep alive, their interest in the health of their organization.

MAINTAINING CONTROL

There are relatively few occupational diseases that can be prevented altogether by stopping harmful exposures. After applying one or more of the control measures outlined, successful prevention depends on two principles: first, environmental monitoring to limit exposures to levels that impose no serious threat to health during a normal lifetime of work; and secondly, by health monitoring or screening to identify persons who are susceptible to particular agents; and where necessary to advise them to avoid further contact with the agent, because such high risk individuals may not be adequately protected even where permissible limits are achieved.

Environmental and medical monitoring should not be regarded as alternatives. They are complementary. Both are needed when dealing with newly recognized hazards for which dose/response relationships are not clearly defined, or with diseases like byssinosis, where the permissible limits do not protect susceptible persons (US Department of Labor, 1978). Health monitoring also evaluates environmental control measures and offers opportunity to detect illness not directly related to work.

Environmental monitoring

Chapters 17 and 18 give a fuller description of environmental monitoring and control procedures.

Permissible levels of chemical substances

The United States of America and the USSR, the two countries which have contributed most to the development of permissible levels of chemical substances for workplaces, have basically different concepts for setting such standards. These differences are best illustrated by their definitions.

In the USSR permissible levels are officially defined as follows: 'Maximum allowable concentrations (MACs) of harmful substances in the air of the working area are those concentrations that in the case of daily exposure at work for eight hours throughout the entire working life will not cause any diseases or deviations from a normal state of health, detectable by current methods of investigation, either during the work itself or in the long term.'

In the USA the term Threshold Limit Values (TLVs) is used. It is defined by the American Conference of Governmental Industrial Hygienists (ACGIH) as follows: 'Threshold limit values refer to airborne concentrations of substances and represent conditions under which it is believed that nearly all workers may be repeatedly exposed day after day without adverse effect. Because of wide variation in individual susceptibility, however, a small percentage of workers may experience discomfort from some substances at concentrations at or

below the threshold limit; a smaller percentage may be affected more seriously by aggravation of a pre-existing condition or by development of an occupational illness.
There are three types of TLV:

1. Threshold Limit Value – Time Weighted Average (TLV–TWA): the time-weighted average concentration for a normal eight-hour workday or 40-hour work-week, to which nearly all workers may be repeatedly exposed day after day without adverse effect.
2. Threshold Limit Value – Short Term Exposure Limit (TLV–STEL): the maximal concentration to which workers can be exposed for a period up to 15 minutes continuously (provided that no more than four excursions per day are permitted with at least 60 minutes between exposure periods, and provided that the daily TLV–TWA also is not exceeded) without suffering from irritation, chronic or irreversible tissue change or narcosis of sufficient degree to increase accident proneness, impair self-rescue or materially reduce work efficiency. The STEL is a maximal allowable concentration, or absolute ceiling, not to be exceeded at any time during the 15-minute excursion period.
3. Threshold Limit Value–Ceiling (TLV–C): the concentration that should not be exceeded even instantaneously.

In 1968 international agreement was reached on safe concentration zones for only twenty-four industrial and agricultural chemicals (WHO, 1969). Differences are due to a number of factors including the definition of what constitutes an adverse health effect, and the extent to which governments, and other bodies such as the ACGIH, consider costs, feasibility and resources in standard setting.

Thus general international agreement on permissible levels is not foreseeable in the near future. What is possible, and probably more desirable, is to achieve agreement among scientists on the dose/ response effects on which both MACs and TLVs are based.

Figure 25.3 is an example of a composite dose response curve for a toxic agent based on behavioural, subclinical and medical indices (Hatch, 1972). The permissible levels adopted will depend on the type of response used as an index. A country can decide on its own standards, which will be influenced by a number of factors such as cultural background, the type of economy and the stage of development of the industries. In developing countries, which inevitably have limited resources, the problems of achieving hygiene standards are further complicated by special factors such as genetic differences which may affect susceptibility, the prevalence of under-nutrition and endemic parasitic diseases, adverse climatic conditions and very long hours (*see* Chapter 2).

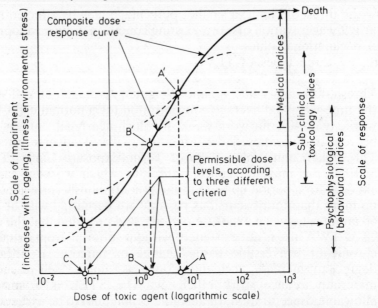

Figure 25.3 Points A and A' at permissible limits to prevent diagnosable illness; points B and B' to keep levels of response below detectable precursors of illness; points C and C' to prevent the earliest demonstrable change from normal behaviour (as is the stated purpose of MACs in the USSR). (Reproduced from Hatch, 1972, by courtesy of the Author and of the Editor of Journal of Occupational Medicine.)

Permissible levels are not sharp dividing lines between 'safe' and 'dangerous' concentrations. The best working practice is to keep concentrations of all airborne contaminants' as low as practicable, whether or not they are known to present a hazard and irrespective of their TLVs or MACs (Health and Safety Executive, 1978). Permissible levels may be based on incomplete information. Previously unsuspected health risks have arisen from substances assumed to be comparatively safe.

Hygiene standards for physical agents

Hygiene standards for physical agents such as ionizing and non-ionizing radiations, noise and heat-stress are published by the ACGIH (1976). They are based on much the same principles as those used for chemical substances, namely to establish dose/response relationships and then decide what is permissible. For heat and noise there are certain limitations in standard setting (WHO, 1977). In heat exposure a distinction has to be made between optimal climatic conditions, which provide thermal comfort, and permissible climatic conditions, which do not ensure thermal comfort but do not cause health impairment. Permissible and not optimal climatic conditions have to be applied in hot industries. They may involve thermal discomfort and impaired performance, but their aim is to avoid heat exhaustion and heat stroke. The criterion for setting permissible levels of noise is hearing loss. In setting standards, little consideration has been given to other effects such as annoyance, interference with communication or work performance, largely for economic reasons (it is costly to eliminate or reduce noise) and because most people accept it as inevitable in modern society.

The strategy of environmental monitoring

The environment has to be assessed in terms which express the dose of the hazardous agent absorbed by the worker. It involves sampling, which should be representative as far as possible of conditions throughout a working shift; choosing the right instruments and analytic methods; and, finally interpreting the results. The strategy of sampling for airborne contaminants and physical agents is discussed in Chapters 17 and 18.

Health monitoring

Health screening and other forms of medical examinations are discussed in Chapters 9, 10 and 11. As they have a special value in the prevention of occupational disease, they will be discussed briefly in this context.

The preplacement medical examination

This has the positive purpose of proper job placement according to the physical and mental capabilities of the worker. It also makes it possible to identify persons likely to be vulnerable to certain exposures and provides base line data that make it possible to measure early adverse effects of exposure.

Periodic health examinations

These have two main functions.

1. They evaluate the effectiveness of preventive measures – a group of workers showing clinical or biological evidence of excessive

exposure may be the first indication of faulty design or break-down in plant or protective devices.

2.   They identify workers showing undue susceptibility to a particular type of exposure who can be recommended for a change of job before the onset of serious or disabling disease.

Biological monitoring

There are many examples of the use of biological tests in the early detection of occupational disease and its precursors. It entails the periodic examination of blood or urine specimens in order to detect excessive absorption of a potentially toxic substance. The assay may be made on the absorbed substance or a metabolite or enzyme affected by it. The term may also include determination of gases or vapours (e.g. carbon monoxide, trichlorethylene) in the breath. It has certain advantages over air sampling. In the case of agents such as lead or mercury which have a long biological half life, i.e. take a long time to be metabolized or excreted, biological assay gives a more reliable measure of exposure than an air sample.

Biological monitoring takes into account absorption through the skin and gastrointestinal tract and effects of workload (breathing rate and depth) and exposure outside the workplace. It can also identify the susceptible worker (Zielhuis, 1975). Thus biological assays may sometimes be more reliable indicators of health risks than environmental measurements (Nelson, 1973).

Biological Limit Values (BLVs) have been recommended for certain agents; for example, metals such as inorganic lead and cadmium, benzene, organophosphorus and organochlorine insecticides (Lloyd Davies, 1970). They should be regarded as useful guidelines in the health surveillance of exposed workers, and complementary to environmental monitoring, which is generally the more effective way of measuring exposure. At present there is no move by governments, or other official bodies, to recommend that BLVs replace TLVs (Calabrese, 1978b).

The susceptible worker

Many of the permissible levels for chemicals are not designed to protect susceptible persons who are at higher risk because of genetic disorders, other diseases, or personal habits such as smoking and excessive consumption of alcohol (Calabrese, 1978a).

Genetic disorders

There are a number of genetic disorders which can be identified by screening.

Atopy Generally, atopic status seems to play a relatively minor role as a predisposing factor in the development of allergic respiratory disease. As atopics, defined by positive skin tests to environmental allergens, may comprise about one-third of the population, it is not reasonable to screen out individuals on this basis, except in high risk industries. Where subjects are heavily exposed to an allergen or substances of high allergenicity such as complex platinum salts, screening for atopy is probably of value. Isocyanates, which are being used increasingly in making foams, lacquers and paints, give rise to a highly variable immunological response among exposed workers. At present there is no satisfactory way of predicting who will become sensitized.

Deficiency in red cell glucose-6-phosphate dehydrogenase (G6PD) This condition increases susceptibility to haemolytic anaemia. Based on existing evidence, Calabrese (1978a) suggests that black males who have a high prevalence (11 per cent) of G6PD deficiency should be screened where they are exposed to welding fumes or haemolytic chemicals.

Serum total αl-antitrypsin deficiency (SAT) This predisposes to alveolar destruction. It has been tried as a screening procedure for persons exposed to respiratory irritants and apparently abandoned because of the low incidence of the deficiency.

Genetic disorders may explain susceptibility to occupational exposures such as carbon disulphide and organophosphorous insecticides. Screening for susceptibility only becomes of practical use where these abnormalities can be reliably detected, have a relatively high frequency and their presence can be shown to increase the risk of occupational disease.

Other diseases

Persons with chronic obstructive pulmonary disease should be excluded from occupations involving exposure to dust or aerosols which are likely to exacerbate their condition. Function tests of various organ systems, for example lung, liver, kidneys, hearing and blood, are valuable where workers are to be exposed to agents which may damage them (*see* Chapters 5, 6, 11 and 12).

Smoking and alcohol consumption

Cigarette-smoking potentiates the carcinogenic activity of asbestos. Among coal workers there is evidence that emphysema and heart disease have a significantly higher incidence in smokers than in non-smokers. In cotton workers, smoking and cotton dust have an additive adverse effect on lung function. High consumption of alcohol may impair ability to detoxify lead, pesticides and polychlorinated biphenyls (PCBs). Where personal habits like smoking and drinking substantially increase the risk of developing occupational disease, the physician or nurse should warn employees of the extra risk, and attempt either to persuade them to change their habits or to transfer to jobs where there is less risk.

STATE RESPONSIBILITY

In many countries the employer has had a long-standing common law duty to take reasonable care for the health and safety of his or her employees. In the nineteenth century this duty proved to be inadequate in protecting workers in factories and mines. The result was that statutory laws were introduced to enable governments to establish and enforce standards of health and safety. This legislation gradually raised general standards of working conditions and reduced the incidence of occupational disease. Laws were made to prohibit the use of several highly toxic substances; for example, high solubility lead glazes in the pottery industry and white phosphorus in making matches. Previous attempts to *control* disease from such substances had failed.

For a time too much reliance was put on government legislation. Nevertheless, it is clear from the experience of many countries that state intervention is essential for the enforcement of minimum standards, especially under conditions of competition and rapid economic growth. The emphasis in industrialized and developed countries is more on the personal responsibility of employers who create the risks, and employees who work with them (Harvey, 1976). Statutory law is limited to a statement of general principles; powers to make regulations and instigate codes of practice; controls on the use and storage of dangerous materials; and provision for dealing with offences and penalties.

The enforcing authorities, the health or labour inspectorate, play an important part in giving information and advice. The state has further responsibility for undertaking or sponsoring research which leads to the early detection of risks and their causes. It sets or approves hygiene standards and codes of practice in consultation with professional associations, employers and trade unions.

REFERENCES

American Conference of Governmental Industrial Hygienists (1979). *TLVs for Physical Agents adopted by ACGIH for 1976.* PO Box 1937, Cincinnati, Ohio

British Medical Journal (1980) 'Hepatitis B vaccines.' *British Medical Journal,* **280,** 203–204

Calabrese, E.J. (1978a) *The Biological Basis of Increased Human Susceptibility to Environmental and Occupational Pollutants.* New York: Wiley–Interscience Series

Calabrese, E.J. (1978b) *Methodological Approaches to Deriving Environmental and Occupational Health Standards.* New York: Wiley–Interscience Series

Faculty of Community Medicine (1978) 'Rubella vaccination reconsidered.' *Faculty News Letter,* **5,** 46–52

Gardner, A. Ward (1967) 'Legge's axioms re-examined.' *Transactions of the Society of Occupational Medicine,* **17,** 74–75

Harvey, B.H. (1976) 'Some legislative problems in the control of toxic substances.' *Annals of Occupational Hygiene,* **19,** 135–138

Hatch, Theodore, F. (1972) 'The role of permissible limits for hazardous airborne substances in the working environment in the prevention of occupational disease.' *Bulletin of the World Health Organization,* **47,** 151–159

Health and Safety Executive (1978) *Threshold Limit Values for 1978.* Guidance note EH 15/78. London: HM Stationery Office

Holmberg, B.O. and Winell, Margaret (1977) 'Occupational health standards: An international comparison.' *Scandinavian Journal of the Work Environment and Health,* **3,** 1–15

Lloyd Davies, T.A. (1970) Annual Report H.M. Chief Inspector of Factories. London: HM Stationery Office

Nelson, K.W. (1973) 'The place of biological measurements in standard setting concepts.' *Journal of Occupational Medicine,* **15,** 439

Printing and Publishing Industry Training Board (1977) *Training for Health and Safety.* Training Guide No. 6. Merit House, Edgware Road, London NW9

US Department of Labor (1978) *Occupational Safety and Health Administration. Occupational Exposure to Cotton Dust.* Federal Register 43, No.122

WHO: World Health Organization (1969) *Permissible Levels of Exposure to Airborne Toxic Substances.* Sixth Report of the joint International Labour Organization/WHO Committee on Occupational Health. Technical Report Series, No.415 Geneva: WHO

WHO (1977) *Methods Used in Establishing Permissible Levels in Occupational Exposure to Harmful Agents.* Technical Report Series 601. Geneva: WHO

Zielhius, R.L. (1975) 'Permissible limits for chemical exposures.' Ch. 25 in Carl Zenz, ed. *Occupational Medicine.* Chicago: Year Book Medical Publishers, Inc.

Index